International Conference of Young Scientists and Students: Topical Problems of Mechanical Engineering (TopME 2018)

IOP Conference Series: Materials Science and Engineering
Volume 489

Moscow, Russia
20 – 23 November 2018

ISBN: 978-1-5108-8484-7
ISSN: 1757-8981

Printed from e-media with permission by:

Curran Associates, Inc.
57 Morehouse Lane
Red Hook, NY 12571

Some format issues inherent in the e-media version may also appear in this print version.

Printed by Curran Associates, Inc. (2019)

For permission requests, please contact the Institute of Physics at the address below.

Institute of Physics
Dirac House, Temple Back
Bristol BS1 6BE UK

Phone: 44 1 17 929 7481
Fax: 44 1 17 920 0979

techtracking@iop.org

Additional copies of this publication are available from:

Curran Associates, Inc.
57 Morehouse Lane
Red Hook, NY 12571 USA
Phone: 845-758-0400
Fax: 845-758-2633
Email: curran@proceedings.com
Web: www.proceedings.com

Table of contents

Volume 489

International Conference of Young Scientists and Students "Topical Problems of Mechanical Engineering 2018"

20–23 November 2018, Moscow, Russian Federation

Accepted papers received: 20 February 2019
Published online: 26 March 2019

Preface

Papers

IOP Conf. Series: Materials Science and Engineering **489** (2019) 011001 doi:10.1088/1757-899X/489/1/011001

PREFACE

We are pleased to introduce the Proceedings of the 30th International Conference of Young Scientists and Students "Topical Problems of Mechanical Engineering 2018" (TopME 2018).

The Conference was held from November 20th to November 23rd, 2018 in Moscow, Russia. It was organized by the Blagonravov Mechanical Engineering Research Institute of the Russian Academy of Sciences (IMASH RAS). The Russian Foundation of Basic Research supported the Conference.

The scientific sections of the Conference covered a wide range of topics on materials science, strength, survivability and safety of machines, tribology, vibration processes and systems, machines mechanics and biomechanics.

IOP Conf. Series: Materials Science and Engineering **489** (2019) 011002 doi:10.1088/1757-899X/489/1/011002

30th International Conference of Young Scientists and Students "Topical Problems of Mechanical Engineering 2018"

Chairman of Program Committee

V.A. Glazunov, director of IMASH RAS

Chairman of Organizing Committee

Prof. A. N. Polilov

Program Committee

R. F. Ganiev, academician of the RAS, scientific director of IMASH RAS
A. Yu. Albagachiev
V. K. Astashev
Yu. I. Bobrovnitskii,
S. M. Kaplunov
O. I. Kosarev
Yu. G. Matvienko
N. A. Makhutov
S. Yu. Misjurin
G. Ya. Panovko
I. A. Razumovsky
A. N. Romanov
V. O. Solov'ev
I. V. Stepanyan

Organizing Committee

A. V. Frolova
A. N. Nikiforov
M. V. Prozhega
M. S. Pugachev
H. F. Saberov
K. B. Salamandra
V. I. Svirin
V. V. Shabalin
A. E. Shokhin
N. A. Tatous'

IOP Conf. Series: Materials Science and Engineering **489** (2019) 011003 doi:10.1088/1757-899X/489/1/011003

Peer review statement

All papers published in this volume of *IOP Conference Series: Materials Science and Engineering* have been peer reviewed through processes administered by the proceedings Editors. Reviews were conducted by expert referees to the professional and scientific standards expected of a proceedings journal published by IOP Publishing.

IOP Conf. Series: Materials Science and Engineering **489** (2019) 012001 doi:10.1088/1757-899X/489/1/012001

Mathematical model of the impact response of a linear viscoelastic auxetic plate

O Ajeneza, Yu A Rossikhin and M V Shitikova

Research Center on Dynamics of Solids and Structures, Voronezh State Technical University, Voronezh, Russian Federation

ajeneza@mail.ru

Abstract. The aim of the present paper is to construct a mathematical model describing the impact of a rigid sphere upon a linear Kirchhoff-Love plate, made of viscoelastic auxetic with fractional viscosity. Auxetic's Poisson's ratio is a time-dependent value changing from negative to positive magnitudes with time. In the case of a linear plate, the solution out of the contact domain is found through the Green function, and within the contact zone via the modified Hertz contact theory. Integral equations for the contact force and local indentation have been obtained.

1. Introduction

Nowadays material engineers are looking for methods rising physical and mechanical characteristics of traditional materials via creating the structures possessing essential abnormal deformational features. As demonstrated by experimental results, materials with negative Poisson's ratio (auxetics), belong to such abnormal materials. When stretched, they become thicker perpendicular to the applied force [1]. This occurs due to their particular internal structure. Auxetics present enhanced mechanical properties, which make them superior to classical materials for many practical applications.

There are a lot of papers devoted to auxetic materials, however, the majority of them deal with the internal structure of auxetics, experimental determination of Poisson's ratios, as well as with the description of features of different auxetics [1-5]. Papers discussing the mathematical models describing the behavior of viscoelastic auxetics are rare, and there are practically no studies devoted to the solution of boundary-value dynamic problems with such materials [6,7]. With the accelerated progress in studies of auxetics made in last few years, there are new potential opportunities for researchers. Auxetic materials are still in developing stage and require intensive research [8,9].

In the present paper, a mathematical model describing the impact of a rigid sphere upon a linear viscoelastic plate is suggested, in doing so the Poisson's ratio is considered to be a time-dependent operator described by the fractional derivative model.

2. Problem formulation and governing equations

Let us consider a rigid spherical impactor of mass m and radius R moving along the z-axis with the velocity V_0 towards a simply-supported linear Kirchhoff-Love plate with the dimensions a and b. The origin of the Cartesian coordinate system x, y is at the plate's center. The plate's viscoelastic properties are described by the Kelvin-Voigt model with fractional derivatives [10]. The impact occurs at the moment $t=0$ at the point N with Cartesian coordinates x_0, y_0.

In this case, the equation of motion of the spherical impactor is

$$m\ddot{z} = -P(t), \tag{1}$$

subjected to the initial conditions

$$\dot{z}(0) = V_0 \quad , \quad z(0) = 0, \tag{2}$$

where $P(t)$ is the contact force, $z(t)$ is the displacement of the spherical impactor, and over dots denote time-derivatives.

The equation of motion of a simply-supported viscoelastic plate has the form

$$\frac{\tilde{E}h^2}{12(1-\tilde{\nu}^2)}\left(\frac{\partial^4 w}{\partial x^4} + 2\frac{\partial^4 w}{\partial x^2 \partial y^2} + \frac{\partial^4 w}{\partial y^4}\right) = -\rho\ddot{w} + h^{-1}P(t)\delta(x-x_0)\delta(y-y_0), \tag{3}$$

where $\tilde{E}$ and $\tilde{\nu}$ are, respectively, time-dependent Young's operator and Poisson's ratio operator, h, $w(x,y,t)$ and ρ are the plate's thickness, deflection and density, respectively, $\delta(x-x_0)$ and $\delta(y-y_0)$ are the Dirac delta-functions.

Equation (3) is subjected to the following boundary and initial conditions:

$$w = 0 \quad \text{at} \quad x = 0, a \ \text{and} \ y = 0, b. \tag{4}$$

$$\dot{w}\big|_{t=0} = 0\ . \tag{5}$$

The values $z(t)$ and $w(x_0, y_0, t)$ are coupled as follows

$$z(t) = \alpha(t) + w(x_0, y_0, t), \tag{6}$$

where $\alpha(t)$ is the value characterizing the local bearing of the plate's material within the contact domain, which is connected with the contact force by the generalized Hertzian law

$$P(t) = \tilde{k}\left[\alpha(t)\right]^{3/2}, \tag{7}$$

where $\tilde{k} = \dfrac{4\sqrt{R}}{3}\tilde{d}$, and $\tilde{d} = \dfrac{\tilde{E}}{1-\tilde{\nu}^2}$.

3. Method of solution

The solution to equation (3) satisfying the initial conditions (4) has the form

$$w(x,y,t) = \sum_{m=1}^{\infty}\sum_{n=1}^{\infty} \xi_{mn}(t)\sin\frac{m\pi x}{a}\sin\frac{n\pi y}{b}, \tag{8}$$

where $\xi_{mn}(t)$ are the generalized displacements, which are yet unknown functions.

Substituting (8) in equation (3) and considering the orthogonality of $\sin\dfrac{m\pi x}{a}$ within the segment $[0,a]$ and $\sin\dfrac{n\pi x}{b}$ within the segment $[0,\mathrm{b}]$, as a result we obtain

$$\ddot{\xi}_{mn}(t) + \frac{h^2\pi^4}{12\rho}\frac{\tilde{E}}{(1-\tilde{\nu}^2)}\left(\frac{m^2}{a^2} + \frac{n^2}{b^2}\right)^2 \xi_{mn}(t) = P(t)F_{mn}, \tag{9}$$

where $F_{mn} = \dfrac{4}{\rho hab}\sin\dfrac{m\pi x_0}{a}\sin\dfrac{n\pi y_0}{b}$.

IOP Conf. Series: Materials Science and Engineering **489** (2019) 012001 doi:10.1088/1757-899X/489/1/012001

In order to find the equation for the Green function $G_{mn}(t)$, we have to substitute the function $P(t)$ in equation (9) with the Dirac delta-function $\delta(x)$, i.e.,

$$\ddot{G}_{mn}(t)+\frac{h^2\pi^4}{12\rho}\left(\frac{m^2}{a^2}+\frac{n^2}{b^2}\right)^2\frac{\tilde{E}}{1-\tilde{\nu}^2}G_{mn}(t)=\delta(t)F_{mn}. \quad (10)$$

We also have to take into consideration that the bulk extension-compression operator $\tilde{K}$ is assumed to be time-independent, i.e., volumetric relaxation is neglected (this assumption is due to the fact that for many viscoelastic materials volumetric relaxation is much smaller than the shear relaxation [11]), i.e.

$$\tilde{K}=K_0,$$

resulting in

$$\frac{\tilde{E}}{1-2\tilde{\nu}}=3K_0, \quad (11)$$

where $K_0=\text{const}$ is the bulk modulus.

In addition we have to assume that the shear operator $\tilde{\mu}$ is preassigned using the fractional derivative Kelvin-Voigt model [10]

$$\tilde{\mu}=\mu_0\left[1+\left(\tau_\sigma^\mu\right)^\gamma D^\gamma\right], \quad (12)$$

where μ_0 is the relaxed shear modulus, τ_σ^μ is the retardation time during shear deformations, γ $(0<\gamma\le 1)$ is the fractional parameter,

$$D^\gamma x(t)=\frac{d}{dt}\int_0^t\frac{x(t')dt'}{\Gamma(1-\gamma)(t-t')^\gamma} \quad (13)$$

is the Riemann-Liouville fractional derivative [12], $\Gamma(1-\gamma)$ is the Gamma-function, and $x(t)$ is a certain function.

To find the Poisson's operator, first it is necessary to find the operator $\tilde{E}$ from formula (11). For this purpose using the Volterra correspondence principle, the following formula could be utilized:

$$\tilde{E}=\frac{9K_0\tilde{\mu}}{3K_0+\tilde{\mu}}. \quad (14)$$

First using formula (12), we write the operator

$$3K_0+\tilde{\mu}=(3K_0+\mu_0)(1+t_\sigma^\gamma D^\gamma), \quad (15)$$

where $t_\sigma^\gamma=\mu_0(\tau_\sigma^\mu)^\gamma(3K_0+\mu_0)^{-1}$.

Then we find the operator reverse to (15), i.e.,

$$(3K_0+\tilde{\mu})^{-1}=(3K_0+\mu_0)^{-1}\Im_\gamma^*(t_\sigma^\gamma), \quad (16)$$

where $\Im_\gamma^*(t_\sigma^\gamma)=(1+t_\sigma^\gamma D^\gamma)^{-1}$ is the dimensionless Rabotnov's fractional operator [13].

Substituting now (12) and (16) in formula (14) and considering [13]

$$t_\sigma^\gamma D^\gamma\Im_\gamma^*(t_\sigma^\gamma)=1-\Im_\gamma^*(t_\sigma^\gamma), \quad (17)$$

we have

$$\tilde{E}=9K_0\left[1-\frac{E_0}{3\mu_0}\Im_\gamma^*(t_\sigma^\gamma)\right], \quad (18)$$

where $E_0=\dfrac{9K_0\mu_0}{3K_0+\mu_0}$.

Now we can calculate Poisson's operator $\tilde{\nu}$ from formula (11), i.e.,

$$\tilde{\nu} = -1 + \frac{E_0}{2\mu_0} \mathcal{Э}^*_\gamma(t^\gamma_\sigma). \tag{19}$$

When operator $\tilde{\nu}$ will act on the unit Heaviside function $\tilde{\nu}H(t)$, as a result we obtain

$$\nu(t) = -H(t) + \frac{E_0}{2\mu_0} \mathcal{Э}^*_\gamma(t^\gamma_\sigma) H(t). \tag{20}$$

Considering that

$$\lim_{t\to\infty} \mathcal{Э}^*_\gamma(t^\gamma_\sigma) H(t) = 1,$$

from equation (20) we have

$$\nu(0) = -1, \quad \nu(\infty) = \frac{E_0 - 2\mu_0}{2\mu_0} = \nu_0. \tag{21}$$

From formulas (21) it is evident that the proposed model (12) with $\tilde{K} = K_0 = \text{const}$ describes the behavior of a viscoelastic auxetic material. In so doing, the Poisson's ratio could vary from -1 to its relaxed magnitude ν_0.

Knowing the form of operators $\tilde{E}$ (18) and $\tilde{\nu}$ (19), we could calculate the operator of rigidity

$$\tilde{d} = \frac{\tilde{E}}{2}\left(\frac{1}{1+\tilde{\nu}} + \frac{1}{1-\tilde{\nu}}\right) = \xi + \eta D^\gamma + \varsigma \mathcal{Э}^*_\gamma(T^\gamma), \tag{22}$$

where

$$\xi = \frac{1}{4}(9\lambda_0 + 10\mu_0), \quad \eta = 3(\lambda_0 + \mu_0) t^\gamma_\sigma, \quad \varsigma = -\frac{1}{4}\frac{(3\lambda_0 + 2\mu_0)^2}{\lambda_0 + \mu_0}, \quad T^\gamma = \frac{4(\lambda_0 + \mu_0)}{\lambda_0 + 2\mu_0} t^\gamma_\sigma.$$

The operator $\tilde{d}^{-1}$, which is reverse to the operator $\tilde{d}$, could be represented in the following form:

$$\tilde{d}^{-1} = a_1 \mathcal{Э}^*_\gamma(\tau^\gamma_1) + a_2 \mathcal{Э}^*_\gamma(\tau^\gamma_2), \tag{23}$$

where the unknown constants a_1, a_2, τ_1, and τ_2 could be found by utilizing the property of the reciprocal operators $\tilde{d}\tilde{d}^{-1} = \tilde{d}^{-1}\tilde{d} = 1$, resulting in the equation for defining the constants

$$1 = \left[a_1 \mathcal{Э}^*_\gamma(\tau^\gamma_1) + a_2 \mathcal{Э}^*_\gamma(\tau^\gamma_2)\right]\left[\xi + \eta D^\gamma + \varsigma \mathcal{Э}^*_\gamma(T^\gamma)\right], \tag{24}$$

the solution of which could be found using the algebra of the dimensionless Rabotnov's operators [13].

Knowing the reciprocal rigidity operators (22) and (23), the following integral equations could be obtained: for the determination of the contact force $P(t)$

$$-\frac{1}{m}\int_0^t P(t')(t-t')dt' + V_0 t = \left\{\frac{3}{4\sqrt{R}}\int_0^t \left[a_1 \mathcal{Э}_\gamma\left(-\frac{t-t'}{\tau_1}\right) + a_2 \mathcal{Э}_\gamma\left(-\frac{t-t'}{\tau_2}\right)\right] P(t')dt'\right\}^{2/3}$$
$$+ \sum_{m=1}^\infty \sum_{n=1}^\infty \sin\frac{m\pi x_0}{a} \sin\frac{n\pi y_0}{b} \int_{0.}^t G_{mn}(t-t')P(t')dt', \tag{25}$$

and for the local indentation of the plate's material $\alpha(t)$:

$$\alpha(t)=V_0t-\frac{4\sqrt{R}}{3m}\xi\int_0^t\left[\alpha^{3/2}(t')+b_1\frac{d}{dt'}\int_0^{t'}\frac{\alpha^{3/2}(t'')dt''}{\Gamma(1-\gamma)(t'-t'')^\gamma}+b_2\int_0^{t'}\ni_\gamma\left(-\frac{t'-t''}{T^\gamma}\right)\alpha^{3/2}(t'')dt''\right](t-t')dt'$$
$$-\frac{4\sqrt{R}}{3m}\xi\sum_{m=1}^{\infty}\sum_{n=1}^{\infty}\sin\frac{m\pi x_0}{a}\sin\frac{n\pi y_0}{b}\int_0^t G_{mn}(t-t') \tag{26}$$
$$\times\left[\alpha^{3/2}(t')+b_1\frac{d}{dt'}\int_0^{t'}\frac{\alpha^{3/2}(t'')dt''}{\Gamma(1-\gamma)(t'-t'')^\gamma}+b_2\int_0^{t'}\ni_\gamma\left(-\frac{t'-t''}{T^\gamma}\right)\alpha^{3/2}(t'')dt''\right]dt'.$$

where $b_1=\eta\xi^{-1}$, $b_2=\varsigma\xi^{-1}$, and

$$\ni_\gamma\left(-t/\tau_i\right)=\frac{t^{\gamma-1}}{\tau_i^\gamma}\sum_{n=0}^{\infty}\frac{(-1)^n(t/\tau_i)^{\gamma n}}{\Gamma\left[\gamma(n+1)\right]} \tag{27}$$

is the Rabotnov's fractional exponential function [14]. Equations (25) and (26) could be solved numerically.

4. Conclusion

In the present paper, the problem on impact of a rigid spherical impactor upon a linear viscoelastic auxetic Kirchhoff-Love plate, made of viscoelastic auxetic with fractional viscosity, has been formulated for the case, when the shear operator is governed by the fractional derivative Kelvin-Voigt model in conjunction with the assumption that the bulk extension-compression operator $\tilde{K}$ is assumed to be time-independent, i.e., volumetric relaxation is neglected.

The behavior of Kirchhoff-Love plate, made of viscoelastic auxetic is different from the behavior of the same viscoelastic plate. This is due to the fact that the auxetic's Poisson's ratio is a time-dependent value changing from negative to positive magnitudes with time.

The solution out of the contact domain is found through the Green's function, and within the contact domain via the Hertz theory. Using the algebra of dimensionless Rabotnov's fractional operators, integral equations for the contact force and the local indention have been obtained.

Acknowledgments

This research has been supported by the Russian Science Foundation, Project No. 19-11-00200.

References

[1] Ren X, Das R, Tran P, Ngo T D and Xie Y M 2018 Auxetic metamaterials and structures: A review *Smart Mater. Struct.* **27** (2)

[2] Liu Q 2006 *Literature review: Materials with negative Poisson's ratios and potential applications to aerospace and defence* (Victoria, Australia: DSTO Defence Science and Technology Organization)

[3] Chen C P and Lakes R S 1993 Viscoelatic behaviour of composite materials with conventional Poisson ration or negative Poisson ratio foam as one phase *J. Mater. Sci.* **28** 4288-98

[4] Chen C P and Lakes R S 1996 Micromechanical analysis of dynamic behaviour of conventional and negative Poisson's ratio foams *J. Eng. Mater. Technol.* **118** 285-8

[5] Scarpa F and Tomlin P J 2000 On the transverse shear modulus of negative Poisson's ratio honeycomb structures *Fatigue Fract. Engng. Mater. Struct.* **23** 717–20

[6] Duc N D, Seung-Eock K, Cong P H, Anh N T and Khoa N D 2017 Dynamic response and vibration of composite double curved shallow shells with negative Poisson's ratio in auxetic honeycombs core layer on elastic foundations subjected to blast and damping loads *Int. J. Mech. Sci.* **133** 504–512

[7] Wang X, Wang B, Wen Z and Ma L 2018 Fabrication and mechanical properties of CFRP composite three-dimensional double-arrow-head auxetic structures *Compos. Sci. Technol. J.* **164** 92–102

IOP Conf. Series: Materials Science and Engineering **489** (2019) 012001 doi:10.1088/1757-899X/489/1/012001

[8] Rossikhin Y A and Shitikova M V 2017 Mathematical models of viscoelastic auxetics *Abst. of the 3d Int. Conf. on Mechanics of Composites (Bologna: Italy)* p 29
[9] Rossikhin Y A, Shitikova M V and Krusser A I 2016 To the question on the correctness of fractional derivative models in dynamic problems of viscoelastic bodies *Mech. Res. Commun.* **77** 44–9
[10] Rossikhin Y A and Shitikova M V 2018 The fractional derivative Kelvin-Voigt model of viscoelasticity with and without volumetric relaxation *Proc. of the 5th Int. Conf. on Topics Problems of Continuum Mechanics (Armenia) IOP Conf. Ser.: J. Phys.* **991** 012069
[11] Rabotnov Y N 1966 *Creep of Structural Elements* (Moscow: Nauka, Engl. transl. by North-Holland, Amsterdam in 1969)
[12] Samko S, Kilbas A, and Marichev O 1993 *Fractional Integrals and Derivatives. Theory and Applications* (Switzerland: Gordon and Breach Science Publishers)
[13] Rossikhin Y A and Shitikova M V 2014 Centennial jubilee of Academician Rabotnov and contemporary handling of his fractional operator *Fract. Calculus Appl. Anal.* **17** 674–83
[14] Rabotnov Y N 2014 Equilibrium of an elastic medium with after-effect *Fract. Calculus Appl. Anal.* **17** 684–96

Microstructure influence on high temperature deformation of low carbon martensitic steel

N V Batanova and V L Danilov

Bauman Moscow State Technical University, 5 2nd Baumanskaya str., Moscow, 105005, Russia

kolebina.nv@mail.ru

Abstract. Shaping of workpieces by the methods of plastic deformation is an advanced technology of fabrication of large dimension parts. In order to optimize the process of manufacturing, the optimal conditions of plastic deformation should be chosen based on knowledge of low carbon martensitic steel specific properties. In the present paper the influence of microstructure on high temperature deformation is investigated based on experimental data for 03Cr13Ni4Mo steel.

1. Introduction

Low carbon martensitic steel has extremely low carbon content and economy alloying, it provides a rare combination of properties of high corrosion resistance, strength and good weldability [1]. These types of steel are widely used in the oil, gas and hydropower industries. For example, they are used for the fabrication of parts of hydraulic turbines. One of the advanced manufacture methods of large dimension parts from this type of steel is the plastic or superplastic deformation. The usual temperature for such methods is from 800°C to 1100°C. However, martensitic steels undergo a phase transformation when heated [2]. It is known that structural transformations lead to substantial changes in the plastic properties of a material. Hence, it is expedient to investigate the influence of the structure of supermartensitic steel on its behavior at elevated temperatures in order to choose the optimal condition of deformation.

Due to the wide application of low carbon steels in welded structures a lot of works, for example, by Kleyner L.M. [1], Ma K.P. [3], Lee Y. S. [4], Courage D. [5] and others [6-7], are dedicated to the special features of microstructure and weldability properties of these steels as well as investigation mechanical properties at room temperature after welding or heat treatment [8-9]. However, only a few works describe the mechanical properties of these steels at high temperatures [10-11]. For example, the work by Bogachev I.N. [11] was devoted to the investigation of the temperature and grain size influence on the elongation of low carbon martensitic steel. In the present paper, the influence of the structure features on the high temperature deformation of low carbon supermartensitic 03Cr13Ni4Mo steel is investigated.

2. Investigation of microstructure special features influence on high temperature deformation

Optical and electron microscopes were used to study the microstructure. Steel 03Cr13Ni4Mo has the martensite structure at room temperature with grain size 2-7μm, figure 1, and the inclusion of delta-ferrite. The presence of delta-ferrite in low carbon steels depends on heat treatment and the manufacturing method and can affect the kinetics and mechanism of phase transformations.

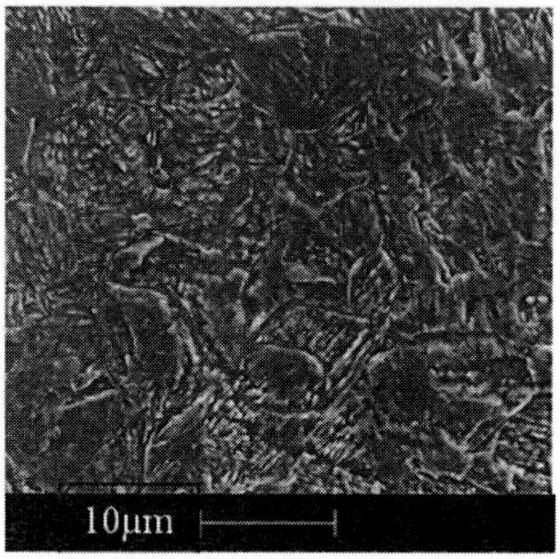

Figure 1. Microstructure of low carbon supermartensitic steel (martensite grains boundaries are highlighted with the red lines).

The dilatometric method was used to study structural transformations with continuous heating [12]. The dependence of austenite quantity on temperature was obtained by the analysis dilatometer curve using lever rule [13]. The morphology of the structure was studied using optical and electron microscopes and is consistent with data from earlier studies [3,5]. The phase transformation of martensite to austenite with continuous heating at a rate of 10°C/min occurs in the temperature range 680°C (A_{c1}) - 850°C (A_{c3}).

Based on the experimental data and the results of previous studies [1,5,9], it was found that when heated over A_{c1}, phase transformation of martensite to austenite begins, when small austenite grains are formed on the boundaries of the martensite plates and grains, figure 2a. Taking into account the specific properties of austenite and martensite at such temperatures, the structure of steel can be considered as dispersion-strengthened. With a further temperature increase to A_{c3}, the main phase is austenite, and the steel structure can be considered as austenitic with inclusions of martensite grains, figure 2b, c. It was also found that the latter undergoes a phase transformation the region around delta - ferrite. This is due to the fact that the decrease in nickel content was observed near delta-ferrite, which affects the rate of phase transformation. At temperatures above A_{c3}, only the austenitic phase exists. Thus, we can identify three types of phase morphology depending on temperature: single-phase with a known grain size (martensite ($<A_{c1}$) or austenite ($>A_{c3}$); dispersion-strengthened martensite matrix (A_{c1} - A_{c3}); austenite phase with inclusions of martensite (A_{c1} - A_{c3}).

The high-temperature tensile tests were performed to determine the effect of the phase composition on the thermo-mechanical properties of supermartensitic steel. The mechanical experiments were conducted before failure at a constant strain rate in the temperature range from 680°C to 950°C. It has been established experimentally that the characters of the dependences of stresses on deformations for the martensitic and austenitic phases are very different. The deformation curves at temperatures of the austenitic phase demonstrate noticeable hardening of material, then a small region of constant stresses is observed, and then a long gradual decrease in stress until final failure, figure 3. Deformation at temperatures of the martensitic phase occurs without appreciable hardening, which indicates different deformation mechanisms. The results obtained are consistent with the results of earlier studies [9-11,14].

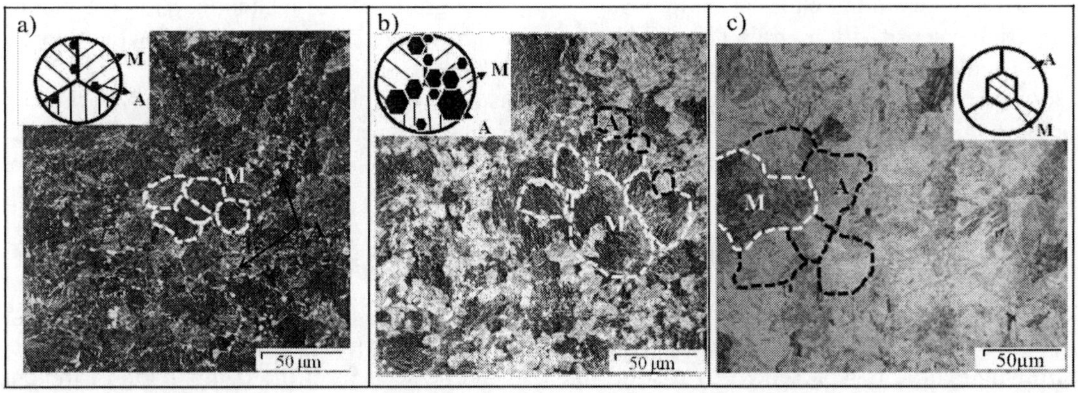

Figure 2. Morphology and microstructure of steel, depending on temperature (with continuous heating at a rate of 10°C/min): a) a martensitic grains with small austenitic grains on the boundaries, b) austenite-martensite structure, c) austenite structure with inclusions of martensite grains.

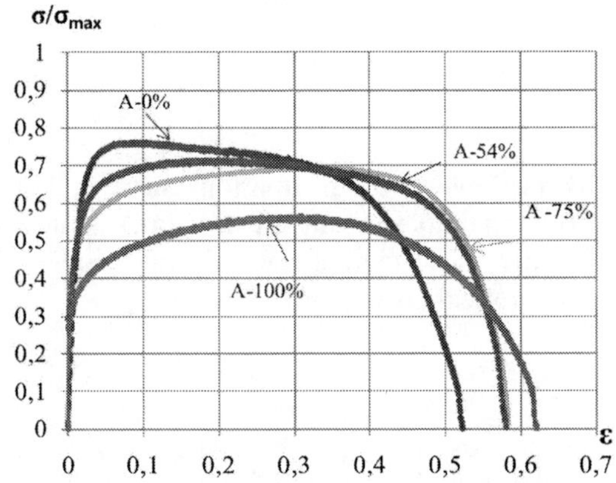

Figure 3. Deformation curves for tensile tests at different temperatures.

Based on experimental data, it was discovered that the phase composition and morphology of the phases significantly affect the mechanical properties. Therefore, for each of the phases (austenite and martensite), it is necessary to determine the state equation. And in the two-phase temperature range we should use such combinations of these equations, which take into account the morphology of the structure.

The most appropriate description of the deformation curves of the austenitic phase corresponds to the theory of creep hardening [15]:

$$\sigma_A = A_A \xi^{m_A} \varepsilon^{n_A} \tag{1}$$

where σ_A is the stress; ε is the creep deformation; ξ is the creep deformation rate; A_A is the experimental material constant; m_A is the parameter, characterizing strain rate sensitivity of the austenitic phase, n_A is the parameter, characterizing hardening of the austenitic phase.

For the most accurate description of steel properties, as well as to determine the mechanism of deformation, the dependence of grain size, temperature, and deformation activation energy on the stress was introduced into the equation:

$$\sigma_A = \left(B_A / d_A^{\,p_A}\right) \xi^{m_A} \varepsilon^{n_A} \exp\left(Q_A m_A / RT\right) \tag{2}$$

where B_A is the experimental material constant, d_A is the grain size of austenite, p_A is the parameter, characterizing sensitivity to grain size of the austenitic phase, Q_A is the deformation activation energy of the austenitic phase, R is the universal gas constant, T is the testing temperature.

However, equation (2) describes only homogeneous deformation of sample. Therefore, the damage parameter, ω, was introduced, to take into account the necking formation and the fracture process. The damage parameter ranges from 0 to 1, where, 0 is no damage, 1 is failure, to take into account the damage of sample.

$$\sigma_A = (1-\omega)^k \left(B_A / d_A^{p_A}\right) \xi^{m_A} \varepsilon^{n_A} \exp(Q_A m_A / RT) \tag{3}$$

where k is experimental constant; ω is damage parameter.

The theoretical curve, calculated by the equation (3), is shown on figure 4 by a dashed line, the experimental curve is represented by a solid line. As can be seen from the comparison of the curves, the theoretical curve is consistent with the experimental one.

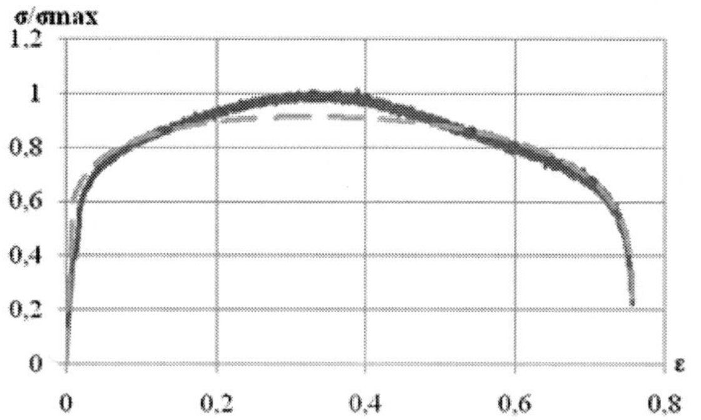

Figure 4. The deformation curve at 880°C (taking into account the fracture): solid line is the experimental data, dashed line is the theoretical curve.

To describe the behavior of the martensitic phase, the expression similar to equation (3) was used, taking into account that the martensite deformation curve shows no hardening of material [16]:

$$\sigma_M = (1-\omega)^k \left(B_M / d_M^{p_M}\right) \xi^{m_M} \exp(Q_M m_M / RT) \tag{4}$$

where B_M is the experimental material constant, d_M is the grain size of martensite, p_M is the parameter, characterizing sensitivity to grain size of the martensitic phase, m_M is the parameter, characterizing strain rate sensitivity of the martensitic phase, Q_M is the deformation activation energy of the martensitic phase.

In the two-phase temperature range, in accordance with morphology, two types of relations were used: the equation describing the dispersion-strengthened structure and the equation of state as the sum of stresses of the austenitic and martensitic phases. In accordance with the experimental data, figure 2a, b, the temperature of the beginning of the phase transformation is characterized by the microstructure, which can be considered as a martensite matrix reinforced with small austenite grains. The behavior of the material can be described by the following equation of state, provided that the second phase does not form a closed structure in the material and does not deform [17-18]:

$$\sigma_1 = (1-f)^{-h(m1)} \sigma_M \tag{5}$$

where f is the volume fraction of austenite at a given temperature (T), $h(m1)$ is the index of the hardening function, σ_M is the stress of martensitic phase, defined by the equation of state (4).

At temperatures at the end of the phase transformation, the microstructure consists of the martensitic and austenitic phases, figure 2c. Since the results of microstructural analysis of the samples after deformation at these temperatures showed that there are no discontinuities at the phase boundary, we can assume that the deformations of martensite and austenite are equal. Based on these facts, the equation of state can be written as the sum of stresses of the austenitic and martensitic phase [19–20]:

$$\sigma_2 = f\sigma_A + (1-f)\sigma_M \tag{6}$$

where f is the volume fraction of austenite at a given temperature (T), σ_A is the stress in the austenitic phase, defined by the equation of state (3), σ_M is the stress in the martensitic phase, defined by the equation of state (4).

Thus, using equations (3) - (6), we can describe the behavior of low carbon supermartensitic stainless steel in the temperature range from 680°C to 950°C. The research results showed that the structure of steel strongly affects the behavior of the material at high temperatures. The influence of temperature on maximum tensile stresses and deformations before failure is shown on figure 5.

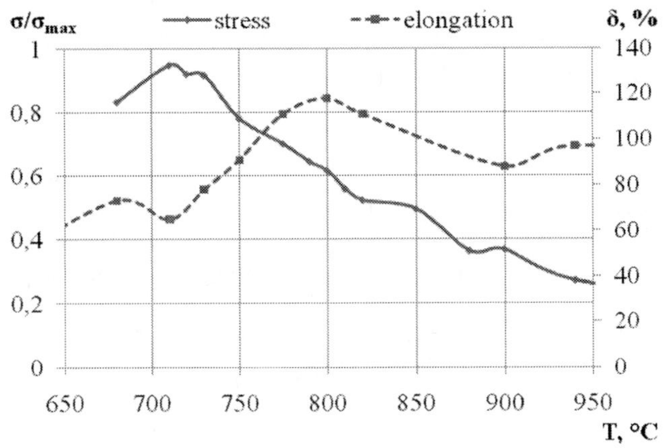

Figure 5. Maximum stresses and maximum strain to failure, depending on temperature.

The maximum stress decreases with increasing temperature, while the maximum deformation before failure increases, figure 5. However, maximum deformation strongly increases only until the end of the phase transformation temperature, A_{c3}. Based on this it can be concluded that the presence of the second phase during deformation allows obtaining the stable fine-grained structure. Therefore, the deformation in the two-phase temperature range may be of practical interest.

3. Conclusion

Experimental studies allowed to identify structural features and show its influence on the high-temperature deformation of low carbon supermartensitic steel, and also specify the equations of state that take into account the morphology of the phase of steel at a given temperature. The results of these studies can be used to determine the optimal condition of workpieces deformation from low carbon supermartensitic steel 03Cr13Ni4Mo.

References

[1] Kleiner L M, Shatsov A A and Melnikov L D 2011 Low-carbon martensitic steels: Alloying and properties *Metal Science and Heat Treatment* **52** 540–44
[2] Gulyaev A P 2011 *Metal Science* (Moscow: Alyans Press) p 643
[3] Ma X P, Wang L J and Liu C M 2012 Microstructure and properties of

IOP Conf. Series: Materials Science and Engineering **489** (2019) 012002 doi:10.1088/1757-899X/489/1/012002

13Cr5Ni1Mo0.025Nb0.09V0.06N supermartensitic stainless steel *Material Science and Engineering* A **539** 271-79
[4] Lee Y S, Kim D W, Lee D Y and Ruy W S 2001 Effect of gain size on creep properties of type 316MN stainless steel *Metals and material International* **7** 107-14
[5] Carrouge D 2002 *Phase transformations in welded supermartensitic stainless steels* (University of Cambridge) p 195
[6] Wang P, Shanping L and Dianzhong L 2008 Investigation on phase transformation of low carbon martensitic stainless steel Zg06Cr13Ni4Mo in tempering process with low heating rate *Acta Metallurgica Sinica* **6** 681-85
[7] Park Eun Seo, Yoo Dae Kyoung and al. 2004 Formation of Reversed Austenite during Tempering of 14Cr-7Ni-0,3Nb-0,7Mo-0,03C Super Martensitic Stainless Steel *Metals and Materials International* **10** 521-25
[8] Delgado Reina S Yu 2014 *Formation of submicrocrystalline structural state during thermomechanical treatment of low carbon steels and steel composites* (Yekaterinburg) p 168
[9] Tong W U 2007 *Experimental and numerical simulation of welding induced damage stainless steel 15-5PH* (INSA-Lyon France) p 186
[10] Ajit K Roy 2005 Tensile Properties of Martensitic Stainless Steels at Elevated Temperatures *Journal of Materials Engineering and Performance* **2** 212-18.
[11] Bogachev I N and Lepekhina L I 1977 Superplasticity of chromium martensitic steels *Fiz. Metal. Metalloved* **44** 1282–85
[12] Gladshtejn L I, Rivanenok T N and Khristo A V 2008 Dilatometer analysis of kinetics of polymorphic transformation during the heating of steel *Zavod. Lab. Diagnost. Mater* **6** 36–9
[13] Leontiev P A, Simonov Yu N and Panov D O 2014 Method of processing the primary dilatometric data obtained under the conditions of high-temperature heating and cooling *Zavod. Lab., Diagnost. Mater* **80** 45-8
[14] Kolebina N V and Danilov V L 2015 Influence of morphology and phase composition on creep properties of low carbon martensitic steel *Nanomechanics Sciense and Technology: An Internation Journal* **6** 1 -9
[15] Rabotnov Y N 2014 *Creep of Structural Elements* (Moscow: Nauka Press) p 752
[16] Roy A K 2005 Tensile Properties of Martensitic Stainless Steels at Elevated Temperatures *Journal of Materials Engineering and Performance* **2** 212-18
[17] Bao G, Hutchinson J W and McMeeking R M 1991 Particle reinforcement of ductile matrices against plastic flow and creep *Acta Metall and Mater.* **39** 1871-82
[18] Hom C L, McMeeking R M 1991 Plastic flow in ductile materials containing a cubic array of rigid *Int. J. Plasticity* **7** 255-74
[19] Ravichandran K S 1994 A simple model of deformation behavior of two phase composites *Acta metall.* **42** 1113-23
[20] French J D, Zhao J, Harmer M and al. 1994 Creep of duplex microstructures *Journal of the American Ceramic Society* **77** 2857-65

IOP Conf. Series: Materials Science and Engineering **489** (2019) 012003 doi:10.1088/1757-899X/489/1/012003

Acousto- and electroplastic effects in alloy with reversible martensitic transformation

A V Frolova[1,3], Y V Tsarenko[2], V V Rubanik Jr[2], V V Rubanik[2] and V V Stolyarov[1,3]

[1] Mechanical Engineering Research Institute of the Russian Academy of Science, Moscow, 101990, Russia
[2] Institute of Engineering Acoustics, Belorussian Academy of Sciences, Vitebsk, 210009, Republic of Belarus
[3] National Research Nuclear University MEPhI (Moscow Engineering Physics Institute), Moscow, 115409, Russia

anna13frolova@gmail.com

Abstract. The article discusses the deformation behaviour of $Ti_{49.3}Ni_{50.7}$ alloy in the austenitic state during tensile tests under pulsed current and ultrasound at ambient temperature and 200 °C. The stress jumps from a combination of ultrasound and / or single pulse current are analyzed. The direction and magnitude of stress jumps are determined by the phase state of the material, the degree of deformation and the presence of electroplastic (EPE) and acoustoplastic (APE) effects. The different direction of the stress jumps in this shape memory alloy is related to the position of the deformation temperature relative to the deformation martensite formation temperature.

1. Introduction

Modern materials in the process of production or operation may experience complex external effects of various natures, for example, associated with mechanical stresses and deformation, electromagnetic field, acoustic waves, and vibrations. The behaviour of materials in such difficult conditions requires an analysis of the influence of external factors on their physical and mechanical properties. The presence of structural phase transformation in materials imposes additional requirements on the prediction of technological and operational responses to external impacts. One of the strategic tasks in the process of production is to increase the deformability and reduce the effort in metal forming process, for example, when cold rolling or drawing long products of thin section.

In [1, 2], the effect of the joint use of Severe Plastic Deformation (SPD) and ultrasound (US) was demonstrated in pure nickel - reduction of flow stresses, relaxation processes and hardening of the material depending on the ultrasound amplitude. The combined use of SPD and pulse current also proved to be a method for improving technological and functional properties, as well as the microstructure of various materials: in TiNi alloy [3], in TRIP and stainless steels [4], in aluminum [5-7] and magnesium alloys [8].

In recent decades, the combined effect of various methods with the aim of obtaining the required technological and functional characteristics has been actively investigated. In pure copper [9] and

IOP Conf. Series: Materials Science and Engineering **489** (2019) 012003 doi:10.1088/1757-899X/489/1/012003

nickel [10] this possibility was shown by the example of the sequential application of ultrasound and pulse current, which led to an increase in the effect of a decreasing flow stresses.

At present time, a special role belongs to structural alloys with the shape memory effect $Ti_{49.3}Ni_{50.7}$, which can undergo thermoelastic reversible martensitic transformations, both during production and during operation processes. A potential method of increasing the deformability of $Ti_{49.3}Ni_{50.7}$ alloys is the use of electroplastic (EPE) and acoustoplastic (APE) effects [11-13], however, the mechanism of such influence is not sufficiently studied.

The purpose of this work is to investigate the electroplastic and acoustoplastic effects during tension under pulse current and ultrasound in $Ti_{49.3}Ni_{50.7}$.

2. Material and methods

The object of the study was an alloy with shape memory effect in the austenitic state at ambient temperature $Ti_{49.3}Ni_{50.7}$ in the form of a wire with a diameter of Ø 1.5 mm after quenching from 800 °C with a grain size of ~ 50 μm. In accordance with the quality certificate, the start temperature of martensitic (M_s) and final temperature of austenitic (A_f) transformations for the $Ti_{49.3}Ni_{50.7}$ alloy are M_s = 6 °C and A_f = 26 °C, respectively, at cooling and heating. Tensile tests were performed on Instron machine (a ram speed of 1.5 mm/min) at different temperatures (23 °C and 200 °C). The specimens were heated with a BOSCH GHG 660 technical hot air gun; the temperature was controlled using a UT321 contact thermometer with a thermocouple attached to the center of the sample. The pulse current and ultrasonic oscillations in various sequences were introduced into the specimens at different stages of deformation and the amplitude of a stress jump from the EPE or APE effects was recorded. Regimes were as follows: frequency of 0.8 - 1 kHz, current density $j = 150$ A/mm^2 and pulse duration $\tau = 1 \times 10^{-4}$ s for pulse current; the frequency of 20 kHz, the duration of 1 s and the amplitude of 5 and 20 μm for ultrasound.

3. Experimental results

3.1. Tensile tests at ambient temperature

The $Ti_{49.3}Ni_{50.7}$ alloy without external impacts is characterized by high ductility ($\delta > 30$ %) and strength (800 MPa) (Figure 1, curve 1). The plateau corresponding to the martensitic transformation is short, practically passing to the inflection point. Single current pulses lead to a decrease in plasticity and multidirectional stress jumps: downward ($\Delta\sigma$ ~ 30–40 MPa) and upward ($\Delta\sigma$ ~ 50-60 MPa) (Figure 1, curve 2). A change in the direction of the jumps approximately corresponds to a strain of 10 %, and the stress peaks oscillate around curve 1. The use of ultrasound (Figure 1, curve 3) or a combination of pulse current and ultrasound (Figure 1, curve 4) leads to sharp embrittlement and destruction in almost elastic area. All jumps caused by ultrasound are directed downwards and have a smaller stress amplitude ($\Delta\sigma$ ~ 5–15 MPa) than jumps from pulse current ($\Delta\sigma$ ~ 35-45 MPa). The combined effect of ultrasound and current is similar in character to curve 3, (Figure 1, curves 3 and 4).

3.2. Tensile tests at 200 ° C

Under similar tests at the temperature of 200 °C, the type of curves and the direction of the stress jumps varies. First of all, this can be seen in the absence of a plateau associated with martensitic transformation, as well as in the same direction of stress jumps “down” from the pulse current and ultrasound (Figure 2). One can note, that all the curves from pulse current and ultrasound lie above curve 1 or practically coincide with it.

IOP Conf. Series: Materials Science and Engineering **489** (2019) 012003 doi:10.1088/1757-899X/489/1/012003

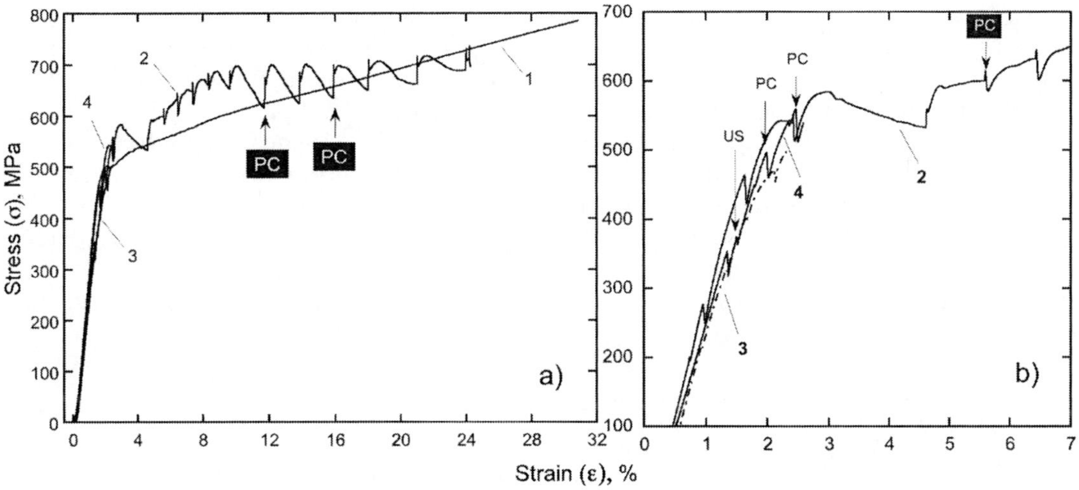

Figure 1. Stress-strain curves at ambient temperature: 1 - without external impacts, 2 - with pulse current (PC), 3 - with ultrasound (US), 4 – pulse current + ultrasound. (a) General view, (b) enlarged area for the curves 2, 3 and 4. Arrows indicate stress jumps from pulse current and ultrasound.

The $Ti_{49.3}Ni_{50.7}$ alloy at 200 °C, as compared with ambient temperature, exhibits less ductility (δ ~17 %) with almost the same level of strength (850 MPa) (Figure 2, curve 1). The impact of current pulses leads to stress jumps downward with an amplitude of Δσ ~ 35-40 MPa, (Figure 2, curve 2). In the case of the ultrasound introduction, material destruction occurs already in the elastic region at stresses below 300 MPa (Figure 2, curve 3). The combined effect of current and ultrasound (Figure 2, curve 4) plasticizes the alloy, the elongation to failure of which increases from 2.5 % to 7.5 % compared with ambient temperature. The amplitude of the stress jumps from the current (Δσ ~ 35-45 MPa) is higher than the amplitude of the jumps from the ultrasound (Δσ ~ 10 MPa).

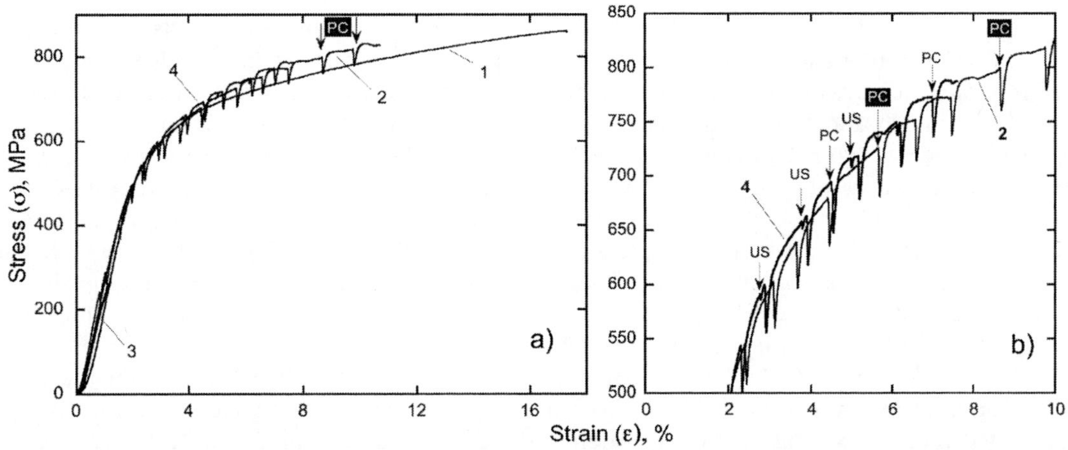

Figure 2. Stress-strain curves at a temperature of 200 ° C: 1 - without external impacts, 2 - with pulse current (PC), 3 - with ultrasound (US), 4 – pulse current + ultrasound. (a) General view, (b) enlarged area for the curves 2 and 4. Arrows indicate stress jumps from pulse current and ultrasound.

4. Discussion

The tensile curves of the alloy at ambient and 200 °C temperatures without external impacts are presented on Figure 3, curve 1 and curve 2. For the $Ti_{49.3}Ni_{50.7}$ alloy in the austenitic phase at ambient temperature, a not typically short plateau is observed, due to the phase transformation A → M, which is probably due to the proximity of the deformation temperature and A_s (26 °C). The difference in elastic modulus before and after the plateau is explained by the existence, respectively, of austenite and martensite, for which the ratio $E_A > E_M$ is true [14]. The difference in the slope of the curves in the elastic deformation region is can be also explained by this fact.

The character of the tensile curve at 200 °C, which according to many literature data is higher than the deformation martensite formation temperature M_d, corresponds to stable austenite, which is not capable of experiencing martensitic transformation. This causes the absence of a plateau from martensitic transformation.

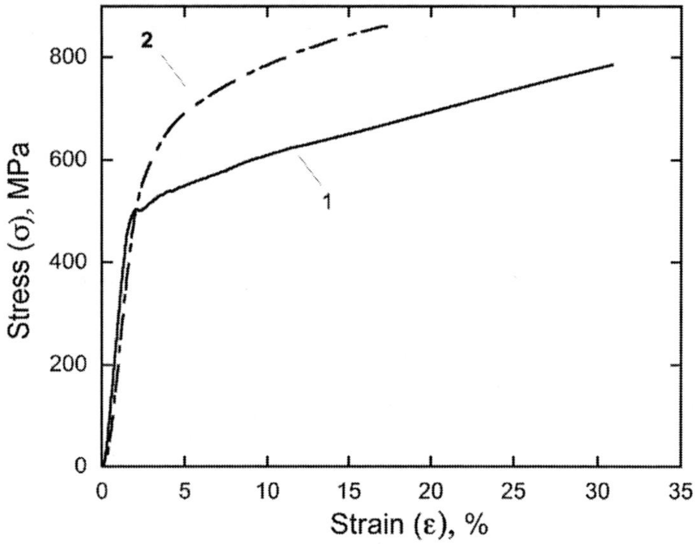

Figure 3. Stress-strain curves without external impacts: 1 - at ambient temperature, 2 - at a temperature of 200 °C.

Now compare the deformation behaviour of the alloy with the introduction of pulse current and / or ultrasound. The introduction of single current pulses at ambient temperature leads to emergence of differently directed stress jumps caused by different physical nature. EPE causes jumps "down" at the stage of martensitic transformation, and reverse phase transformation M → A leads to stress jumps "up" at the stage of work hardening [11]. The introduction of ultrasound causes the emergence of stress jumps "down" associated with the APE, the value of which, in this case, is noticeably less than the EPE. Note that, despite the general thermal nature of the ultrasound and pulse current, they have features of the deformation effect: these are the elastic deformation of the lattice (APE), and the plastic deformation due to dislocations (EPE). Therefore, hardening due to martensitic transformation prevails under pulse current (EPE), and a relaxation effect prevails under ultrasonic action (APE), which causes softening (Figure 1, curves 2 and 3).

At 200 °C, the stress peaks have the same direction downward regardless of the deformation region. The direction of the jumps "down" indicates the absence of martensitic transformation, which is associated with the existence of stable austenite at the deformation temperature $T_d > M_d$. At the same time, EPE and APE are still present.

The combined effect of both, ultrasound and pulse current, did not reveal any noticeable differences in the deformation behaviour of the alloy, except for the increased brittleness ($\delta < 5\%$) due to low-cycle fatigue during introduction of ultrasound. However, the combined effect from APE and

IOP Conf. Series: Materials Science and Engineering **489** (2019) 012003 doi:10.1088/1757-899X/489/1/012003

EPE requires additional research with a relatively equal level of input energy or other types of metal forming.

5. Conclusions

An increase in the deformation temperature from ambient temperature to 200 °C in the $Ti_{49.3}Ni_{50.7}$ alloy leads to an increase in the flow stress. The plateau disappears due to stabilization of austenite, as well as decrease in ductility.

The direction of the stress jumps from the introduction of ultrasound (APE) and / or pulse current (EPE) is determined by the phase state of the material and the strain degree. At 200 °C, stress jumps have the same direction (downward), regardless of the strain degree. At ambient temperature, prior to the onset of strain hardening, stress jumps also have the same direction (downward), and then they become multidirectional ("down" from APE and "up" from EPE).

The different direction of the stress peaks in the shape memory alloy is related to the position of the deformation temperature regarding to the deformation martensite formation temperature. At the deformation temperature $T_d < M_d$, the direction of the jumps upwards is due to the reverse martensitic transformation caused by the thermal effect of the current. On the contrary, if the deformation temperature $T_d > M_d$, the direction of the jumps "down" corresponds to the absence of martensitic transformation.

Acknowledgments
The study was carried out with the financial support of the Russian Foundation for Basic Research in the framework of the research project No. 16-58-48001 IND_omi and the BRFFR project No. T16P-152.

References

[1] Samigullina A A et al. 2018 *Ultrasonics* **82** 313–21
[2] Zhilyaev A P, Samigullina A A, Medvedeva A E, Sergeev S N, Cabrera J M and Nazarov A A 2017 *Mater. Sci. Eng.,A* **698** 136–42
[3] Stolyarov V V 2015 *Mater. Sci. Technol.* **31 (13a)** 1536-40
[4] Beridze E, Gennari E C, Michieletto F and Forzan M 2015 *Appl. Mech. Mater.* **698** 264-72
[5] Ghiotti A et al. 2018 *CIRP Annals – Manuf. Technol.* **67 (1)** 289-92
[6] Yuryev V A, Baranov Yu V, Stolyarov V V, Shulga V A and Kostina I V 2008 *Bull. Russ. Acad. Sci. Phys.* **72 (9)** 1248–50
[7] Shirinkina, I G, Brodova I G, Astaf'ev V V, Yablonskikh T I, Stolyarov V V, Potapova A A, Frolova A V, Mukhgalin V V and Bel'tyukov A L 2014 *Phys. Metals Metallogr.* **115 (12)** 1221-30
[8] Xu Z, Tang G, Tian S, Ding F and Tian H 2007 *J. Mater. Process. Technol.* **182** 128–33
[9] Kozlov A V, Mordyuk B N and Chernyashevsky A V 1995 *Mater. Sci. & Eng. A* **190** 75-79
[10] Samigullina A A, Mukhametgalina A A, Zhilyaev A P, Nazarov A A, Rubanik V V and Tsarenko Yu V 2017 *Ultrasound: problems, developments, prospects. Materials of the Int. Sci. Conf.* (Ufa: Bashkir State University) p 71-73 (in Russian)
[11] Potapova A A, Stolyarov V V, Bondarev A B and Andreev V A 2012 *Machine Build. Eng. Education* **2** 33-38 (in Russian)
[12] Rubanik V V, Rubanik V V Jr, Dorodeiko V G and Miliukina S N 2013 *Mater. Sci. Forum* **738-739** 362–66
[13] Misochenko A A, Tsarenko Yu V, Rubanik V V and Stolyarov V V 2017 *Ultrasound: problems, developments, prospects. Materials of the Int. Sci. Conf.* (Ufa: Bashkir State University) p 41-43 (in Russian)
[14] Brailovski V, Prokoshkin S, Terriault P and Trochu F 2003 *Shape Memory Alloys: Fundamentals, Modelling and Applications* (Quebec, Montreal) p 844

IOP Conf. Series: Materials Science and Engineering **489** (2019) 012004 doi:10.1088/1757-899X/489/1/012004

Aluminum alloy matrix composite wire reinforced by continuous carbon fibers

S N Galyshev[1], A I Gomzin[2], R F Gallyamova[2], A Yu Nazarov[3], E L Vardanyan[3], U Sh Shayakhmetov[4] and F F Musin[5]

[1]Institute of Solid State Physics of RAS, Chernogolovka, Moscow distr., Russia
[2]Department of Material Science and Metal Physics, Ufa State Aviation Technical University, Ufa, Russia
[3]Mechanical Design Technology, Ufa State Aviation Technical University, Ufa, Russia
[4]Department of Engineering Physics and Materials Physics, Bashkir State University, Ufa, Russia
[5]Ufa Chemistry Institute of RAS, Ufa, Russia

andre-gomzi@yandex.ru

Abstract. In this study, samples of composite wire with a matrix of aluminum alloy, reinforced with continuous carbon fibers were obtained by the method of liquid-phase infiltration with ultrasonic treatment. Mechanical tests for three-point bending were carried out and the structure of the composite was studied. The qualitative relationship between the strength of the CF/Al composite and the relief of the fracture surface is shown.

1. Introduction

Fabrication of metal matrix composites reinforced with carbon fiber is primarily complicated by the fact that carbon materials are practically not wetted physically by liquid metals [1], as a result the negative capillary effect prevents the spontaneous impregnation of carbon fiber by melt. The solution to this problem is to apply external pressure to the molten metal that compensates for the capillary effect [2]. However, the processes (liquid stamping, die casting, etc.) that implement this solution, in most cases, imply long-term contact (at least 60 seconds) of the melt with carbon fiber. This leads to the second technological problem of metal matrix composites at the interface of the matrix/fiber. As a rule, this is the cause of the destruction or complete transformation of the fiber in the molten metal, as in the case with carbon/titanium composites, the technology of which remains at the development stage today. It should be noted that the formation of carbides at the interface of the matrix/fiber in most cases plays a negative role, deteriorating the mechanical properties of the composite with a metal matrix.

An alternative to the aforementioned technological methods is the use of ultrasonic treatment of melt in the process of carbon fiber infiltration. Under the influence of ultrasonic vibrations in the matrix melt acoustic cavitation occurs. This phenomenon is the formation and collapse of cavitation bubbles, which locally causes a significant increase in pressure in the metal melt (up to several GPa [3]), which in turn compensates for the capillary effect and contributes to the infiltration [4].

In this paper, a scheme has been implemented for obtaining composite wire by pulling a bundle of carbon fibers through an AMg6 aluminum alloy melt subjected to ultrasonic processing. This method has several advantages, firstly, during its exploitation, the contact time of liquid aluminum with carbon fiber is only a few seconds. Secondly, the method allows to obtain a composite wire of any length, which can be used as a finished product, for example, as the basis of a conductive core for high-voltage power lines. Another application of the composite wire with a metal matrix is its use for building composites with a hierarchical structure, where the reinforcing component is composite wire, and as a matrix, metal or alloy.

2. Experimental materials and methods

2.1. Materials

In this work, AMg6 alloy as a matrix material was used. The chemical composition of the alloy is presented in Table 1. The choice of matrix alloy is due to the presence of magnesium, which suppresses the formation of aluminum carbide at the matrix melt / carbon fiber interface [5, 6].

Table 1. Chemical composition of AMg6, %

Al	Si	Fe	Cu	Mn	Mg	Zn	Be	Ti	Other
≥91,1	0,4	0,4	0,1	0,6	6,3	0,2	0,005	0,06	0,1

PAN-based carbon fiber in the form of a bundle of UMT49 and UMT430 grades was selected as a reinforcing component, the strength and modulus of elasticity of which are 4.9 GPa, 260 GPa and 4.5 GPa, 430 GPa, respectively (UMATEX Group, Russia).

2.2. The fabrication method

Samples of the composite wire were obtained by drawing carbon fibers through a molten aluminum alloy using a method described in the paper [7]. During the process of drawing, ultrasonic processing of the melt was performed (Figure 1). The fiber was passed through a 5 mm diameter hole in an ultrasonic niobium alloy horn, the lower part of which was immersed into the melt. The input power of the ultrasound unit was 1000 watts. The melt temperature was 800 ° C.

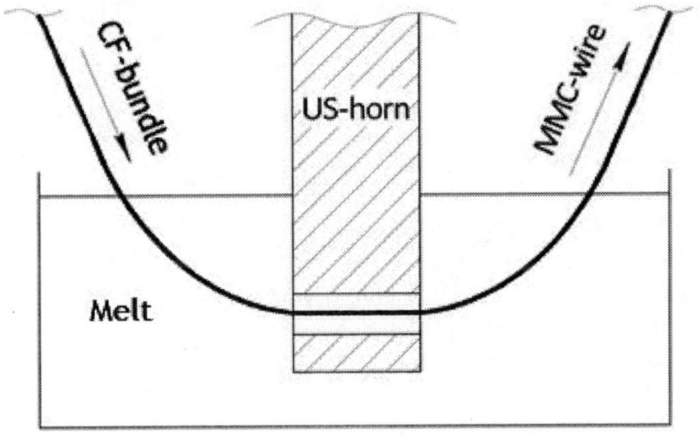

Figure 1. Scheme of composite wire fabrication

2.3. Bending tests and SEM

Mechanical tests for three-point bending were carried out according to the scheme close to GOST R 56810-2015. The cross-sectional shape of the wire was taken as an ellipse, the maximum strength values were obtained by the following equation (for ellipsoidal cylinders):

$$\sigma = \frac{8FL}{\pi ab^2} \quad (1)$$

where F is the maximum load preceding failure, *a* is the major axis of the ellipse in the cross section of the sample, *b* is the minor axis of the ellipse in the cross section of the sample, L is the distance between the supports; L is 15 mm.

The microstructure analysis of the CF/Al composite specimens was conducted using scanning electron microscopy (JSM-6490LV, JEOL, Japan). The material was shot in the mode of secondary electrons with an accelerating voltage of 20 kV at magnification up to 500 times.

3. Results and discussion

As a result of experiments on the carbon fibers drawing through an AMg6 melt, samples of CF/Al composite wire were obtained. All obtained samples had a similar specific structure (Figure 2a). One must note that the carbon fiber bundle is completely infiltrated with aluminum alloy. The fibers in the volume have nonequilibrium distribution, in the central part the volume fraction of fiber in the composite was approximately 60%, the average volume fraction of fiber over the whole cross-section was about 30%.

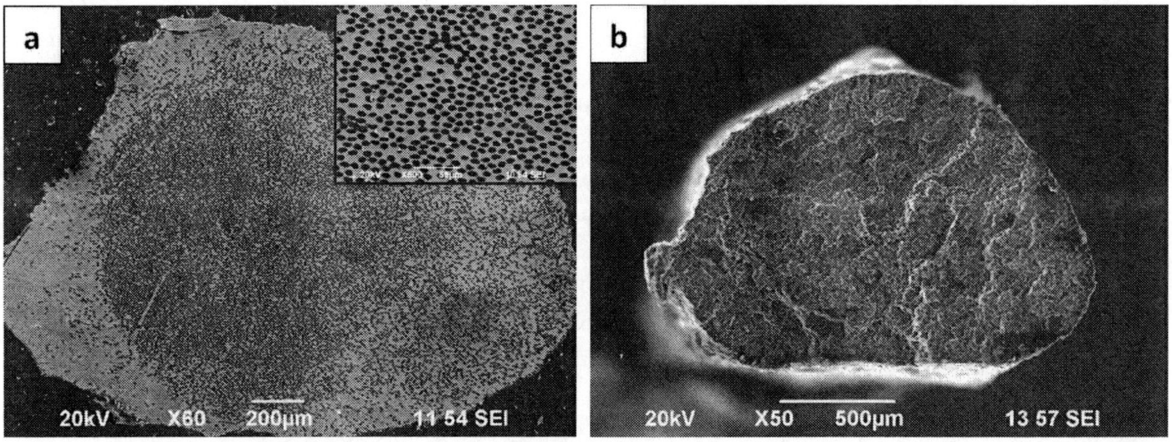

Figure 2. Cross-section of a CF/Al composite wire: microstructure (a) and the fracture surface after strength tests (b).

The bending strength of the obtained samples of CF/Al composite wire ranged from 315 to 1445 MPa. The surface relief of the tested samples fractures correlates with the value of their strength (Figure. 3). Samples with greater strength have a more rough fracture surface, which indicates a high level of crack energy dissipation. This character of destruction is more typical for the composites of the ceramic/ceramic system type with "weak" matrix/fibers interfaces [8], however, similar situation was observed in the CF/Al composite [9]. It should be noted that for maximum strength of the CF/Al composite, the matrix/fiber interface must not be "weak" - it has to be at the optimal level.

The fracture surface of the weakest specimen, on the contrary, is practically flat, which indicates easy crack propagation. The authors of [10, 11] showed that the observed phenomenon is associated with the formation of aluminum carbide Al_4C_3 at the matrix/fiber interface. The causes of the different types of composite failure will be considered in further studies.

The composite having the maximum strength (1446 MPa) was reinforced with high-modulus UMT450 fiber, while the remaining samples of composites whose strength did not exceed 625 MPa were reinforced with standard UMT49 fiber. Despite the fact that the high modulus fiber has lower strength compared to the standard (4.2 GPa and 4.9 GPa, respectively), the composite reinforced with UMT450 fiber has maximum strength.

IOP Conf. Series: Materials Science and Engineering **489** (2019) 012004 doi:10.1088/1757-899X/489/1/012004

Most likely, this is due to the orientation of atomic planes of carbon, which are the structural elements of fiber surface layer that were known as the basic structural units (BSU) [12]. High modulus fiber is produced under conditions, which are different from conventional low modulus fiber. It can be assumed that the resulting high-modulus fiber surface is different from the surface of conventional fiber in a greater orientation of BSU along the fiber, whereas in a standard fiber they are less oriented. This means that the edges of unoriented BSU come to the surface of the fiber, thus creating on the surface areas with a higher reactivity, where later the formation of the first aluminum carbide crystals occurs, which affects the crack propagation [8].

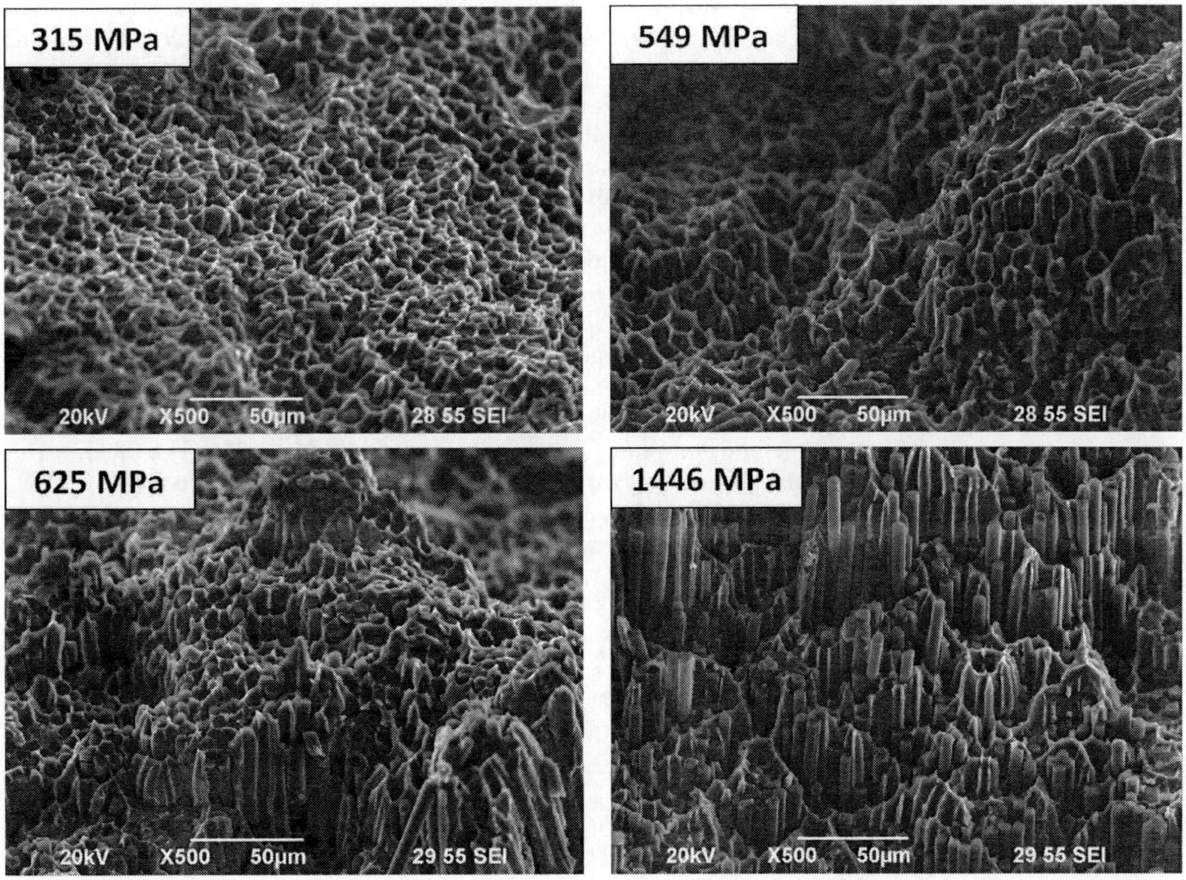

Figure 3. Results of mechanical testing of composite specimens and corresponding images of

4. Conclusion

Aluminum alloy composite wires reinforced by continuous carbon fiber were fabricated using the liquid-phase infiltration method with ultrasonic treatment. The maximum strength of the composite on a three-point bending amounted to 1446 MPa.

The relationship between the strength of the CF/Al composite and the relief of the destruction surface is established. The flat fracture surface indicates more easily propagated crack, resulting in a low strength of the composite.

Acknowledgments

IOP Conf. Series: Materials Science and Engineering **489** (2019) 012004 doi:10.1088/1757-899X/489/1/012004

This work has been performed with the financial support of the Ministry of Education and Science of the Russian Federation within the framework of the base part of the state task №11.6454.2017/BP and within the state assignment framework of the Ufa Institute of Chemistry Russian Academy of Sciences on the subject №AAAA-A17-117001191002l-8.

References

[1] Chernyshova T A, Kobeleva L I, Shebo P, Panfilov A V 1993 *Vzaimodeystviye metallicheskikh rasplavov s armiruyushchimi napolnitelyami* [Interaction of metallic melts with reinforcing fillers] (Moscow: Nauka) p 272

[2] Galyshev S N, Gomzin A I, Musin F F 2018 On infiltration pressure in the liquid-phase fabrication technology of metal matrix composites *Composites & nanostuctures* **10** (2) p 85 – 91

[3] Tzanakis I, Xu W W, Eskin D G, Lee P D, Kotsovinos N 2015 In situ observation and analysis of ultrasonic capillary effect in molten aluminium *Ultrasonics Sonochemistry* **27** p 72-80

[4] Matsunaga T, Ogata K, Hatayama T, Shinozaki K, Yoshida M 2006 Infiltration mechanism of molten aluminum alloys into bundle of carbon fibers using ultrasonic infiltration method *Journal of Japan Institute of Light Metals* **56** (4) p 226–32

[5] Revzin B, Fuks D, Pelleg J 1996 Carbide formation in aluminium-carbon fibre-reinforced composites *Compos. Sci. Technol.* **56** p 3-10

[6] Pelleg J, Ashkenazi D, Ganor M 2000 The influence of a third element on the interface reactions in metal-matrix composites (MMC) Al-graphite system *Mater. Sci. Eng. A* **281** p 239-47

[7] Matsunaga T, Matsuda K, Hatayama T, Shinozaki K, Amanuma S, Jin P, Yoshida M 2006 Development in manufacturing of carbon fiber reinforced aluminum preform wires using ultrasonic infiltration method *Journal of Japan Institute of Light Metals* **56** (1) p 28–33

[8] Mileiko S T 1997 *Metal and Ceramic Based Composite* (Amsterdam: Elsevier) p 690

[9] Vidal-Se´tif M H, Lancin M, Marhic C, Valle R, Raviart J-L, Daux J-C, Rabinovitch M 1999 On the role of brittle interfacial phases on the mechanical properties of carbon fibre reinforced Al-based matrix composites *Mater. Sci. Eng. A* **272** p 321-33

[10] Wang X, Jiang D, Wu G, Li B, Li P 2008 Effect of Mg content on the mechanical properties and microstructure of Grf/Al composite, *Mater. Sci. Eng. A* **497** p 31-6

[11] Yunhe Z, Gaohuia W 2010 Comparative study on the interface and mechanical properties of T700/Al and M40/Al composites *Rare Metals* **29** (1) p 102-7

[12] Guigon M, Oberlin A, Desamort G 1984 Microtexture and Structure of Some High-Modulus, PAN-Base Carbon Fibres *Fiber. Sci. Technol.* **20** p 177-98

IOP Conf. Series: Materials Science and Engineering **489** (2019) 012005 doi:10.1088/1757-899X/489/1/012005

The analysis of applicability of the FlowVision software for capillary effect simulation

S N Galyshev[1], L A Ishbulatov[2] and A A Aksenov[3]

[1]Institute of Solid State Physics of RAS, Chernogolovka, Moscow distr., Russia
[2]Department of Material Science and Metal Physics, Ufa State Aviation Technical University, Ufa, Russia
[3]Joint Institute for High Temperatures of RAS, Moscow, Russia

blacksail772@mail.ru

Abstract. The article discusses the modeling of hydrodynamic processes in the FlowVision software package, taking into account the influence on the capillary effect of changing various parameters of a liquid, such as capillary diameter, wetting angle, surface tension force and fluid viscosity. Authors solved test tasks about the effect of the above parameters on the difference between the values obtained by mathematical methods and modeling. A qualitative and quantitative analyses of the results were made, and conclusions were drawn about the possibility of using the FlowVision software to simulate the infiltration process of the frame of reinforcing fibers with a binder when forming products from metal-matrix composites.

1. Introduction

Over the past decades, composites have actively entered our life and replaced traditional materials in such areas as energy, transport, electronics and others.

Nevertheless, there is a high scatter of their stiffness and strength characteristics, the source of which is the instability of source materials, deviations in the implementation of technological processes, insufficient production culture, the choice of mode, and the calculation of the time of infiltration. Because of this, it is necessary to introduce additional safety coefficients when determining the projected load, which affects the cost of the manufacture [1].

To solve the problem of choosing the mode and calculating the infiltration time, modern modeling methods can be used in software products such as ANSYS Multiphysics, PAM-RTM, FlowVision, etc., which will reduce cost and time for the manufacture of test products.

Let us analyze the capabilities of the FlowVision software. It is intended to simulate the dynamics of the flow of viscous liquids and gases, taking into account the effects of wettability, surface tension forces, hydraulic resistance, etc. The advantages of this software package are relatively low requirements for computer resources and a relatively low cost in comparison to major global competitors, such as ANSYS, PAM-RTM, ProCAST.

To confirm the suitability of FlowVision for solving the tasks related to the infiltration of the frame of reinforcing fibers with a binder, a number of test tasks were solved. The results confirm the possibility of taking into account the capillary effect in the FlowVision software package. Qualitative pictures of the distribution of a fluid in the both problems correspond to the reality, but a comparison of the simulation results and theoretical calculations showed that the quantitative discrepancy in the data was

IOP Conf. Series: Materials Science and Engineering **489** (2019) 012005 doi:10.1088/1757-899X/489/1/012005

about 30%. [2] To obtain a more accurate result, reduce the error, with the assistance of technical support service, some corrections were made and additional test tasks were solved to determine the effect of the capillary diameter, wetting angle, surface tension force, and fluid viscosity on the flow of a fluid in the capillary.

2. Results and discussion

With the assistance with technical support service, some changes in the program were made, and in the new version of the program we achieved the following results: we changed the parameters of capillary radius, surface tension, wetting angle and viscosity.

The parameters were not chosen randomly, almost all of them are contained in the Jurin's formula for calculating the height of a liquid:

$$h = \frac{2\sigma \cdot \cos\Theta}{r(\rho - \rho_0)g}$$

where
r is the tube radius;
ρ is the liquid density;
ρ_0 is the external medium (air) density;
Θ is the wetting angle;
σ is the surface tension force.

By changing these parameters in the simulation we obtained different heights of the fluid and compared them with the mathematical calculations. Various parameters were used to ensure that we could fairly say that the Jurin's formula is correct with all the differences in the FlowVision software package and the sofware can be used to simulate more complex fluid flow processes, such as modeling the infiltration of the frame of reinforcing fibers with a binder.

1. Under standard conditions, two different radii were used: 0.5 and 1 mm. In figures 1 and 2, one can see a line to which they have risen, these lines are the estimated capillary height, in the both cases the error was within 5%.

Figure 1. Tube radius is 0.5 mm.

IOP Conf. Series: Materials Science and Engineering **489** (2019) 012005 doi:10.1088/1757-899X/489/1/012005

Figure 2. Tube radius is 1 mm.

2. Then, the radius was left equal to 1 mm and the surface tension of water equal to 0.0589 N m^{-1} at 100 ° C was used. At the same time, the height should have been lower than that under standard conditions (figure 3). A red line shows the calculated height, the capillary, after small oscillatory movements, stopped at the required height. The error was less than 5%.

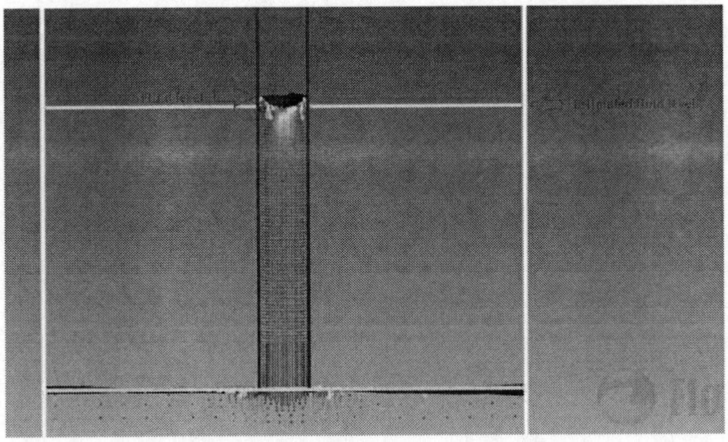

Figure 3. Surface tension is 58.9 N m^{-1}.

3. The height of a liquid column also depends on the material of a tube itself and its interaction with the liquid. To indicate the nature of this interaction in FlowVision, one can set a wetting angle for the surface. Under standard conditions, the value of 5 degrees was used, in the subsequent tasks we used the values of 70 and 145 degrees. At a wetting angle of 70 degrees, we observed a full agreement with the mathematical calculation (figure 4).

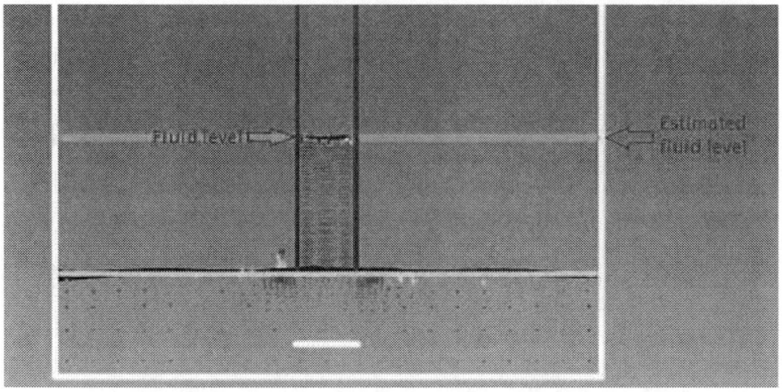

Figure 4. Wetting angle is 70°.

4. In the case of a "non-wetting" capillary surface, we observed a qualitatively correct picture (figure 5), but quantitatively there were discrepancies, the error with the mathematical model was 10%, and the capillary with a wetting angle of more than 90 degrees needs optimization.

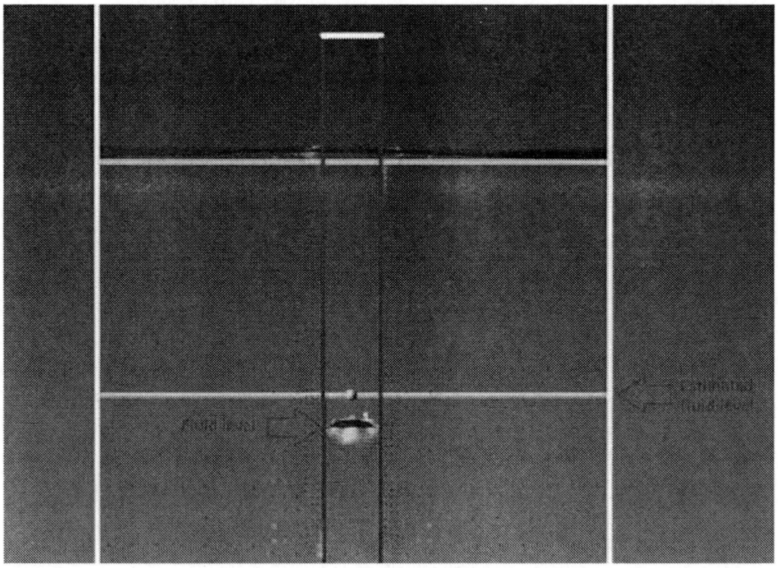

Figure 5. Wetting angle is 145°.

3. Conclusion

The FlowVision software package shows good results in matching the mathematical and computed visualized models, the discrepancy for most tasks was only about 5%. However, for a task with a non-wetting surface, the discrepancy is about 10%, and it is necessary to optimize for tasks with a wetting angle of more than 90 degrees. The qualitative and quantitative pictures of modeling correspond to the reality, which enables us to use it for the further simulation of the flow of liquids in more complex structures, taking into account wettability, surface tension, viscosity, capillary effect, etc. This makes possible to solve the task of impregnating reinforcing fibers with a binder when forming metal-matrix composites in the FlowVision design package.

4. References

[1] Skvortsov Yu.V. Mechanics of composite materials [Lecture text] – Samara: Publishing center

IOP Conf. Series: Materials Science and Engineering **489** (2019) 012005 doi:10.1088/1757-899X/489/1/012005

SSAU, 2013.
[2] Lenar Ishbulatov, Sergey Galyshev, Pavel Solovev, Andrey Aksenov, Simulation of the capillary action via FlowVision software, MATEC Web of Conferences 224, 02092 (2018)

Acknowledgments
This work has been performed with the financial support of the Ministry of Education and Science of the Russian Federation within the framework of the base part of the state task №11.6454.2017/BP.

IOP Conf. Series: Materials Science and Engineering **489** (2019) 012006 doi:10.1088/1757-899X/489/1/012006

Composite materials cohesion parameters

K S Kotova[1] and G S Slavcheva[2]

[1]Department of Design of Buildings and Constructions named after N.V. Troitskogo, Voronezh State Technical University, Bldg 84, 20-letiya Oktyabrya St., Voronezh 394006, Russia
[2]Department of Building Materials Technology and Structures, Voronezh State Technical University, Bldg 84, 20-letiya Oktyabrya St., Voronezh 394006, Russia

kottova-k@yandex.ru

Abstract. The article presents the study results of parameters of the macroporous composite cohesion with glass composite parameters, basalt composite and metal reinforcement elements by the pull-out of the bar from a concrete cube. It has been established that the best joint operation of reinforcement element with macroporous composite can be ensured by the use of glass composite reinforcement element. Ultimate strength of its cohesion with macroporous composite of B7.5 (D1400), B10 (D1600) classes is 20-45% higher than the values of cohesion strength of basalt composite and metal reinforcement elements for all options of combining the classes by strength and diameter of reinforcement elements (∅ 6,8,10 mm). With increased strength class of the composite, ultimate strength of all the studied types of reinforcement elements is growing consistently. The change in strength of reinforcement elements cohesion with macroporous composite through variation of their diameter has different patterns for the studied types of reinforcement elements.

1. Introduction

The value of ultimate strength of cohesion ensures reliable operation of constructive elements consisting of heterogeneous composite materials. This feature is the one defining the construction rigidity on the whole. Cohesion of reinforcement elements with composite is cooperation of three elements: composite, reinforcement, and contact layer area. The structure and characteristics of macroporous composite influence the adhesion together with geometric and physical-mechanical properties of reinforcement.

It should be admited that joint operation of metal reinforcement element with cement composites of different classes has been studied rather thoroughly. In addition to studying the cohesion of steel reinforcement elements with the composite, we have also obtained the results of experimental and theoretical research of fiberglass parameters and basalt plastic reinforcement elements and structures reinforced with them [1- 6]. Issues of cohesion methodology tests have been studied and recommendations on the change of the composite protective layer thickness of reinforcement element in the composite embedment ratio were suggested [7, 8]. The obtained results of the ultimate strength values of reinforcement elements cohesion with thick and light composites were received through numerical modeling of their behavior or experimentally by the method of axial pull-out of reinforcement element from the body of the composite, as the most popular one [9-11]. At the same time, influence of the type, geometry, reinforcement element surface, kind and strength of composite as well as its age and hardening conditions were evaluated [12, 13]. As a result, requirements for reinforcing elements securing the best cohesion with thick high-strength composites are well-known.

At the same time, there is no reliable information on parameters of macroporous composites cohesion with the reinforcement elements, which are considerable factors of constructions bearing capacity. This information is necessary for calculating and designing constructions based on macroporous composites [14-21].

Principal conditions ensuring reliable joint operation of composite and reinforcement element in the constructions are the following:

- cohesion of reinforcement element with composite along their connection preventing the pull-out (shift) of reinforcement element in the composite,
- the value of composite linear strain coefficient and reinforcement element.

The structure of macroporous composite will have a heterogeneous contact layer with the surface of reinforcement, which will logically influence the value of cohesion. Secondly, macroporous composite has lower strength and increased deformability compared to the other types of composites [22]. Therefore, there is a high probability of pull-out (shift) of high-strength metal reinforcement made of macroporous composite. The result is that construction potential of metal reinforcement cannot be fully used in constructional parts made of macroporous composite. Thirdly, macroporous composite does not always ensure protection of reinforcement from corrosion in contrast to thick composite.

For the mentioned above reason, the main problem of the study is the definition of requirements to reinforcement element (type, physical-mechanical properties, profile, diameter) which secure optimal joint operation of reinforcement element with macroporous composite.

This paper presents the studies results of the metal glass composite cohesion value and basalt composite reinforcing elements, different in physical-mechanical properties, with macroporous composites of density 1400-1600 kg/m^3 considered as materials for compression and bending structural elements.

2. Materials and methods

2.1. Properties of materials and factor space of the research

The following factors varied in the experiments during the study of the cohesion parameters in "macroporous composite-reinforcement" system:

1. Type and diameter of reinforcement element. In addition, all reinforcement elements had similar periodic profile.
2. Average density of macroporous composite. That said, composite mix remained the same as for mass ratio of cement to sand; average density was regulated by variation of the water amount and SAS chemical additives.

The method of pull-out the reinforcement bar from the body of composite was used to evaluate the joint operation of reinforcement with composite. Embedment length depended on the diameter and accounted for 5d. Total length of reinforcement elements, in compliance with the requirements of European Standards [23, 24], was determined by the embedment length, structure of testing machine, and length of testing clutch structure so that the protruding bars could be captured by the testing machine and the pull-out sensor could be installed. Macroporous composite mixtures with property packages optimised in our previous studies were used to manufacture the samples [25] (table 1). Reinforcement elements properties were used in the experiment and parameters of their embedment into composite samples are presented in table 2.

Table 1. Properties of macroporous composite mixtures

Average density, kg/m^3	Ratio C:S	W/C ratio	Foaming agent dosage, % from the mass of cement	Superplasticiser dosage, % from the mass of cement

1400	1:1.75	0.30	0.04	0.4
1600	1:1.75	0.31	0.05	0.5

Table 2. Properties of the used reinforcement elements and parameters of their embedment into macroporous composite samples

Type of reinforcement	Production factory	Nominal diameter d, mm	Modulus of elasticity, MPa	Ultimate strength under tension, σ_b MPa	Yield strength, σ_τ H/mm ^{2}	Values of design compression resistance, MPa	Specific elongation, %	Unrestrained bar length, mm	Embedment length, mm	Bar length, mm
Glass composite	Voronezh composite materials factory	6	50 000	800	none	300	2.2	300	30	600-800
		8						400	40	
		10						500	50	
Basalt composite	Yaroslavl composite materials factory	6	50 000	800	none	300	2,2	300	30	600-800
		8						400	40	
		10						500	50	
Metal profiled A500	OOO «Metallinvest Plus»	8	200 000	600 (ultimate strength)	500	350	Not less than 14	300	40	750-850
		10						600	50	

Cohesion with all types of reinforcement described in table 1 was evaluated in the experiments for macroporous composites of each average density.

2.2. Description of experimentation conditions

Series of sample cubes 100×100×100 mm in size were used in cohesion tests. There were 6 samples in the series for each option of type combination, reinforcement diameter, and average density of macroporous composite. The samples were manufactured in metal forms with one opening for reinforcement in each parallel side. The openings served for fixing the bar horizontally in the centre of each cube form where the reinforcement bar was mounted perpendicularly to the sides and parallel to the direction of composite laying.

Macroporous composite was manufactured by two-stage method: at the first stage water was mixed with air-entraining and superplasticing SAS additives, at the second stage bonding agent and filler were added. They were mixed in a turbine mixer at a speed of 1300 min^{-1} for 4 minutes. Once the mixture was put in the form, short time vibration was used to eliminate the air stuck near the walls of the form.

Testing of cohesion parameters was conducted through pull-out the reinforcement bar out of the sample cube using a universal static electromechanical testing system INSTRON 5982. The sample was put in a special metal frame fixed from one side in the gripping device of the testing machine. The sample in the frame was supported by a metal bearing plate and rubber interlayer between it and the frame. A displacement sensor with precision of indication of 0,01 mm was mounted on the loose end of the bar in a special carrier frame to measure the slipping of reinforcement in the composite. To measure longitudinal deformations of the reinforcement bar under tension, an extensimeter mounted in the centre of operational height of embedded bar was used. The size of the frame ensured the required distance for the installation on the protruding end of the displacement sensor bar. During the pull-out tests the value of the applied load with indications of 0,05 mm, 0,1 mm and 0,25 mm was recorded together with the value of the bar displacement under maximum load. Before the test the samples were examined and measured. The scheme of testing the samples under axial reinforcement pull-out is presented in figure 1.

In order to control the strength and density of macroporous composite, control series of unreinforced sample cubes 100×100×100 mm in size were manufactured at the same time in the volume of 24 samples for each average density.

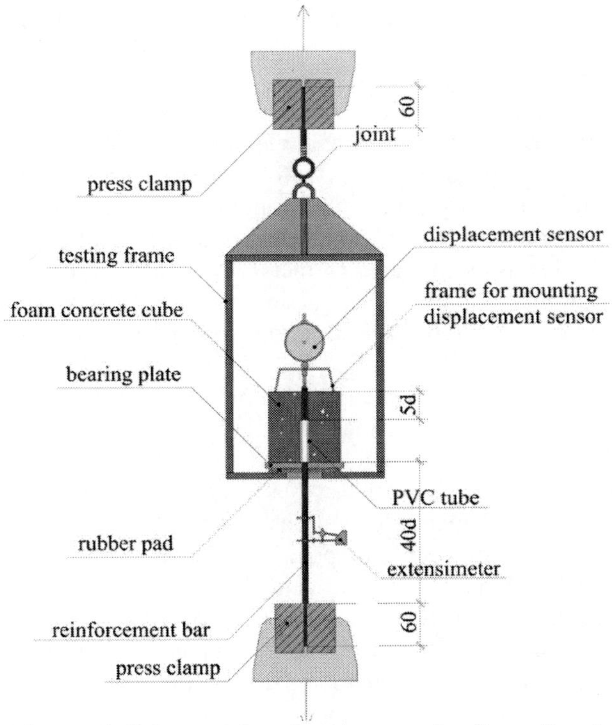

Figure 1. Scheme of testing the samples for pull-out

The samples hardened in normal conditions for 28 days. Strength tests were performed by standard method using universal 4-column static hydraulic testing system 1500HDX by INSTRON. Test results defined statistically-valid class by strength and brand by average density according to regulated standard procedures [18, 19].

2.3. Cohesion value calculation method

Strength of reinforcement cohesion with macroporous composite was evaluated by the value of resistance of pull-out the reinforcement bars from the composite. The calculation is based on defining the values of shear stress along the boundary of reinforcement cohesion with the composite realised under maximum load obtained through stretching a sample until it is destroyed regardless of the place of destruction – along the bar or along the boundary of reinforcement cohesion with the composite. Cohesion strength, according to International Standard [26], was determined by the formula (1).

$$\tau = \frac{P}{cL_{fb}}, \tag{1}$$

P is applied load, N;
c is nominal circumference of the bar, mm;
L_{fb} is length of bar embedment into composite, mm.

Processing of the test results included designing “cohesion stress – slipping” charts for each diameter. Cohesion stress for each sample was calculated for the values of applied load under the indications of bar slipping sensor at marks of 0.05; 0.1; 0.25 mm, as well as under maximum load causing the slipping of the loose end of the bar. Then, average cohesion value was calculated for each series of the samples.

3. Results

Complex test results analysis (table 3, figures 2,3,4) allowed to establish the following. During the tests, slipping of all the studied reinforcement types and diameters in the composite was determined. Pull-out of reinforcement elements of all types was accompanied by the damage in "composite – reinforcement" contact area, rupture of reinforcement element was not observed (see figure 2).

3.1. Reinforcement element type influence

The highest cohesion values were recorded for glass composite reinforcement element. Ultimate cohesion strength of this type of reinforcement element is 1.3-1.8 times higher than for basalt composite element and 1.2-1.9 times higher than for steel element for all the studied diameters of reinforcement elements and classes by composite strength. In addition, for macroporous composite of B7.5 and B10 classes the pull-out value of basalt composite reinforcement elements in control displacement points of the bar is comparable to metal reinforcement element under the condition of equal nominal diameter. Comparison of the destruction nature with "cohesion stress – slipping" charts show that it is conditioned by the nature of destruction when pulling the reinforcement element out of the composite body.

As you can see on the presented "cohesion stress – slipping" characteristic charts (figure 4), pull-out of glass composite reinforcement element is accompanied by significant slipping deformations, which leads to the increase of the effort required for its pull-out. As for fiberglass reinforcement element, local layer separation of coiling from the bar was observed when pulled out (figure 2a). This allowed assuming that this type of reinforcement element has higher cohesion of coiling with composite than with the body of the bar. Pull-out of basalt composite reinforcement bar from macroporous composite was done exactly along "composite – reinforcement element" contact area (figure 2b) without breaking the coiling, as in this type of reinforcement element the coiling is partly sunk in longitudinal fibre. Pull-out of metal reinforcement element was accompanied by partial destruction of composite near the area of contact with reinforcement element (figure 2c).

3.2. Influence of reinforcement element diameter

Change of cohesive strength of reinforcement element with macroporous composite under the variation of its diameter has different patterns for the studied types of reinforcement elements (table 3, figure 3).

Glass composite reinforcement element of B7.5 class has shown the best result of cohesive strength for the bars of diameter 6, then the value reduced by 25% and remained the same with the increase of cross section. Basalt composite reinforcement element within class B7.5 has not shown clear cohesion dependence on the change of diameter (figure 3a). As for class B10, reduction of cohesion with macroporous composite as the diameter is increased was recorded for composite reinforcement element of two types.

However, metal element shows an increase of cohesive strength by 20% for B7.5 class composite with the change of its diameter from 8 to 10 mm (figure 3b). Deterioration of cohesive strength values is noticed in class B10.

3.3. Influence of composite strength

With the increase of composite compression strength class, ultimate strength limit of cohesion of reinforcement elements with macroporous composite naturally grows (table 3). As the composite class is increased, ultimate strength limit of cohesion grows to a variable degree depending on the diameter of reinforcement element:

- by 1.3 times for glass composite reinforcement element for bars ∅ 6 mm, by 1.6 times for bars ∅ 8 mm, by 1.3 times for bars ∅10 mm;

- by 1.7 times for basalt composite reinforcement element for bars ∅6 mm, by 1/3 times for bars ∅8 mm, by 1/1 times for bars ∅10 mm;
- by 1.8 and 2.2 times for metal reinforcement element, for bars ∅8 mm and ∅10 mm correspondingly.

Figure 2. Nature of reinforcement elements cohesion failure with macroporous composite

Table 3. Test results of determining the value of reinforcement elements cohesion with macroporous composite

Strength class	Density grade	Type of reinforcement element	Diameter	MAX pull-out value P, kN	Displacement of A for P_{max}, mm	Cohesive strength limit with concrete, MPa
B 7.5	D 1400	Glass composite	6	4.01	3.28	7.21
			8	5.05	2.55	5.36
			10	8.63	1.63	5.36
		Basalt composite	6	1.38	1.40	3.08
			8	3.94	1.55	4.05
			10	5.84	1.88	3.81
		Metal	8	2.90	0.85	2.80
			10	5.68	0.35	3.36
B 10	D 1600	Glass composite	6	5.08	2.80	9.12
			8	8.46	2.93	8.98
			10	11.57	1.38	7.19
		Basalt composite	6	2.29	3.30	5.12
			8	4.91	2.20	5.04
			10	7.42	1.80	4.84
		Metal	8	6.34	0.48	6.11
			10	10.08	0.33	5.97

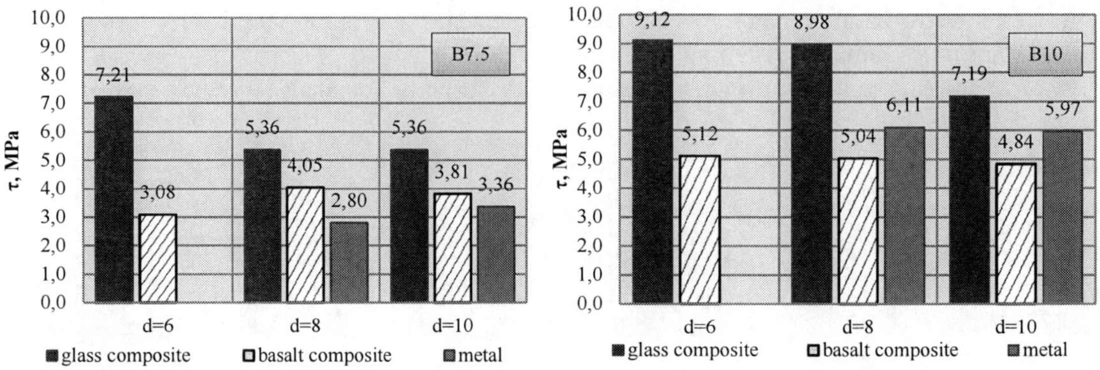

Figure 3. Cohesive strength of different types of reinforcement elements with macroporous composite of classes

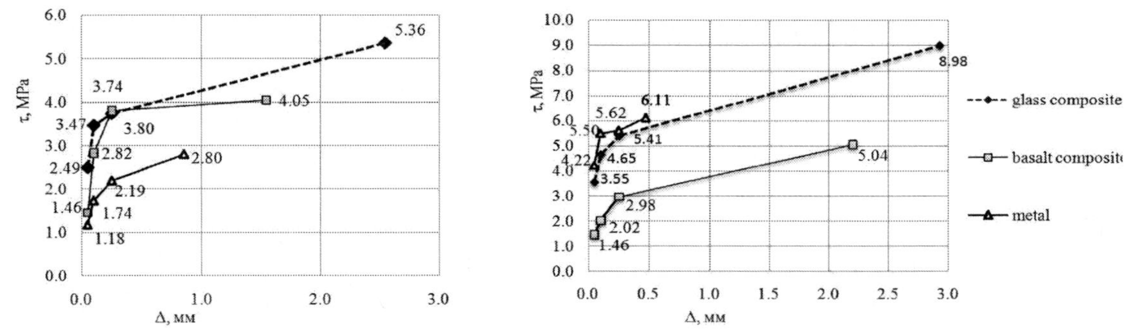

a) Strength class B7.5 b) Strength class B10

Figure 4. Dependence on reinforcement elements cohesion of the 8 mm diameter with macroporous composite of classes from slipping

4. Conclusion

It has been established that best joint operation of reinforcement elements with macroporous composite can be ensured by using glass composite reinforcement element. Ultimate strength of its cohesion with macroporous composite of classes B7.5 (D1400), B10 (D1600) is 20-45% higher than the values of cohesive strength of basalt composite and metal reinforcement element for all types of combining the classes by strength and diameters of reinforcement (∅6,8,10 mm). It is conditioned by the nature of destruction at the pull-out of reinforcement elements from the composite body. As for fiberglass plastic reinforcement element, cohesive strength in the "composite – reinforcement element" contact area is higher than coiling cohesion forming the profile of the reinforcement element with the body of the bar. Basalt composite reinforcement element has the lowest cohesive strength with macroporous composite, when pulled out it is destroyed right along the "composite – reinforcement element" contact area. Pull-out of high-strength metal reinforcement element is accompanied by partial destruction of the composite near the contact area with the reinforcement element.

With increased strength class of composite, ultimate strength of all the studied types of reinforcement elements is growing consistently. However, the change in strength of reinforcement elements cohesion with macroporous composite through variation of their diameter has different patterns for the studied types of reinforcement elements. As for composite reinforcement element, the strength of cohesion with composite is inclined to decrease as its diameter grows, and it is exactly the opposite for metal reinforcement element.

The prospects of further research are related with the solution of the following tasks:

IOP Conf. Series: Materials Science and Engineering **489** (2019) 012006 doi:10.1088/1757-899X/489/1/012006

- comprehensive experimental-theoretical studies and tests of constructive macroporous composite elements reinforced with different types of reinforcement elements;
- development of recommendations for calculations and design of macroporous composite constructions based on the results of experimental research.

References

[1] Dong-Uk C, Sung-Chul C and Sang-Su H 2012 *Eng. Struct.* **34** 303-13
[2] Baena M, Torres L, Turon A, Barris C 2009 *Compos. Part B: Eng.* **40**(8) 784-97
[3] Jabbar A A S, Farid B H S 2018 *Karbala Int. J. Mod. Scien.* **4(2)** 216-27
[4] Okelo R, Yuan R L 2005 *J. Compos. for Constr.* **9(3)** 203-213
[5] Bazli M, Ashrafi H, Oskouei AV 2017 *Constr. Build. Mater.* **148** 429-443
[6] Aydin F 2018 *Constr. Build. Mater.* **163** 1-8
[7] Hao Q, Wang Y, Ou J 2008 *Eng. Struct.* **30(11)** 3239-46
[8] Carvalho E P, Palhares M M, Fernandes D S G and Alves G V 2018 *Constr. Build. Mater.* **83** 243-52
[9] Phan TS, Rossi P, Tailhan J-L 2015 *Cem. Concr. Compos.* **59** 1-9
[10] Baena M, Torres L, Turon A, Barris C 2009 *Compos. Part B: Eng.* **40(8)** 784-97
[11] Caro M, Jemaa Y, Dirar S, Quinn A 2017 *Eng. Struct.* **147** 448-57
[12] Rolland A, Quiertant M, Khadour A, Chataigner S, Benzarti K and Argoul P 2018 *Constr. Build. Mater.* **173** 136-148
[13] Li T, Zhu H, Wang Q, Li J and Wu T 2018 *Constr. Build. Mater.* **185** 545-54
[14] Bilek V, Bonczková S, Hurta J, Pytlík D and Mrovec M 2017 *Proc. Eng.* **190** 243-247
[15] Al-Shannag M, Charif A 2017 *J. King Saud Univ. – Eng. Sci.* **29(4)** 365-372
[16] Novidis D, Pantazopoulou S 2008 *4th Int. Conf. on FRP Composites in Civil Engineering* (July 22-24, Zurich: Switzerland) pp. 323-333
[17] Song X, Wu Y, Gu X and Chen C 2015 *Constr. Build. Mater.* **94** 209-217
[18] Sancak E, Simsek O , Apay A 2011 *Int. J. Phys. Sci.*, **6 (14)** 3437-54
[19] Shang H-S, Cui F-K, Zhang P, Zhao T-J, Ren G-S 2017 *Constr. Build. Mater.* 150 529-537
[20] Yoo D-Y, Shin H-O 2018 *Cem. Concr. Compos.* **93** 206-217
[21] Hassan AA, Hassan K M A and Lachemi M 2010 *Constr. Build. Mater.*, **24** 520-530
[22] Novikov M V, Chernyshov E M, Slavcheva G S 2016 *Constr. Mater.* **11** 26-31
[23] EN 10080, 2005 Steel for the reinforcement of concrete - Weldable reinforcing steel – General European Standards
[24] EN 12390-3 2009 Testing hardened concrete - Part 3: Compressive strength of test specimens European Standards
[25] Slavcheva G S, Chernyshov E M, Novikov M V 2017 *Constr. Mater.*7 20-24
[26] ISO 10406-1:2015 Fibre-reinforced polymer (FRP) reinforcement of concrete-Test methods-Part 1: FRP bars and grids International Standards

IOP Conf. Series: Materials Science and Engineering **489** (2019) 012007 doi:10.1088/1757-899X/489/1/012007

Influence of the inclusion shape on the effective elastic properties of composites

I V Magnitsky[1] and E S Sergeeva[1,2]

[1]JSC "Kompozit", Pionerskaya, 4, Korolev, Moscow Region, 141070, Russia
[2]Bauman Moscow State Technical University, 2-ndBaumanskaya, 5, Moscow, 105005, Russia

info@kompozit-mv.ru

Abstract. The elastic properties of composites are mainly determined by the characteristics of reinforcing inclusions. This work addresses the dependence of the elastic properties of inclusion-reinforced composites on the inclusion shape. Specifically, ball-shaped, spherical and cubic inclusions are considered.

1. Introduction

Currently, composite materials are widely used in the technology and science. This can be explained by their unique properties, specifically, their ability to withstand simultaneous high-intensity mechanical, physical and thermal loads, while having smaller specific weight as compared to pure metals. In view of this, these materials are of particular importance in the aerospace industry [1].

In aircraft engineering there is a tendency to apply increasingly the carbon-ceramic composites (CMCM). An important feature of this class of composites is a combined matrix composed of a carbon base and inclusions of individual particles that can form one or more ceramic-nature substances. These ceramic inclusions enhance the erosion/oxidation resistance of the composite, while influencing its physical and mechanical properties. In the CMCM fabrication process, as well as during design, development and operation of CMCM-based structures, the influence of these inclusions on the carbon matrix characteristics should be taken into account, particularly, the elastic behavior as being the subject addressed in this paper. For the composite matrix, solution of this problem represents micro-level modeling in terms of building a three-level model of the composite [2].

One of the most widely used and mature methods to introduce ceramic inclusions in a material is the liquid-phase siliconizing that involves formation of particles composed of silicon and silicon carbide. The inclusion shape depends on the specificity of the porousness pattern of the initial carbon-carbon piece, as well as on the siliconizing process mode. Therefore, the inclusion shape can vary widely even for the same material. Thus to build a micro-level model that provides correct representation of the material matrix behavior, it is necessary to investigate the degree and nature of the inclusion shape influence on the matrix effective properties.

There is a number of papers focused on the analytical approaches to investigation of the CMCM elastic behavior on the micro-scale level [3-9]. The advantages of these approaches are the simplicity of relationships and the minimum requirements to the computing capacity. However, these methods have a number of shortcomings, which makes them inapplicable for solving the problem stated above.

The analytical methods are based on assumptions that are often unrealistic for the actual material; moreover, most of these methods don't take into account the shape of the inclusions.

Therefore, this paper focuses on the use of the finite-element method (FEM) implemented using the ANSYS software to investigate the inclusion shape influence on the inclusion-reinforced matrix elastic behavior.

2. Problem definition

As the target of our investigation, we take the 4DL -reinforced ceramic matrix carbon material [7]. As can be seen in Figure 1, the matrix inclusions have various shapes and dimensions. When designing a composite, it is impossible to consider the actual arrangement of the ceramic inclusions in the material; therefore, to simulate the combined matrix behavior, we consider some typical shapes of ceramic inclusions that can be found in the actual composite. For example, the simplest type is a ball-shaped inclusion that consists of silicon carbide (type I) or, mainly, of pure silicon (type II). However, given the specific form of the material's reinforcing element cross section and the reinforcement pattern, the ceramic inclusions are often polygon-shaped, which can be simulated by a cubic form (type III) when making computations. Since in the carbon matrix, formation of silicon carbide takes place mainly on the surface of the pores that may be inaccessible during the saturation process until the moment of complete saturation, there are also thin-wall shell-shaped inclusions in the composite that can be simulated as a full (type IV) or fragmented (type V) sphere.

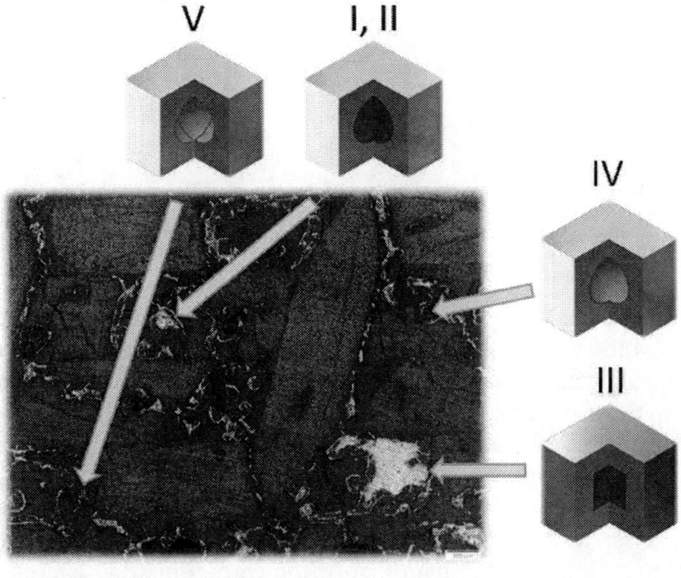

Figure 1. Types of inclusions in the composite material.

Here we analyze what is the influence of the ceramic inclusion shape on the carbon matrix elastic behavior for these five typical shapes of various dimensions. As a dimension measure, the volume ratio of the filled pore is taken, as related to the matrix material, irrespective of what part of the matrix is particularly filled with the inclusion. This value is called conditional reinforcement ratio μ. Analysis of the modern CMCM structures shows the practical importance of range $\mu \in [0;\ 0,35]$, as being characteristic for the most widely used siliconizing processes and, consequently, determinative for the requirements to the carbon-carbon semi-products.

To simulate the elastic behavior of the combined matrix, its cubic volumes with different-type inclusions at their centers were considered. For the volumes, kinematic boundary conditions were set to provide their uniform strain states as per [8]. Note that for the problem in this formulation, these volumes are representative in terms of [9] where tension of volumes along three orthogonally-related axes, as well as shear in the planes of supposed orthotropic properties, are considered. The 3D elastic problems were solved by the finite element method in the ANSYS software environment. To determine the effective elastic properties of the combined matrix, the approach described in [10] was applied. The elastic properties of the inclusion materials corresponded to the data given in [11], while those of the matrix – to the results described in [12].

3. Results

Dependencies of effective elasticity modulus E and shear modulus G on conditional reinforcement ratio μ for the combined matrix with inclusions of different types are shown in Figures 2 and 3, correspondingly.

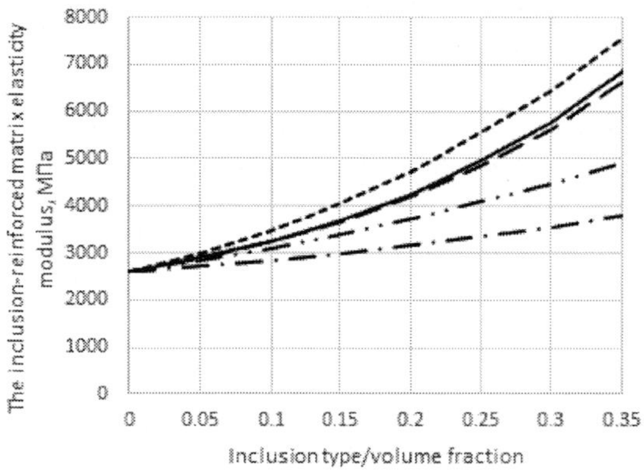

Figure 2. The inclusion-reinforced matrix elasticity modulus vs. the inclusion type/volume fraction.

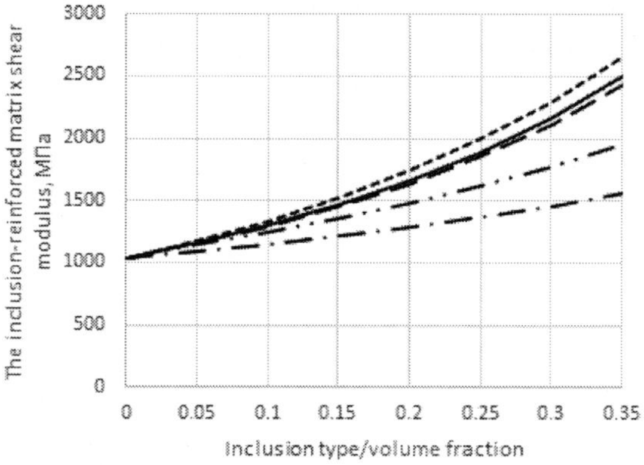

Figure 3. The inclusion-reinforced matrix shear modulus vs. the inclusion type/volume fraction.

Here and elsewhere, the dash-dot single-point curves correspond to the results obtained for the fragmented SiC sphere, while the dash-dot double-point curves – to those obtained for the SiC sphere, the dotted ones – to those obtained for the Si ball, the solid ones – to those obtained for the SiC ball, and the dashed ones – to those obtained for the SiC cube.

From the curves shown above it can be seen that the influence of the solid ceramic inclusion does not exceed 10%, which is negligibly small within the reasonable design calculation accuracy. Even less is the difference between the results obtained for the matrix with silicon carbide inclusions and that with pure silicon ones, which can be explained by the fact that both types of inclusions have stiffness two orders of magnitude higher than that of the carbon matrix. The hollow inclusions (type IV and V) have less influence on the matrix properties; however, it is still large enough to be taken into account when estimating the composite characteristics. The fragmented structure of the spherical inclusion reduces considerably the apparent stiffness of the matrix.

The value of Poisson's ratio ν of the matrix depends only slightly on the presence of inclusions of types IV and V, while the presence of solid ceramic inclusions results in a considerable reduction of this parameter, which can be seen in Figure 4.

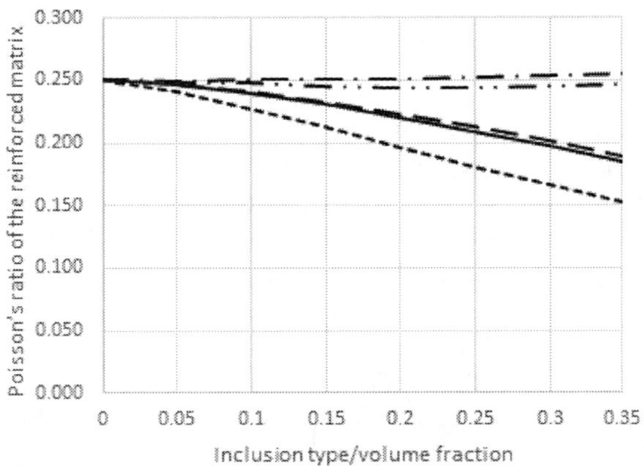

Figure 4. Poisson's ratio of the reinforced matrix vs. the inclusion type/volume factor.

Generally, the degree of inclusion influence on the composite's elastic properties was estimated by the relative change in the effective elasticity moduli along X and Y axes and the shear modulus in XZ plane, given the conditional reinforcement ratio of 35%. For the estimation, the 4DL-reinforced elastic model described in [13] was used. The estimation results are shown in Figure 5. It can be seen from the results that the influence of ceramic inclusions in the composite's elasticity moduli along the reinforcement direction does not exceed 10%, while that on the shear modulus can range to one-and-a-half times, depending significantly on the inclusion type. Consequently, the presence and the shape of ceramic inclusions should be taken into account when making design calculations for the composite elastic properties.

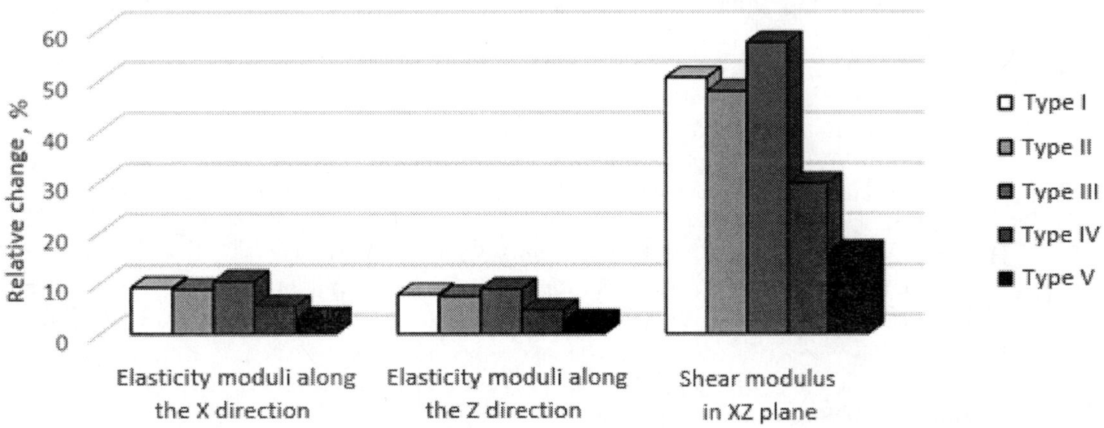

Figure 5. Relative change in the composite effective elastic properties resulting from inclusions in the matrix of the reinforcement factor of 0.35.

As the matrix reinforcement pattern in the composite involves cubic packing of inclusions, it seems reasonable to determine the degree of anisotropy in the resulting matrix material, which is obviously the case, given the random arrangement of ceramic inclusions in the matrix at its larger scales. As the anisotropy measure, relative deviation of the apparent shear modulus value from that obtained by the following well-known formula [14] is taken:

$$G = \frac{E}{2(1+\nu)}. \quad (1)$$

Analysis of the results showed that the larger the conditional reinforcement factor, the greater the deviation from isotropy, with its considerable values of 15-25% obtained for inclusion types I, II and III, while in case of hollow inclusions, this effect is negligibly small. This fact should be taken into account when modeling the characteristics of materials reinforced by discrete randomly-arranged inclusions, as the design case of the symmetry of properties is quite different from that expected in the actual materials. To balance out the effect of inclusion arrangement, stochastic models of materials can be used, as well as larger representative volumes containing multiple randomly-arranged different-type inclusions can be considered. The latter option would make it possible to take into account the presence of ceramic inclusions of various dimensions and configurations in the actual materials.

4. Conclusion

From the results obtained, the following conclusions were derived. Ceramic inclusions of different shapes and dimensions have a considerable impact on the elastic properties of the combined carbon-based matrix, depending on the inclusion type. Introduction of ceramic inclusions in the matrix influences considerably some elastic properties of the composite.

When considering the cubic packing of inclusions, the modeling predicts disturbance of isotropy of the matrix elastic properties, which can be avoided in practice through random arrangement and orientation of ceramic particles. This fact should be taken into account when predicting performance of composite materials under development.

Acknowledgments
The work was performed within the framework of implementation of the basic part of the governmental task of the Ministry of Education and Science of the Russian Federation (Project 9.7784.2017/BP).

References

[1] Biksha J 2014 *Bulletin of Electronics* **1** 24–7
[2] Tatarnikov O V 2012 *All materials. Encyclopaedic Guide* **7** 21–6
[3] Shermegor T D 1977 *Theory of elasticity for micro-heterogeneous media* (Moscow: Nauka). p 400
[4] Hashin Z and Strikman S *J.* 1962 *Mech. Phys. Solids* **10** 335–43
[5] Eshelby J 1963 *Continuous dislocation theory* (Moscow: IL) p 248
[6] Skudra A M and Bulavs F Ya 1978 *Structural theory of reinforced plastics* (Riga: Zinatne) p 192
[7] Tarnopolsky Yu M, Zhigun I G and Polyakov V A 1987 *Spatially reinforced composite materials* (Moscow: Mashinostroyeniye) p 224
[8] Pobedrya B E 1984 *Mechanics of composite materials* (Moscow: Publishing House of Moscow State University) p 336
[9] Hashin Z 1962 *J. Appl. Mech.* 29 pp 143–50
[10] Zarubin V S and Sergeeva E S 2018 *Mathematical Models and Computer Simulations* 10 pp 288–98
[11] Grigoryev I S and Meylikhov E Z 1991 *Physical values* (Moscow: Energoatomizdat) p 1232
[12] Vagin V P, Dvoretsky A E, Magnitsky I V et al 2015 *Non-destructive testing of composite materials: Collected papers of the 1st distant scientific and technical conference NKKM-2014* (St. Petersburg: Sven) p 228.
[13] Magnitsky I V 2013 *New technology* In: Proceedings of the Xth All-Russian Conference **2** (M.: RAS) 21–32.

IOP Conf. Series: Materials Science and Engineering **489** (2019) 012007 doi:10.1088/1757-899X/489/1/012007

[14] Zarubin V S, Kuvyrkin G N 2008 *Mathematical Models in Continuum Mechanics and Electrodynamics*(Moscow: The Bauman University Publishing House) p 512.

IOP Conf. Series: Materials Science and Engineering **489** (2019) 012008 doi:10.1088/1757-899X/489/1/012008

Measuring control of construction materials parameters in order to increase reliability of engineering objects

E D Pozdnyakova, E V Maslennikova, A S Komshin and S R Orlova

Department of Metrology and Interchangeability, Bauman Moscow State Technical University, Moscow, 105005, Russia

katya2110m@yandex.ru

Abstract. The use of string and string elements find widespread practical application in technology. In the field of measuring technology, there are many devices in which the sensitive elements are elements in the form of strings: strain gauges, accelerometers, crack meters, etc. String sensors are widely used due to its simplicity and low cost. The present paper demonstrates that preliminary studies of structural materials for the parameters of structural materials caused by a group of factors, such as surface properties, chemical composition, processing technology, can be achieved using modern methods of scanning probe, confocal, electron raster microscopy, spectroscopy, and x-ray.

1. Introduction

Strings and string elements are widely used in the technique. In the field of measurement technology, there are many devices in which the sensitive elements are elements in the form of strings: strain gauges, accelerometers, crack gauges, etc. String sensors are widely used due to their simplicity and low cost. The paper presents studies of the structural materials properties using modern methods of scanning probe, confocal, electron scanning microscopy, spectroscopy, radiography [1].

2. Samples of the hypothesis

The concept of the proposed study is based on the hypothesis that the rate of degradation can be significantly influenced by factors such as the surface topography, especially in the elemental and chemical compositions, which also need to be taken into account. The use of phase-chronometric method with the theory of measurements reduction opens up new opportunities to study the behavior and degradation properties of structural materials, including elastic and rheological [2,3].

Many scientists conduct comprehensive researches of the structural materials properties. The study of materials is of particular interest that undergoes mechanical stress during operation [4]. Elastic modules in this case are dependent on the frequency of the applied force, and the mechanical characteristics of the construction materials are not constant values, as is commonly considered, even for a specific product, but vary in time under the influence of external influencing factors [5-8].

Strings and string elements which are widely used not only in musical instruments but also in the other fields, for example, there are many different string strain gauges, string accelerometers, crack meters, etc. in measuring equipment, were chosen as a constructive element for the present research. String sensors are widely used due to their simplicity and low cost.

Oscillations of systems on the example of strings in various environments are widely considered. The questions of studying natural frequencies are widely covered and the features of changes in natural frequencies, numerical-analytical modelling of strings free oscillations problems and rods in various media, etc. are established in the scientific literature.

A mathematical model of the degradation process of an elastic element (string) using the example of a linear oscillator with variable parameters is proposed, and the possibility of observing changes in its properties is demonstrated.

Parameters that directly relate to the properties of the structural material from which the string under study is made are estimated using mathematical modeling of the string oscillation process taking into account the degradation of the structural material using the example of a linear oscillator with variable parameters in the form of:

$$\ddot{x} + 2\beta(t)\dot{x} + {\omega_0}^2 x = 0 \tag{1}$$

where x - is the deviation from the equilibrium position;
$\beta(t)$ - attenuation coefficient, slowly varying in time, directly related to changes in material properties;
ω_0 - natural angular frequency.
The known variable replacement

$$x = v(t)e^{-\int \beta(t)dt} \tag{2}$$

allows converting the original equation to the form

$$\ddot{\upsilon} + \Omega^2(t)\upsilon = 0 \tag{3}$$

here $\Omega^2(t) = {\omega_0}^2 - \beta^2(t) - \dot{\beta}(t)$.

Application of the generating function

$$\varphi = \frac{1}{2} m_{\Delta}(t) v^2 ctgQ \tag{4}$$

provides a transition using the Hamilton formalism from variables p, v to canonical variables action - phase (P, Q).

The achievement of this result is guaranteed by the fact that the information capacity of gating, determined by quantization errors of time intervals δτc and high-speed recording of discrete samples of the coordinate δz or phase δφ, is significantly higher than that of the existing approaches. At the same time, the gain in the value of information capacity on the basis of achievements in the field of fast processes metrology, radio electronics and optics compared with the variant caused by the "mechanical" increase in accuracy is estimated by the expression

$$I_2' - I_1 = \log_2\left[\frac{\Delta\varphi}{\delta\varphi} \cdot \frac{2\sigma_{\tau 0}}{\delta\tau_c}\right] \tag{5}$$

where δτc - errors in the quantization of time intervals, $\delta\varphi$ - errors in the registration of discrete phase samples; $\Delta\varphi$ - phase increment; $\sigma_{\tau 0}$ - root-mean-square deviation of the measurement of the time instant τ0, corresponding to the triggering of the chronometric system.

The proposed measuring system is implemented in the form of a test bench, an analogue to which is the installation that provides measurement of the vibrations addition and the demonstration of the mechanical resonance phenomena, given in paper.

The design provides the sensitivity of phase relationships, due to the measurement of which it is possible to control the degradation of the slowly changing properties of structural materials in time. Such engineering implementation of the measurement setup makes it possible to measure some peculiarities of string vibrations, for example, the dependence of the spectral composition of the vibrations on the tension and conditions of the string excitation, etc.

The principle of the stand operation is the next: the string is fixed on the one side motionless, and on the other it is necessary to provide adjustable tension through the feedback system. To register the oscillation period, the measuring system emitter-receiver is used. Free damped oscillations are excited automatically. The period of oscillation is measured using the photovoltaic method. The laser diode and the receiving photodiode are matched in wavelength. The luminous flux is modulated by passing

IOP Conf. Series: Materials Science and Engineering **489** (2019) 012008 doi:10.1088/1757-899X/489/1/012008

the zero position of the string. The error of measurement of time intervals is no more than $5 \cdot 10^{-8}$ seconds.

As the example of the implementation of such measuring systems installation, providing a measurement of the oscillations addition and demonstration of the mechanical resonance phenomena can be considered. The proposed design provides the sensitivity of phase relationships, due to the measurement of which it is possible to control the degradation of the slowly varying properties of structural materials in time.

A whole group of factors can influence the results of a measurement experiment and the degradation of the structural materials properties: surface properties, chemical composition, type of treatment, etc.

This paper presents the results of research on guitar strings, as the most widely distributed and industrially manufactured using the atomic-force microscopy, scanning electron microscopy, providing a study of the topography and surface structure, as well as the results of elemental and chemical analysis using spectroscopy.

3. Results

In figure 1, strings are presented in a cut and it can be seen that coatings from D'addario and Elixir manufacturers reflect electrons differently, which indicates that manufacturers use different materials as protective coatings. At the first measurements of features in the structure of the strings' surface were not detected. Two weeks later repeated studies were conducted. At the same time, during a break between studies, the strings were not subjected to any mechanical or chemical influences. But, as it can be seen from figure 2, the covering of the D'addario string, even before exploitation, began to peel intensively.

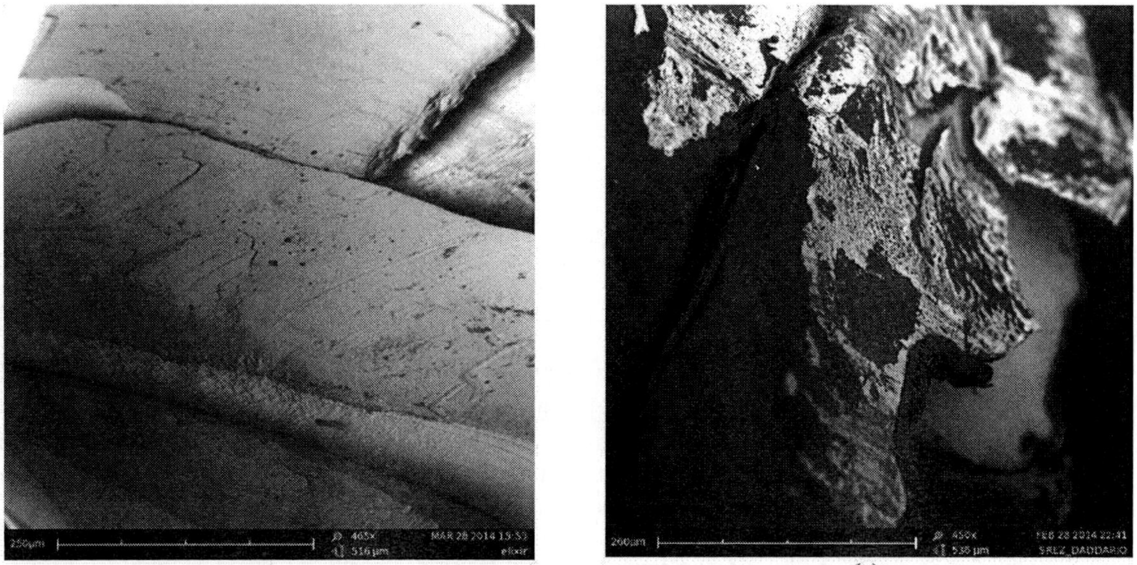

a) b)

Figure 1. (a) SEM image of the ElixirPolyWeb (USA) string cut surface; (b) SEM image of the D'addario EXP26 (USA) string cut surface

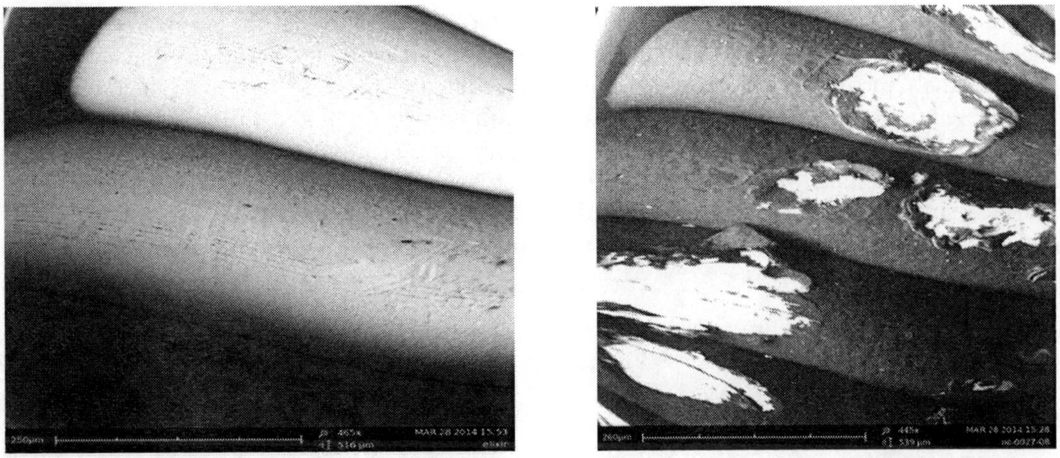

a) b)

Figure 2. (a) SEM image of the Elixir PolyWeb (USA) string surface; (b) SEM image of the D'addario EXP26 (USA) string surface

Figure 3 (a) shows the surface relief of the strings D'addario EXP26, the scan area is 40x40 μm. After a series of measurements, it was found that vertical bands are not image defects, but represent a feature of the surface structure. The investigated string has a winding and a protective coating.

Strings with braid are considered. The strings are presented in a cut, which shows that the coatings of the manufacturers D'addario and Elixir reflect electrons differently, which indicates the different properties of the protective coatings. Figure 3 (b) shows the relief of the Elixir PolyWeb string surface, the scanning area is 40x40 μm. It can be noted that the surface structure of this sample differs markedly from the previous one, the surface is more uniform, therefore at this stage it can be noted that the manufacturers D'addario and Elixir apply different protective coatings for their strings.

Since the degradation of a structural material properties is significantly influenced by a group of factors such as surface properties, chemical composition, type and duration of processing, measuring the parameters of the strings under investigation prior to testing them as a harmonic oscillator provides a priori information for accurately determining the residual life of the structural material. The presented results of the string surfaces studies by atomic force microscopy make it possible to determine the initial state of the parameter β(t) of expression (1) more accurately.

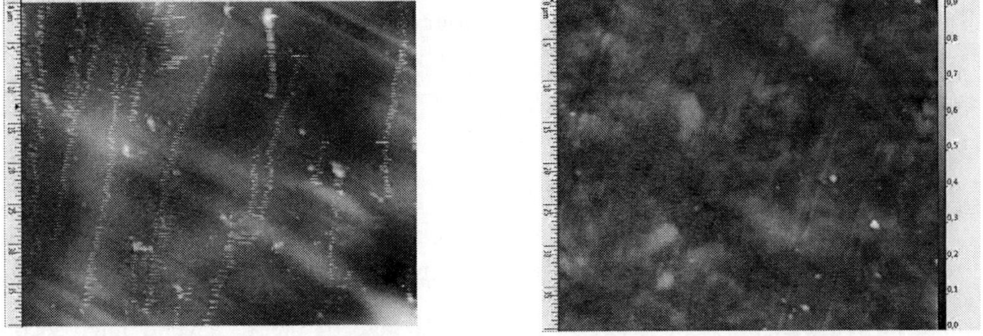

a) b)

Figure 3. (a) AFM image of D'addario EXP26 (USA) string surface; (b) AFM image of Elixir PolyWeb (USA) string surface

IOP Conf. Series: Materials Science and Engineering **489** (2019) 012008 doi:10.1088/1757-899X/489/1/012008

Figure 4 shows the relief of the D'addario EXP26 strings surface, the scan area is 30x30 μm. Strings do not have any winding or coating. On the AFM image of the surface of the strings, one can see longitudinal "furrows", which are a technological feature of the strings' manufacture. In order to obtain the required shape and size of the section, the string is pulled through the spinnerets. Mr. Musician "Bronze Age" has the scan area of 30x30 microns. Strings do not have winding and coating. The difference in heights in comparison with foreign analogues differs by 2-2.5 times and in general the surface structure is very heterogeneous. It can also be noted that the traces of the broach are practically invisible, which indicates that when making strings of the domestic producer, another technology is used.

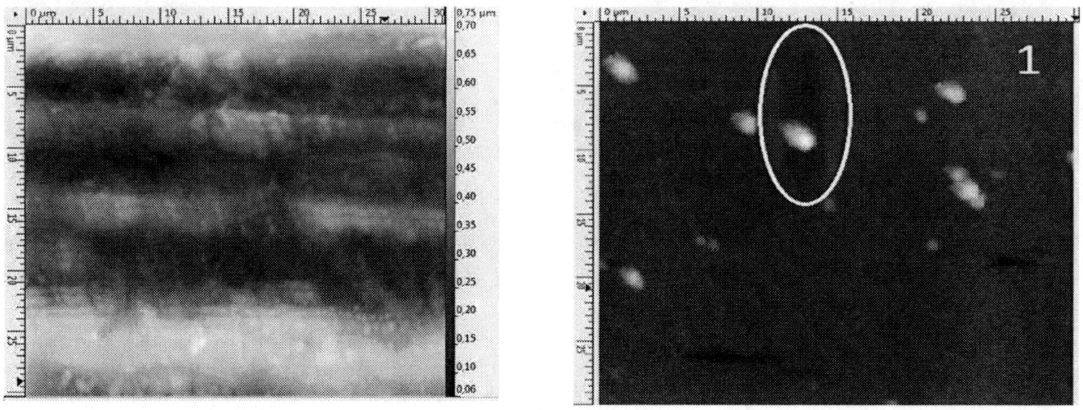

a) b)

Figure 4. (a) AFM image of D'addario EXP26 (USA); (b) AFM image of Mr. Musician "Bronze Age" (Russia).

The selected area (1) shows a surface defect of 1x9 μm, which is an indicator of a serious violation of the manufacturing technology.

The time intervals (period and its multiple fractions, corresponding to the passage of the corresponding phases of oscillations) measuring result obtained with a Ntegra Spectra microscope is influenced by such groups of factors as the change in the dimensions of the string as a result of stretching, the surface roughness, and the tension of the sample. Estimates of the effect of these parameters on the measurement result are given below.

Table 1 shows the result of calculating the effect of a change in the diameter of a string caused by tension and a continuous force action for the period of its oscillations.

Table 1. The effect of the change in the diameter of a string on the period of its oscillation

Measurement number	Δd = 5 μm, increment of the period, μs.
1	49,6 (2,23%)
2	49,7 (1,67%)
3	49,5 (1,33%)
4	40,1 (0,79%)
5	38,3 (0,57%)
6	39,7 (0,44%)

A change in diameter of 5 μm leads to a period change of 0.2 to several microseconds. The influence of the roughness of the string surface on the periods of oscillations is associated with a change in the outer diameter. It is shown that if the period of free damped oscillations is T1 = 0.006648184 s, then a two times decrease in the roughness of the string surface to Ra = 1.3 μm will lead to a change in the period of free damped oscillations within ± 2 μs.

The spectra obtained from the surface of GHS strings and the Mr. Musician "Bronze Age" with the help of Auger electron spectroscopy are shown in figure 5.

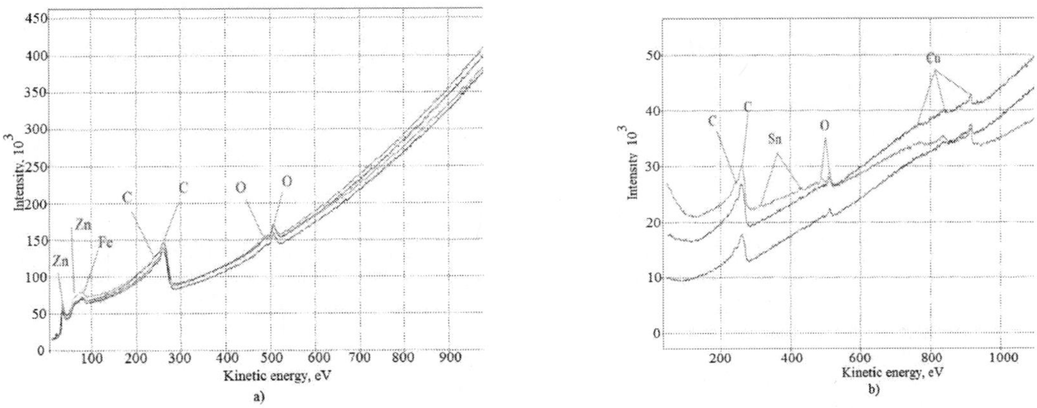

Figure 5. (a) Auger electron spectroscopy energy spectrum of GHS (USA) string; (b) Auger electron spectroscopy energy spectrum of Mr. Musician "The Bronze Age"(Russia) string

4. Conclusion

The presented results of the measuring control of the structural materials properties' degradation open fundamentally new possibilities for evaluation in the process of functioning, increasing the accuracy of determining the physical parameters and properties of structural materials by several orders of magnitude, and as a result, there are reduction of time and increase of reliability of design and technological development.

The paper shows that preliminary studies of structural materials in order to control the parameters of structural materials caused by the influence of a group of factors, such as surface properties, chemical composition, processing technology, can be provided by using modern methods of scanning probe, confocal, electron scanning microscopy, spectroscopy, X-ray.

The results of the strings' surface investigations by the methods of atomic force microscopy (AFM) and scanning electron microscopy (SEM), which provide research of topography and surface structure are presented. The results of elemental and chemical analysis by Auger electron spectroscopy are presented.

Acknowledgments

The results of the work were obtained within the framework of implementing the initiative scientific projects of the basic part of the State Assignment (NIR 9.4968.2017/BCh).

References

[1] Komshin A S and Orlova S R 2016 *Meas. Tech.* **59** 589 – 94
[2] Komshin A S 2015 *All materials. Enc Guide* **6** 2 – 7
[3] Postnikov V S 1954 *Physics-Uspekhi* **LIII** 87 – 108
[4] Afanasev K N, Ozerin S A, Rodionov I A et al. 2017 *IEEE Transactions on Nanotechnology* **16** 274 – 80
[5] Rodionov I A, Baburin A S, Zverev A V, Philippov et al. 2017 *Metamaterials, Metadevices, and Metasystems* **10343** 10343371-7
[6] Malysheva G V, Tumasova M S, Tepishkina E S et al. 2016 *Polymer Science. Series D* **9** 223 – 7
[7] Braginsky V B 1967 *JETP* **53** 1434 - 41
[8] Amstislavsky YaE and Mamakov A S 1966 *Physics-Uspekhi* **551** 710 – 4

IOP Conf. Series: Materials Science and Engineering **489** (2019) 012009 doi:10.1088/1757-899X/489/1/012009

Improvement of the machine determining technological capabilities method

A V Sakharov and N A Rodionova

Mechanical Engineering Research Institute of the Russian Academy of Science, Moscow, 101990, Russia

n.rodionova@mail.ru

Abstract. The article presents the method of determining the position of the modules of surfaces that manufactured on the machine in its working space. The method is based on the use of rectangular coordinate systems of the machine working space, the module of surfaces and establishment of dimensional connections between them.

One need to know the technological capabilities of the machine when forming the enterprise production program, determining the possibility of performing individual orders for production, purchasing machines, as well as developing technological processes for manufacturing parts.

The technological capabilities of the machine is a list of items of production, which are the surfaces, manufactured on the machine, with certain ranges of dimensions, level of accuracy and roughness, with certain dimensions of the parts.

The technological capabilities of the machine should be reflected in the assignment wording. Assignment wording analysis of machines with different degrees of universality showed that they do not have sufficient information about what surfaces can be made on these machines and with what characteristics. As an example, table 1 shows the assignment wording of a screw-cutting lathe and a milling machining center with an indication of their shortcomings.

Table 1. The assignment wording of a screw-cutting lathe and a milling machining center.

Machine	Assignment wording	Remarks
Screw-cutting lathe model CDE6140A	Designed to perform a variety of turning operations, including cutting metric, inch, modular and pitching thread, allows usage of modern high-speed processing modes with coolant-cutting fluid and with usage of advanced tools and accessories.	– there is no list of surfaces manufactured on the machine; – the sizes of the produced surfaces, the level of their accuracy and roughness are not indicated; – range of dimensions of manufactured parts is not specified

Milling machining center model 69П80МФ4-4	The high-precision milling machining center with a tool changer, interchangeable heads and an additional 4th coordinate has wide technological capabilities when performing milling, drilling, countersinking, reaming and threading operations. The machine consists of a rigid bed, along which the work table moves in the longitudinal direction. A column is rigidly connected to the bed, along which the console moves. In the transverse direction of the console moves the headstock. At the end of the headstock spindle head is rigidly fixed.	– there is no list of surfaces manufactured on the machine; – the sizes of the produced surfaces, the level of their accuracy and roughness are not indicated; – range of dimensions of manufactured parts is not specified

Analyzing the assignment wording of other machines, it was found that the higher the machine versatility, the greater the formulation uncertainty of its technological capabilities. The most complete information about the technological capabilities of the machine is given in the assignment wording of the specialized and operational machines used in large-scale and mass production.

Thus, in the assignment wording of the machines the technological capabilities are represented vaguely. The uncertainty in the description of the machines technological capabilities makes it difficult to choose machines when designing technological processes for manufacturing parts, purchasing new machines, causes difficulties in conducting a technological audit of an enterprise's machine park, assessing the possibilities of executing a production program and orders.

The problem of the machine technological capabilities describing was to choose the subject of production. In the existing machines assignment wording as the subject of production is taken a surface, a set of surfaces or a part. However, the part as an object of production is characterized by an almost unlimited variety of possible designs and its characteristics. The combination of surfaces is characterized by a large variety of surfaces forming it and their characteristics. Surfaces also have almost unlimited variety of geometric shapes, sizes, and accuracy.

In order to eliminate these shortcomings, it was proposed to adopt the module of surfaces as a subject of production. The module of surfaces (MS) is a combination of surfaces united by the joint performance of the service function of a part [1]. The main advantages of MS are their limited variety and consistency of structures. According to the classification, all MSs are divided into three classes: basing, working and binding. Each MS has its own number of standard designs, and each design has its own classification by size, accuracy, and roughness.

When using MS as an object of machine production, the wording of the machine technological capabilities will change, and will contain a list of manufactured MSs with ranges of dimensions, accuracy and surface roughness, overall dimensions and material of the parts being manufactured.

To determine the MSs manufactured on the machine, a special method was developed [2], the initial data of which are the processing methods implemented on the machine, the processing tool used and the technical characteristics of the machine. The method includes the following steps:

1) Determination of surfaces manufactured on the machine;
2) Determination of MS by the composition of the surfaces manufactured on the machine;
3) Determination of the surfaces positions compliance in the MS construction with their positions in the machine working space during manufacture;
4) Determination of the size ranges of MS, obtained on the machine;
5) Determination of achievable accuracy of MS manufacturing on the machine.

For the correct choice of machine for technological operation the great importance has the position of the MS surfaces in the machine working space during manufacture. It is so because the MS construction is a combination of surfaces having a certain relative position. After installation of the workpiece on the machine according to the technological bases, a situation may arise when one of the MS surfaces takes a position relative to the tool in which it cannot be made. To avoid such cases, when determining the technological capabilities of the machine, each MS is checked for positions compliance of the surfaces in its structure with the positions of the same surfaces in the working space of the machine during manufacture.

The disadvantage of such a test is that it does not take into account the possible positions of the entire MS structure in the machine working space. This may lead to the fact that the MS as the composition of the surfaces can be made on the machine, but after the workpiece is based on the machine, the MS structure may take such a position in the working space of the machine that its manufacture will be impossible. The following is an example of such a case: after basing the shaft on a lathe mod.16A20Φ3, the MSB312 module that consisting of a cylindrical surface produced on this machine, is offset from the axis of the centers and cannot be manufactured on this machine.

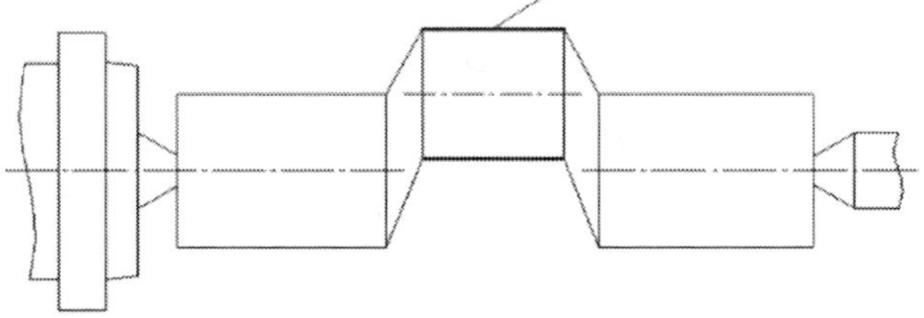

Figure 1. Position MSB312 in the working space of the machine mod.16A20Φ3.

Therefore, it is necessary to determine not only the conformity of the surfaces positions in the MS structures and in the working space of the machine, but also the permissible position of the entire MS structure in the working space of the machine.

To solve this problem it is necessary to develop a method which would help to determine the admissible MS structure position in the machine working space. The basic data are: the scheme of machine working space and the MS structure, which in terms of the composition and the location of the surfaces can be produced on the machine.

The initial stage will be constructing of a rectangular coordinate system of the machine working space on the working body surface, on which the workpiece is based. For example, in a milling machine, this is the surface of the working table, in a lathe - the spindle flange and the axis of the centers, etc. In some cases, like with a lathe, it is necessary to first install possible basing patterns of the workpiece on the machine, and then build the coordinate system of the machine working space for each scheme.

The second stage is constructing of a rectangular coordinate system on the MS. For example, if the MS consists of cylindrical and flat surfaces, then one of the axes is located on the cylindrical surface axis, and the second is on a flat surface.

The admissible position of the MS structure in the machine workspace will be determined by three linear and three angular dimensions that bind the machine coordinate systems and the MS. Therefore, the task is to establish admissible values of six sizes that bind two coordinate systems. In connection with the above, the following method is proposed for determining the MS structure permissible position in the machine working space:

1. Construction of a rectangular coordinate system on the MS.
2. Construction of a rectangular coordinate system of the machine.
3. Determination of permissible values of three linear and three angular dimensions that bind the machine coordinate systems and the MS.

Using the proposed method one can determine the allowable position of the MSB212 module, consisting of the threaded outer surface and the end face in the working space of a 16A20Φ3 lathe while basing the workpiece in the cartridge.

First, we construct the coordinate system on MSB212. The $0_M X_M$ axis will pass through the axis of the threaded outer surface, and the $0_M Z_M$ axis will pass through the flat outer surface (Figure 2). The

$0_C X_C$ axis of the lathe mod.16A20Ф3 coordinate system, when basing the workpiece in the three-jaw chuck, is located on the rotation axis of the cartridge, and the $0_C Z_C$ axis - on the right end of the cartridge (Figure 3).

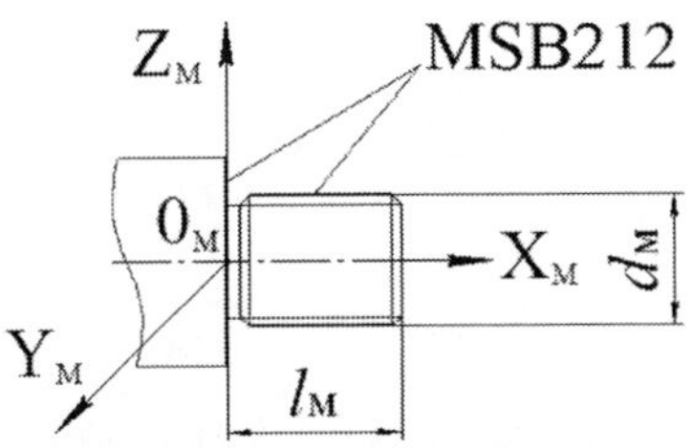

Figure 2. The MSB212 coordinate system.

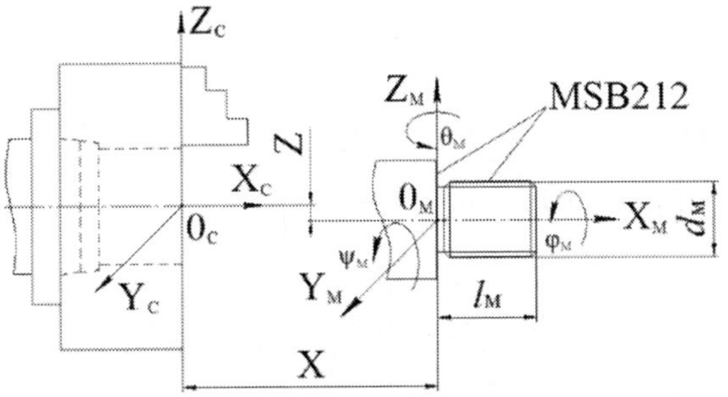

Figure 3. The MSB212 position in the coordinate system of the lathe model 16A20Ф3.

The position of the MSB212 coordinate system in the coordinate system of a lathe mod.16A20Ф3 is described by three linear (X, Y, Z) and three angular coordinates (Figure 3). The angular coordinates include:

φ_M - the MSB212 rotation coordinate around the $0_C X_C$ axis;
ψ_M - the MSB212 rotation coordinate around the $0_C Y_C$ axis;
θ_M - the MSB212 rotation coordinate around the $0_C Z_C$ axis.

If one analyze the shaping movements of the machine, one see that the ψ_M angle can have two values: 0° and 180°, the φ_M angle can vary from 0° to 360°, and the θ_M angle is zero. The linear coordinates Z and Y must also be zero. The value of the linear coordinate X (Figure 3) can vary in the range from 0 to 900 mm.

The relative position of the MSB212 coordinate system parameters in the lathe mod.16A20Ф3 coordinate system are summarized in Table 2.

Table 2. Dimensional relations of coordinate systems.

MS coordinate system	Machine coordinate system	Relative position parameters					
		X	Y	Z	φ	ψ	θ
		mm			degrees		
$X_M 0_M Y_M Z_M$	$X_C 0_C Y_C Z_C$	0-900	0	0	0-360	0, 180	0

In a similar way, the required positions of the other MS structures that manufactured on a lathe mod.16A20Ф3 are determined.

Thus, the proposed method, based on the construction of the machine coordinate systems and the MS, allows definitely determine the parameters of the possible MS parts position in the machine working space during manufacture.

References

[1] Bazrov B 2001 Modular technology in mechanical engineering (Moscow: Mashinostroenie) p 368
[2] Bazrov B and Sakharov A 2016 The technological capabilities determination of the enterprise machine park Stankoinstrument number 2 pp 29-34

IOP Conf. Series: Materials Science and Engineering **489** (2019) 012010 doi:10.1088/1757-899X/489/1/012010

High nitrogen steel powder formation for additive technologies

A S Zhukov, I V Shakirov, V N Klimov, E A Samodelkin and A D Bykova and M A Markov

NRC "Kurchatov Institute" – CRISM "Prometey", Saint-Petersburg, 191015, Russia

npk3@crism.ru

Abstract. Today, when manufacturing and restoring bulk products of almost any shape and complexity degree, additive technologies are in demand, which use metal-powder compositions of various alloys as a building material. In the present work, the possibility of obtaining powder by gas atomization from batch products of high-nitrogen 04Cr20Ni6Mn11Mo2NVNb steel used for selective laser melting is shown, its structure and properties are evaluated.

1. Introduction

The production technology of high nitrogen steels has been developed most intensively over the last 20 years [1-3]. Nitrogen is an austenite stabilizing element. Doping with nitrogen helps to increase static and cyclic strength, hardness, wear resistance, hardening of steel during plastic deformation. The main area of application for nitrogen steels are: the chemical and oil and gas industries, applications where alloys with high resistance to corrosion cracking in chlorides, halides, pitting corrosion, and marine corrosion (heat exchangers, absorbers, condensers, stop valves) are required. Steel with a high nitrogen content can be obtained, including powder metallurgy methods [4]. To do this, different schemes for obtaining powder are used. For example, the methods of plasma rotational spraying, gas atomization are applied. These and other production schemes have proven themselves in powder metallurgy, which makes them [4, 5] the most promising methods for obtaining materials with a high nitrogen content.

The use of high nitrogen powders in additive technologies by selective laser melting (SLM) can provide a high (possibly even non-equilibrium) nitrogen content in the 3D-modeling and at the same time can ensure uniform nitrogen distribution throughout the product section, but at the moment there are no powdered nitrogen steels used in the SLM technology [6]. The aim of this work is to study the possibility of obtaining powders with a high nitrogen content for additive technologies by melt spraying using melting ingots of high-nitrogen 04Cr20Ni6Mn11Mo2NVNb steel as an example.

2. Materials and experimental methods

All experiments to obtain powders of nitrogen steel were carried out using HERMIGA 75/3VI gas atomizer with induction heating of a 3 kg volume crucible. This type of atomizer has wide potential for the production of spherical metal powders, which makes it the most flexible tool for conducting highly efficient research activities and the production of small batches of special purpose metal powders. HERMIGA 75/3VI is a high-performance closed-type monoblock gas sprayer that meets modern requirements:

a) To provide greater purity and smaller particle size than the existing systems;

b) Cooling rate with enhanced fast crystallization (RSP).

Powder of high-nitrogen steel 04Cr20Ni6Mn11Mo2NVNb with the following melt spraying parameters:

- Melt temperature – 1500 ° C;
- Excess pressure in the melting chamber – 0.25 bar;
- Exposure time of the molten metal to spray – 40 minutes.

Melting was carried out by induction heating in a protective gas. Both nitrogen (pressure at the nozzle 50 bar) and argon (pressure at the nozzle 30 bar) were used as a spray and at the same time as a protective gas.

Table 1 presents the chemical composition of the powders and the initial charge, determined by X-ray fluorescence analysis on a Niton XL3t unit. Table 1 shows that the chemical composition of the obtained powders under various technological conditions corresponds to the chemical composition of the ingot.

Table 1. Chemical composition of powders.

Material	Basic alloying elements, wt. %					
	Fe	Cr	Mn	Ni	Mo	others
Ingot	60.5	20.8	10.5	5.9	1.7	0.6
Powder (30 bar argon)	61.1	19.3	10.1	7.3	1.4	0.8
Powder (50 bar nitrogen)	59.7	20.6	10.0	6.7	2.0	1.0

The granulometric composition of the powders was measured by laser diffraction using the Fraunhofer approximation by Malvern Mastersizer 2000. According to the results of particle size analysis, powder fractions of more than 80 μm were excluded, which are not suitable for using in SLM technology. The share of raw materials suitable for fusion was 70%, in which the 0-20 μm fraction was 1%, the 20-40 μm fraction was 40%, the 40-63 μm fraction was 10%, and the 63-80 μm fraction was 5%. Powder sieving by fractions used a sieve system on a shaker table. To determine the content of nitrogen and oxygen in the powder, the complex of gas analyzers LECO CS-230, TC-400 was used. According to the results of the nitrogen content measure before spraying, the value was 0.39%. Structural studies were performed by a TESCAN Vega 3 scanning electron microscope (SEM). The non-destructive testing of the elemental composition was performed by an Inka X-ray fluorescence attachment on a TESCAN Vega 5136-LM scanning electron microscope.

3. Experimental results and discussion

The particles morphology of the obtained high-nitrogen steel powder of 04Cr20Ni6Mn11Mo2NVNb is demonstrated in Figure 1. It can be concluded that the particles have a shape close to spherical, with no particles of irregular shape, which makes the powder suitable for using in the SLM technology. The selected pressure values at the nozzle provide particles of the specified form with a maximum proportion of available powder. In some powders production methods [7], on the surface of the powder particles, one may see foreign inclusions that the quality characteristics of the final product worse. Figure 2 shows the characteristic appearance of particles after gas atomization: it is clear that the surface of the powder particles does not contain satellites.

IOP Conf. Series: Materials Science and Engineering **489** (2019) 012010 doi:10.1088/1757-899X/489/1/012010

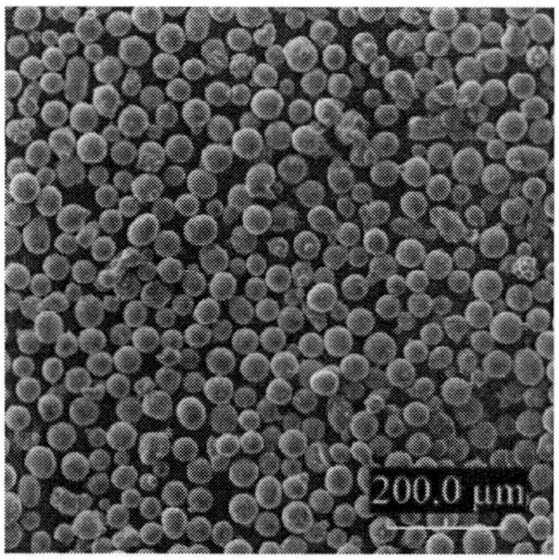

Figure 1. The morphology of the particles of the obtained high-nitrogen steel powder of 04Cr20Ni6Mn11Mo2NVNb

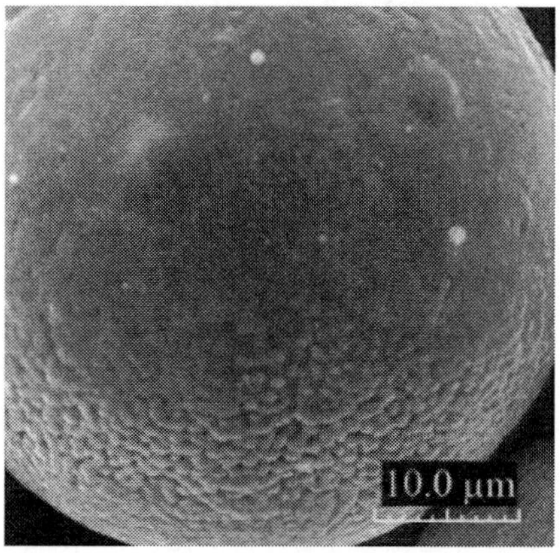

Figure 2. The characteristic appearance of high-nitrogen steel particles after gas atomization

The measurement results made by gas analyzers of nitrogen and oxygen in the powders obtained in different technological conditions are presented in table 2.

Table 2. The content of nitrogen and oxygen in the obtained powders of steel 04Cr20Ni6Mn11Mo2NVNb.

Pressure, Bar	Melt gas	Tmelt., ° C	Dispersity, µm	Oxygen,%	Nitrogen,%
30	Ar	1600	50-100	0.020	0.250
30	Ar	1600	50-100	0.020	0.270
50	N	1500	0-20	0.064	0.771
			20-32	0.061	0.500
			32-40	0.060	0.450
			40-50	0.062	0.420
			50-63	0.050	0.390
			63-71	0.050	0.410
			71-80	0.060	0.370
			>80	0.040	0.350

It is known [8] that at low pressure above the melt nitrogen is removed from steel. Therefore, an optimal pressure was chosen that impedes this process. Argon was used to study diffusion processes taking place with the melt in the crucible before spraying (Table 2 shows that the use of argon leads to the nitrogen content decrease in the melt and in the powder). The argon pressure above the melt was 0.25 bar, at the nozzle at a spray of 30 bar, with the nitrogen content decrease in the powder compared to the initial ingot (~ 0.26% versus 0.39%, respectively). It enables to suggest that the process of nitrogen diffusion from the melt proceeds. When nitrogen is used (at excessive pressure above the melt of 0.25 bar, as in case of using argon), the melt is saturated with nitrogen, and during the subsequent spraying (50 bar at the nozzle), the melt drops are further saturated.

The nitrogen content in the fractions of the nitrogen sprayed powder obtained by sieving through a sieve system, is graphically represented (Figure 3). A regular occurrence has been revealed: in particles of low dispersion, the nitrogen content exceeds that value even in the initial ingot. The decrease in nitrogen content depending on the particle diameter (fractional composition of the powder mixture) can be explained by diffusion processes, namely, the different depth of nitrogen diffusion into the powder particle from its surface at the moment of spraying, when the powder looks like metal droplets (in the liquid phase). Accordingly, in a particle with a smaller diameter, the volume fraction of nitrogen is higher than in particles with a larger diameter. In this case, the time interval of the diffusion process is insignificant.

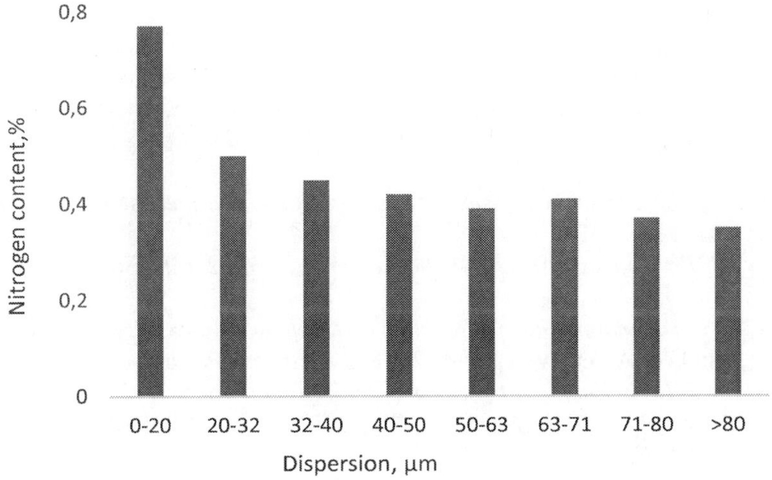

Figure 3. Nitrogen content in various powder fractions

The issue of high-nitrogen steels is to maintain a non-equilibrium nitrogen content (heat treatment modes, melt cooling rates). Therefore, obtaining the final part from high-nitrogen steels is associated with a number of challenges. Additive technologies provide prerequisites for solving this problem: at each stage, when the alloy undergoes a transition to a liquid state (sputtering in the atomizer and subsequent selective laser melting), it is possible to control the partial pressure above the melt, the cooling rate of the melt.

A further research continues with a more detailed study of the structure and properties of the additive samples (according to the SLM technology) from obtained powder materials. The nitrogen content and the phase composition of the melted sample are of importance. Samples are supposed to be grown in nitrogen to maintain (or even increase) their nitrogen content. The phase composition is determined by the ratio of alloying elements; therefore with nitrogen (nickel equivalent) and chromium contents one can obtain a fully austenite or an austenite-ferrite structure.

4. Conclusions

Powder of high-nitrogen steel 04Cr20Ni6Mn11Mo2NVNb of necessary dispersion (less than 80 microns) and morphology (sphericity) that meets the requirements of equipment used for selective laser melting of metallic powder materials was obtained by melt spraying in a nitrogen atmosphere by HERMIGA 75/3VI atomizer.

It was determined that when argon is used during atomization of the high-nitrogen steel ingot 04Cr20Ni6Mn11Mo2NVNb, the nitrogen content in the final product (powder) decreases in comparison with the initial ingot.

The nonequilibrium content of nitrogen in a metallic powder material with dispersity of 0–20 µm with about 0.7% of nitrogen (in the initial ingot – 0.39% of nitrogen) is obtained. The increasing of nitrogen content in steel in powders production for additive technologies by HERMIGA 75/3VI gas atomizer is investigated. Thus, a number of technological challenges to obtain high nitrogen samples are addressed.

Acknowledgments

IOP Conf. Series: Materials Science and Engineering **489** (2019) 012010 doi:10.1088/1757-899X/489/1/012010

This work was supported by the grant of Russian Scientific Fund №15-19-00210. Experimental studies were performed on the equipment of the collective use center "Composition, Structure and Properties of Structural Materials" NRC "Kurchatov Institute" – CRISM "Prometey", Saint-Petersburg, Russia, supported by the Ministry of Education and Science of the Russian Federation (unique subsidy identifier – RFMEF159517X0004).

References

[1] Sedlaka J, Rican D, Piska M and Rozkosny L 2015 *Procedia Engineering* **100** 1232-41
[2] Barakhtin B K, Bobyr V V, VoznIuk A V et al. 2017 *Problems of Materials Science* **2(90)** 146-52
[3] Shishkovsky I V 2009 *Laser synthesis of functional-gradient mesostructures and bulk products* (Moscow: Fizmatlit) p 424
[4] Rashev Ts 1995 *High-nitrogen steels: metallurgy under pressure* (Sofia: Prof. Marin Drinov) p 270
[5] Lieberson G A 1975 *Fundamentals of powder metallurgy* (Moscow: Metallurgy) p 200
[6] Kuznetcov P, Zhukov A, Deev A et al. 2018 *Structure and Properties of the Bulk Standard Samples and Cellular Energy Absorbers* (Rieka: IntechOpen) p 22
[7] Rodionov A I, Efimochkin I Yu, Buyakina A A and Letnikov M N 2016 *Aviation materials and technologies* **S1 (43)** 60-4
[8] Averin V V, Revyakin A V, Fedorchenko V I and Kozina L N 1976 *Nitrogen in metals* (Moscow: Metallurgy) p 224

IOP Conf. Series: Materials Science and Engineering **489** (2019) 012011 doi:10.1088/1757-899X/489/1/012011

The structure and properties of the coating based on particles of copper and zinc deposited by gas-dynamic spraying

V E Arkhipov, A F Londarsky, G V Moskvitin and M S Pugachev

Mechanical Engineering Research Institute of the Russian Academy of Science, Moscow, 101990, Russia

vearkhipov@mail.ru

Abstract. The results of the study of the structure and properties of the coating based on a particulate mixture of Cu and Zn deposited by the method of cold gas-dynamic spraying are presented. The presence of diffusion during the deposition of a coating with the formation of a solid solution of copper in zinc is shown. The process of predominant diffusion of copper into zinc with a slight exposure time in the furnace with the formation of intermetallic compounds $Cu_5 Zn_8$ and CuZn , the mass fraction of which reaches 44% and 28%, respectively, is revealed. The diffusion coefficient of copper to zinc is not less than $1.56x10^{-13}$ $m^2 \cdot s^{-1}$. An increase in the exposure time leads to a decrease in the content of beta (β ′) and gamma (γ) phases to ≈ 15% and the formation of a solid solution of zinc in copper (α phase, 58%) with a different ratio of initial components. The formation of electronic compounds reduces the cohesive strength of the coating metal by almost two times, and an increase in the proportion of the phase leads to an increase in the cohesion above the initial value.

1. Introduction

Copper and brass coatings are applied using a variety of chemical and physical processes, such as flame spraying methods. In the last decade, considerable attention has been paid to the technology of applying coatings using the method of gas-dynamic spraying, which is distinguished by substantially less heating of the substrate and, as a result, the absence of deformation of the product and changes in the mechanical properties of the substrate [1]. In foreign scientific centers, they are actively engaged in the research of gas dynamic coating processes for tribotechnical purposes using brass powders, copper powders and a mixture of copper particles with zinc particles. Positive results were obtained on the deposition of a brass coating and an assessment was made of the effect of heat treatment of powders and coatings on the quality and mechanical properties of the applied metal layer [2,3,4,5,6].

In the Russian Federation, manufacturers of the "DYMET" series equipment used for gas-dynamic spraying of various functional coatings, in the nomenclature of powder materials, offer a mechanical mixture of copper and zinc particles to obtain a coating of brass [1]. Testing of the coating based on copper and zinc particles under dry friction conditions at room temperature showed very low values of sample wear with the coating and especially the counter body compared to copper coating, which is caused by the mass transfer of copper and zinc to the surface of the counter body [7]. However, an increase in the test temperature to 250°C significantly increases the wear of the coating and the counter body, which may be due to a change in the structure and properties of the coating.

The purpose of this work is to study the sequence of structural-phase transformations in the coating based on a mechanical mixture of copper and zinc particles deposited by cold gas-dynamic spraying and their influence on the cohesive strength of the coating metal.

2. Materials, methods, equipment.

The coating on the 40X steel substrate was sprayed using the DYMET – 404 gas-dynamic installation at an air flow temperature of 450°C, a moving speed of the replaceable nozzle relative to the sample surface 10 mm/s and a distance from the nozzle section to the surface 10 mm. Since the width of the applied coating layer when moving the nozzle in one direction does not exceed 6 mm, the deposition on the entire surface of the substrate is carried out with a successive displacement of the nozzle relative to the previously applied metal layer by 3 mm (overlap 50%). The mechanical mixture of particles of copper and zinc (C – 01 – 11) is used in the delivery condition and the ratio of ingredients by weight is – Cu : Zn : $Al_2 O_3$ = 35%:35%:30% [1].

Heat treatment of the coating with the substrate is carried out in a SNOL laboratory furnace at a temperature below the melting point of zinc (405-415 ° C) and exposure times of 3, 5, 10, 20, 40, 60 and 180 minutes.

The phase constitution of the Cu – Zn system coatings is investigated on a multifunctional X-ray diffraction meter Rigaku Ultima IV using $Cu_{K\alpha}$ radiation and a parallel beam. To form a parallel beam, X-ray optics in the form of a multilayer parabolic mirror is used. Diffraction patterns are taken in a symmetric mode, the scan range 2θ from 20 to 120, step 0.05°, the speed of the detector 3°/min.

Analysis of diffractograms is carried out on the basis of the PDXL software product Rigaku using the PDF-2 database. Quantitative analysis is carried out using the Rietveld method that is implemented in the PDXL software package Rigaku.

The broadening of diffraction lines was analyzed to determine the parameters of the fine crystal structure (block size, micro strain) using the PDXL software Rigaku using the Rietveld method. Accounting for instrumental broadening was carried out using a standard survey — lanthanum hex boride (LaB_6), which does not have physical broadening.

X-ray fluorescence analysis is performed on a Rigaku spectrometer Primus II using an X-ray source an X-ray tube with Rh-anode in vacuum and range of elements from Ca to U. Quantitative analysis is carried out after recording and decoding the experimental spectra from the samples under study using special software from Rigaku ZSX using the SQX fundamental parameters method (without using reference samples).

The study of the structure and elemental composition of the coating is carried out on an FEI "Quanta-650" scanning electron microscope with an energy dispersive X-ray spectrometer EDAX. The study uses a detector of back reflected electrons, which allows obtaining an image with atomic number contrast and visualizing different phases on the surface of the samples. The electron source in the microscope is a tungsten cathode; accelerating voltage is 25 kV.

Coated samples are tested using the SHIMADZU universal servo hydraulic testing machine with a constant displacement rate of 0.002 mm/s.

The study of the cohesive strength of the metal coating is carried out using the method of "ring separation" [8]. The cohesive strength of the coating metal is calculated as the ratio of the ultimate load preceding the destruction of the coating to the area of the coating. Based on the calculated data, the dependence of the cohesion of the copper-zinc coating on the heat treatment time is constructed.

3. Results

The mass content of copper and zinc in the applied coating is 64.8% and 35.0%, respectively, which differs significantly from the initial ratio of components in the mechanical mixture of powders. In the structure of the coating, a rather uneven alternation of particles of zinc and copper is observed – areas of light and light gray, respectively figure 1. In addition, uneven in shape and size of the inclusion of corundum (dark areas) are clearly visible.

IOP Conf. Series: Materials Science and Engineering **489** (2019) 012011 doi:10.1088/1757-899X/489/1/012011

Figure 1. Coating structure based on copper and zinc particles: scanning electron microscopy using a back-reflected electron detector.

X-ray diffraction phase analysis reveals copper (≈ 3.61A^0), zinc (≈ 2.66A^0) and zinc-based solid solution (≈ 2.75A^0) in the initial coating structure (table 1).

Table 1. Phase composition of the coating based on copper and zinc.

Name of phases and compounds	Mass fraction of elements, chemical compounds and phases,%							
	Holding time, min							
	0	3	5	10	20	40	60	180
Cu	58	48	38	26	26	19	22	1
α – tv. rr	-	1	14	16	17	30	34	58
CuZn (β')	-	-	-	28	29	27	17	15
Cu_5 Zn_8 (γ)	-	44	44	22	20	14	17	14
Zn	23	7	-	-	-	-	-	-
ZnO			4	8	8	10	10	12
η - phase	19	-	-	-	-	-	-	-
Total content,%	100	100	100	100	100	100	100	100

Six phases can form in solid state of copper and zinc alloys: α – solid solution of Zn in Cu; β' – solid solution of electronic type based on CuZn compound; γ is an electronic solution based on Cu_5 Zn_8; ε is a solid solution of electronic type based on $CuZn_3$; δ is a solid solution (nature is not established); η is a solid solution of Cu in Zn. At normal temperature, the brass used in practice is single-phase and consists of the α-solid solution or the β'-phase or two-phase on the basis of α and β' [9].

Thus, it can be argued that in the process of spraying a coating based on a mixture of particles of copper and zinc at a temperature of 450°C the deposited metal layer undergo initial diffusion processes, which leads to the formation of η-solid solution of Cu in Zn.

The following factors can influence the diffusion process and the formation of a solid solution of copper in zinc during the deposition of a coating. The results of the study of electrical resistivity and X-ray analysis of the copper coating show the presence of plastic deformation of particles with an increase in the number of vacancies [8]. However, the most important factor should be considered to increase the temperature of the coating (used metals) to a temperature close to the temperature of the

IOP Conf. Series: Materials Science and Engineering **489** (2019) 012011 doi:10.1088/1757-899X/489/1/012011

airflow (≈450°C) [8]. In this case, zinc particles are in a state close to melting, when the bonds between the lattice atoms weaken and the number of vacancies increases significantly, which leads to the possibility of diffusion passing through the vacancy mechanism [10].

A slight exposure of the coating based on copper and zinc particles is accompanied by the beginning of the formation of a solid solution based on copper with a period different from pure copper (a-phase; ≈3.72A^0), a decrease in the content of pure zinc (up to 7%) and the formation of an electronic compound – gamma phase Cu_5 Zn_8 , the mass fraction of which is 44% table 1. It can be seen from the copper-zinc state diagram that the gamma phase is formed when the zinc content in the melt is from 50% to 78% (β′+γ, γ, γ+ε) [9].

Spectral analysis of the surface of the coating was carried out along the route, which includes a zinc particle (≈15 μm), limited to copper particles figure 2.

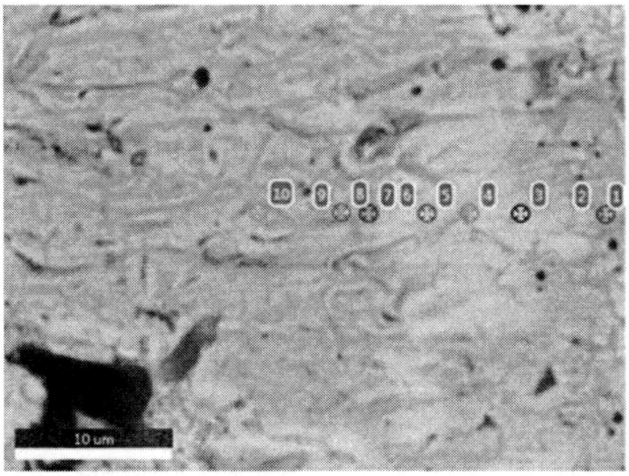

Figure 2. The trace of the analysis of the chemical composition of the surface area of the coating after heat treatment for 3 minutes.

At the edge of dark tint particles, the copper concentration is within 95.21-96.82 at.% (error 2.05%), and the rest is zinc. The result obtained indicates the presence of diffusion of zinc into copper and the formation of a solid solution of zinc in copper (α phase). The diffusion of copper atoms into a zinc particle takes place on both sides fairly evenly (throughout the volume); the copper concentration decreases to the center of the zinc particle and reaches 26 at.% figure 3.

Diffusion coefficient [10], which is ≈$1.56x10^{-13}$ m^2/s is calculated by the simple relation of $D_x = x^2/2t$, using copper diffusion distance (half-size zinc particle) and a heat treatment time (180 s).

In zinc particles in different parts of the surface of the coating, 19 measurements were made of the content of the starting components. Based on the copper-zinc state diagram: 12 (63.2%) measurements are in the region of existence of the γ phase; 3 (15.8%) measurements on the concentration of copper and zinc correspond to the region of existence of the γ+ε phases; 2 (10.5%) measurements correspond to the existence region of the γ+β ′phases and 2 (10.5%) measurements on the atomic content of the components correspond to the phase of existence of the α+β′ phases.

IOP Conf. Series: Materials Science and Engineering **489** (2019) 012011 doi:10.1088/1757-899X/489/1/012011

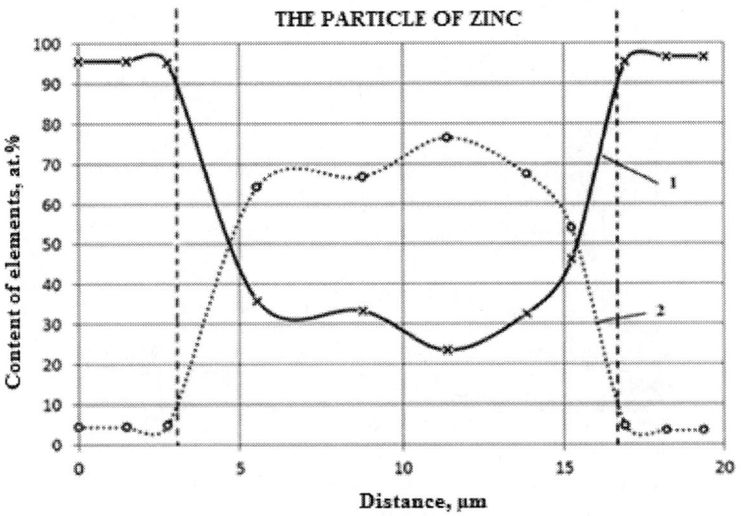

Figure 3. Changes in the concentration of copper (1) and zinc (2) in the coating after heat treatment for 3 minutes

Thus, based on the results of micro spectral analysis, the main phase formed as a result of the diffusion of copper into zinc is $Cu_5 Zn_8$, which coincides with the results of X-ray structural analysis. In this case, after heat treatment of the coating for 3 minutes, a four-phase structure is formed on the basis of copper (≈48%), zinc (≈7%), gamma phase (≈44%) and a-phase (≈1%).

Heat treatment of the coating for 5 min is accompanied by a decrease in the copper content from ≈48% to ≈38% and an increase in the content of the solid solution based on copper from 1% to ≈14%. The amount of gamma phase does not change, but zinc oxide is additionally detected in the amount of ≈4% (table 1). The gas-dynamic coating of copper and zinc has an open porosity in the range of 3-4%, which can contribute to the oxidation of zinc over the depth of the deposited metal layer when it is heated in a furnace with the formation of zinc oxide [8].

An increase in the heat treatment time of the initial coating to 10 min leads to a further decrease in the copper content from ≈38% to ≈26% and to a slight increase in the amount of solid solution based on copper (up to 16%). In the same time interval, the content of the γ - phase decreases by half (22%) with an increase in the amount of zinc oxide to ≈8% (table 1). In this case, X-ray phase analysis shows the presence of a new e-compound of β' phase CuZn in an amount of ≈28% (table 1).

A further increase in the exposure time of the coating in the furnace to 180 minutes is accompanied by a decrease in the copper content to ≈1%, an increase in the a-phase content to 58% and a decrease in the β' and γ phases to 14% and 15%, respectively (table 1). An increase in the solid solution of zinc in copper and a decrease in the content of electronic compounds indicates a predominant diffusion of zinc into copper. At the same time, the zinc oxide content increases from 8% to 12%.

Microspectral analysis of elements in the coating after 180 minutes of heat treatment was carried out at points and on the area of plots (grains, particles), which are limited to dark tint lines and differ in size (from 10 µm to 20 µm) figure 4.

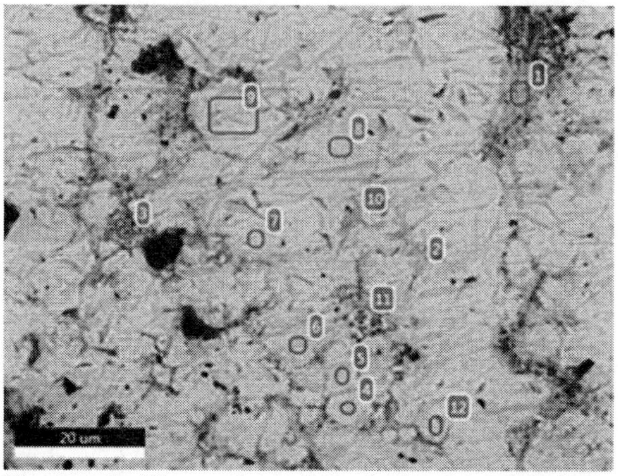

Figure 4. The order of analysis the chemical composition of the surface area of the coating after heat treatment for 180 minutes

Analysis of the results shows that the grains (particles) of an insignificant size (8–12 μm) contain zinc in the range of 37.21–38.24 at.%. In this case, in the structure of the coating after heat treatment with a maximum exposure time, there are areas that, by mechanical properties, correspond to single-phase brass L63, that is, they have high strength with good ductility [9]. In particles (grains) of a larger size, zinc content from 44.82 at.% to 47.14 at.% is observed, which according to the copper-zinc state diagram can be interpreted as the presence of two-phase brass, where α solid solution and β′ phase are present. Two-phase brass with such a zinc content has the highest strength with low ductility. In addition, zones with a zinc content of about 8.59 – 17.99% are noted in insignificant amounts, which indicates the presence of structural components that, by mechanical properties, correspond to single-phase brass of type L90 (low brass) and L80. The identified distribution of components in a solid solution of zinc in copper (α phase) is undoubtedly due to the uneven distribution of copper and zinc particles in the coating figure 1. There are also areas with zinc content in the range of 66.55-70.53 at.%, Which according to the copper-zinc state diagram corresponds to the region of existence of the γ phase. It should be noted and the presence of a certain number of points where the zinc content within the measurement error is close to zero. Microspectral analysis of the surface areas of the coating in general confirms the results of the study of the phase composition of the coating obtained by X-ray structure analysis.

In the areas (lines) of dark gray color, an elevated oxygen content of 31.19–40.79 at.% was revealed and it can be said with confidence that it is zinc oxide that forms along the boundaries of metal particles and in structure defects (pores) figure 4.

The analysis of the results obtained shows that the cohesive strength of the coating metal in the initial state has a value of 78 MPa. Heat treatment for 3 minutes reduces the cohesion to ≈64.6 MPa, which is due to the formation of the electronic compound – gamma (γ) of the Cu_5 Zn_8 phase, figure 5 (table 1). With an increase in heat treatment time up to 10 minutes, the cohesive strength of the coating metal increases slightly to ≈75.4 MPa, which is associated with an increase in the content of solid solution of zinc in copper (α phase) (figure 5, table 1). A further increase in the exposure time in the furnace (20 min) is accompanied by a maximum decrease in cohesion to the level of ≈42.6 MPa, which is caused by the formation of another electronic compound beta (β′) phase based on Cu Zn (figure 5, table 1). It is known that the formation of an electronic compound in alloys of copper and zinc (brass) is accompanied by a sharp decrease in strength and ductility [9]. Increasing the heat treatment time to 60 minutes or more increases the cohesive strength of the coating metal, which reaches a value of 93 MPa.

IOP Conf. Series: Materials Science and Engineering **489** (2019) 012011 doi:10.1088/1757-899X/489/1/012011

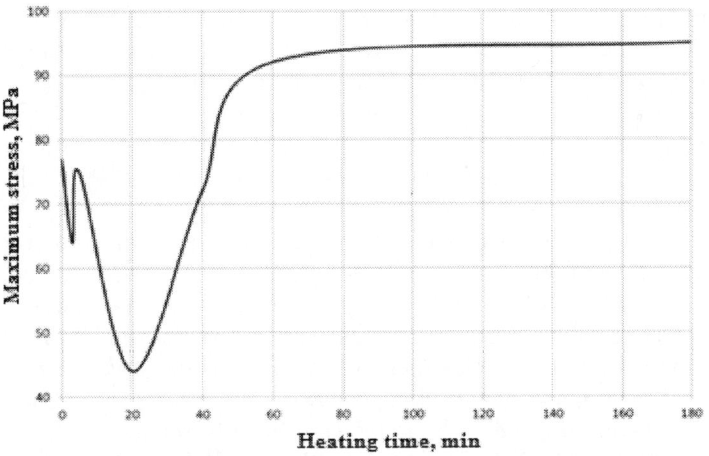

Figure 5. The dependence of the cohesive strength of coatings based on a mixture of copper and zinc particles on the heat treatment time at a temperature of 405-415 °C.

With an increase in the exposure time in the oven from 40 minutes. up to 60 min. content of the main structural components, namely copper, and α, β', γ phases varies slightly, which should not have a significant effect on improving cohesive strength of the coating metal (figure 5, table 1). However, in this time range of heat treatment of the metal of the coating, the size of the substructures of the main components of the coating significantly changes figure 6. Therefore, a sharp increase in the size of substructure fragments, especially the alpha phase, which becomes the dominant component in the structure of the coating with an increase in the exposure time, has an effect on increasing the cohesive strength of the coating metal (figure 6, table 1).

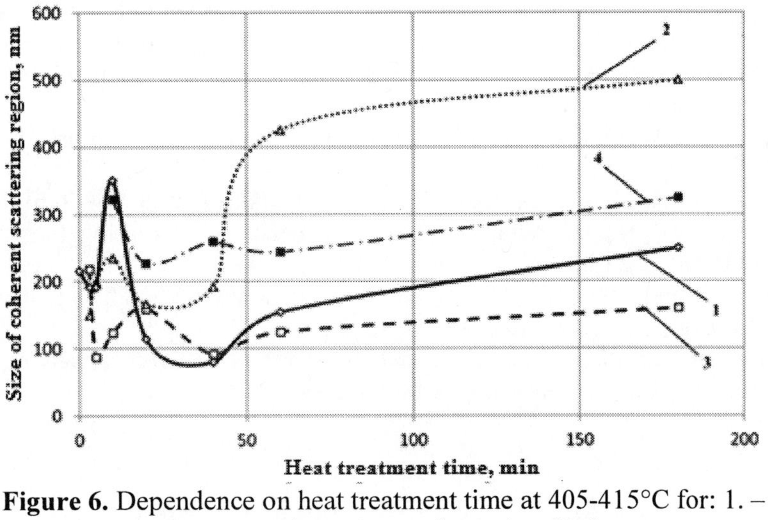

Figure 6. Dependence on heat treatment time at 405-415°C for: 1. – copper, 2 – α phase; 3. – γ phase; 4. – β′ phase.

4. Conclusion

1. Gas dynamic spraying of a coating based on a mechanical mixture of copper and zinc particles is accompanied by the formation of a metal layer based on copper (58%), zinc (23%) and solid solution based on zinc (η – 19%).

IOP Conf. Series: Materials Science and Engineering **489** (2019) 012011 doi:10.1088/1757-899X/489/1/012011

2. Heat treatment of the coating at a temperature of 405-415°C for 3 minutes is accompanied mainly by the diffusion of copper into zinc and the formation of an electronic type compound, the gamma phase of $Cu_5 Zn_8$ (44%). The share of solid solution of zinc in copper (α – phase) is at the level of ≈1%. The diffusion coefficient of copper to zinc is not less than 1.56 x 10^{-13} $m^2 \cdot s^{-1}$.

3. Increase the exposure time to 10 min. leads to an increase in the proportion of the a-phase (16%), a decrease in the content of the γ phase (22%) and the formation of an electronic compound — the β′ phase of Cu Zn (28%).

4. Heat treatment for 180 minutes is accompanied by the formation of a solid solution of zinc in copper (58%), which has a different ratio of components. Furthermore, in the coating are present intermetallic compound CuZn and $Cu_5 Zn_8$ (β′ and γ phase), the mass fraction of which is ≈15%.

5. Along the boundaries of the particles (grains) and in the defects of the structure (pores), zinc oxide is formed, the mass fraction of which increases to 12% at the maximum exposure time.

6. Heat treatment of the coating based on copper and zinc particles at a temperature of 405-415°C and a holding time of 3 minutes and 20 minutes significantly reduces the cohesive strength of the coating metal from 78 MPa to 64.6 MPa and to 42.6 MPa, which is due to loss of γ and β′ phases.

7. An increase in the size of the substructure of the main structural components of the coating after heat treatment for 60 minutes and more contributes to an increase in the cohesive strength of the coating metal.

References

[1] DIMET *The use of technology and equipment* [Electronic resource] – Access mode: http://www.dimet-r.narod.ru/ (Date of access 23.11.2018)

[2] Shinichi NISHIMURA, Yukitaka HAMADA 2017 Feasibility study of High-Strength Brass Coatings by the Cold Spray Process *Journal of the Society of Powder and Powder Metallurgy* **64** 661-6

[3] Gershman Iosif S et al 2016 *Description of Seizure Process for Gas Dynamic Spray of Metal Powders from Non-Equilibrium Thermodynamics Standpoint, Entropy* p 315

[4] Marple William J 2012 *The cold gas-dynamic spray and characterization of microcrystalline and nanocrystalline copper alloys* (United States Naval Academy, NAVAL POSTGRADUATE SCHOOL December Graduate Thesis for Master of Science in Mechanical Engineering)

[5] Kowalski M and Spencer P J 2015 Thermodynamic Re-evaluation of the Cu-Zn System *J. Phase Equi* **14** 432–8

[6] Warren B E and Warekois E P 1955 Stacking Faults in Cold Worked Alpha-Brass *Acta Metallurgica* **3** 473–9

[7] Sachek B Ya, Mezrin A M, Arkhipov V E and Londarsky A F 2018 Modifying the surfaces of tribo -spraying by spraying metal coatings as a means of increasing their fretting resistance *Friction and Wear* **39** 24-8

[8] Arkhipov V E, Londarsky A F, Moskvitin G V, Pugachev M S 2017 *Gas-dynamic spraying: structure and properties of coatings* KRASAND p 240

[9] Efremov B.N. BRASS 2016 *From phase structure to structure and properties* INFRA-M p 314

[10] Bokshtein B.S. 1978 *Diffusion in metals* Metallurgy p 248

IOP Conf. Series: Materials Science and Engineering **489** (2019) 012012 doi:10.1088/1757-899X/489/1/012012

Track vehicles self-extrication mechanism

N V Belov, D S Kiper and I S Chabunin

Moscow Higher All-Arms Command School, Golovacheva street, 21, 09380, Moscow, Russia,

tchabunin@rambler.ru

Abstract. The analysis of extrication of stuck track vehicles methods available led to the conclusion that in case of light or moderate sticking it is justified to employ the self-extrication method using the drive wheels of the vehicle as winch heads. However, it is seldom resorted to taking into account high labor intensity and complexity of the efforts necessary to accomplish it. To exclude this drawback the authors propose a structure containing a drum with a wire wound on it that is attached to the drive wheel externally. To increase the efficiency rate and to reduce the load on the wire and the head with the wheel a fixed block is used. Essentially, this results in a winch with a tackle system whose ratio is 2. For self-extrication it is necessary to mount the mechanism on the drive wheels of the track vehicle (without removing the tracks), to fasten the blocks on fixed bearing(s), such as a stump, a tree or an anchor log, located at a distance from the vehicle, to pass the wire from the head through the block and to fasten it on the vehicle's towing shackle. The authors use the example of a 14-ton heavy vehicle to determine the main geometrical parameters of the mechanism and provide the fields of equivalent stress distribution calculated using the finite element method.

1. Introduction

It is well-known that the application of track running gear facilitates the vehicle's off road capability and enhances its towing performance when moving across the terrain characterized by weak supporting capacity or in total road impassability.

The off-road capability is one of the key features of combat vehicles; it is mainly determined by the contact pressure, the area of the running gear contact with the supporting face and their adherence. The distribution of the weight of the vehicle across a much larger contact surface than that of the wheel the unit pressure affecting it is significantly lower, whereas the availability of lugs provides for their advanced adherence. However, there may be cases when weak supporting capacity of the terrain (mud, swamp, snow, etc.) or because of insufficient clearance of the vehicle, the latter is stuck. Thus, when accomplishing combat missions, track combat vehicles may fail not only because of breakages received during the operation, but because of getting stuck as well. Its extrication is one of the integral parts of the technical support of the military.

2. Task setting

According to the Guidelines on Armored Vehicle Extrication [1], there exist the following methods of extricating stuck track vehicles: self-extrication, extrication by tractors, extrication by the recovery winch and tackle gear. It is natural that the employment of a particular method is determined by the type of sticking, the technical condition of the vehicle and the extrication means available. If a mobile vehicle is stuck because of insufficient adherence power between the tracks and the surface, and the towing effort it can potentially develop using its engine power exceeds the motion resistance force, it is preferable to use self-extrication. This method does not require any tractor, which constitutes its main advantage.

Lepeshinsky et al. [2] provide the description of the following self-extrication techniques: using a log attached to the tracks of the stuck vehicle (figure 1); using a wire attached to the tracks of the stuck vehicle (figure 2); attaching a tow wire to the tracks of the stuck vehicle and using the drive wheels of the stuck vehicles as the winch head (figure 3). The techniques based on attaching the wire to the tracks prove to be easier, however, if self-extrication attempts are made without proper preparation, this may lead to more severe sticking. Secondly, the distance that the vehicle may be moved to is estimated as less than a half of the track length as after the points of wire attachment are released from beneath the supporting surface, the means of self-extrication of the vehicle itself may be damaged. Thirdly, practically all the load is concentrated in one or two track links. The technique that consists in using the drive vehicles of the stuck vehicle as winch heads proves the most efficient in the majority of casesas it allows to obtain the highest momentum on the wire, which will enable the extrication of the stuck vehicle. However, this is a labor-intensive and, therefore, time-consuming procedure, which results in its rare employment, regardless of its efficiency.

Figure 1. Self-extrication of a track vehicle using a log

3. Solution

As the main difficulty in using drive wheels of the stuck vehicle as winch heads consists in the need to remove and then put back the tracks (e.g. the weight of each track of IFV 2 is over 600 kilos) that impede the attachment and winding of the wire onto the wheel, it is expedient to fasten the head on the outer part of the drive wheel.

Figure 2. Self-extrication of the vehicle using wire attached to the tracks and to the fixed bearing

IOP Conf. Series: Materials Science and Engineering **489** (2019) 012012 doi:10.1088/1757-899X/489/1/012012

Figure 3. Self-extrication of a track vehicle by winding the wire onto the drive wheels

The idea of attaching the head to the drive wheels was long ago developed for cars. For example, Patent № 74622 entitled 'Self-extrication mechanism for a skidding car' published on March 31, 1949 describes the mechanism designed as a head attached to the drive wheels of the car 'onto which the rope is wound that is anchored to the fixed object located in front'. The research in the public domain did not allow to discover the description of the mechanism used to accomplish this method for track vehicles. The only outcomes were two technical improvement proposals developed at Novosibirsk Higher Military Command School. The first one by A.N. Chernenko and A.A. Klepikov is entitled 'IFV self-extrication mechanisms using the drive wheel' and the second one by S.A. Boyarinstev, A.N. Chernenko and A.I. Tsarenko is entitled 'IFV self-extrication mechanisms using the drive wheel without the disconnection of the track'. With the view to reducing the force extending the wire as well as enhancing the efficiency rate of the mechanism, we suggest upgrading the structures described in the technical improvement proposals by additional blocks. This will result in actual winch with a tackle system whose ratio is 2. To conduct self-extrication it is necessary to mount the mechanisms onto the drive wheels of the track vehicle (without removing the tracks), to fasten the blocks on fixed bearing(s) – for example, a tree, a stump, or an anchor log – located at a distance from the vehicle (see figure 2), the wire from the head must be passed through the block and fastened at the towing shackle of the vehicle.

Let us determine the main geometrical parameters of this mechanism. To exemplify our calculations we take a track vehicle with a full weight of 14 tons. In this case the value of the pulling force F with light or medium sticking should not exceed 54 kN on each drive vehicle considering the efficiency rate provided that the force is evenly distributed between the two vehicles. For further calculations we accept the pulling force to equal 61kN.

We take into account that during self-extrication works, the employment of the wire tends to be rather short-term. Therefore, to facilitate it in field conditions we need to take its assurance factor at 1,7 – 2 [2], which leads us to choosing the wire matching the following characteristics: 12.5-200-B-Л-OGOST 3077-55 with a breaking stress of no less than 101.01 kN.

For design purposes we take the outer diameter of the head at 320mm, the diameter of the limiting wall that prevents the wire slipping lengthwise at 410 mm, the distance from the rotational axis of the head to fixation holes at 220 mm; the length of the head at 150 mm.

The wall of the drum will be in the condition of combined stress [3], however, the compression strain will be the dominant factor. Let us calculate the approximate value of its thickness on the basis of the values of these compressions:

$$\delta = 0.5 \cdot \left(D - \sqrt{D^2 - \frac{2pD^2}{[\sigma]}} \right),$$

where D is the outer diameter of the head;

$[\sigma]$ is working stress;

IOP Conf. Series: Materials Science and Engineering **489** (2019) 012012 doi:10.1088/1757-899X/489/1/012012

$p = \frac{2F}{Dt}$ is the distributed load on the wire;

t is the wire thickness.

Then $p = \frac{2 \cdot 61000}{0.32 \cdot 0.0125} = 30.5\, MPa$;

$$\delta = 0.5 \cdot \left(D - \sqrt{0.32^2 - \frac{2 \cdot 30.5 \cdot 10^6 \cdot 0.32^2}{350 \cdot 10^6}} \right) = 0.015\, m, \text{ if}[\sigma] = 350\, MPa.$$

As apart from standard compression strains, the wall of the head will experience the strain from the bending moment and tangent stress from the rotative moment, we take $\delta = 16\, mm$. In this case the compression strain from the distributed load p is calculated as follows:

$$\sigma_c = \frac{2pD^2}{D^2 - (D - 2\delta)^2} = \frac{2 \cdot 30.5 \cdot 10^6 \cdot 0.32^2}{0.32^2 - (0.32 - 2 \cdot 0.016)^2} = 321\, MPa;$$

strain from the bending moment stands at:

$$\sigma_b = \frac{32 M_b D}{\pi \left(D^4 - (D - 2\delta)^4 \right)} = \frac{32 \cdot 61 \cdot 10^3 \cdot 0.15 \cdot 0.32}{3.14 \cdot \left(0.32^4 - (0.32 - 2 \cdot 0.016)^4 \right)} = 8.3\, MPa.$$

Tangent stresses caused by the rotative moment are:

$$\tau = \frac{16 T_t D}{\pi \left(D^4 - (D - 2\delta)^4 \right)} = \frac{16 \cdot 61 \cdot 10^3 \cdot 0.22 \cdot 0.32}{3.14 \cdot \left(0.32^4 - (0.32 - 2 \cdot 0.016)^4 \right)} = 6.1\, MPa.$$

According to the strength theory of maximum tangent stresses, the maximum equivalent stress equals:

$$\sigma_{eq} = \sqrt{(\sigma_c + \sigma_b)^2 + 3\tau^2} = \sqrt{(321 + 8.3)^2 + 4 \cdot 6.1^2} = 329.5\, MPa.$$

As the calculation object is represented by rather a complex structure that finds itself in complex strained conditions, the outcomes of the calculations obtained using material strength methods can be considered approximate. Therefore, more defined calculations involving the finite element method will be conducted at the Nastran software. Figure 4 represents the model of the mechanism of self-extrication of a track vehicle with the weight of 14 tons; figure 5 gives the equivalent stress distribution (where figure 5,*a* gives the picture at the upper surface and figure 5,*b* the one in the horizontal cross-section). It was predictable that the maximum equivalent stresses in the wall that arose due to the increased edge stiffness of the envelopment slightly increase the values calculated and amount to approximately 385 *MPa*, rather than 329.5 *MPa*. These stresses exist at the inner surface. The yield limit for 12XH3A steel, for instance, stands at 700 *MPa*. Therefore, the assurance factor ratio for yield equals 1.82, which is rather acceptable in our case.

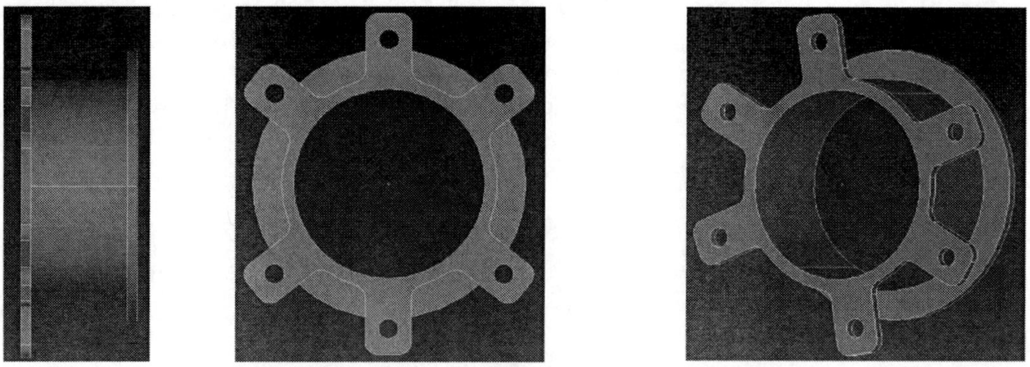

Figure 4. Self-extrication mechanism for track vehicles

IOP Conf. Series: Materials Science and Engineering **489** (2019) 012012 doi:10.1088/1757-899X/489/1/012012

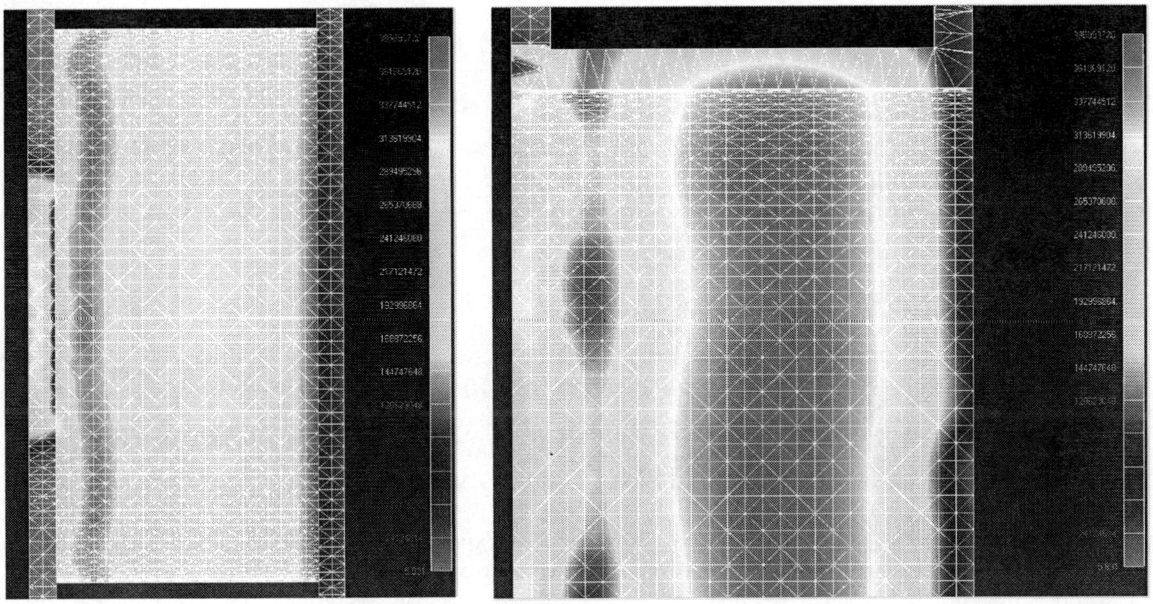

a) *b*)

Figure 5. Equivalent stress distribution

Conclusions

Thus the research has determined the main geometrical parameters of the mechanism for self-extrication of track vehicles with the weight of 14 tons (obviously, for extricating a particular vehicle specific calculations are required). The weight of this device is 30.2 *kilos,* however, with 30-meter long metal wire and the shackle it will increase to 60 *kilos*, which is a drawback as it will impede manual fastening on the drive wheel. Therefore, the next stage of the research will consist in searching for designing solutions to facilitate the mounting of the mechanism.

References

[1] Group of authors 1981 *Guidelines on Armored Vehicle Extrication* (Moscow: USSR Ministry of Defence Publishing House) p 60

[2] Lepeshinsky IU, Chikirev OI, Varlakov PM, Anufriev EV and Pogodaev VP 2013 *Tank extrication procedure* (Omsk: OSTU Publishing House) p 66

[3] Aleksandrov MP 2001 Lifting machines (Moscow: Publishing House of N. Bauman MSTU – Higher School) p 124

The authors express their gratitude to Ekaterina N. Luchinina, head of the Foreign Languages Department of Moscow Higher All-Arms Command School for the translation of this article

IOP Conf. Series: Materials Science and Engineering **489** (2019) 012013 doi:10.1088/1757-899X/489/1/012013

Small wind-driven power plant operating experience

A A Belsky and V S Dobush

Electromechanical Department, Saint-Petersburg Mining University, Vasilevsky Island, 21th Line, 2, Saint-Petersburg, 199106, Russia

Belskij_AA@pers.spmi.ru

Abstract. This paper addresses the operating experience of a small wind-driven power plant at Saint Petersburg Mining University. The paper describes the wind power generator operation, the deficiencies, revealed during the exploitation, as well as provides the recommendations for the plant improvement.

1. Introduction

According to the World Wind Energy Association (WWEA) as of 2017, the installed capacity of wind-driven power plants is 539.3 GW [1]. According to recent data in the field of small wind-power engineering, presented by WWEA, in 2015 the number of wind-driven power plants (WDPP) in the world reached 990 thousand (a unit capacity of a plant is less than 100 kW) [2].

In Russia, the potential owners of small WDPP are the inhabitants of sparsely populated areas that are beyond the reach of the centralized power supply system, as well as separate facilities of the mineral resources sector, located on the northern sea shelf and in the remote north-eastern regions of the country [3, 4].

Currently, there are favorable conditions to develop wind-power engineering in Russia: there are enough regulatory and legal framework and technical rules and regulations [5]. Nevertheless, at the present time, the operating experience of low-power WDPP is insufficient and problems arising during its operation are not disclosed in a proper manner [6–8].

In 2012, the Saint-Petersburg Mining University acquired an electro-technical complex based on a small WDPP Briz of domestic manufacture by CJSC Wind Power Production Company (St. Petersburg). The power complex is designed for conducting scientific research [9–11], assessing the performance and the residual life of such systems [12, 13], as well as for providing outdoor illumination autonomous power supply within the territory of the training and experimental base Neftianik, which serves for educational purposes (figure 1).

During the operation of the wind-powered complex, a few deficiencies and observations have been revealed that affect the ease of operation and maintenance of the complex, safety, and reliability of power supply to consumers.

In this article, the authors have analyzed the ways to solve and eliminate the revealed deficiencies and formulated recommendations to improve this energy complex based on the WDPP Briz, which are given below.

IOP Conf. Series: Materials Science and Engineering **489** (2019) 012013 doi:10.1088/1757-899X/489/1/012013

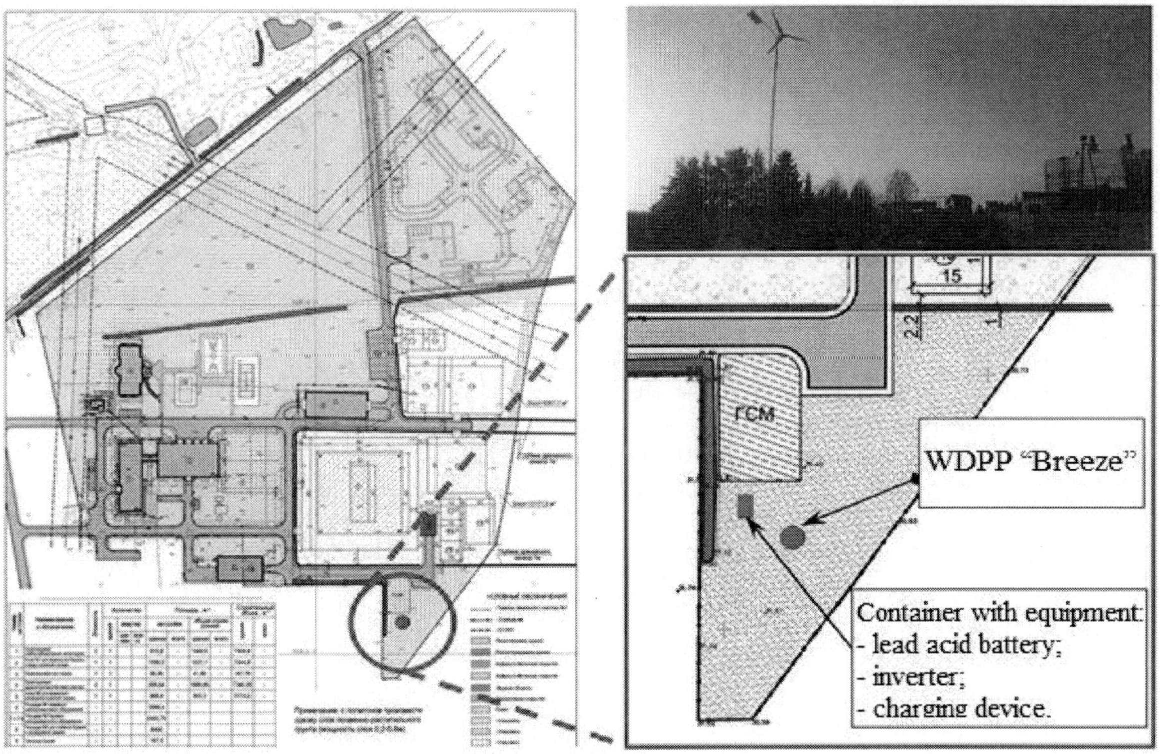

Figure 1. The WDPP installation site within the territory of the training and experimental base

2. Description of WDPP

The main elements of the electro-generating complex (Figure 2): the WDPP Briz: sealed lead-acid rechargeable batteries (SB); inverter; charging device.

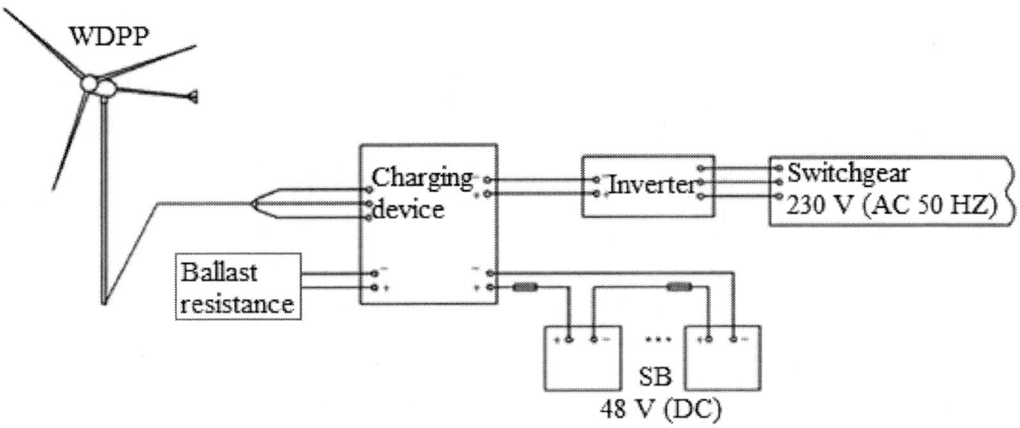

Figure 2. The scheme of the electro-generating complex with the WDPP

The WDPP Briz consists of a wind wheel with three blades of fiberglass, which is attached directly to the generator shaft. The generator is mounted on a supporting and turning assembly, which has an extended tail swiveling vane, to orient the wind-powered generator with the wind. The supporting and turning assembly hosts a current collector to transfer energy from the generator to the cable line, the

IOP Conf. Series: Materials Science and Engineering **489** (2019) 012013 doi:10.1088/1757-899X/489/1/012013

assembly itself is attached to an assembled tubular mast, which is held in a vertical position via two levels of stretching wires.

Basic technical specifications of the power-generating complex under consideration:

- nominal rating power: 4 kW;
- cut-in-wind speed: 3 m·s^{-1};
- nominal (rated) wind speed: 12 m·s^{-1};
- cut-out-wind speed: 45 m·s^{-1};
- rotor diameter (spread of blades): 5 m;
- number of blades: 3 pcs.;
- mast type: steel pipe with tension cables;
- operating temperature: from – 40°C to + 50°C;
- orientation to the wind: swiveling vane;
- method of power control: taking the wind wheel away from the wind;
- generator: synchronous three-phase energized by constant magnets;
- nominal speed of the wind wheel: 400 rpm;
- foundation blocks (L x W x D): 0.8 x 0.8 x 1.8 m, 5 pcs.;
- wind-driven power plant's operational lifetime: 20 years;
- DC link voltage: 48 V;
- storage battery (SB): lead acid, 12 V, 200 Ah, 4 pcs.;
- parameters of the inverter: U_{input} = 40–60 V (DC); U_{output} = 230 V (AC 50 Hz); 6 kW;
- ballast resistance power: 6 kW.

3. Analysis of WDPP

The authors have analyzed the ways to solve and eliminate the revealed deficiencies (Figure 3) and formulated recommendations to improve this energy complex based on the WDPP Briz, which are given below.

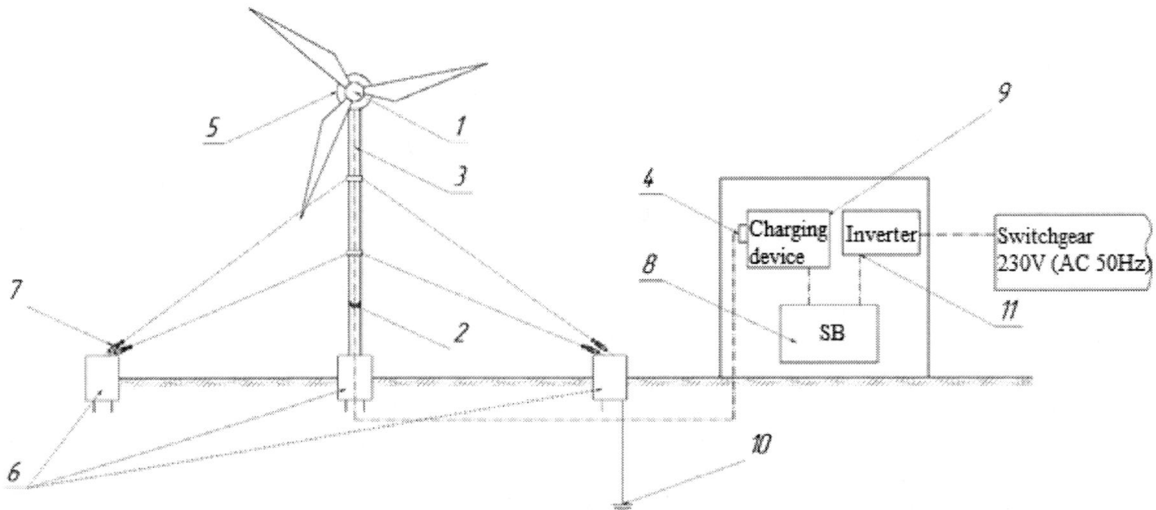

Figure 3. The scheme of the energy complex with the WDPP Briz with an indication of the revealed deficiencies' numbers

Observation No. 1: The attaching point of the blades to the wind wheel hub and the generator shaft.

IOP Conf. Series: Materials Science and Engineering **489** (2019) 012013 doi:10.1088/1757-899X/489/1/012013

Existing implementation (a complete set), manufacturer's solution: Open threaded connection of elements on the wind wheel hub.

Proposed implementation (a complete set), the authors' solution: Equip the wind wheel hub with a protective aerodynamic housing.

Brief comment: Under the effect of precipitation the open threaded joints get rusty. The protective housing will reduce the impact of precipitation, decrease the aerodynamic factor of wind flotation along the WDPP body, and also increase the aesthetic appeal of the complex.

Observation No. 2: Wind wheel emergency brake.

Existing: None.

Proposed: Place on the mast (at a height of 1.8 m from the ground level) a cutoff switch (short-circuit of the WDPP generator windings), which serves as the wind wheel emergency electric brake.

Brief comment: According to national standard GOST R 51991–2002, a wind-power unit must have a manually actuated brake. Brake control must be available to the operator, standing at the ground level.

Observation No. 3: Power cable.

Existing: The rubber-insulated cable is used.

Proposed: Use the PVC-insulated cable.

Brief comment: The rubber-insulated cable service life, according to the technical specifications for manufacturing is 4 years, which does not comply with to the equipment service life — 20 years. The service life of the PVC-insulated cable is at least 25 years.

Observation No. 4: Plug connector, for connecting a charge regulator, in the power supply line, coming from the WDPP.

Existing: Power plug connector 2P + PE, 32 A is installed.

Proposed: Replace the plug connector with the closed terminal connection.

Brief comment: The presence of a plug connector in the power supply line reduces the safety of the WDPP operation. Disconnection of this connector will result in the WDPP operation under the idle mode with a subsequent uncontrolled speeding-up of the wind wheel.

Observation No. 5: Open current-carrying parts, on a supporting and turning arrangement of the WDPP generator.

Existing: Open terminal blocks are installed for connecting wires, coming from the generator to the slip rings inside the tilting device.

Proposed: Install terminal blocks in closed junction boxes IP67.

Brief comment: Under the effect of precipitation, open terminal connections are oxidized, get rusty, which contributes to an increase in transient resistance at the connection point.

Observation No. 6: Grounding device.

Existing: Reinforced concrete foundation blocks are installed separately of each other, complicating their use as natural grounding electrodes.

Proposed: Connect the reinforcement of separate reinforced concrete foundation blocks with a steel strip of 40x4 mm under the ground in a trench at a depth of at least 0.7 m.

Brief comment: As a result, it will allow equalizing the potentials of the electric plant; performing the potentials grading (to reduce step voltage) in case of thunderstorm damage to the WDPP; in case of autonomous power supply it will allow using the existing grounding loop for other equipment (inverter).

Observation No. 7: Restraining wire ropes tension control.

Existing: A "ring-hook" lanyard M16 is used, the range of adjustment is about 140 mm.

Proposed: Use an M24 "male-male" lanyard, the range of adjustment is about 300 mm.

Brief comment: the range of adjustment of the used lanyards is not enough to compensate for the natural extension of wire ropes while in operation, especially in summer-time.

Observation No. 8: Lead-acid rechargeable batteries installation site (as a condition).

Existing: Lead-acid rechargeable batteries are placed on open shelves in a container without heating.

Proposed: Place lead-acid rechargeable batteries in a separate closed cabinet (housing) with an electric heating element installed therein (25–50 W), to maintain a positive temperature and a fan that provides the required number of air changes.

Brief comment: When using lead-acid rechargeable batteries, in order to prolong lead-acid rechargeable battery lifetime, it is necessary to prevent them from operation at negative ambient temperatures.

Observation No. 9: Lead-acid rechargeable batteries charging control method.

Existing: Electromechanical relay is used, without redundancy and emergency shutdown.

Proposed: Use a solid-state relay with an emergency shutdown of the WDPP [14, 15].

Brief comment: This relay connects/disconnects the ballast resistance to control the process of lead-acid rechargeable batteries charging from the WDPP. The electromechanical relay has a low threshold frequency of switching.

Observation No. 10: The grounding of lead-acid rechargeable batteries.

Existing: The manual does not contain information on the necessity to ensure the grounding of lead-acid rechargeable batteries.

Proposed: Ensure the correct operation of the inverter; it is necessary to provide the grounding of the battery.

Brief comment: This requirement is present in the manual for the inverter, which is not provided to the customer of the WDPP.

Observation No. 11: Protection of autonomous inverter against overvoltage at the input.

Existing: None.

Proposed: Use a voltage control relay at the input of the inverter.

Brief comment: In case of failure of lead-acid rechargeable batteries charge control relay (see Observation No. 9), overcharge of lead-acid rechargeable batteries may occur and the voltage at the inverter input will exceed the permissible value, which may cause the inverter breakage.

4. Conclusions

The above observations are not critical but can complicate the process of the energy complex operation and should be taken into account when purchasing and mantling the WDPP. The updated list of observations is submitted to the manufacturer.

References

[1] Wind Power Capacity reaches 539 GW, 52.6 GW added in 2017 (World Wind Energy Association)

[2] WWEA released latest Global Small Wind Statistics (World Wind Energy Association)

[3] Abramovich B N and Sychev Yu A 2017 *J. of Mining Inst.* **217** 132–9

[4] Abramovich B N, Sychev Y A, Kuznetsov P A and Zimin R Yu 2018 *IOP Conf. Ser.: Earth Environ. Sci.* **194** 052001

[5] Kulakov A V and Nazarova Yu A 2015 *J. MPEI Vestnik* **5** 65–9

[6] Kvitko A V, Semenov Ya A and Otmahov G S 2015 *Polythematic Online Sci. J. Kuban State Agrarian University* **112** 1003–15

[7] Yanhong Wang and Shunbin Hu 2012 *Proc. Fifth Int. Conf. Advanced Computational Intelligence* (IEEE) 937–40

[8] Peter Kadar 2013 *Proc. Int. Conf. Clean Electrical Power* (IEEE) 421–5

[9] Kopteva A V and Malarev V I 2018 *Proc. Conf. Russian Young Researchers in Electrical and Electronic Engineering* (IEEE) 678–81

[10] Belsky A A and Dobush V S 2016 *Proc. Int. Conf. Dynamics of Systems, Mechanisms and Machines* (IEEE) 1–4

[11] Belsky A A and Dobush V S 2017 *IOP Conf. Ser.: Earth Environ. Sci.* **87** 032006

[12] Korolev N A and Solovev S V 2017 *IOP Conf. Ser.: Materials Sci. and Engineering* **177** 012007

IOP Conf. Series: Materials Science and Engineering **489** (2019) 012013 doi:10.1088/1757-899X/489/1/012013

[13] Zhukovskiy Y L, Korolev N A, Babanova I S and Boikov A V 2017 *IOP Conf. Ser.: Earth Environ. Sci.* **87** 032056
[14] Shklyarskiy J E and Bardanov A I 2018 *Proc. 11th Conf. Russian-German Raw Materials* (Potsdam: St Petetersburg Mining University) **1** 241–7
[15] Demidovich V B, Chmilenko F V and Rastvorova I I 2018 *IOP Conf. Ser.: Earth Environ. Sci.* **194** 062029

Acknowledgments
The reported study was funded by RFBR according to the research project №18-38-00530.

IOP Conf. Series: Materials Science and Engineering **489** (2019) 012014 doi:10.1088/1757-899X/489/1/012014

Aspects of strength calculation for the differential variable speed drive body of a hybrid bicycle

V V Bogdanov[1,2], B V Demin[2] and E V Metscheryakova[1]

[1]Moscow Higher Military Command University, Moscow, Russia
[2]Limited production company Gibridnye Tekhnologii, Skolkovo, Moscow, Russia

vvbogdanov@t-online.de

Abstract. The article describes some aspects of strength calculation for the differential variable speed drive body of a hybrid bicycle in bodywork of Skolkovo residents' implementation of the design on the construction of hybrid bicycle for the civil and military mission.

1. Introduction

It is considered that a bicycle is an alternative form of urban and suburban transport. Precisely in these directions generally the development and evolution of its' design proceeds. Now the conversion tendency from mechanical bicycles to the so-called hybrid bicycles combining the muscular drive and the electric transmission is observed [1].

It is known that the use of bicycles as an alternative form of urban and suburban transport has a significant shortcoming – low speed and significant physical effort spent on movement, which adversely affects the comfort of travel and does not allow using fully this kind of transport for people with disabilities (or underdeveloped physical abilities). The existing available designs of electric bicycles urged to solve the above-mentioned issues due to the technical solutions used now have either insufficient cruising range/low-power auxiliary motor, or high weight and speed parameters transferring it into a scooter class or environmentally unsafe battery.

The hybrid bicycle capable to solve the abovementioned issues provides a competitive advantage and can be employed for both civil and military purposes, for example, in a group of bicycle infantries. The last direction has more than 200 years history [2; 3] from the moment of appearance of the first serial designs of bicycles (see figure 1) to up-to-date equipment of the army with the electric bicycles (see figure 2).

Figure 1. The first bicycles for army needs

Figure 2. The modern army cycling groups of NATO (left) and Russian electric bicycle Gran (right)

2. Issue formulation and main results

The main issue of the technical component when developing designs of hybrid drive structures on electric bicycles is an insufficient reasonableness of the optimizing process of human effort and battery charge consumption, expressed in the non-optimal and uneven load of the electric motor. Equipping the bicycle with a hybrid variable speed drive based on summing-up differential forms for two power flows – from the muscular cyclist power and (main) from the electric motor – will solve the abovementioned issues. The hybrid variable speed drive keeps to constants the rotational speed of pedals, smoothly and dynamically distributes the loading applied by the cyclist. Two electric motors plus combined transmission in one body, automatic stepless speed change and the possibility of software upgraded up to the power scooter class distinguishes this development from analogues. Due to the possibility of changing and controlling the rotation's speed of the main electric motor, it is possible to change the gear ratio between the pedal speed and the speed of the bicycle wheel. This design of hybrid variable speed drive also allows using hybrid bicycle as the electric generator from muscular force or as emergency / reserve power source. The possibility of charging the storage battery both the electric bicycle and other batteries or emergency power supply for communication facilities is provided. Of course, the power of such generator is small, but it does not require fuel reserves, does not make noise and is mobile.

Today product development and manufacture of completely hybrid bicycle have occupied such well-known companies as Bosch, Shimano and some smaller companies, including development for a military mission. From the technical point of view, the electric bicycle designed for military use differs little from the civilian version, with the only difference of no limit on the ride maximum speed, which, in particular, according to modern European standards, transfers e-bicycles in the category of scooter class if a certain high-speed limit is exceeded.

One of the main elements of the electric bicycle hybrid is the differential variable speed drive (see figure 3), which in part one of the basic elements is the body combined with the electric drive. Its strength directly influences the safety and functioning of the whole design. In addition, the selection of the optimal thickness of the body promotes to weight reduction of the construction and therefore increases the average speed of movement.

At the first stage, the design participants were given the following task: to conduct a metric synthesis of the differential mechanism based on the power balance. This has been done (not present in this article). According to the results of this stage, the next task of the design second stage was set: to determine the stress state of the variable speed drive body. The results of the strength calculation, in turn, are the initial data for the next third stage: to reduce the thickness of the variable speed drive body without losing the general strength of the design for implementation of the body weight reduction at constant dimensions, material and known given loads.

IOP Conf. Series: Materials Science and Engineering **489** (2019) 012014 doi:10.1088/1757-899X/489/1/012014

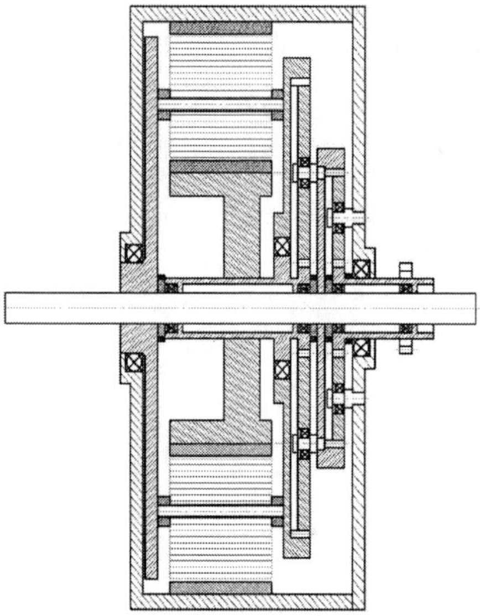

Figure 3. Drawing of the differential variable speed drive

The Femap with NX Nastran software package to solve the second stage problem was used [4]. Due to the complex geometrical features of the design the three-dimensional Solid element with an intermediate node was used. The results of strength FEM-calculation in the form of equivalent stress distribution fields are shown (see figure 4).

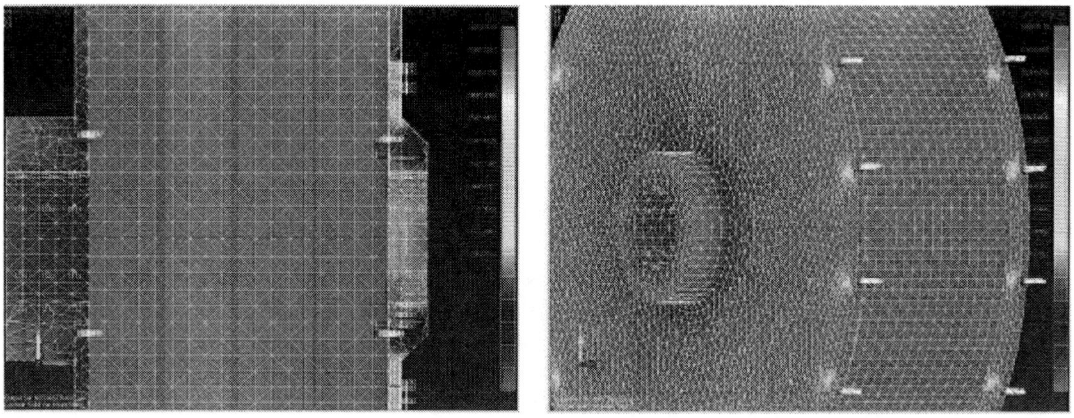

IOP Conf. Series: Materials Science and Engineering **489** (2019) 012014 doi:10.1088/1757-899X/489/1/012014

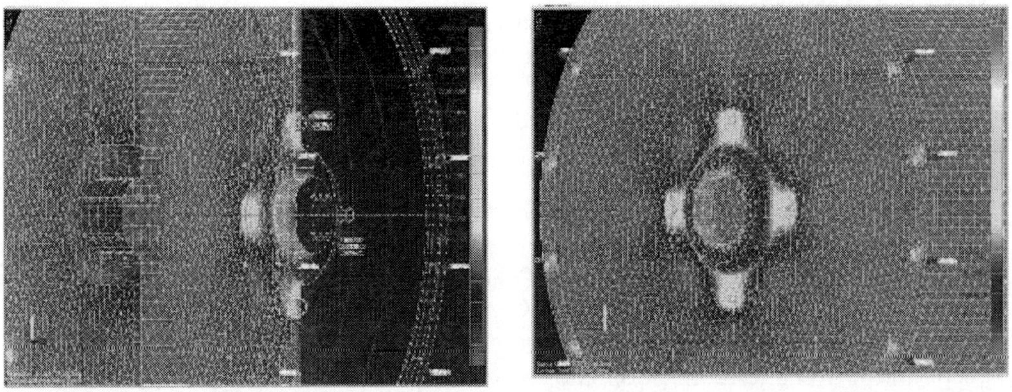

Figure 4. Examples of fields distribution of equivalent stresses for variable speed drive body of an electric bicycle

3. Conclusions and recommendations

It is possible to give the following general conclusions and recommendations based on results of FEM-calculation. Different forms of the body of differential variable speed drive from the point of view of ensuring the maximum strength and eliminating the concentration of stresses in local zones are analyzed; the most optimum form of the variable speed drive body is chosen. The stress-strain state of the body under influence of loads for the main prototype of the design is determined. The main directions for motivated weight reduction of the body and other structural elements of the variable speed drive without losing the general strength of the design at constant dimensions, material and effective load spectrum are selected. Subtasks for further optimization of the real design are set.

The complexity of this project and the advisability of collaboration with teams interested in this topic should be noted. As shown by 3D-model (see figure 5), the design includes not only strength calculations and kinematic calculations, but also electromechanical calculations. Moreover, this project involves the electronic control system development of the electromechanical part of the hybrid variable speed drive, which is no less complex and includes such areas as electrical engineering; electronics; programming.

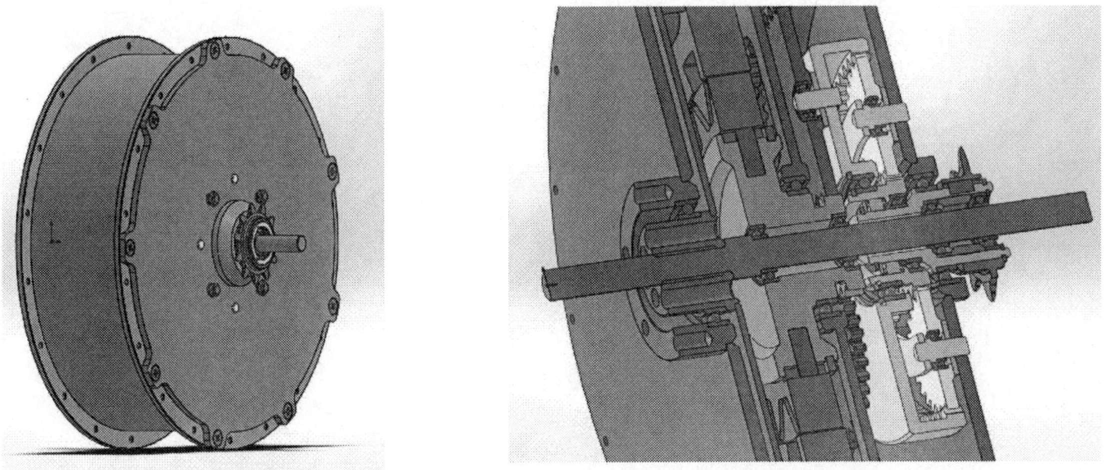

Figure 5. 3D-model of the differential variable speed drive of electric bicycle

IOP Conf. Series: Materials Science and Engineering **489** (2019) 012014 doi:10.1088/1757-899X/489/1/012014

As noted above the received results of the strength calculation is the initial data for the next, third stage of structural design. The general difficulty of the next stage is the static redundancy of the structure, so to make the target function of reducing the weight of the design is the exceptional complexity task. Now this issue is being solved.

References

[1] Bogdanov V V, Chabunin I S and Afanasyev S S 2018 Features of strength calculation by FEM based on the example of differential variable speed drive body of hybrid bicycle. *Proc. Scientific and Technical Conf. on "Use of computer technologies in educational activity"* (Moscow: Moscow Higher Military Command University Press) pp 57-64

[2] Link: http://army-news.ru/2015/05/voennye-velosipedy-v-21-veke

[3] Link:http://trenager.com/publ/stati_o_velosipedakh/voennoe_primenenie_velosipeda/6-1-0-90

[4] Rychkov S P 2013. *Modeling of structures in the Femap with NX Nastran environment.* (Moscow DMK Press) p 784.

IOP Conf. Series: Materials Science and Engineering **489** (2019) 012015 doi:10.1088/1757-899X/489/1/012015

On assessment of crack-type defect sizes with using results of modal analysis

A S Chernyatin and I A Razumovsky

Blagonravov Mechanical Engineering Research Institute of the Russian Academy of Sciences, 4 Maly Kharitonyevsky Pereulok, Moscow, 101990, Russia

murza45@gmail.com

Abstract. This paper presents the methodogical approach to the assessment of geometrical and localization parameters for crack-type defects (including hidden subsurface flaws) by means mathematical treatment of experimental information represented by the amplitude displacement fields of researchable object registered by optic-interference methods.

1. Introduction

The goal of research, the results of which are presented in this paper, is to assess the capabilities of finding the geometrical parameters for crack-type defect and its localization zone based on the data of experimental analysis of oscillation natural modes and frequencies (simulation analysis).

As of today, numerous experimental methods (as well as appropriate devices) that provide finding capabilities of oscillation modes and frequencies applied to technical systems have been developed, e.g. holographic interferometry, electronic digital speckle interferometry, method of digitized image correlation, digital holography. These methods and devices make it possible to register the displacement field pictures of the examined object that are obtained at any point of time during dynamic process or the oscillations corresponding to definite natural mode of oscillation to a high precision (~0.1 μm/interference band) [1].

It is shown in this paper that the parameters defining defect sizes and location can be determined on the basis of minimizing the complex discrepancy between experimentally registered fields of vibration displacements and their computed values obtained from simulation analysis with FEM invocation at current values of required parameters. Similar approach has been applied successfully in [2,3] with the purpose of estimating the parameters of surface and internal defects in construction components on the basis of mathematical treatment of deformation response fields registered at static loads.

2. Problem definition

The model of a gas turbine blade (Figure 1a) in the form of the profiled plate permanently attached at the bottom (in section $x = 0$) is considered as a subject of inquiry; this plate has the following dimensions: length $L = 0.1$ m; width $W = 0.05$ m; maximal profile thickness $t = 0.01$ m. Plate material (steel): modulus of elasticity $E=2\times10^5$ MPa; Poisson ratio $\mu = 0.3$; density $\rho = 7800$ m^3/kg. The blade has a defect in the form of semi-elliptical flat surface crack with length $2a$ and depth b (a, b – axes of semi-ellipse). Location of the crack which is orthogonal to the lower flat surface of the plate is defined by localization parameters x_c,y_c (Figure 1b).

IOP Conf. Series: Materials Science and Engineering **489** (2019) 012015 doi:10.1088/1757-899X/489/1/012015

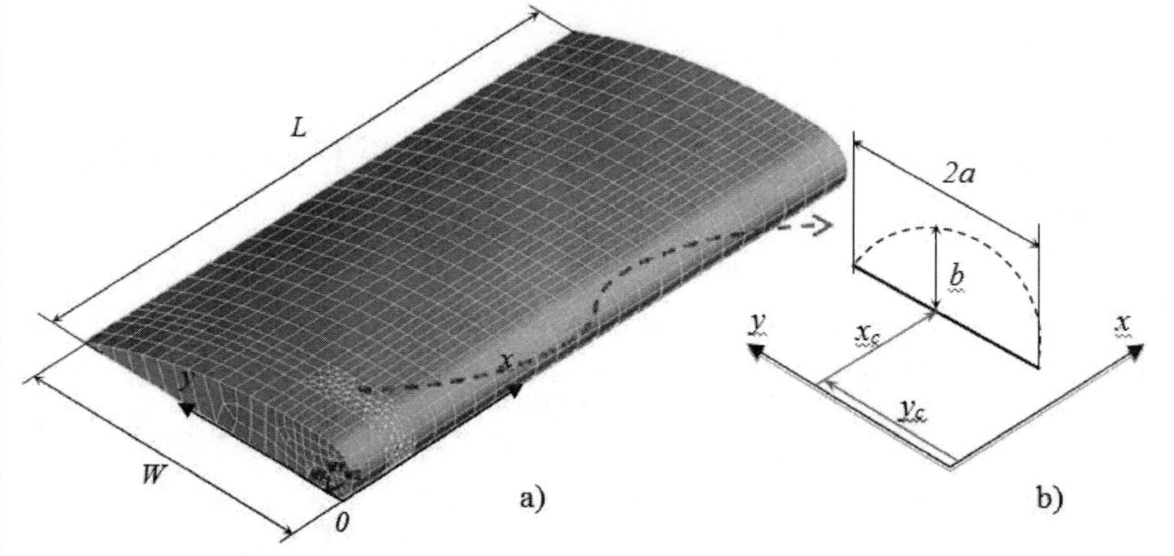

Figure 1. Finite difference model of the researchable object (a) with surface crack (b)

The objective is to assess the capability to determine defect parameters a, b, x_c, y_c (forming vector of unknowns $P\ \{a, b, x_c, y_c\}$) through mathematical treatment of amplitude displacement field pictures for plate points $w\ (x,y)$ (w values at N points form the vector of $e = \{w_j\}\ j = 1,...N$) corresponding to oscillation modes of the examined object at natural frequencies. It should be pointed out, that modal analysis does not immediately allow obtaining of real values of point oscillation amplitudes (they are defined up to a constant). In view of this several variants for normalizing the displacement fields corresponding to natural modes were developed by authors of this paper.

The finite-element model of the examined object in ANSYS circumference built using 3D elements with quadratic approximation is shown in Figure 1a. The model is divided into two zones: a zone with defect and the basic part. Along the crack front the grid has a crowding, the average size of elements is $(L+W)/100$. The finite-element model is parametric, and a special APDL macro was developed for its generation and subsequent calculation of natural frequencies and modes.

3. Calculation procedure

The proposed approach for determining $P=\{a, b, x_c, y_c\}$ vector on the basis of processing of displacement fields recorded experimentally consists of following stages:

- Handling the problem of modal analysis for the examined object – with the purpose of generating the so-called "response database" [2].
- Handling the problem of determining the required parameters of defect based on mathematical treatment of the interference images pertaining to the oscillation natural modes of the object as a multiparameter nonlinear optimization problem [2,3].
- Assessment of the effect of experimental data inaccuracies on the precision of determining the parameters characterizing the defect sizes and location.

The modal analysis carried out with the purpose of determining the natural frequencies and modes of oscillations based on solving of the eigenvalue problem is the "point of departure" for more complicated inquiries on dynamic system responses, which, in particular, can be used for defect identification. For the purpose of determining natural frequencies and modes of oscillations, the preconditioned conjugate gradient (PCG) method coupled with Lanczos algorithm was applied. This method is preferable to the direct methods of solving equations with sparse matrices (as it requires far less resources) and possesses a superlinear rate of convergence – even in the problems of high dimensionality and with ill condition sine qua non.

Figure 2 represents the calculation oscillation modes corresponding to the first five frequencies for the blade with the defect with parameters of P_0^* ={10;-5;10;5} mm.

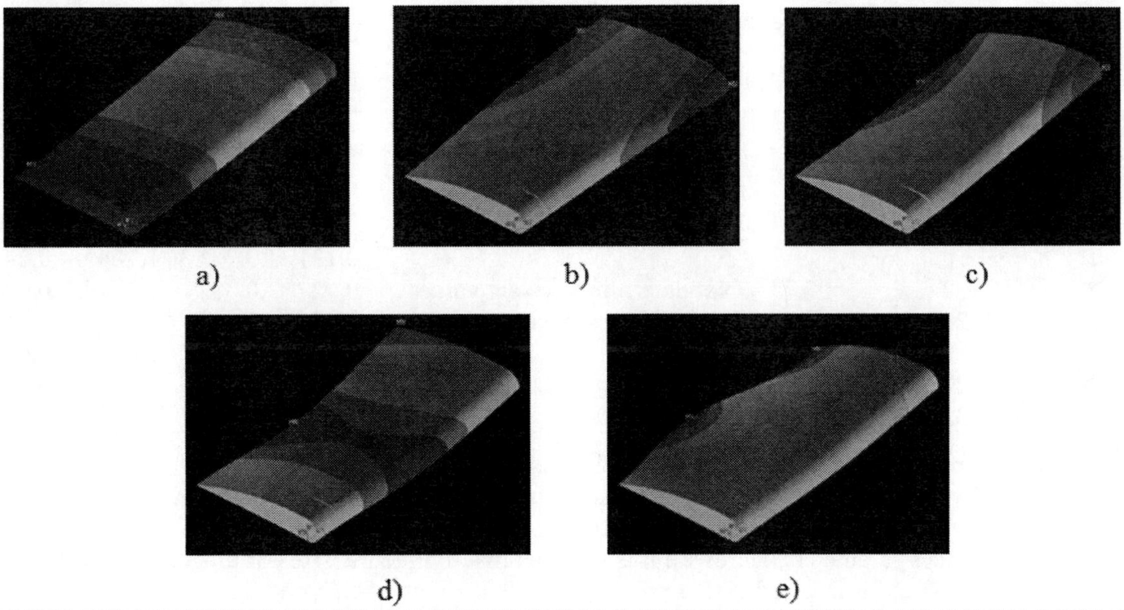

Figure 2. Natural modes of object oscillations under the frequencies: 672 Hz (a), 2814 Hz (b), 3236 Hz (c), 4091 Hz (d), 8345 Hz (e).

For the purpose of solving the problem of mode analysis for the object in question, a series of problems of finding the amplitude displacement for surface points of the examined object $w(x,y)$ under variation of vector P components is solved for the "provisioning" of the response database. This procedure is repeated for each natural mode. Based on the hyper-spline surfaces obtained in such a manner for each j-th component of e vector the approximation of the $e_i = F(P)$ form is set forth.

It should be noted, that the normalized module of the relative gradient g_w is used in the calculation, rather than the w displacement in the points of the examined object; this allows to circumvent the problem of uncertain scaling for displacement fields.

In the next step the vector P components are determined according to the displacement field $w^*(x,y)$ available from experiments and presented through the e^* vector of g_w gradient values in actual points of measurements on the basis of minimizing the objective function I, which reproduces the integrated divergence between e^* and e, which, in turn, is calculated by dint of well-formed response database. It should be noted that within the confines of this work the numerical experiment is performed when the w^* displacements are determined by the numerical finite element simulation. The defect with P_0^* parameters (see above) was adopted as the definable one, and the measurement points in which w^* value is calculated are located in the nodes of a rectangular mesh corresponding to the undamaged (inverse) surface of the blade (see Figure 1a). This way, the defect is "undetected" in relation to the surface of measurements. Mean-square deviation was adopted as an objective function. Taking into account the fact that optic-interference methods allow to obtain a practically unlimited volume of experimental information, the number of N points of displacement "measuring" was imputed as $17 \times 17 = 289$.

For the purpose of examining the effect of the errors of the w values measuring on the precision of determining the P defect parameters, a computed procedure of e^* vector "noise masking" was conducted by introducing random departures out of $\pm \Delta e$ range to each vector component with subsequent use of the provided approach of P vector determination. This procedure was repeated multiple times in order to obtain statistical parameters.

4. Results

The results of the numerical experiment for determining the defect parameters in the form of mathematical expectation M and dispersion D for P vector components which are classified as valid (i.e. corresponding to P_0^*) values are presented in table 1.

Table 1. Results of defect parameters determination.

Calculation conditions		Δe	x_c		y_c		a		b	
№№	Oscillation mode		M	D	M	D	M	D	M	D
1	1	±15%	Not determined, exact values were given				0,131	0,231	1,124	0,240
2	1, 2						0,927	0,082	1,0314	0,048
3	1, 2, 3						0,964	0,007	1,023	0,002
4	1, 2, 3, 4						0,995	0,003	1,002	0,002
5	1, 2, 3		1,000	0,000	1,016	0,005	0,952	0,007	1,037	0,006
6	1, 2, 3, 4, 5		1,000	0,000	1,000	0,006	0,996	0,005	1,002	0,003

The following conclusions can be made on the basis of the results obtained.

- Under exceptionally wide adopted scatter band of experiment uncertainties the usage of registration results for the oscillation mode corresponding to one of natural frequencies does not assess the values of defect parameters. Herewith it is obvious that, unlike the sizes of hidden defect, the zone of its localization can be assessed with a precision eligible for practice based on visual examination of the oscillation mode.
- In the case of oscillation modes obtained for several natural frequencies, the proposed approach allows to determine both the localization and the geometric defect parameters reliably and precisely. One can expect that the precision of problem solving with the use of displacement fields registered at the поверхности выхода дефекта, will be higher, compared to the provided example, when the defect emerges to a surface that does not allow to register the displacement fields.

Acknowledgments
This work was financially supported by the Russian Science Foundation, grant No 14-19-00776-П

References

[1] Igor A. Rasumovsky 2011 Interference-optical Methods of Solid Mechanics / Series: Foundations of Engineering Mechanics. Springer. 270 p (doi 10.1007/978-3-642-11222-5)
[2] Chernyatin A S and Razumovskii I A 2013 Methodology and Software Package for Assessment of Stress-Strain State Parameters of Full-Scale Structures and its application to a Study of Loading Level, Defect Rate, and Residual Stress Level in Elements of NPP Equipment Strength of Materials **45** 506–11
[3] Razumovskii I A, Chernyatin A S and Fomin A V Experimental-Computation Methods for Determination of Stress-Strain State of Structural Components Inorganic Materials **50** 57-64

IOP Conf. Series: Materials Science and Engineering **489** (2019) 012016 doi:10.1088/1757-899X/489/1/012016

Evaluation of the operational life of direct current motors

O A Filina and A N Tsvetkov

Kazan State Power Engineering University, Krasnoselskaia str., Kazan, 420066, Russia

olga_yuminova83@mail.ru

Abstract. This article is devoted to the issue of technical diagnostics in urban electric transport. To resolve the issue, it is necessary to translate the qualitative determination of the TS on some quantitative basis. The formalization of qualitative definitions is a necessary condition for the construction of formal (computable) diagnostic algorithms.

1. Introduction

The relevance: The operational reliability of the traction electric-motor decreases suddenly after the depletion of the operational life. Life cycle extension methods should maintain the operational reliability of a direct current traction motor (DCTM) equal or close to the nameplate data for a number of years. On the one hand, the life cycle extension of an electric machine should have an economic effect, and on the other hand, technical measures of operational reliability maintaining cause an increase in material expenditures. Therefore, a position is required where the difference between the result obtained and the cost is positive. In this case, the life cycle extension is an appropriate and cost-efficient task.

2. Tasks evaluation of the engines operating resource

It can be said that maintaining the operational reliability of an electric machine that has reached the end of its service life at an appropriate level entails an increase in technical and economic effect. Therefore, scientific research is aimed at determining the economically justified level of reliability of electric machines for a given year in certain areas of their use.

The work poses the following tasks:

- research how the deviations of the design and technological parameters of the electromagnetic and mechanical nature influence the quality of functioning of the collector-and-brush assembly unit (DCTM);
- determine the effective set of parameters for diagnosing the technical state of the collector-and-brush assembly unit and monitor its performance with the most informative value and distinctiveness;
- explain the selection of diagnostic signals and their parameters for assessing the sparking rate and the surface condition of the collector of the traction motor, research the statistical characteristics of these parameters;
- develop a method and device for monitoring the state of the collector' surface taking into account the influence of the collector' profile on the sliding contact stability.

A direct current traction motor is the most important element for rolling stock that determines its reliability in general.

IOP Conf. Series: Materials Science and Engineering **489** (2019) 012016 doi:10.1088/1757-899X/489/1/012016

At the present time, there is a trend towards the development of automated diagnostic systems for electric actuators without putting them into repair. It can significantly reduce the standby time of rolling stock (RS) and improve the accuracy of troubleshooting.

Traction motor (TM) is a complex multi-element technical system. Its individual elements are combined into numerous assemblies related to mechanical and electromagnetic systems. Issues with them are related to the non-identity of operating sequences.

Comprehensive solution of these issues: Mechanical and electromagnetic non-identity of operating sequences meets the modern requirements of scientific and technological progress, since one of the priority areas is to reduce losses associated with the components and parts wear, as well as the introduction of modern tools, methods, diagnostics, and repair of the TM collector-and-brush assembly unit. There are also external conditions affecting the TM along with internal issues. These factors include the environment, such as humidity, temperature, dust. With a decrease in pressure, heat removal from the elements deteriorates, which leads to overheating.

Thus, the topic of scientific research related to the development of means of DC motors state operational diagnostics is an important task.

Currently, rolling stock-operating organizations widely use the system of scheduled-preventive repair (SPR) in the operation of unrecoverable or disposable parts (electrical brushes, capacitors, resistors, devices, etc.). But it is unacceptable for modern rolling stock in general. It entails an increase in standby time of RS with DCTM during repair and the overall rise in the cost of operation, which negates the profitability of this type of actuator and further complicates the situation with asynchronous and synchronous motors. It does not take into account various developments, unequal load factors and, consequently, different wear of RS equipment. Obviously, the same equipment overhaul period at different wear leads to unnecessary consumption of material resources.

The systematic collection and analysis of statistical information about faults and failures of the RS, its weak assemblies, the researches of parts and assemblies wear common factors in the function of operating time are necessary conditions for improving the efficiency of production processes, reliability, and quality of maintenance and repair. The work on pre-repair defects and damage to electrical machines is called troubleshooting.

A significant number of rejections occur during the inter-repair period or are detected during preventive maintenance. Well-trained personnel, provided with a sufficient amount of automatic control features for the electric actuator, ensures the operation of the equipment with less expenditure of effort and means. A production base has been created in the rolling stock-operating organizations to ensure the traction rolling stock repair and its maintenance during operation. Most of the TM malfunctions over the research period are associated with damages at anchor and collector (Fig. 1).

As can be seen from the diagram, the main issues are insulation breakdowns and shifting. The overlap on the collector often goes with a circular fire with the transfer to the body and the burning out of machine parts that fall into the burning area. The cause of shifting is usually pollution and oiling of the collector, accumulation of coal dust in bar-to-bar grooves. To eliminate the consequences of breakdown and reruns, the failed coil or section is replaced, the insulation is restored. Vibration is of traction motor. Increased vibration is the cause of burning without a common factor, causing brushes to break away from the collector, the brush sparking, and increased collector heating. These features cause rapid abrasion of brushes in height and the formation of abundant graphite dust. Vibrations and noise with high-frequency components appear when rubbing brushes against a collector or contact wheels in an electric machine.

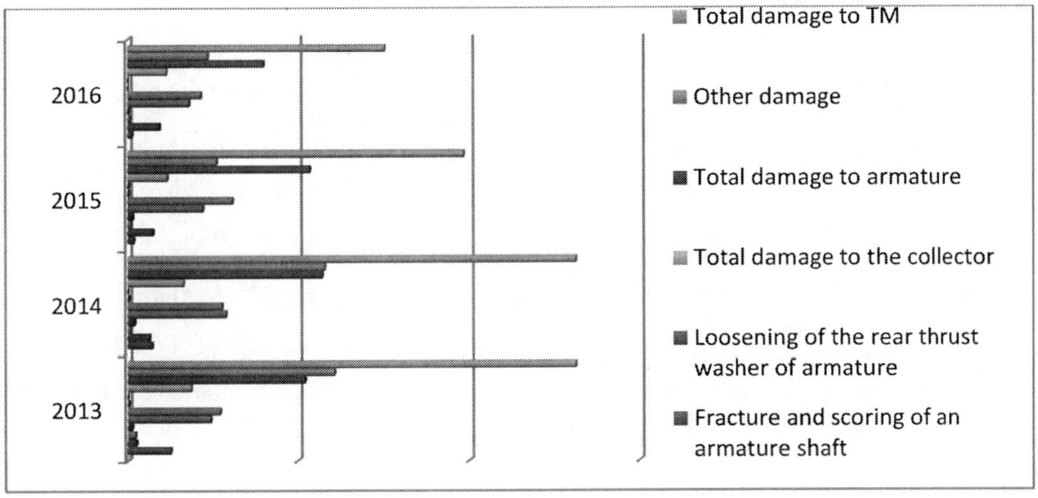

Figure 1. Cases of all damages

Damage to the TM collectors may occur in the collector non-cylindrical surface formation due to its unequal brushing wear during long-term work, breaking the polishing of the collector surface with scratches on it due to unequal pressing of individual brushes on the collector and their hardness, burning plates by commutation failure, their flashing in the case of a circular fire on the collector during machine overload, etc.

The causes of rejections of anchor bearings are the allowable bearings heating temperature rise, their contamination during assembly or the presence of a contaminated lubricant, its excess, wear or destruction of bearing parts if the latter being imbalanced, small radial clearance, friction in bearing seals [1]. The most likely causes of low-frequency oscillations (from 0 to 300 Hz) are an eccentricity, imbalance of rotating masses, imperfection of rolling bearings, violation of the geometry of slider bearers [2]. Brushes EG-2AF have the friction coefficient equal to 0.35 when idling, and operating under load, it decreases to 0.014. In these conditions, great importance appears to be in the vibration resistance of brushes, their ability to resist intense wear and destruction (chipping) with increased vibration. At present, a set of vibro-acoustic devices, which are used to diagnose inter-shaft, inter-rotor and other rolling bearings, is used quite effectively to control the vibration state of the motor. On the other hand, the vibration itself is a source of serious damage and accidents.

A further reduction in the number of rejections may be due to better repair, improvement of the existing technology. The system is divided into blocks based on the unity of the product functioning and the physical processes occurring in it.

Therefore, the introduction of maintenance and repair system is the right resolution to the issue of ensuring the facilities durability.

According to statistical indicators, it is possible to determine the availability factor of an RS for the considered period of operation by the SPR. It is expressed by the formula:

$$A_i = \frac{M_{UT}}{M_{UT}+M_D T_{CM}} \tag{1}$$

where M_{UT} is the average work time;

$M_D T_{CM}$ is the average non-work time due to the implementation of unscheduled types of maintenance and repair performed to eliminate all faults and failures for the same period of operation (except for cases of vandalism, third-party culpability, incorrect operation, etc.).

Considering the probability of no-failure (PNF) at all time intervals for each of the TM assemblies, the probability of failure-free operation during time t means the probability of an event $\{\tau > t\}$, when the product (TM) will not fail during time t. I.e. symbolically it is written like this:

$$p(t)=P\{\tau>t\}, \tag{2}$$

where t is the specified time;

τ is the time before product failure, a random value.

The function $p(t)$ is often called the reliability function. It is believed that $p(0)=1$ and with increasing time t, this function monotonically decreases, tending to zero.

The probability that after time t, the assembly will work smoothly for a time x,

$$P^{*}(t)=1-\frac{n(t)}{N},\ n(t)\leq N \tag{3}$$

The value of $\sigma=|\sqrt{D\tau}|$, where the mean-square deviation shows the deviation of the service life τ from its average $M\tau$.

It is seen to be considered that τ_i (i=1, 2, ...) are independent and equally distributed random values with the distribution function $F(t)$.

The coefficients "$C_к$" form a random amplitude probability distribution. The reliability functions will change depending on the flows of elements failures over time. In particular, according to the law of Weibull, it can be represented as an equation:

$$P(t)=exp[-\lambda t^{\alpha}] \tag{4}$$

where λ, α – parameters.

If vectors of durability of ri reflect properties in a reference state, and the index of R(t/r) is equal to the conditional probability of failure-free operation at a stage of operation of t, then probability of failure free operation for any system taking into account change of a condition of the U(t/r) system and an index of its quality of V (t/r) can be expressed by means of the corresponding functions.

The following types of electric stock clusters failures stock are most widespread and demonstrate that 20-25% of the total number of failures fall to the share of traction engines. Charts of malfunctions of TED are provided on Fig. 2 and 3.

Experimental studies carried out by the authors to assess the influence of external vibrations on the work of the current-collecting assembly were performed on an experimental setup where experiments have been carried out. The developed test method contains a list of reliability indicators to be monitored, as well as the following data for each specific reliability indicator: acceptance $P\alpha$ and failure $P\beta$ levels of the failure-free operation probability, risks of consumer β and manufacturer α, the test method, the test plan, the list of parameters characterizing the state of products, test conditions (values of influencing factors, their sequence, duration), and the decision rule. A feature of the applied method of sequential analysis is that the number of test items is not fixed in advance for given α and β but depends on the observations outcome.

A mechanical wear of the collector depends on the area of contact with the electrical brushes and their state. Repair of a restored technical system (a brush with a detachable lower section) developed by the authors (patent No. 182855 of the Russian Federation for the brush assembly of an electric machine) is less expensive than a new one. Knowing the nature, the rate of change of the output process parameter, and its limiting value, one can determine the operational life of the collector-brush assembly before the next maintenance or repair. The pressure of the brush on the collector is also measured.

The TM collector-brush assembly as a subject of diagnosis can be in one of the states that form the set of states W, which can be divided into the following subsets: A – good technical states when the values of all structural and output parameters are within acceptable limits and, therefore, the collector-and-brush assembly unit is functioning normally; B^t – faulty technical state when at least one structural or output parameter is beyond the permissible limits. On the other hand, the set of the collector-and-brush assembly unit states can be divided into subsets: E – operational states when most of the main structural and output parameters are within acceptable limits and the collector-and-brush assembly unit performs specified functions; D – rejections of the collector-and-brush assembly unit, the loss of its performance.

IOP Conf. Series: Materials Science and Engineering **489** (2019) 012016 doi:10.1088/1757-899X/489/1/012016

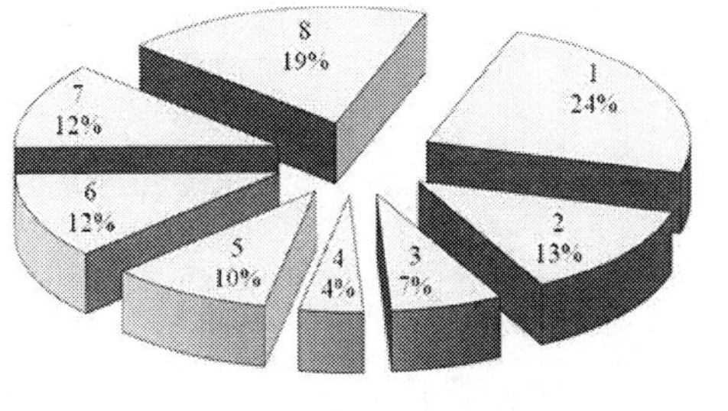

Figure 2. Fault diagram of tram traction motors: 1 – oiling; 2 – damage to the anchor bearers; 3 – shifting, flashing, burnings, tightening of the collector bars; 4 – collector scoring; 5 – the collector surface runout; 6 – melting of brazes from collector lugs; 7 – breakdown and inter-turn closing of the armature winding; 8 – other faults.

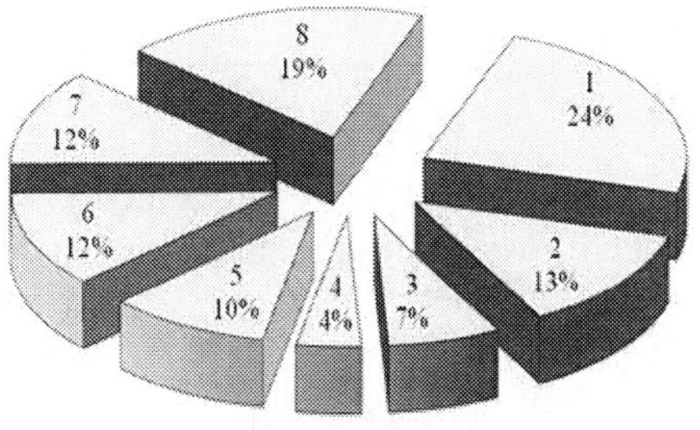

Figure 3. Fault diagram of the trolley bus traction motors: 1 – insulation breakdown and inter-turn armature closing; 2 – melting of brazes from collector lugs; 3 – oiling in the core; 4 – collector surface runout; 5 – damage to the anchor bearers; 6 – low winding insulation; 7 – insulation breakdown and inter-turn closing of the main and additional poles; 8 – other faults.

The connection of the considered states of the collector-and-brush assembly unit can be illustrated by the equations:

$$A + B' = W/(E\backslash D) = W\backslash AB_t = 0;\ E/D = 0; \tag{5}$$

$$B_t/E = C;\ A\,JG\backslash JD = W, \tag{6}$$

where G is a subset of faulty but operational states.

In case a state belongs to the subsets G or D, it is necessary to resolve the issue of diagnosing, i.e. defect recognition.

Accept the smallest values:

$$S = \sum_{i=1}^{n}\left(aN_i^{\frac{1}{4}} - \frac{в}{N_i} - R_{nep\,i}\right)^2 \tag{7}$$

System natural frequency

$$\omega_0 = \sqrt{k/M_Д} \tag{8}$$

Oscillatory acceleration is determined by a formula

$$a = \omega^2 X = (2\pi F)^2 X = \frac{4\pi^2}{T^2}X \text{ (cm/sec. }^2\text{, m/s}^2\text{).} \tag{9}$$

The current has a harmonic proportional to the speed of rotation of the armature. Then, for example, when modulating the rotation harmonics of the rotor (the inner wheel of the bearer) at the first type combination frequency [i.e. if we take take $\omega_1 = 2\pi(qf_{H.K} \pm mf_B)$, $\omega_2 = 2\pi f_B$], the vibration spectrum may have got lateral components at frequencies

$$kf_B \pm n(qf_{н.к} \pm mf_B), \tag{10}$$

and if the signal is modulated at a frequency $kf_{H.K}$ by a second combinational frequency [i.e. $\omega_1 = 2\pi(q_1 f_B \pm m_1 f_C)$, $\omega_2 = 2\pi f_{H.K}$] at frequencies

$$kf_{н.к} \pm n(q_1 f_B \pm m_1 f_C), \tag{11}$$

where q=1, 2, 3, …, m=1, 2, 3, … - are harmonic numbers.

IOP Conf. Series: Materials Science and Engineering **489** (2019) 012016 doi:10.1088/1757-899X/489/1/012016

The frequency of the harmonics resulting from the engine operation and previously called "speed" ones-, are defined as follows:

$$f_г = \frac{N_L n}{6}, \tag{12}$$

where N_L is the number of collector bars on the DC motor, n is the rotor speed, U/min, 60 is a number that allows counting minutes in seconds.

The frequency spectrum is limited to a certain frequency fg. The center frequency of the f_0 range is selected since there is no more than one maximum of the function spectral density $S_w(f)$ in each subrange.

If the function $f^*n(n)$ within the boundaries of possible values does not have jumps that correspond to the available information about Rn, namely:

$$f_0^*(r) = \begin{cases} \frac{R}{R_в - R_а} & \text{при } R_а < r < R_в \\ 0 & \text{при } R_а > r;\ r < R_в \end{cases} \tag{13}$$

where the value of R is distributed in the interval $[R_a, R_в]$ evenly, so the distribution in the number (4...10) of indicators does not give an accurate implementation of R within the interval $[R_a, R_в]$.

These functions can be determined using the experiment planning theory by implementing the planning matrix of a full factorial experiment for four factors (matrix type 2^4). Thus, we have in this example: fault A_1 occurs in the case of simultaneous going beyond the standard parameters S_1 and S_3, fault A_2 with parameters S_2 and S_4, and fault A_3 with parameters S_3 and S_4 (table 1).

Table 1. Diagnostic matrix

Diagnostic parameters	Possible damages		
	A_1	A_2	A_3
S_1	1	0	0
S_2	0	1	0
S_3	1	0	1
S_4	0	1	1

This issue of defects detection and distinguishing comes down to a spectral analysis of their vibration. The quantitative and high-quality changes can lead to sudden failure both in contact couple-, and beyond it (corrosion, erosion, excessive increase in current or temperature, blows, pushes, vibration). The correct interpretation of such range allows not only to recognize defects-, but also to establish their interrelation and nature (depth) of their interactions. Calculations should be carried out by means of the COMPUTER.[2]

Wear rate of the collector α is affected by a large number of random factors-, diverse by nature, therefore distribution α is considered normal. These malfunctions occur usually after 16•103 km of a run, the wear on height is only 20 mm and it is unsuitable to operation. Influence of misalignment degree on the level of vibration can be reduced by application of couplings. The increased vibration caused by the insecure shaft interface, defects of bearings or fans, turn-to-turn short- circuits on an anchor, a double rigidity of an anchor, resonances of separate clusters etc cannot be eliminated with balancing of an anchor. Therefore they constitute the greatest danger from all mechanical influences to which isolation on stream is exposed (owing to the accumulation of an extremely large number of small deteriorations in isolation). In places where rod stock pull out of grooves the bigger number of breakdowns occurs (in process and at preventive tests), than in other parts of rod stock. It is promoted by gradual weakening of winding fastenings, both in grooves, and in front parts. Vertical vibrations cause in hydraulic generator rotor elements alternating stresses which are imposed on tension from centrifugal forces and the rotating moment. The measuring technique of a wear in the course of the electric motor tests is described in GOST P 51667-2000 and involves the measurement of the electric brush length an before and after a test cycle (each 20 or 50 h).

The distinction of statistical characteristics estimates of of vibration loading of each brush holder gives the chance to receive the functional dependence between the change of transfer resistance of brush and collector assembly and size of the maximum acceleration regardless of brush type. The main issue of monitoring and rationing of machines vibration consists of the parameters selection for the intensity of vibration process characteristic.

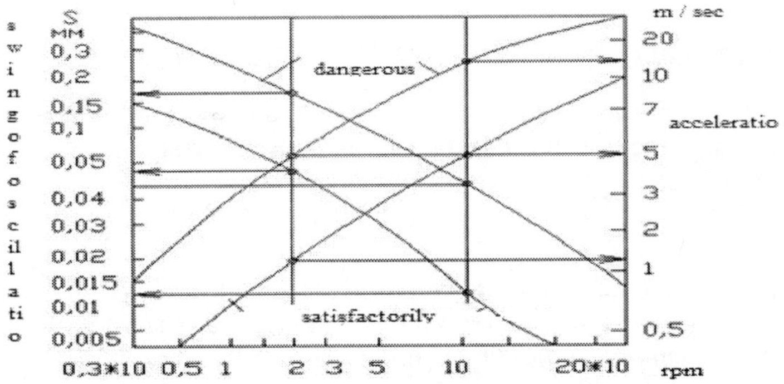

Figure 4. Vibration rates of machines, depending on the frequency of vibrations and accelerations

Eventually in the course of machine operation the rigidity of vibrating parts fastening can weaken, not only further increase of this vibration, but also some decrease in resonant frequencies is result of it that is especially dangerous when the working frequency is on the site of characteristic where vibration increases with the increase in frequency, therefore, of rather small shift of characteristic that vibration sharply increased.

According to the schedule of figure 4.- of machine base vibrations with a frequency of 2000 fluctuations a minute, a range of S=0.05 of mm and acceleration up to 1.2 $m \cdot s^{-1}$ a satisfactory assessment is given to the machine operation.

Increase in the vibration of the electrical machines shown in Fig. 4 demonstrates the emergence of damages though vibration is a source of serious damages and accidents. As a result of a wear of electric insulating materials at operation, the shorting voltage decreases and the probability of system survival decreases respectively. The physical nature of elements and communications can be any. [3]

The decrease in a shorting voltage of isolation does not depend on the frequency of vibrations (within 500-1000 vibrations a minute), but depends on the amplitude of vibrations. The device the single acceleration sensors is shown in Fig. 2. Sometimes the user should install vibration sensor to inaccessible places permanently, and in such cases the user can either remove cables to available places and to measure vibration in them by the device, or to unite these sensors in the stationary system of diagnostics.

Therefore it is expedient to consider a question of inertness of a system and to reveal influence on its quantitative characteristics of operational factors of work of brush and collector assembly on EPS. In the modern control and diagnostics devices, at least, of the rotating inventory, vibration becomes a main type of the analyzed processes, actively forcing out many other processes including thermal.

Purpose of vibration monitoring (Fig. 5) is the detection of brush and collector assembly vibration condition changes in operational process for which defects of an electric brush are the reason in many cases.

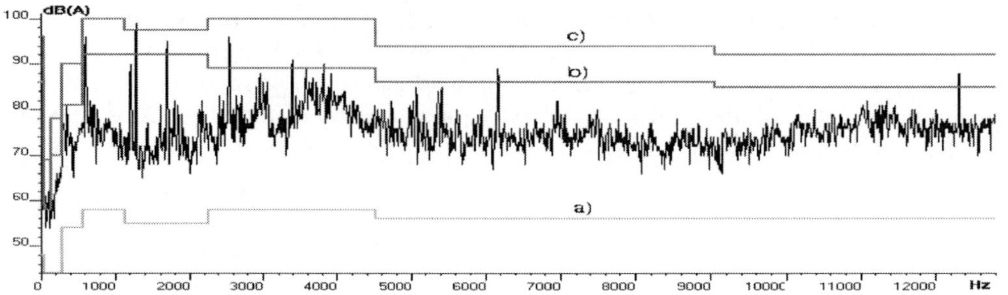

Figure 5. Monitoring the vibration state of the machine. a) – the threshold of a weak vibration signal; b) – the threshold of the average vibration signal; c) – the threshold of a strong vibration signal.

3. Conclusion

Thus, the development of automated monitoring systems for the rolling stock electric motor with the development of the DC motors state operational diagnostics means is an important task. The research of the brush-collector assembly reliability represents a typical example of the non-disassembly diagnostics of vibration tests method application for the study of the individual elements reliability. The above static equation makes it possible to predict the normal operation of the electric brush on the rolling stock in the future.

References

[1] Filina O A 2017 Research of the operational resource of electric brushes of the electric motor of the direct current of the rolling stock *News of higher educational institutions. Power engineering issues* **19** 133-9.

[2] Filina O A, Idiyatullin R G 2015 Approximation of the current- voltage characteristic of brush contact of the compound brush with the raised resource *The current trends of an advancement of science and technologies* **8-1** 134-6.

[3] Filina O A 2014 Technical condition of the product and its non-disassembly diagnostics *In the collection: Science today the collection of scientific works on materials of the international scientific and practical conference. Disput scientific center*. 74.

[4] Filina O A 2010 Increase in effectiveness of brush holders *In the world of discoveries* **6-1** 227-30.

IOP Conf. Series: Materials Science and Engineering **489** (2019) 012017 doi:10.1088/1757-899X/489/1/012017

The study of automotive vehicles technical operation indicators

I N Kiryushin and V N Retyunskikh

Ryazan Institute (branch) of Moscow Polytechnic University, Ryazan, Russia

aitts@rimsou.ru

Abstract. Nowadays automobile transport plays the main role nationwide. That's why transport efficiency increasing is the pressing issue of technical automobile transport operation. To assess and compare the different enterprises' operation vehicle (automobile) fleets efficiency, the production coefficients, automobile vehicles and vehicle (automobile) fleets technical readiness and other. The proposed type of expressions facilitates the calculation and relationship understanding for determining the technical readiness coefficient and determines the main factors relationship characterizing the automobile vehicles and vehicle (automobile) fleets operation efficiency.

Nowadays automobile transport plays the main role not only in agricultural passenger transfer and cargo transportation but nationwide. This mean of transport can change railway and river transport, ensuring reliability and regularity of product delivery from provider to consumer [1]. That's why transport efficiency increasing is the pressing issue of technical automobile transport operation [2]. To assess and compare the different enterprises' operation vehicle (automobile) fleets efficiency, the production coefficients, automobile vehicles and vehicle (automobile) fleets technical readiness, as well as the non-working days coefficient is used.

In modern scientific and educational sources, these indicators have the following definition and calculation formula[3]:

The output coefficient(α_{out}) determines the calendar proportion time during which the vehicle (automobile) or fleet performs transport work on the line:

$$\alpha_{out} = \frac{D_e}{D_e+D_r+D_n} = \frac{D_e}{D_c}; \quad (1)$$

where D_e is the number of vehicle (automobile) exploitation days;

D_r is the number of automobile vehicle idle time in repair and maintenance service days;

D_n is the number of downtime in good condition for organizational reasons days;

D_c is the number of cycle days.

The technical readiness coefficient (α_t) determines the proportion of working time during which the vehicle (automobile) or fleet is operational and can be used in the transport process:

$$\alpha_t = \frac{D_e}{D_e+D_r}; \quad (2)$$

The non-working days (α_n) coefficient defines the calendar time share during which the serviceable automobile (automobile group) is not used in the transport process for organizational reasons (days off, vacation, lack of work, etc.):

$$\alpha_n = \frac{D_n}{D_c}; \tag{3}$$

In determining these coefficients for vehicle (automobile) fleets, days (D) are replaced in vehicle (automobile) days (AD) formulas.

It is worth noting that in determining the technical readiness (α_t) coefficient they determine the working time proportion and the other two factors – the calendar time share. Because of this difference, there are often discrepancies and errors in these indicators calculation, as it turns out that the automobile vehicle idle time in good condition for organizational reasons is not taken into account in the technical readiness coefficient (α_t) calculation. And in many literary sources, it is accurately specified that automobile vehicles stand idle in technically serviceable condition [3]. Therefore, in order to find the total number of days when the automobile vehicle is in good technical condition, the number of downtime days for organizational reasons D_n should be added to the number of operation days D_e [4].

The expression (2) denominator is the days sum when the vehicle is in a technically sound condition D_e and the days when the vehicle is in a technically defective condition D_r. As it was established, the technically serviceable car condition is characterized by the operation days sum and idle time in good condition for organizational reasons days, this means that in the denominator for the calculations accuracy it is necessary to add the number of idle time for organizational reasons days D_n. In this case, the denominator will be given to the number of days in the D_c operating cycle [4].

Then expression (2) takes the form:

$$\alpha_t = \frac{D_e+D_n}{D_e+D_r+D_n} = \frac{D_e+D_n}{D_c} \tag{4}$$

where D_e+D_n is the number of days per cycle when the automobile vehicle is in good technical condition.

In most of the literature the ratio α_{out} to α_t is considered as one of the operation vehicle (automobile) fleets efficiency indicators:

$$\frac{\alpha_{out}}{\alpha_t} = \frac{D_e+D_n}{D_e+D_r+D_n} = \frac{D_c-D_n}{D_c} = 1 - \frac{D_n}{D_c} = 1 - \alpha_n; \tag{5}$$

From the expression (5) it can be concluded that the ratio α_{out} to α_t is in direct connection with the non-working days coefficient α_n. And the expression for determining the output coefficient has the form:

$$\alpha_{out} = \alpha_t(1 - \alpha_n) \tag{6}$$

In our case, this ratio α_{out} to $\alpha_t$will have the following form:

$$\frac{\alpha_{out}}{\alpha_t} = \frac{\frac{D_e}{D_e+D_r+D_n}}{\frac{D_e+D_n}{D_e+D_r+D_n}} = \frac{D_e}{D_e+D_r+D_n} \times \frac{D_e+D_r+D_n}{D_e+D_n} = \frac{D_e}{D_e+D_n} \tag{7}$$

If we change the numerator and denominator in the formula (7), we get a more convenient expression for the analysis:

$$\frac{\alpha_t}{\alpha_{out}} = \frac{D_e+D_n}{D_e} = 1 + \frac{D_n}{D_e}; \tag{8}$$

Based on the formulas (1) and (3) we get:

$$\frac{\alpha_n}{\alpha_{out}} = \frac{D_n}{D_c} : \frac{D_e}{D_c} = \frac{D_n}{D_e};$$

Then expression (8) takes the form:

$$\frac{\alpha_t}{\alpha_{out}} = 1 + \frac{\alpha_n}{\alpha_{out}}; \tag{9}$$

As a transformation expression result (9), we get the final expression form to determine the main coefficients relationship characterizing the automobile vehicle and vehicle (automobile) fleets operation efficiency:

$$\alpha_t = \alpha_{out} + \alpha_n; \tag{10}$$
$$\alpha_{out} = \alpha_t - \alpha_n; \tag{11}$$

To check the formula correctness (10), substitute the coefficients values into it (1), (3), (4):

$$\frac{D_e + D_n}{D_c} = \frac{D_e}{D_c} + \frac{D_n}{D_c} = \frac{D_e + D_n}{D_c}; \tag{12}$$

Expression (12) confirms the main indicators relationship efficiency theoretical analysis correctness of automobile vehicles and vehicle (automobile) fleets operation among themselves.

Now it is necessary to check the results in practice. To do this, let us resolve the issue of determining the automobile vehicle and vehicle (automobile) fleets operation efficiency during the calendar year in two ways: standard and using the derived expressions.

Task 1. Truck ZIL-130 had carried out transport work during the calendar year (D_c =365 days). With a five-day working week, the non-working days' number for the year was D_n =118 days, while in maintenance service and repair the automobile vehicle stood idle D_r =25 days. Determine the release coefficients values, technical readiness and non-working days.

Standard solution:

$$\alpha_{out} = \frac{D_e}{D_e + D_r + D_n} = \frac{D_e}{D_c} = \frac{D_c - D_r - D_n}{D_c} = \frac{365 - 25 - 118}{365} = \frac{222}{365} = 0{,}608;$$
$$\alpha_t = \frac{D_e}{D_e + D_r} = \frac{222}{222 + 25} = 0{,}899;$$
$$\alpha_n = \frac{D_n}{D_c} = \frac{118}{365} = 0{,}323;$$

Proposed solution:

$$\alpha_{out} = \frac{D_e}{D_e + D_r + D_n} = \frac{D_e}{D_c} = \frac{D_c - D_r - D_n}{D_c} = \frac{365 - 25 - 118}{365} = \frac{222}{365} = 0{,}608;$$
$$\alpha_t = \frac{D_e + D_n}{D_e + D_r + D_n} = \frac{D_e + D_n}{D_c} = \frac{222 + 118}{365} = 0{,}932;$$
$$\alpha_n = \frac{D_n}{D_c} = \frac{118}{365} = 0{,}323;$$

It is visible that the non-working days account increases technical readiness coefficient value as automobile vehicles on storage and their idle time waiting for transport work statement is allowed only in technically serviceable condition.

Task 2. Truck ZIL-130 had carried out transport work during the calendar year (D_c =365 days). With a six-day working week, the non-working days' number for the year was D_n =65 days, while in maintenance service and repair the automobile vehicle stood idle D_r =50 days. Determine the release coefficients values, technical readiness and non-working days.

Standard solution:

$$\alpha_{out} = \frac{D_e}{D_e + D_r + D_n} = \frac{D_e}{D_c} = \frac{D_c - D_r - D_n}{D_c} = \frac{365 - 50 - 65}{365} = \frac{250}{365} = 0{,}685;$$
$$\alpha_t = \frac{D_e}{D_e + D_r} = \frac{250}{250 + 50} = 0{,}833;$$
$$\alpha_n = \frac{D_n}{D_c} = \frac{65}{365} = 0{,}178;$$

Proposed solution:

$$\alpha_{out} = \frac{D_e}{D_e + D_r + D_n} = \frac{D_e}{D_c} = \frac{D_c - D_r - D_n}{D_c} = \frac{365 - 50 - 65}{365} = \frac{250}{365} = 0{,}685;$$

$$\alpha_t = \frac{D_e + D_n}{D_e + D_r + D_n} = \frac{D_e + D_n}{D_c} = \frac{250 + 65}{365} = 0{,}863;$$

$$\alpha_n = \frac{D_n}{D_c} = \frac{65}{365} = 0{,}178;$$

Let us check the main indicators value, based on the expressions (6) and (11):.

$$\alpha_{out} = \alpha_t(1 - \alpha_n) = 0{,}833(1 - 0{,}178) = 0{,}685;$$

and

$$\alpha_{out} = \alpha_t - \alpha_n = 0{,}863 - 0{,}178 = 0{,}685;$$

Based on the calculations, it can be concluded that using the presented technical readiness coefficient calculation version is possible to determine this indicator, like the rest, for the entire operating cycle without additional calculations. It should be noted that the calculation result is more important than the conventional calculation version. This is due to the fact that the downtime days for organizational reasons imply the downtime of technically serviceable rolling stock.

The proposed expressions type for determining the technical readiness coefficient and to determine the main coefficients relationship characterizing the automobile vehicle and vehicle (automobile) fleets operational efficiency facilitates their calculation and relationship understanding.

References

[1] Uspensky I A et al 2013 *The tractor-transport units movement studies Rural mechanic No.* 5 (51), pp 36-7

[2] Kolupaev S V et al 2017 *Actual problems of vehicles operation". The XIX International scientific and practical conference materials,* pp. 102-5

[3] Anikin N V et al 2012 *Motor transport enterprises designing: the textbook for the course projects* (Ryazan: Publishing house FGBOU VPO RGATU), p.115

[4] Kolotov A S et al 2011 *Scientific works collection of the Ryazan state agrotechnological University named after P. A. Kostychev* pp. 255-6

IOP Conf. Series: Materials Science and Engineering **489** (2019) 012018 doi:10.1088/1757-899X/489/1/012018

Finite element methods application for calculation of defective composite leaf spring

A M Kokurov and N A Tatus'

Blagonravov Mechanical Engineering Research Institute of Russian Academy of Sciences, 4 Maly Kharitonyevsky Pereulok, Moscow, 101990, Russia

nikalet@mail.ru

Abstract. One of the most demonstrative examples of fibrous composites application in the vehicle power structures is a spring. Fiberglass springs are not uncommon, the existing analytical design methods allow calculating its overall dimensions under the conditions of strength and stiffness, but the behavior of leaf springs with defects under load is the point of interest. At the same time, the enormously approach is extremely complicated, but modern calculation methods allow to model defects, their growth and the process of destruction, taking into account the specifics of fibrous composites. In the paper considered the process of modeling the progressive destruction of fiber reinforcement plastic (FRP) leaf spring. The here presented calculation procedure is for one-leaf spring. The features of the calculation using MSC Patran are described.

1. Introduction

Despite the long history of springs, they continue to be used to this very day.. The first applications of springs probably were in the carriages, where elliptic or semi-elliptic multi-leaf springs were used. Semi-elliptic multi-leaf springs have remained unchanged to this day. Traditional spring is made of high strength steel and it applies in suspension of heavy trucks. The field of our interest is appliance of composite materials for heavily loaded elastic elements.

Glass-fiber reinforced plastic (GFRP) or fiberglass (one of the most inexpensive FRP) due to the low elasticity modulus are the most effective for the elastic elements, because the elastic energy is stored at a predetermined force in reverse proportional to elasticity modulus. Unidirectional fiberglass composite leaf-springs are applied in some modern passenger and sport cars and SUV. The aim of this article is description of defective GFRP leaf spring calculation by FEM. Also the analytical method of profiled leaf spring calculation is shortly mentioned.

2. Analytical method of leaf spring calculation

The rational design of the console profiled beam is shown in figure 1. Section dimensions, width $w(x)$ and thickness $t(x)$ м vary according to power laws:

$$w(x)=w(0)\left(1-\frac{x}{l}\right)^{\alpha},\ t(x)=t(0)\left(1-\frac{x}{l}\right)^{\beta} \quad (1)$$

IOP Conf. Series: Materials Science and Engineering **489** (2019) 012018 doi:10.1088/1757-899X/489/1/012018

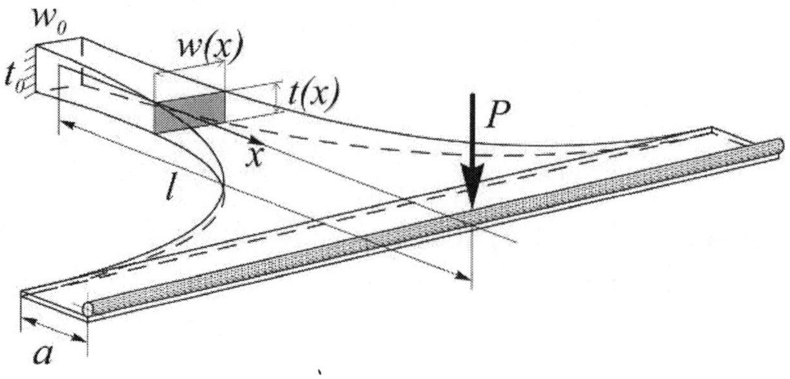

Figure 1. Design scheme of console profiled beam.

Choosing the laws of variation in width and thickness along the length, one can achieve strength uniformity- the same stresses in each section $\sigma_{max}(x)=const$: which implies a linear variation of the bending moment along the length of the spring.

When designing bending elastic elements, it is necessary (knowing the material properties: elasticity modulus E (x, y, z) and flexing strength [σ]) to find the dimensions of the root section w (0) and t (0) and the laws of their change (α-β) for simultaneous fulfillment of the conflicting requirements:

- the requirement for stiffness:

$$c=\frac{P}{v} \tag{2}$$

- the requirement for strength:

$$\sigma_{max}\le[\sigma] \tag{3}$$

Strength uniformity condition:

$\sigma_{max}(x)=\dfrac{6P(l-x)}{w(x)t(x)^2}=\dfrac{6Pl}{w(0)t(0)^2}$ is resulting from (1) to a linear function:

$$\alpha+2\beta=1 \tag{4}$$

The sag v of shaped in accordance with (1) console beam with the load P is determined s determined by the traditional method using the equation for the accumulated elastic energy $\frac{1}{2}Pv$:

$$v(l)=\int_0^l\frac{P(l-x)^2}{E(x)I(x)}dx=\frac{Pl^3}{3EI(0)}\frac{1}{1-\alpha/3-\beta} \tag{5}$$

where the moment of inertia is

$$I(x)=\frac{w(x)t(x)^3}{12}=\frac{w(0)t(0)^3}{12}\left(1-\frac{x}{l}\right)^{\alpha+3\beta} \tag{6}$$

In (6), for simplicity, the elasticity modulus is assumed to be constant. But it, of course, changes due to a change in the misorientation angle of the fibers.

The shape factor for a sag (6):

IOP Conf. Series: Materials Science and Engineering **489** (2019) 012018 doi:10.1088/1757-899X/489/1/012018

$$\delta_v = \frac{v(l)}{v_0(l)} = \frac{1}{1-\alpha/3-\beta}$$

It equals to the ratio of the maximum sag $v(l)$ of the profiled beam to sag $v_0(l)$ of a rectangular beam with the same dimensions of the root section.

The simultaneous fulfillment of the conditions (2) and (3) allows from (4) and (6) to find the rational sizes of root section of the profiled beam:

$$t(0) = \frac{2c[\sigma]l^2}{3EP_{max}} = t_0\delta_v,\ w(0) = \frac{27E^2P_{max}^3}{2c^2[\sigma]^3 l^3\delta_v^2} = \frac{w_0}{\delta_v^2}$$

where t_0, w_0 are cross-sectional dimensions of the rectangular beam which satisfy the conditions (2) and (3).

The cross-sectional dimensions of the end section $w(a)$, $t(a)$ and its lengths a (figure 1) are found from criterion of shear stresses fracture for GFRP:

$$w(l-a)\times t(l-a) = t(0)\times w(0)\left(\frac{a}{l}\right)^{\alpha+\beta} = \frac{3P_{max}}{2\tau^*}.$$

3. Design method of leaf spring progressive fracture

The step-by-step modeling method of analysis the general and local strength is used for the modeling and calculating the progressive fracture of a composite spring. The modeling algorithm is shown in figure 2.

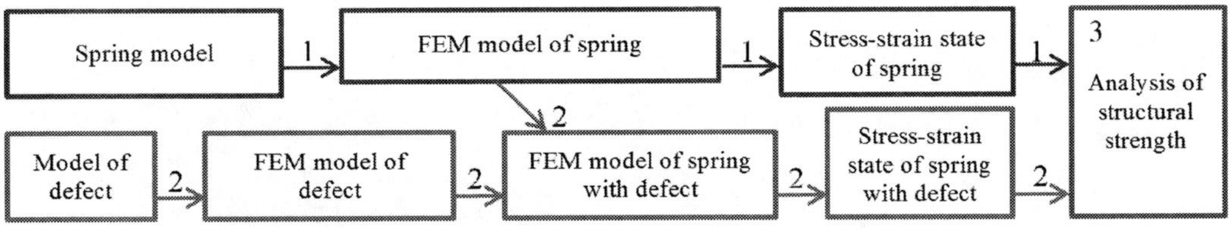

Figure 2. Calculating algorithm of the progressive fracture of a composite leaf spring.

Description of the method presented in figure 2 provided below.

4. Step 1 - calculation of spring designs

First of all, it is necessary to model the geometry of the spring and build a FEM model. Depending on the required detail, different FEM elements (2D or 3D) can be used. Features of model are to be set according to anisotropic and orthotropic composite materials. It is possibly to set individual characteristics in each material layer.

The choice of fracture criterion depends on operating conditions and the types of loading. Criteria for the destruction of Hill, Hoffman, Tsai-Wu, Hashin, maximum stresses, maximum strains, and others can be given.

The specification of boundary conditions and loads depends on the type of problem - this can be either static or dynamic calculations.

To determine the stress-strain state a computational MSC Patran finite-element model of the leaf spring was created see figure 3.

The obtained results of the stress-strain state of construction are presented on figures from 3 to 6, the results are used for determination of the most probable destruction areas. The local coordinate system is used for calculation: the X axis is directed along the spring, the Y axis is perpendicular to the plane of the final element, the Z axis is across the spring.

Figure 3. Finite element model of the leaf springs.

Figure 4. Sag orthographic epure.

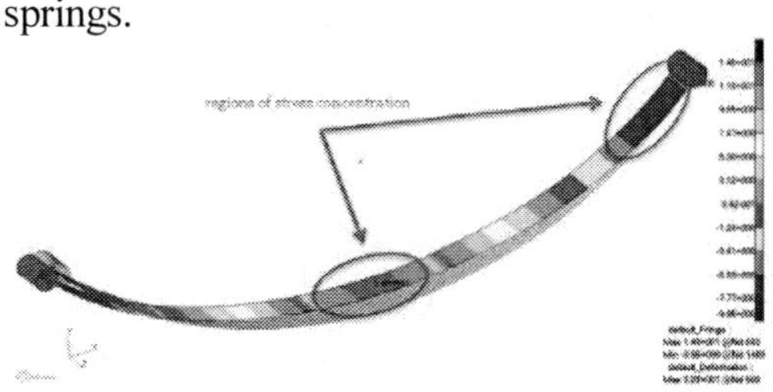

Figure 5. The spring stress- strain state. Stresses along the X axis

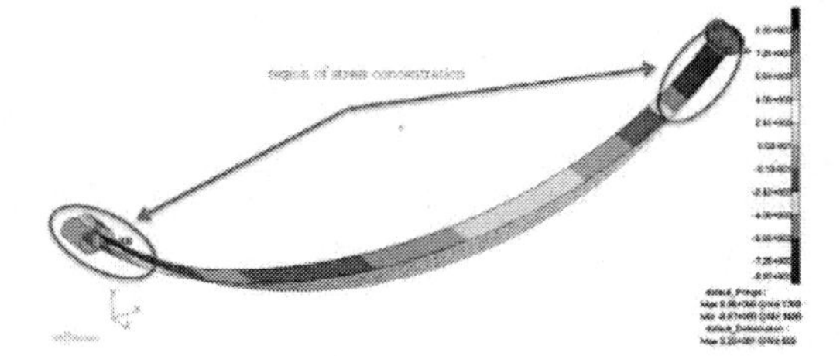

Figure 6. The spring stress- strain state. Stresses along the Y axis

The stress concentration areas are located near the spring eye (highlighted with ellipses in figures 5 and 6). Fracture probability in this area is high and we pay more attention to the eye area.

5. Step 2 – leaf spring progressive fracture calculation

The defect that was "laid" - the delamination or absence of glue. Geometric dimensions are set at the start of modeling, after which a finite element model of the area with the defect is created (figure 7).

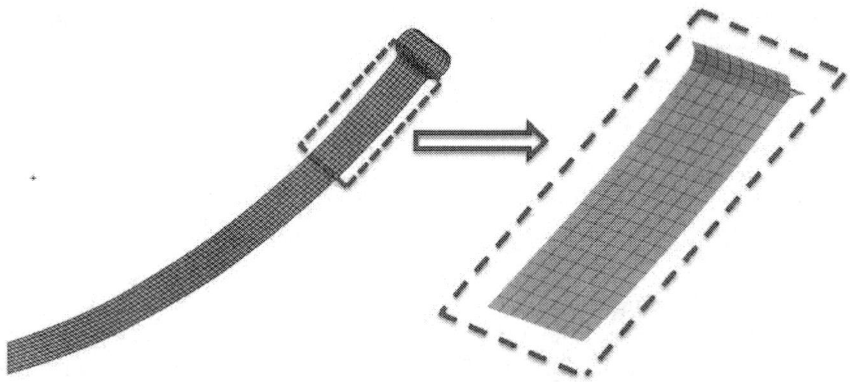

Figure 7. Finite element model of the area with defect.

The layers connection was modeled using the adhesive interface. The process of progressive fracture is modeled by the disruption of the links between the nodes of elements. Adhesive layer fracture criteria are set in the form of maximum permissible stresses in polymer matrix and criteria for the monolayers fracture. For other areas (in which the delamination is not considered) ordinary finite elements were used instead of an adhesive layer.

Before calculation the obtained model is integrated into the initial finite element model by replacing the corresponding array of finite elements.

In total, the results were obtained for 30 load steps (Figure 8). The illustrations of crack progress are presented in figure 9.

6. Step 3 - analysis of results

The process of progressive fracture of the leaf spring is divided into characteristic stages:

Stage of damage accumulation – during the loading steps 1 and 2 the spring deformed as a whole, the values of stresses and deformations in the local area near the eye have not reached critical values yet, the stress patterns have only insignificant differences.

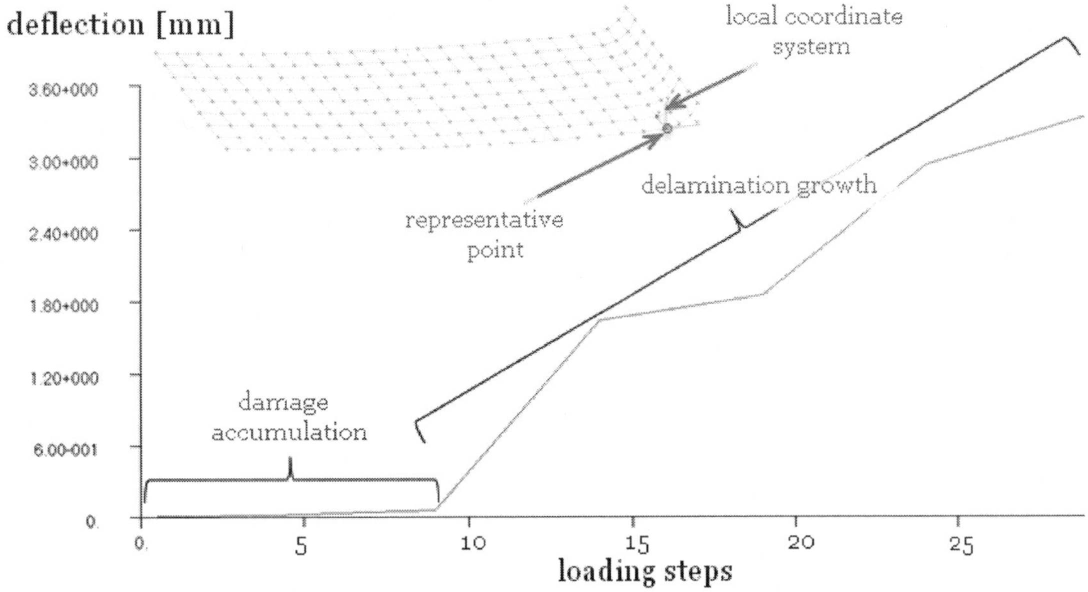

Figure 8. Deflection of the representative points of the spring during the loading process.

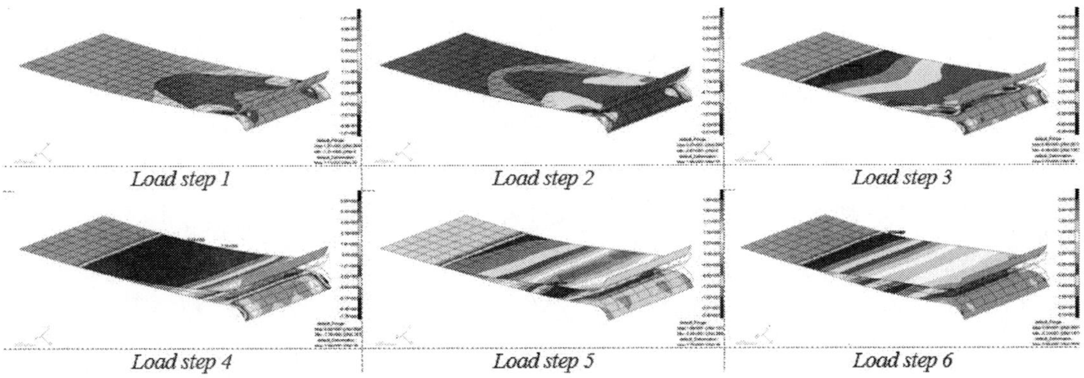

Figure 9. Step 2. The spring stress- strain state. Stresses along the X axis.

With a further increase in load (at loading step 3, figure 9), the stress values reach critical values, which causes local destruction of the binder material between the layers. This effect is modeled by the break between the nodes of the elements in the corresponding area. There is an abrupt increase in the size of the damage, the values of displacements increase sharply.

Then, when the load is increased (steps 4-5, figure 9), the deformation of the spring is accompanied by a simultaneous increase of the area of the defect, as well as the magnitude and distance between the banks of the break.

At the loading step 6, the growth of the delamination is suspended, the spring is deformed as whole again. It is connected with the effects of increasing the spring thickness, the delamination will continue after the values of the stresses again become equal to the critical ones.

7. Conclusion

IOP Conf. Series: Materials Science and Engineering **489** (2019) 012018 doi:10.1088/1757-899X/489/1/012018

The proposed analytical calculation and the finite element model of the defect make it possible to reproduce quite accurately the process of progressive delamination of the composite leaf spring.

Acknowledgments
This work was financially supported by the Russian Foundation for Basic Research, grant No. 18-08-00372 A

References

[1] Polilov A N and Tatus' N A 2018 Evaluation of fiber's misorientation effect on compliance and load carry capacity of shaped composite beams *New Materials 3 IOP Publishing IOP Conf. Series: Materials Science and Engineering* 347. doi:10.1088/1757-899X/347/1/012021

[2] Polilov A N and Tatus' N A 2017 Designing branching or shaped composite elements by analogy with the structure of treetops *Journal of Machinery Manufacture and Reliability* (vol 46) pp 385–93

[3] Kokurov A M and Odintsev I N 2017 Analysis of structural elements made of composite materials with defects *Russian Aeronautics* (vol 60 issue 1) pp 21–6

IOP Conf. Series: Materials Science and Engineering **489** (2019) 012019 doi:10.1088/1757-899X/489/1/012019

The study on external influence on the phasechronometric profile of asynchronous electric motors

A A Kransutskaya, A B Syritskii, E V Tumakova and D D Boldasov

Department of Metrology and Interchangeability, Bauman Moscow State Technical University, 5, 2-nd Baumanskaya, Moscow, 105005, Russia

kransutskaya@bmstu.ru

Abstract. The article deals with the main points of the research on the electric motors influenced by impact and imbalance. Equipment for securing load determining the imbalance was manufactured to obtain asymmetry on the motor shaft. The results of the study on external influence on the phasechronometric profile of the asynchronous electric motors operation are obtained and processed.

1. Introduction

The application of phasechronometric approach for electric motors is proposed for the development of a technical condition assessment system. The system will reduce the number of accidents and reduce potential economic losses. The asynchronous motor model AIR56B2 was chosen as the study subject. Angular displacement sensor (angular encoder) LIR 158A is used as a measuring transducer in the measuring system implementation process to create measurement marks. The sensor produces measuring pulses corresponding to the shaft angular displacements. The stand block diagram is shown in figure 1. The sensor is mounted on the motor shaft (figure 2) with the help of a coupling and special equipment [1].

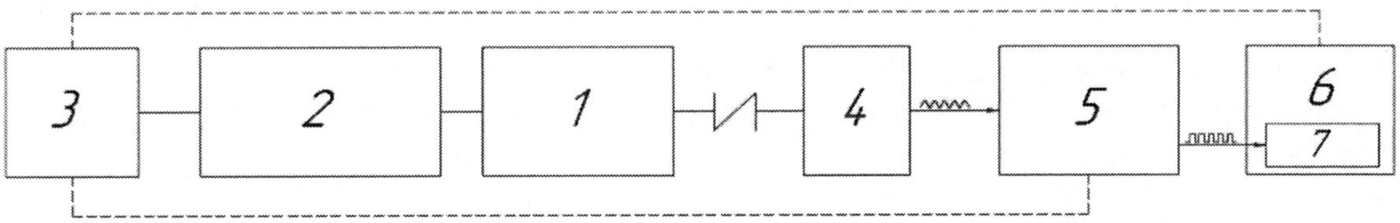

Figure 1. Structural scheme of the stand: 1 – electric motor, 2 – frequency converter, 3 – power source, 4 – angular displacement sensor, 5 – information processing unit, 6 – personal computer, 7 – special software

Figure 2. Fastening the sensor to the motor shaft: 1 – angle encoder, 2 –equipment for fastening the sensor, 3 –coupling, 4 – transition pin, 5 –studs for fastening the snap-in,6 –electric motor

The signal from the sensor in the digital signal form enters the information processing unit, and then in the time intervals format is transmitted to a computer, where the data is processed using special software. The mathematical model provides for the interrelation of measurement results and measurement information processing with the subject design (for example, the spectrum of torsional vibrations, natural frequencies, rotation chronogram, etc.). The mathematical model allows the transition from experimental data to individual system characteristics (inertia moments, stiffness, overall-mass, electrical parameters, etc.) [2-4].

Imbalance characterizes the unbalance of the machine rotating parts. Unbalance occurs when the rotation axis does not coincide with the inertia main axis. The combination of these axes is achieved by balancing. The duralumin sleeve was made to create asymmetry on the motor shaft and conduct an imbalance study. The sleeve serves as an additional load on the shaft, as well as a tool for securing the steel weight, which determines the imbalance. Type of installation for the experiment is shown in figure 3.

According to the Huygens-Steiner theorem [5]

$$J = J_C + md^2 \tag{1}$$

J is the required moment of inertia towards the parallel axis,
J_c is the known moment of inertia towards an axis passing through the body mass center,
m is the body mass,
d is the distance between the specified axes.

According to the formula (1), it is obtained that the steel weight of 28 grams introduces an additional moment of inertia equal to 26.218 kg · m^2.

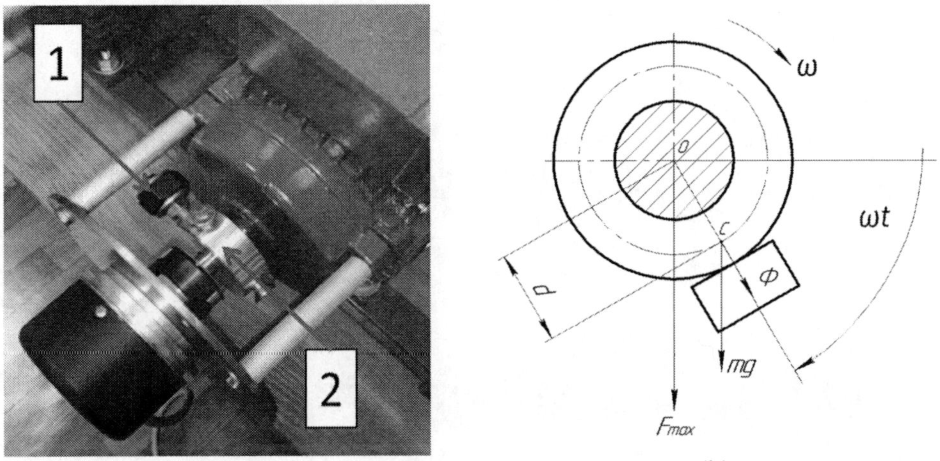

(a) (b)

Figure 3. Realization of unbalance: a – installation on the electric motor shaft of the sleeve 1 with the load 2, b – scheme of forces effect

2. Results

Rotation chronograms are constructed according to the time intervals increments sets obtained during the measurement process. Chronograms in the phasechronometric method [6]) are graphs of changes in the rotation phase duration depending on the phase number. The numbers of measurements (phases) are indicated on the abscissa axis, the phase duration in seconds is indicated on the ordinate axis. Such chronograms allow estimating the uneven rotor rotation. Three graphs reflecting the electric motor idling (figure 4) and external influences on the shaft: an impact (figure 5) and an imbalance (figure 6) are shown for comparison. 5000 measurements are made for one rotor revolution in this experiment.

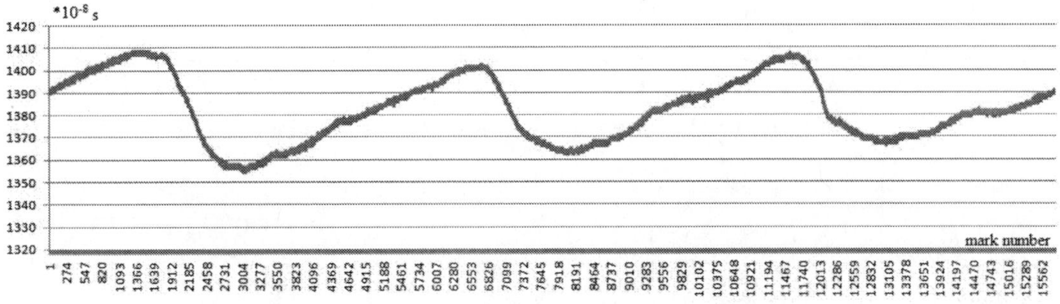

Figure 4. The chronogram of rotation at idle, 15 Hz

Instant speed change occurs upon impact. This is displayed on the chronogram by peaks. The material point makes a fading motion in case of resistance. The applied external effect caused by the impact of three screws fixed symmetrically on the sleeve is reflected in the chronogram in the triple peaks form of superimposed damped oscillations characteristic per rotation (figure 5).The operation of an electric motor with an imbalance introduced into the shaft on the chronogram (figure 6) is reflected as an impact, but the single one, at the point where the superposition of gravity and inertia reaches its maximum.

IOP Conf. Series: Materials Science and Engineering **489** (2019) 012019 doi:10.1088/1757-899X/489/1/012019

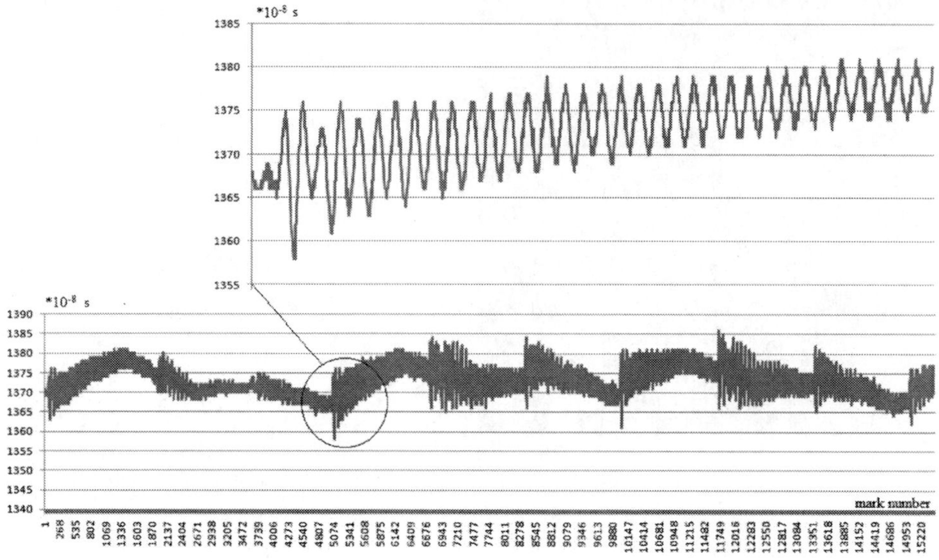

Figure 5. The chronogram of rotation upon impact

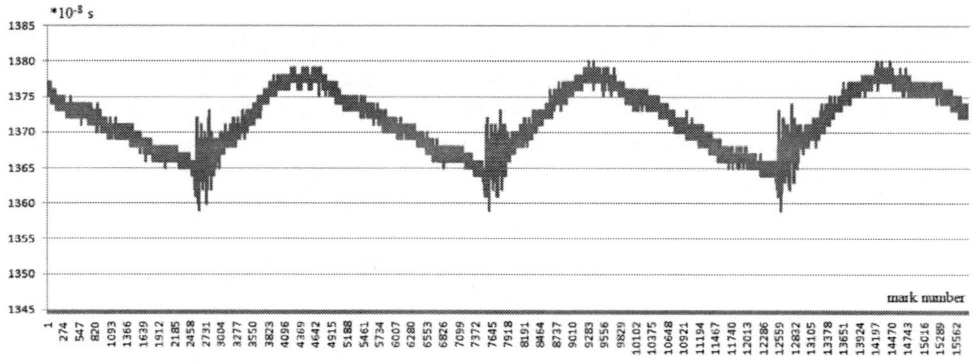

Figure 6. The chronogram of rotation for imbalance

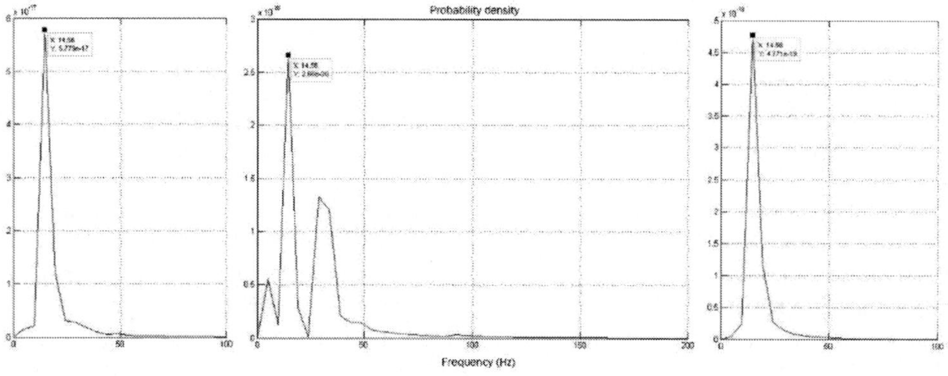

(a) (b) (c)

Figure 7. Spectral analysis: a – the spectrum of the natural frequencies of the system at idle, 15 Hz; b – spectrum of natural frequencies of the system upon impact; c– the spectrum of the natural frequencies of the system when the imbalance

IOP Conf. Series: Materials Science and Engineering **489** (2019) 012019 doi:10.1088/1757-899X/489/1/012019

3. Discussion of results

Spectral analysis of time series showed the presence of repetitive peaks in the range of 10 Hz and 30 Hz. Only the main harmonic appears at idle. The main harmonic coincides with a given frequency on the frequency converter. For example, we select the frequency of 15 Hz for an experiment; on the spectrogram we see the corresponding peak of 14.56 Hz. The difference is due to the use of insufficiently accurate engine frequency converter as part of the stand. Upon impact, the tripling of peaks occurs, i.e. two additional peaks appear. Subharmonics are absent in the imbalance. Due to the fact that the imbalance manifests itself as an impact with rapidly damped oscillations, and due to smaller impact energy, it slightly distorts the spectrogram general view. Thus, with the help of spectral studies we can only identify impacts. The presence of motor shaft oscillation frequencies, independent of the shaft rotation and load options, indicates the presence of stable signs of the mechanism operation, which can be used as diagnostic ones. Carried-out research provides additional information to verify the electromechanical system mathematical model parameters.

4. Conclusion

The obtained results of phasechronometric analysis of the simplest effects on a rotary machine prove the effectiveness of the described approach. The sensitivity shown by the developed measuring equipment together with the innovative phasechronometric approach allows us to make a conclusion not only about the possibility of transferring developments to the field of diagnostics of industrial facilities based on electric motors (drives, pumps, etc.), but also about the applicability of the method in the field of study transients in rotary machines caused by more complex external influences.

Acknowledgments

The results of the work were obtained within the framework of implementing the initiative scientific designs of the basic part of the State Assignment (NIR 9.4968.2017/BCh).

References

[1] Syritsky A B 2017 MATEC Web of Conferences **129** 01049
[2] Tumakova E V 2016 *Meas. Tech.*.**59**:600
[3] Kiselev M I, Pronyakin V I, Tulekbaeva A K 2018 *IOP Conf. Ser.: Mater. Sci. Eng.***312** 012012
[4] Metelkina E D, Komshin A S 2017MATEC Web of Conferences **129**01048
[5] Kane T R, Levinson D A 1985 Dynamics: Theory and Applications (New-York: McGraw-Hill Education) p 379
[6] Boldasov D D, Potapov K G, Syritskii A B 2016 Russ. Eng. Res.**36**:668

IOP Conf. Series: Materials Science and Engineering **489** (2019) 012020 doi:10.1088/1757-899X/489/1/012020

Calculation and experimental method for determination of stress state of the NPP coolant input units into NPP vessel elements

V S Kravets[1] and S V Maslov[2]

[1]Bauman Moscow State Technical University, 5 ul. Baumanskaya 2-ya, Moscow, Russia
[2]Blagonravov Mechanical Engineering Research Institute of the Russian Academy of Sciences, 4 Maly Kharitonyevsky Pereulok, Moscow, Russia

maslovsv@inbox.ru

Abstract. The method of experimental and calculation determination of a stressed state in the NPP cases coolant input units in the conditions of rapid temperature changes is considered. A monitoring algorithm using strain gauge control and an algorithm based on solving direct and inverse problems of experimental mechanics is proposed.

1. Introduction

The estimated life of the NPP equipment contains reserves related to the difference in the real history of changes in the stress-strain state (SSS) of the critical structural elements compared with the design estimates, suggesting the maximum possible impact of operational loading factors. Despite the progress made in the possibilities of numerical methods for calculating the SSS of complex structures, their use at the design stage does not take into account the actual loading history because it is impossible to predict multivariate processes of temperature change, coolant flow, changes in mechanical loads and combinations of various parameters of operation. To determine the real history of SSS changes, monitoring systems based on measurements of the equipment parameters state are created, by which stresses, strains, and displacements of elements are determined using design models. The greatest difficulty is the creation of such models for zones located on the internal surfaces of the coolant circulation circuits. To determine the SSS in these zones, especially for operation modes with rapid changes in the coolant temperature, it is possible to use experimental and calculation methods based on processing the measurement results for accessible construction points [1] and algorithms based on solving direct and inverse problems of experimental mechanics [2].

2. Research methods

Unfortunately, at present, there are no strain gauges in the standard monitoring systems for NPP parameters, which remain operable during the operational period. Experimental data on the SSS can be obtained during startup and commissioning tests (SCT) of nuclear power plants, carried out with the parameters of the power unit that are different from the normal operating ones. At the same time, during the SCT period, it is possible to obtain experimental data necessary for the formation of computational algorithms that allow determining the SSS from standard measured parameters over the entire life cycle. In this paper, we consider the construction of a similar algorithm for the zones of the coolant input nozzles into the vessel elements of heat exchangers, in particular for the emergency water supply branch pipe (EWBP) for VVER-1000 steam generators. EWBP is designed to supply water to the steam generator (SG) in water supply cessation mode in the steam generator by means of main branch pipe during de-energization of nuclear power plant and design basis accidents associated with a significant decrease in the water level. The feed water to the EWBP is supplied through emergency pumps; its temperature (about 40 ° C) is significantly lower than the temperature of the vessel and the internal components of the SG.

The data on temperature differences of the coolant in EWBP for the most critical loading conditions causing significant damage to the metal of the constructions is shown in Table 1.

Table 1. Critical loading SCT modes of the Kozloduy NPP power unit No. 6

Number in order	The cause of the impact load	Temperature drop in EWBP Δt, ^0C
1	Scheduled pumps TX10,20D01 test	100
2	Emergency protection system action during load shedding to idle	255
3	Emergency cooldown in conditions of the power unit de-energization	210

During operation, the nozzle is loaded with pressure of the secondary loop (P = 6.46 MPa), self-compensation efforts from the external pipeline of emergency feedwater and the intracasing collector, as well as uneven temperature field causing significant temperature stresses, especially in areas of geometric inhomogeneities. Figure 1 shows the variants of the branch pipe design, differing constructive options for the supply of water to the steam generator. Initially, the design of the EWBP received the form shown in figure 1a). To reduce the thermal stresses during the supply of emergency feedwater, it was proposed to exclude the direct contact of cold water with the heated bottom by using an intra- collector, as shown in figure 1b).

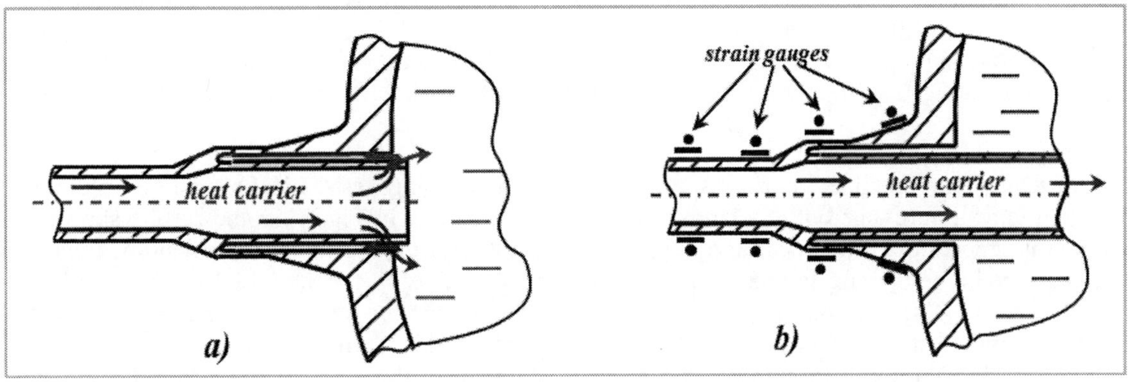

Figure 1. Coolant supply schemes: a) directly into the internal volume; b) with a diversion of the incoming stream.

The strain gauges shown in fig. 1 on the right, were part of the measuring instruments of the SPNI system of the Kozloduy NPP 5th (head) power unit, and were not installed on subsequent serial units. As the analysis performed in this work has shown, it is the experimental data obtained from these strain gages (the SPNI measurement points also included thermocouples installed near the strain gages), allow solving the problem of forming an algorithm for calculating and experimental monitoring of the SSS when the feedwater flowing through EWBP. The standard means of measuring NPPs (temperature sensors of the coolant at the inlet of the nozzle and the temperature of the outer surface of the nozzle zone) have significant thermal inertia, and at realizable rates of temperature change do not provide sufficient information to form the boundary conditions for determining the temperature fields. In addition, as a result of the lag effect of the measured temperatures of the external surfaces in the SSS calculation, the processes of changing temperatures, pressures and self-compensation efforts of pipeline systems are not synchronized. Therefore, when using only the values of the standard sensors, it is necessary to use the maximum calculated estimates of many input parameters (heat transfer coefficients, compensation efforts, etc.), which leads to overestimate of the stresses that arise. Use of indications shown in fig. 1 strain gage allows to create the following updated algorithm.

1. The signals of the strain gauges installed on the pipelines horizontal section (4 pairs in section, left in the figure) determine the bending moment in the pipeline and its change over time (there is no thermal inertia for the strain gauges).
2. The same strain gauges and, additionally, those installed in the second left section, allow to establish the exact time of the beginning of the thermal process according to the change in intensity caused by the heat flow on the inner surface.
3. The readings of the strain gauges in the left section and the thermocouples installed next to them make it possible to form the initial data for solving the inverse problem of thermoelasticity, the solution of which is the function of temperature variation of the internal surface of the pipeline [3]. The solution of the inverse problem is possible if the components of the stress tensor are caused by the loading of the pipe fragment only by the radial temperature field and do not contain components related to other factors. This means that from the measured strain values it is necessary to exclude the components caused by the internal pressure and the components of paragraph 1, associated with the self-compensation of the pipelines.
4. Using the temperature change functions obtained in paragraph 3, the functions of changing the coolant temperature and heat transfer coefficients are selected that are most appropriate for the data obtained in paragraph 3. For this, direct problems of heat conduction are solved and the results are compared with the results obtained in paragraph 3.
5. The determined functions of the coolant temperature changing and the values of the heat transfer coefficient are used to calculate by the FEM methods the temperature fields of and then - the SSS of the analyzed area of the structure.
6. The obtained calculated results are compared with the readings of the strain gauges shown in the right sections in fig. 1. If necessary, the intermediate results of paragraphs 3 and 4 are corrected.

3. Results

Experimental data obtained during startup and commissioning tests (SCT) made it possible to calculate temperatures for both variants of pipe designs, shown in Fig.1. As the results showed, the maximum temperature differences across the wall thickness for the section in the transition zone of the external pipeline to the nozzle body arose in mode No. 3 (Table 1): for the initial variant a) - 149 ° C (at the time of 9.4 s); for option b) - 129 ° C. Thus, the temperature stresses for the upgraded version should be lower than for the original one, which was confirmed by further analysis.

The obtained temperature fields are used to determine the SSS of the EWBP zone under the most dangerous loading mode (No. 1, Table 1), taking into account the loads from the internal pressure, connected pipelines and temperature gradients. Fig. 2 shows the results of determining the SSS for both variants of the structure at the points in time at which the stresses are maximum (9.4 s and 7.4 s).

IOP Conf. Series: Materials Science and Engineering **489** (2019) 012020 doi:10.1088/1757-899X/489/1/012020

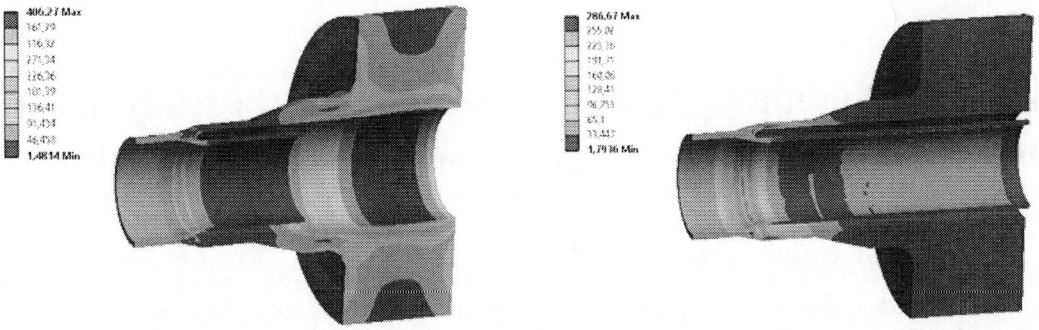

a) *b)*

Figure 2. The results of the calculation and experimental determination of the SSS:a) original and b) upgraded EWBP versions

A comparison of SSS for both EWBP versions shows that the upgraded construction is optimally designed in terms of the resulting thermal stresses, but is more vulnerable to loads from the attached pipelines.

References

[1] Mikhalev Y K, Fomin A V and Maslov S V 2008 *Journal of Machinery Manufacture and Reliability* **5** 121-6

[2] Machutov N A, et al 1992 *Methods and results of reactor VVER-1000 stressed state investigation during the exploatation* (ed International Centre for Scientific and Technical Information, Institute of Machine Science named after A.A. Blagonravov RAS) Moscow, **36** 116

[3] Razumovskii I A, Chernyatin A S, Fomin A V. 2014 Experimental-computational methods for determination of the stress-strain state of structural components. *Inorganic Materials.* **15** 1528-36

IOP Conf. Series: Materials Science and Engineering **489** (2019) 012021 doi:10.1088/1757-899X/489/1/012021

Compensation for rigid body displacements in study of local deformations using electronic speckle pattern interferometry

Igor N Odintsev and Taras P Plugatar

Blagonravov Mechanical Engineering Research Institute of Russian Academy of Sciences, 4, Maly Kharitonyevsky Pereulok, Moscow 101990, Russia

ino54@mail.ru

Abstract. The paper addresses techniques of optical compensation for rigid body displacements of objects undergoing deformation in measurement of in-plane displacements using electronic digital speckle pattern interferometry. Results of demonstration tests are presented. The proposed approaches may be useful, in particular, in study of mechanical properties of materials involving samples with stress concentrators.

Electronic speckle pattern interferometry (ESPI) is used to measure individual components of displacement vectors on surface of solids [1, 2]. The technique is based on recording of images of objects undergoing deformation with laser illumination. Images of objects with rough surface are formed by multiple chaotic speckles. Let two images of an object in different deformation states be recorded using a digit video camera and one picture area be subtracted from the other (e.g. using a proper Photoshop option). As a result, we have a 'difference' image of the object covered with a set of fringes. The emergence of these fringes may be explained as follows. Displacement of solids is accompanied by periodical changes in speckle intensity that causes their chaotic scintillation when the object shifts. This is due to changes in optical difference in motion of rays R illuminating individual points on the object. However speckle intensity distribution on a current picture area will be fully reconstructed in sites where the change in the motion difference is equal to the whole number of laser wave lengths λ, i.e. $R = \lambda N$, where $N = 0, \pm1, \pm2$. Such sites formed by correlated random speckle patterns will have zero intensity after subtraction of two picture areas recorded. Each of these sites looks like an individual dark fringe. Displacement value in a point of the object on such a fringe may be calculated by its interference order N. Direction of the displacement depends on optical arrangement of the measurement set.

Let the object be illuminated with two symmetrical laser beams simultaneously, as shown in figure 1. (The conventional arrangement type has no glass plate in front of the camera lens. The purpose of the glass plate will be described further). In this case visualized fringe patterns will carry information about in-plane displacements U(x, y) along the X axis only. Interpretation of such patterns (interferograms) is done by formula

$$U(x, y) = \frac{N(x,y)\,\lambda}{2\sin(\alpha)} \quad (1)$$

where α is the angle of incidence of illuminating beams.

IOP Conf. Series: Materials Science and Engineering **489** (2019) 012021 doi:10.1088/1757-899X/489/1/012021

This fringe formation principle implies that individual similar speckles have the same (or next to the same) position in two images recorded. (The term 'similar' means here that these speckles are generated by the same microscopic site on the object surface). However displacement of the object under deformation is accompanied by displacement of the speckle pattern on its image in proportion to the scale factor. Therefore, in conventional speckle-pattern interferometry the upper range of displacement measurement is limited by speckle average size. The less the overlap area after superposition of pictures, the lower the fringe contrast in the difference image. If local displacement is greater than the speckle size, no fringes are formed.

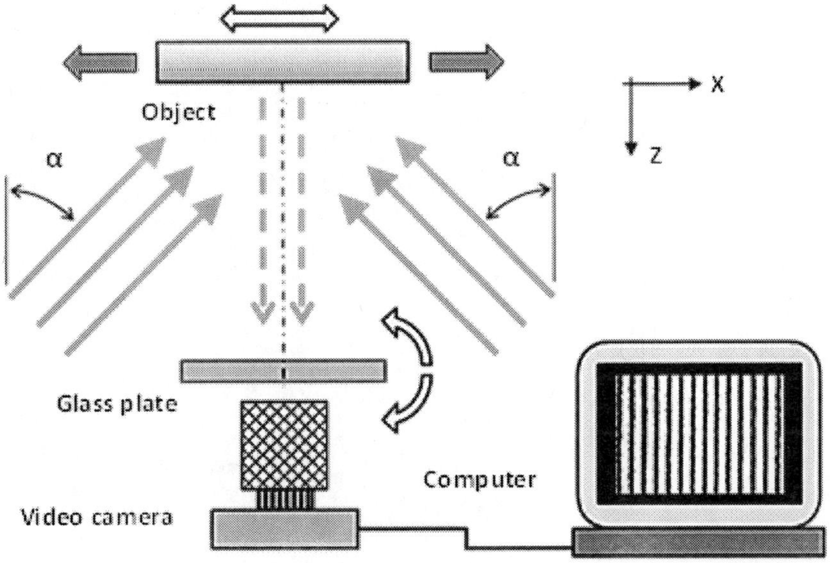

Figure 1. Interferometer optical set-up to measure in-plane displacements of object undergoing deformation

This effect significantly reduces ESPI capacity. In particular, it makes difficult study of deformation state of an object against the background of its displacements as a rigid whole. Such situation is observed, for instance, in tensile testing of samples with stress concentrators (figure 2a). A zone around the concentrator is here of special interest. However, loading of the sample in a test machine causes displacement of this zone due to deformation of a part of the sample beginning from the site of its fixation in a stationary gripper. The axial displacement picture will have satisfactory contrast under low load (figure 2b). However the fringe contrast decreases gradually till complete disappearance of the fringes as the load increases (figure 2c). Note that residual displacements after elastic-plastic deformation and sample unloading concentrate mainly in the concentrator area which results in a rather high fringe contrast on the corresponding interferograms. See, for example, residual displacement fields at the start of plastic deformation in figure 2d. The interferogram in figure 2e was obtained after emerging of cracks during sample cyclic loading.

A special experiment was developed to demonstrate this phenomenon in a greater detail. A plate making a small rotation Ω around normal Z in its plane (XY) was taken as the experimental object, figure 3a. The rotation center was within the observation area. It is easy to show that expected fields of in-plane components of displacements U and V are described by linear relationships, i.e.

$$U = \Omega\, y, \qquad V = -\Omega\, x \tag{2}$$

Figures 3b-e show typical interferograms of such displacement fields. They are obtained using two identical optical set-ups situated at a 90° with respect to each other around the axis Z. As seen, equidistance fringe spacing is decreasing as Ω increases, which is in accordance with formulas (1) and (2). The observed pictures are of a good quality at the entire operation surface at small rotation angles (figures 3b, c), but the contrast at the periphery is decreasing significantly to eventual complete fringe disappearance with increase in Ω. This is due to rise in magnitude of the plate point displacement as the distance from the rotation center becomes longer (figures 3d, e). Note, that the conventional border of the area with satisfactory fringe contrast is described here by a circumference as a line of equal values of displacement vector modulus. Therefore, it is this very characteristic that influences the loss in fringe contrast rather than values of individual components U or V.

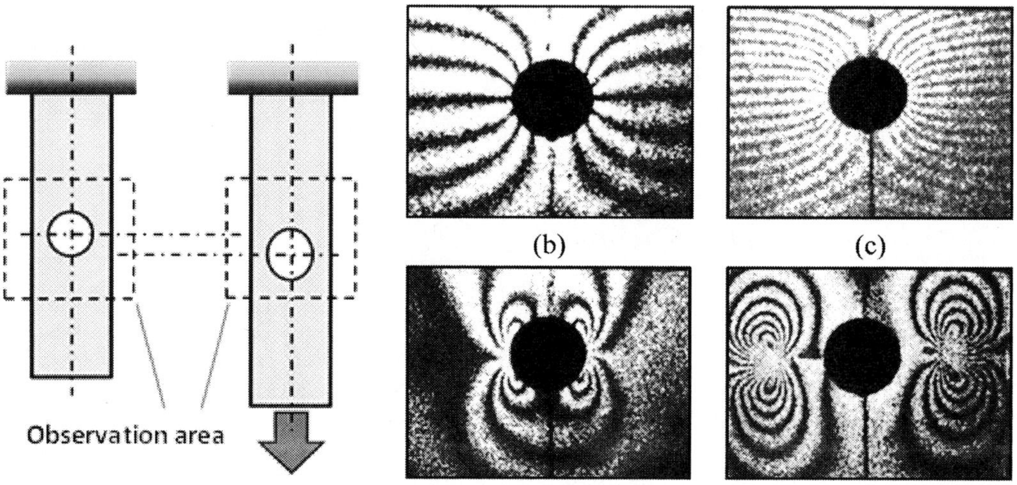

(a) (d) (e)

Figure 2. Tension of a plane sample with a circular stress concentrator (a), and speckle pattern interferograms of increments in elastic (b), (c) and residual (d), (e) displacements U along the loading axis

The rigid body displacements may be compensated for using special computer software through reciprocal shifting of two digitized picture areas before subtraction. (Directions of illumination of all points of the object should remain unchanged when the object is moved. Light beams with a plane wave front are used for this purpose, as demonstrated in figure 1). However, this approach limits the picture area shifts to discrete steps of no less than one pixel, which hampers achieving maximal possible fringe contrast in interferograms. Relationship between the value of the compensated displacement and the pixel physical size is a determinant factor here, the desired value being a whole number. There is a simple technical expedient to avoid the above mentioned limitation. Interferometer conventional optical set-up includes an additional optical element as a plane-parallel glass plate in front of the camera lens, figure 1. Smooth rotation of this plate around some axis in its plane leads to continuous optical displacement of the object image to be recorded. The image distortion may be neglected, if magnitude of the rotation is small. The image displacement direction is evidently orthogonal to the plate rotation axis. This approach helps to achieve maximal fringe contrast, if interferogram visualization is done in the real time mode. Note, that the required direction of the compensated rigid body displacements is often known beforehand. In particular, this is true for the case of tensile testing of samples with a stress concentrator described above.

The following experiment demonstrates potentials of the above-described procedures to compensate for rigid body displacements. As before, a plate rotating in its plane was taken as the experimental object. However, here its center of rotation was outside (to the left of) the picture area to

IOP Conf. Series: Materials Science and Engineering **489** (2019) 012021 doi:10.1088/1757-899X/489/1/012021

be recorded. Figure 4a shows initial speckle interferogram of field U. One can easily see the border of speckle pattern decorrelation as a circular arch. Figure 4b demonstrates the result of one-pixel computer-assisted displacement of the subtracted picture area along the horizontal axis. As seen, the zone of acceptable fringe contrast makes a stick-slip movement to the right. However, we managed to have an interferogram with a high fringe contrast at the entire observation area using the smooth optical compensation procedure, figure 4c. One might say that the plate rotation axis was conventionally moved to the center of the picture area due to the optical compensation, while it was located outside (to the left or right of) the picture area in both example interferograms obtained by computer-assisted image displacement.

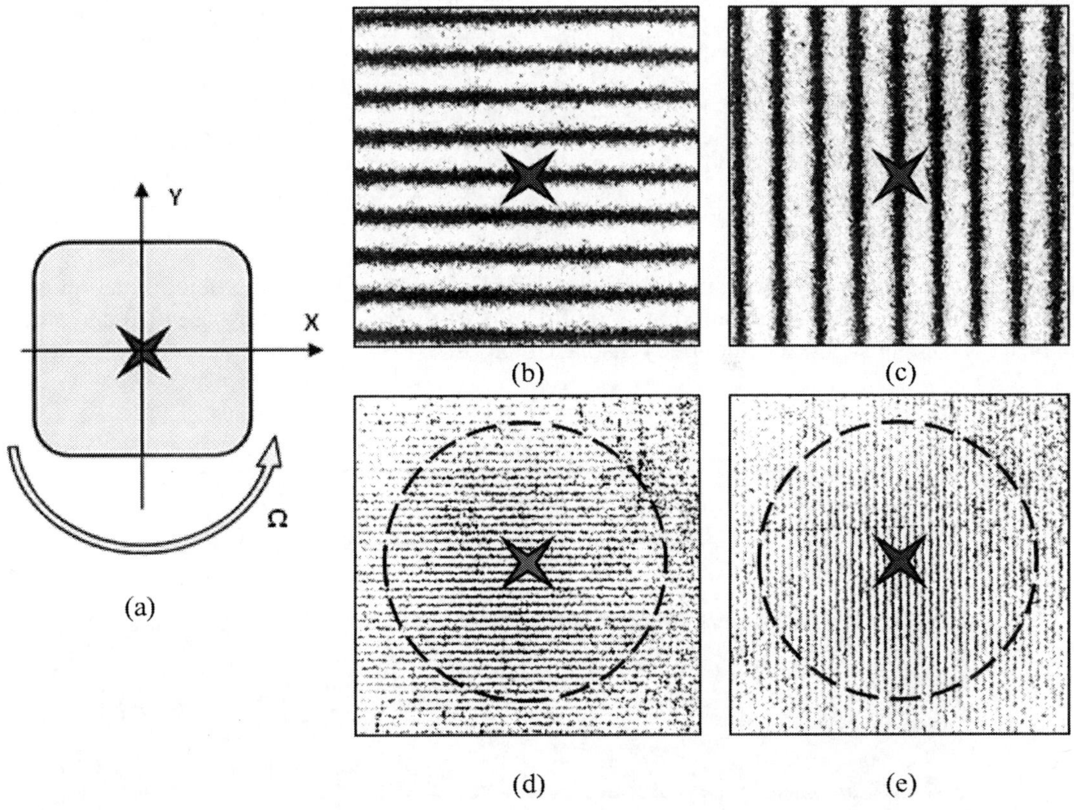

Figure 3. (a) Rotation of the test object in its plane, and (b)-(e) speckle pattern interferograms of horizontal U (left) and vertical V (right) displacements at a small (upper) and a large (bottom) angle Ω

Note that in practice combination of both approaches may be reasonable. The computer-assisted one-pixel shift of subtrahend picture areas is used for rough compensation, while the optical shift is done only within the pixel physical size and serves to visualize interferograms with a maximal possible fringe contrast.

In conclusion, let us demonstrate capacity of optical compensation for rigid body displacement as an actual experiment. A beam with a stress concentrator as a V-notch was tested. The beam was loaded according three-point bending scheme, figure 5a. The concentrator zone of interest was shifted vertically. Under the condition of no displacement compensation in-plane displacement fringes rapidly lost contrast with increase in load. When the sample deflection arrow reached ~30 microns, it became impossible to obtain any information about local deformation, figures 5b, c. However, compensation

IOP Conf. Series: Materials Science and Engineering **489** (2019) 012021 doi:10.1088/1757-899X/489/1/012021

for rigid body displacements helped to obtain high quality interferograms of local deformations in the concentrator zone, figures 5d, e.

One should mention the existence of the upper limit of rigid body displacements fit for compensation by this approach. This upper limit is due to defects of optical elements that lead to appearance of parasite local phase variations in illuminating beams. A series of special tests using optical elements of rather poor quality gave satisfactory results with rigid body displacements reaching ~300 microns, i.e. several-fold greater than average speckle size.

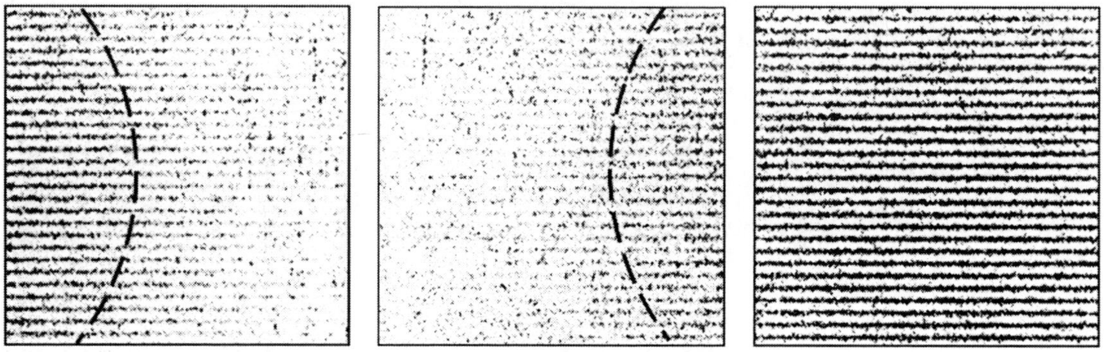

(a) (b) (c)

Figure 4. (a) Initial interferogram of horizontal displacements U with the object rotation center placed to the left of the observation area, and results of discrete computer-based (b) and smooth optical (c) compensation for rigid body displacements

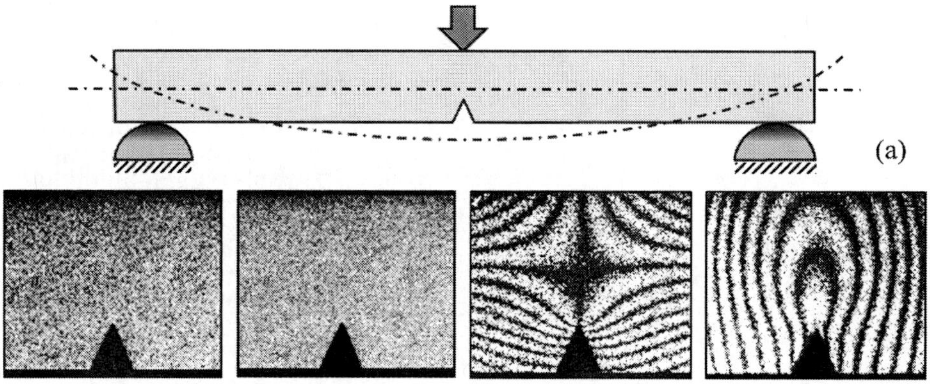

(b) (c) (d) (e)

Figure 5. (a) Arrangement of three-point bending of a beam with a stress compensator, and pair interferograms of horizontal U and vertical V displacements: (b, c) initial, (d, e) after compensation for rigid body displacements

The described approach seems helpful in development of special-purpose measurement ESPI-systems as components of test equipment sets intended to study mechanical properties of materials.

Acknowledgments
This work was financially supported by the Russian Foundation for Basic Research (grant no 18-08-00572).

IOP Conf. Series: Materials Science and Engineering **489** (2019) 012021 doi:10.1088/1757-899X/489/1/012021

References

[1] Jones R and Wykes C 1989 *Holographic and Speckle Interferometry* (Cambridge, UK: Cambridge University Press)

[2] Rastogi P K 2000 *Digital Speckle Pattern Interferometry & Related Techniques* (New York, USA: John Wiley & Sons, Inc.)

IOP Conf. Series: Materials Science and Engineering **489** (2019) 012022 doi:10.1088/1757-899X/489/1/012022

Biaxial elasto-plastic strain-stress state implementation in the case of the simple tension

V V Pashchenko[1], A V Tsarkov[1] and S V Kheylo[2]

[1] Bauman Moscow State Technical University, Kaluga Branch, Kaluga, Russia
[2] The Moscow State University of Design and Technology, Moscow, Russia

vasiliy.pashchenko@gmail.com

Abstract. Biaxial stress state obtaining in the flat tensile specimen is the main subject of this paper. Numerical simulation is the main method of investigation. All calculations were performed in ANSYS. The specimens with different geometry and location of the pairs of raisers stresses are considered. The choices of small plastic strains are shown. The interrelation of actual geometry of the specimen and the shape and size of advanced biaxial area is investigated. The experimental verification of the numerical simulation results is presented.

1. Introduction

It is a matter of common knowledge that the thin wall metal constructions are often used in the biaxial stress state operation conditions. The metal durability in this case is different from the uniaxial loading metal durability. There is no reliable academic failure criterion for plane stress state of metal. For this reason the in situ method of material testing is the best. This method must be based on the direct examination of material properties. Existing testing equipment for biaxial stretching is of the limited availability and very expensive. The unique specimen designing is another tentative solution of the above mentioned problem.

There are many researchers who designed special specimens for simple tension in the recent years [1-5 et al.]. As a rule the plane stress state in these specimens was formed due the stress raisers location. In some cases the specimen elastic strain-stress state displayed the state that take place in the tube, i.e. situation with principal stresses ratio close to 0.5 [5]. The general form of these specimens is shown in figure 1. The desired results are obtained due to corresponding geometry and location of the two symmetrical pairs of stress raisers. It must be emphasized that all results were obtained for the elastic loading zone. At the same time we must realize that in some case the actual stress level may be larger than flow point, as it is shown in figure 1.

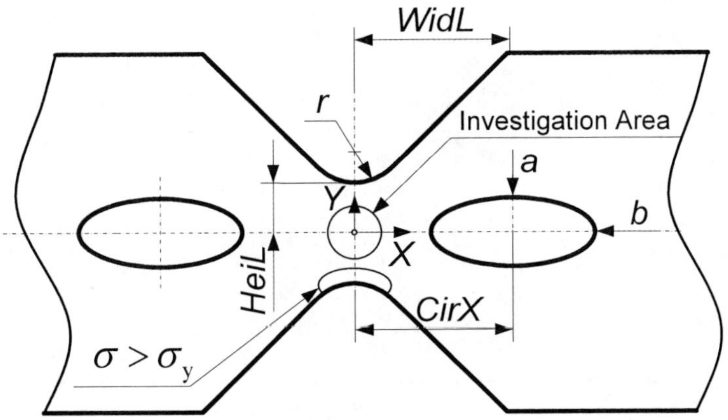

Figure 1. Specimen geometry and advanced strain area.

The main goal of this article is to obtain the specimen, which will represent two-dimensional stress state under the conditions of small elasto-plastic deformations.

2. Method and material

Numerical simulation method was offered as a prime tool for task solution. Numerical experiment was conducted under ANSYS [7, 8]. PLANE182 four node finite elements were used. There were two degrees of freedom in each mode. Five millimeters steel 20 plates were used for numerical simulation and experimental verification. The yield stress of this steel was 320 MPa. For every specimen configuration the principal stress ratio fields were plotted (figure 3).

Variation of shape and size of the stress raisers was performed as specified in the articles [4, 5]. $HeiL, WidL, CirX, r, a, b$: are the main varying parameters (figure 1).

3. Results and discussion

As it appears from the results of simulation the distance between U-shape and elliptical stress raiser has the greatest effect on the forming of ratio=0.5 area (Figure 2). The shape of distribution of the second principle stress does not depend on the specimen geometry. Redistribution of stresses in passing from elastic to plastic deformation depends on the first principle stresses distribution.

Let us consider the influence of different specimen parameters. The advanced plane area (one where principle stresses ratio is close to 0.5) is contracted and moved from the centre of specimen in the case of decreasing of the distance between specimen and elliptical cut centre. The plastic deformations are reformed and concentrated along the line, which connects stress raiser peaks. The value of the plastic strain is increased from 0.5% to 0.7%.

The U-cut radius rounding from 5 mm to 15 mm is modifies the distance between raisers as well as reformation of plastic strains in centre part of specimen and in the elliptical raiser pick area. Plastic deformation is decreased from 1% to 0.5% in the U-cut peak.

Mutual increasing of *WidL* and *R* from 47 mm to 52 and from 20 mm to 25 mm respectively at a constant *CirX* (67mm) leads to reducing the inter raiser distance. The strains are collected along the line that connects raiser peaks. Simultaneously the maximum plastic deformation is decreased from 0.35% to 0.29%. The advanced biaxial area is drifted to the centre and reduced. Providing that the inter raiser distance is going on to decrease the area with principle stress ratio close to 0.5 disappears.

It has been argued that plastic strains redistribution and radical changes of plane state of the centre part of specimen occurs when distance between stress raisers approaching the half of specimen width.

IOP Conf. Series: Materials Science and Engineering **489** (2019) 012022 doi:10.1088/1757-899X/489/1/012022

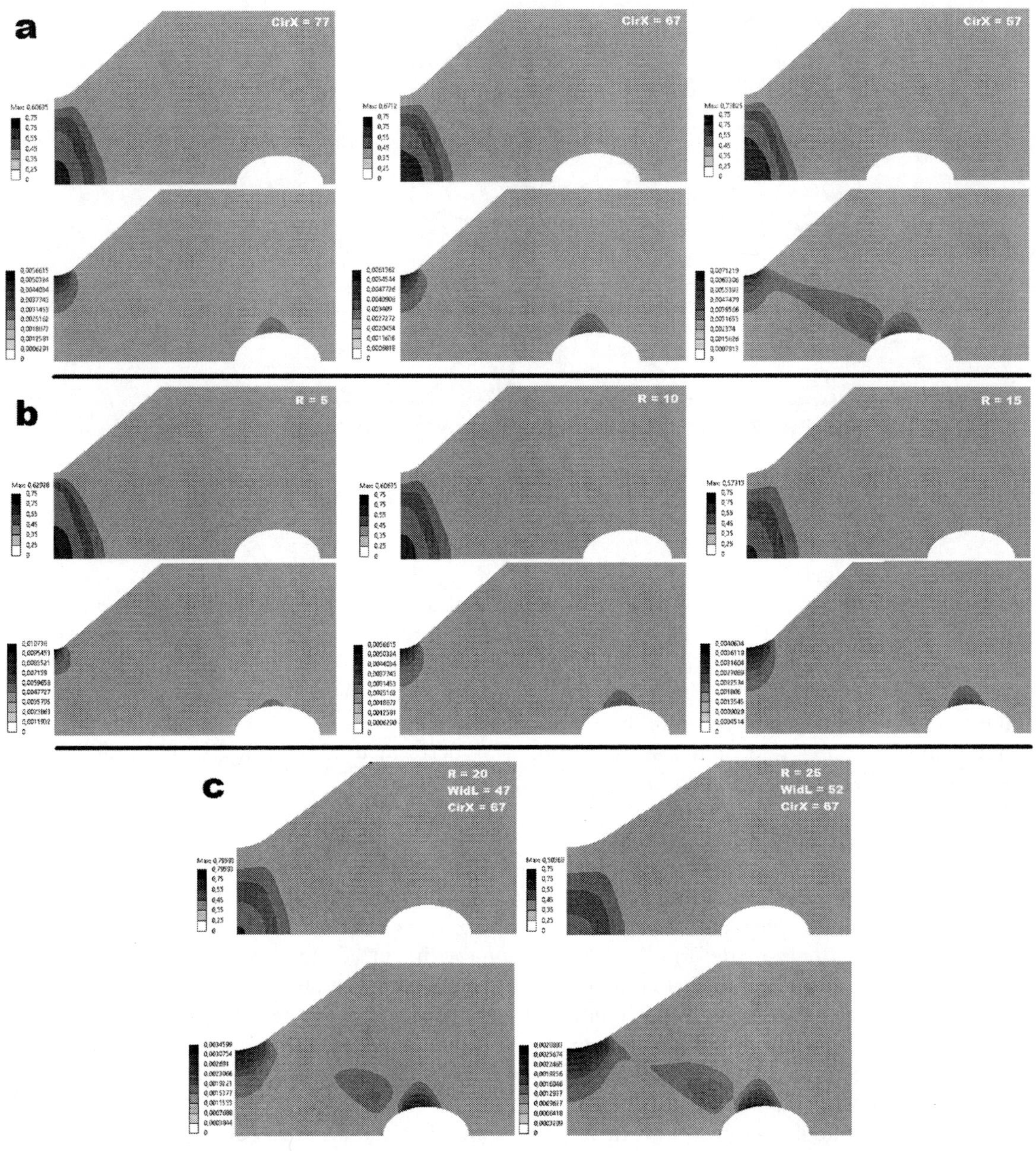

Figure 2. Biaxial stress-strain area (in red) and elastic strain in the different geometry specimens

4. Experimental verification

To confirm the results of simulation the set of natural experiments was performed. In-situ method of strain measurement was used (Figure 3). Rosette-type strain gages were arranged on the both sides of the spacemen that to minimize the bend effect in view of inaccuracies of specimen positioning.

Figure 3. Test specimen.

The test specimens were loaded up to the 50 kN with step of 500 N. Measurement average deformations v.s. step of loading were: $\varepsilon_1 = 20.31$, $\varepsilon_2 = 7.23$, $\varepsilon_3 = 10.23$. Then the principle stress ratio is:

$$\frac{\sigma_1}{\sigma_2} = \frac{\varepsilon_1 + \mu\varepsilon_2}{\varepsilon_2 + \mu\varepsilon_1} \quad (1)$$

With respect to the obtained results, the principal stresses and proper ratio value are:

$$\sigma_1 = 305.69 \text{ MPa}, \sigma_2 = 167.98 \text{ MPa}, \quad (2)$$

$$k = \frac{\sigma_2}{\sigma_1} = 0.55 \quad (3)$$

The principle stresses and plastic strains distributions are presented in Figure 4. These results were obtained as results of numerical simulation. Two principle stresses and their ratio are:

$$\sigma_1 = 329.61 \text{ MPa}, \sigma_2 = 152.59 \text{ MPa}, \quad (4)$$

$$k = \frac{\sigma_2}{\sigma_1} = 0.46 \quad (5)$$

5. Conclusion

1. On the advanced the biaxial stress area with plastic strains or without, difference tensile specimen geometry is needed.
2. In the case of small plastic deformation such tensile specimens are required that the ratio between two stress raisers to specimen width will be less than 0.5.

IOP Conf. Series: Materials Science and Engineering **489** (2019) 012022 doi:10.1088/1757-899X/489/1/012022

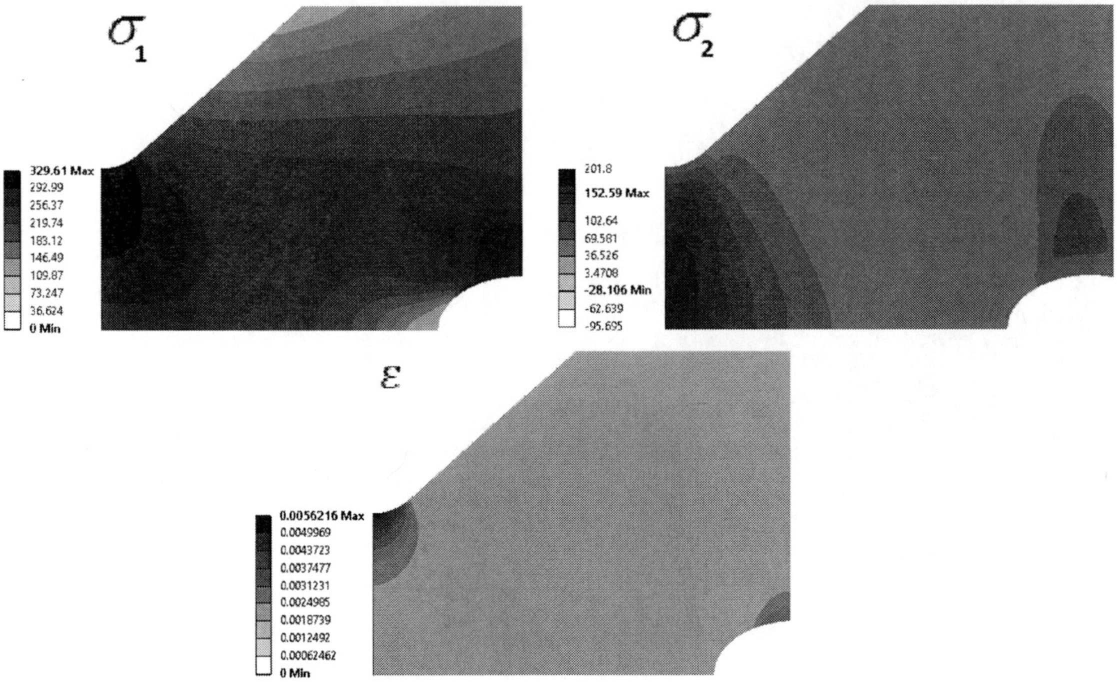

Figure 4. First, second principal stresses and plastic strain in test specimen.

References

[1] Lyapichev D M 2015 Ocenka vliyaniya napryazhyonnogo sostoyaniya podzemnyh gazoprovodov na ih stojkost' k korrozionnomu rastreskivaniyu *PhD thesis* Moscow p 146

[2] Podhalyuzin S Z 1986 Razrabotka metodov povysheniya rabotosposobnosti magistral'nyh truboprovodov *PhD thesis* Moscow p 160

[3] Bellet D, Morel F, Morel A and Lebrun J L 2011 A biaxial fatigue specimen for uniaxial loading. *Int. J. Strain* **47** 227-40

[4] Tsarkov A V and Pashchenko V V 2016 Chislennoe modelirovanie ispytaniya na prochnost' listovyh obrazcov v usloviyah dvuhosnogo napryazhyonnogo sostoyaniya *Svarka i diagnostika* **6** 11–5

[5] Tsarkov A V, Pashchenko V V and Zinoveva O I 2014 Issledovanie vliyaniya koncentratorov napryazhenij na NDS v ploskih obrazcah trub pod davleniem *Inzhenernyj vestnik Dona* **4** 10

[6] Tsarkov A V and Pashchenko V V 2014 Metodika provedeniya chislennyh ehksperimentov pri issledovanii NDS trub pod davleniem *State sci. conf. proc. Prikladnye problemy mekhaniki* pp 160–7

[7] Chigarev A V, Kravchuk A S and Smalyuk A F 2004 *ANSYS dlya inzhenerov: spravochnoe posobie* (Moscow Mashinostroenie) p 512

[8] Zienkiewicz O C and Taylor R L 2000 *The Finite Element Method* (Butterworth-Heinemann)

Features of composite beams bending analysis by energy methods

A N Polilov and N A Tatus'

Blagonravov Mechanical Engineering Research Institute of Russian Academy of Sciences, 4 Maly Kharitonyevsky Pereulok, Moscow, 101990, Russia

nikalet@mail.ru

Abstract. Simple analytical calculations of rectangular-sectioned composite console beams loaded with concentrated force, spread load or dead load are presented. The simplicity of originate expression for maximum deflections by energy methods is shown.

1. Features of strength analysis by set level of stored elastic strain energy

Figure 1 shows the diagram of half of a few-leaf spring of n equal leaves loaded with end force P. The term "few-leaf" is applicable to a spring compiled of equal-length leaves $2l$, uniform strength of which is provided by shaping and not by leaves lengths change according to linear law as in typical steel multi-leaf springs with fixed dimensions of leaves cross sections (w – width, t – thickness).

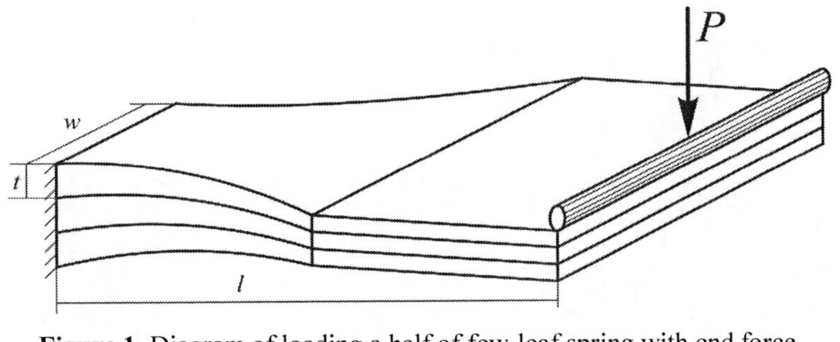

Figure 1. Diagram of loading a half of few-leaf spring with end force.

Elastic members – energy storage unit such as springs, leaf springs, torsional spring – should meet three main conflicting requirements: by strength, by stored energy, and by durability. The inconsistency of these requirements lies in the fact that by set length $2l$ for the provision of requirements on strength and durability, a beam should be as thicker than t as possible, for the provision of requirements on stored energy – as thinner than t as possible. To sort out these differences leaves n multiplication is typically applied, but as it will be shown further, this is not an exclusive and better way as applied to composite materials of quasi-unidirectional glass-fiber-reinforced plastic type. The requirement on stored energy should be met accurately, i. e. a spring loaded by the set static end force P_{st} should have the set deflection v_{st}, and the requirements on strength and durability should be met on the safe side. The lesser stresses σ_{max} occur by the set maximum load P_{max} and the greater

IOP Conf. Series: Materials Science and Engineering **489** (2019) 012023 doi:10.1088/1757-899X/489/1/012023

number of cycles N a spring will bear at the set range ΔP_c of cyclic loads, the greater will be fault-free haul as applied to an automotive vehicle.

At the Technical assignment for the spring design calculation, initial camber and its change are typically set: static v_{st}, maximum v_{max}, and cyclic Δv_c deflections. Stiffness $C=P_{st}/v_{st}$ is defined via a static deflection, and maximum and cyclic loads are defined via stiffness:

$$P_{\max} = k_d P_{st} \approx C v_{\max}; \; P_c = v_{st} \pm C\Delta v_c,$$

where k_d is the dynamic factor depending on operational conditions, for example, of a light motor vehicle or load-carrying vehicle. For SUV, k_d = 1.4, for a mine dump truck, k_d =2.7. As can be seen from the above, for the benchmark related to shaping influence analysis, to set the requirements by load-bearing capacity (maximum load P_{max}) and by stored energy U_{max} (or stiffness C) is enough. The estimation of cyclic stresses in beam approximation is carried out by the same formulas with P_{max} replaced by P_c.

To generalize the assignment, let us assume that console beam is loaded by bending moment changing along x coordinate as a power function (figure 2):

$$M(\bar{x}) = M(1)\bar{x}^{\gamma}; \; \bar{x} = \frac{x}{l} \in [0;1]. \qquad (1)$$

the shearing force is expressed by a derivative of a moment, and spread load – by a derivative of the shearing force:

$$P(x) = \frac{dM}{dx}; \; p(x) = \frac{dP}{dx}.$$

Consequently, loading by end force corresponds to γ=1, spread load – γ=2, linearly changing spread load – γ=3 etc.

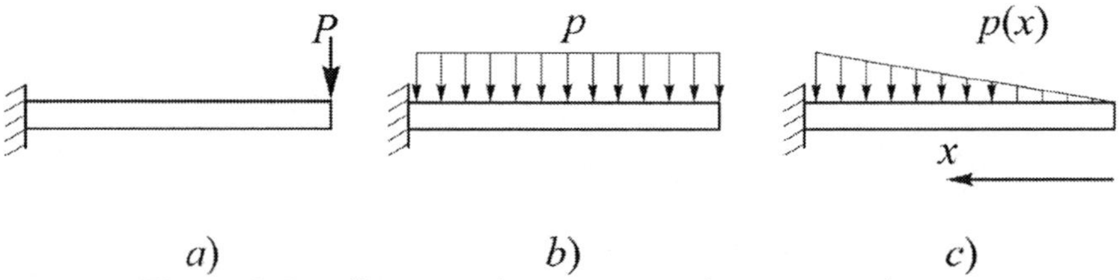

Figure 2. Loading types by power-mode moment change:
$a)\ \gamma = 1;\ b)\ \gamma = 2;\ c)\ \gamma = 3$.

According to relation (1) for the provision of equal stress (strength uniformity), section modulus should also be changed as a power function. Let us assume that thickness t and width w of rectangular sections are changed as a power function:

$$w(\bar{x}) = w(1)\bar{x}^{\alpha}; \; t(\bar{x}) = t(1)\bar{x}^{\beta}; \; \bar{x} \in [\bar{a};1]. \qquad (2)$$

Strength uniformity condition from the expressions (1) and (2) will be as follows:

$$\frac{6M(1)\bar{x}^{\gamma}}{nw(\bar{x})t^2(\bar{x})} = \frac{6M(1)}{nw(1)t^2(1)} \Rightarrow \alpha + 2\beta = \gamma. \qquad (3)$$

The requirement on strength for loading by end force is expressed as follows:

$$\frac{6M(1)}{nw(1)t^2(1)}=\frac{6P_{\max}l}{nw(1)t^2(1)}=\sigma_{\max}\le\sigma_* . \quad (4)$$

The requirement on the maximum stored energy $U_{\max}$ can be expressed by Castigliano theorem:

$$U_{max}=\frac{6ln}{E}\int_0^1\frac{[M(\bar{x})/n]^2}{w(\bar{x})t^3(\bar{x})}d\bar{x}=\frac{6M^2(1)l}{Enw(1)t^3(1)}\int_0^1\frac{\bar{x}^{2\gamma}}{\bar{x}^{\alpha+3\beta}}d\bar{x}=U_0\delta_U,\ \delta_U=\frac{1+2\gamma}{1+2\gamma-\alpha-3\beta}, \quad (5)$$

where U_0 is the energy stored by the same moment in n equal rectangular leaves with the same fixed dimensions of root section $w(1)$, $t(1)$; $\delta_U=U_{\max}/U_0=v_{\max}/v_0$ – shape factor by energy in case of end force is the ratio between the deflection at the end of shaped beam to the deflection at the end of the rectangular beam.

By accurate carrying out the requirement (5) on energy, the necessary thickness of root section (in case of loading by end force) is found:

$$t^3(1)=\frac{6M^2(1)l}{Enw(1)U_{max}}\delta_U=\frac{6P_{max}^2l^3}{Enw(1)U_{max}}\delta_U,$$

and by accurate carrying out the condition (4) on strength – the necessary number of leaves:

$$n=\frac{6P_{max}E^2U_{max}^2}{\sigma_*^3w(1)l^3\delta_U^2}. \quad (6)$$

The relation (6) shows that even considering the strength of unidirectional glass-fiber-reinforced plastic equal to steel strength (although it can be higher), and not considering the useful contribution of shaping (the increase of δ_U), it is possible to reduce the number of leaves by a factor of 20 by elasticity modulus reduction only (210 GPa in steel, 45 GPa in glass-fiber-reinforced plastic). At which point, virtually any steel multi-leaf spring can be replaced by glass-fiber-reinforced plastic single-leaf spring, as it will come into account further considering n=1.

2. The possibility justification of approximate estimates

Let us explain the possibility of passing by the influence of low inter-laminar shearing modulus and low shearing strength for the sufficiently long beams of glass-fiber-reinforced plastic. The presence of shear stresses results in an additional deflection v_2, which can be estimated considering that work done by force at this deflection is spent on shear strain elastic energy:

$$\frac{1}{2}Pv_2=\frac{1}{2}\int_0^l\int_0^w\int_{-t/2}^{t/2}\frac{\tau_{xz}^2(z)}{G_{xz}}dxdydz=\frac{1}{2}\frac{\alpha_0P^2l}{4G_{xz}wt},$$

where $\alpha_0=6/5$ for the transverse bending of a beam with a rectangular section.

The distribution of shear stresses by the beam height is considered as parabolic:

$$\tau_{xz}=\frac{3P}{th^3}\left(\frac{h^2}{4}-z^2\right),$$

and independent of the beam length and axial coordinate x. Across the width, along the axis y, the distribution of the shear stresses is considered uniform at the first approximation. Numerical analysis by finite-element method (FEM) showed that for beams with the relation of span length to the height more than 5, a hypothesis for such type of distribution diagram of shear stresses introduces at most 3% error.

As can be seen from the above, the general deflection can be expressed as follows

$$v = v_1 + v_2 = \frac{Pl^3}{4E_x wt^3} + \frac{\alpha_0 Pl}{4G_{xz} wt} = v_1\left(1 + \alpha_0 \frac{E_x}{G_{xz}} \frac{h^2}{l^2}\right).$$

By relation of longitudinal Young's modulus E_x to inter-laminar shearing modulus G_{xz} of glass-fiber-reinforced plastic around five, a correction to deflection does not exceed 1% for sufficiently long beams $l/t > 20$. Consequently, by further calculations of the stored elastic energy we were restricted to accounting only the deflections from normal stresses, i. e. without sacrificing the plane-sections hypothesis.

By bending short composite beams, failure in the form of delamination by the action of normal and shear stresses combination is possible:

$$\sigma_x + m\tau_{xz} = c\,,$$

where m, c are the experimentally determined material parameters.
In addition, conditional shearing strength becomes dependent on the ratio of span to beam thickness.

$$\tau_0 = \frac{cm}{m^2 + l^2/t^2}.$$

But analytical estimations and experiments show that delamination occurs in glass-fiber-reinforced plastics at $l/t \leq 12$ only, and for sufficiently long beams $l/t > 20$, the influence of shear stresses on strength can be neglected, which is considered in this article.

The glass-fiber-reinforced plastic effect in a leaf spring

Let us analyze how modulus reduction allows increasing cyclic durability for carrying out one of the requirements. For the set cyclic load and stored energy, from the relation (6), the fulfilment of conditions on strength and stored energy results in the following expression for cyclic stresses range:

$$\sigma_c = \frac{1}{l}\left(\frac{6P_c E^2 U_{max}^2}{nw(1)\delta_U^2}\right)^{1/3}. \tag{7}$$

Formula (7) reflects the main features of strength analysis at the set elastic energy. For example, for the given maximum load P_{max} (or P_c), a long (!) beam occurs to be less stressed than a short one. To increase the operational life, it is required to reduce maximum stress in the cycle σ_c for the set cyclic load P_c. According to expression (7), this can be achieved by five methods:

1) increase leaves number n (typical approach);
2) increase width $w(1)$ (that is why wide cross springs are effective);
3) increase length l (let us remember old-fashioned carriages with a spring exceeding the carriage sizes);
4) increase shape factor δ_U by shping of uniform strength;
5) use glass-fiber-reinforced plastic with elasticity modulus E lower than elasticity modulus of steel.

Wohler curves describing the cyclic durability are often considered as straight lines in semilog coordinates: $\sigma_c = \sigma_*(1 - k_c \lg N)$, from which

$$lg\,N = \frac{1 - \sigma_c/\sigma_*}{k_c}. \tag{8}$$

As an example, let us assume that $k_c = 0.1$, which approximately corresponds to Wohler curves slope for unidirectional composites and for a variety of steels. Let the cyclic stress amounts 40% from static strength: $\sigma_c/\sigma_* = 0.4$ is a national norm adopted for composite aeronautical structures, so saying "ignorance factor". Then, from relation (8), the critical number of cycles $N = 10^6$. According (7), the elasticity modulus reduction by 4 times under otherwise equal conditions will result in the cyclic stresses reduction by 2.5 times, which from expression (8) is equivalent to the durability increase by more than 100 times:

$$lg\,N = \frac{1 - 0.4/2.5}{0.1} \approx 8.4.$$

This conclusion is based on data close to reality and given for illustration purposes only. Consequently, it can not be recommended for use as a strict result.

For the single-leaf version, from relations (3), (4) and (5), sizes of root section by the strict fulfilment of conditions on strength and stored energy can be specified:

$$t(1) = \frac{\sigma_* M(1) l}{U_* E(1 + 2\gamma - \alpha - 3\beta)} = t_0 \delta_U;$$
$$w(1) = \frac{6U_{max}^2 E^2 (1 + 2\gamma - \alpha - 3\beta)^2}{\sigma_*^3 M(1) l^2} = w_0 \delta_U^2. \tag{9}$$

Where t_0, w_0 are the sizes of rectangular beam root section meeting the same conditions (4)-(5).

According to relation (2), the volume of a shaped beam is defined by integration:

$$V = l\int_0^1 w(\bar{x})t(\bar{x})d\bar{x} = lw(1)t(1)\int_0^1 \bar{x}^{\alpha+\beta}d\bar{x} = V_0\delta_V;$$
$$\delta_V = \frac{1}{1 + \alpha + \beta}, \tag{10}$$

where δ_V is the shape factor by volume equal to the relation of volumes of shaped and rectangular beams with similar sizes of root section. Now, from expression (9) and (10), it is possible to estimate the required weight of spring leaf via density ρ:

$$m = \rho w(1) t(1) l \delta_V = \rho w_0 t_0 l \frac{\delta_V}{\delta_U} = m_0 \delta_S;$$
$$\delta_S = \frac{\delta_V}{\delta_U}, \tag{11}$$

where δ_s is the aggregate coefficient factor of weight reduction; m_0 is the rectangular leaf weight meeting the same conditions on strength and energy; its section sizes differ from $w(1)$, $t(1)$. From relations (11), it follows that by low density (2500 kg/m^3 in glass-fiber-reinforced plastic; 7800 kg/m^3 in steel) and much lower elasticity modulus, the weight of glass-fiber-reinforced plastic spring (ideally) can be at 15 times lesser than the weight of steel spring with the same service properties. This makes the direct effect evident. The constructive effect follows from relation (6) and shows the possibility of leaves number reduction by a factor of 20. Finally, technological effect involves the possibility to shape composite beams and, as it follows from formulas (5), (10) and (11), by carrying out the condition of uniform strength (3) $\alpha = \gamma - 2\beta$:

IOP Conf. Series: Materials Science and Engineering **489** (2019) 012023 doi:10.1088/1757-899X/489/1/012023

$$\delta_S = \frac{\delta_V}{\delta_U} = \frac{1+2\gamma-\alpha-3\beta}{(1+\alpha+\beta)} = \frac{1+\gamma-\beta}{(1+\gamma-\beta)(1+2\gamma)} = \frac{1}{1+2\gamma}. \quad (12)$$

As can be seen from the above, according to relation (12), the efficiency factor of shaping of uniform strength depends on the velocity of bending moment change (1): at loading by end force, any "ideal" equistrong beam is 3 times lighter than rectangular one, at spread load – 5 times lighter, at linear load (see figure 2, *c*) – 7 times lighter.

A real beam can have neither zero nor infinite sizes at the end. Consequently, due to shearing force resistance conditions, it is required to leave the end area of some length *a* with fixed dimensions (figure 3).

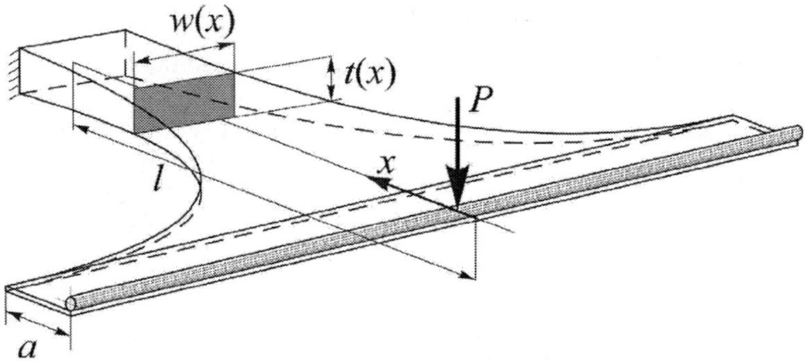

Figure 3. Constant-area beam with cross section fixed area and end area with fixed dimensions of sections.

In calculating of energy (5) and volume (10), the integration should be carried out by two areas: with variable and fixed sizes of the section, and as a result, the expressions for shape factor will be as follows:

$$\delta_U^* = \delta_U + (1-\delta_U)\bar{a}^{3/\delta_U};$$
$$\delta_V^* = \delta_V + (1-\delta_V)\bar{a}^{1/\delta_V};\ \bar{a} = a/l. \quad (13)$$

Formula (13) should be used for the calculation of the required section sizes (9) and weight (11), but the result will no longer be that sort of decisive as (12).

Four design factors α, β, $w(1)$, $t(1)$ can not be found definitely from three conditions: strength (4), uniform strength (3) and stored energy (5). Consequently, one more relation linking the section sizes is required. Fig. 4 gives the most common types of beams of uniform strength for the most important case of loading by end force. This is either triangle beam of uniform thickness: $\alpha = 1,\ \beta = 0,\ t(\bar{x}) = const,$ or parabolic beam of uniform width: $\alpha = 0,\ \beta = 1/2,\ w(\bar{x}) = const,$ or constant-area beam with the cross section fixed area:

$$w(\bar{x}) \times t(\bar{x}) = const \Rightarrow \alpha + \beta = 0 \Rightarrow из\ (3)\ \alpha = -1, \beta = 1. \quad (14)$$

IOP Conf. Series: Materials Science and Engineering **489** (2019) 012023 doi:10.1088/1757-899X/489/1/012023

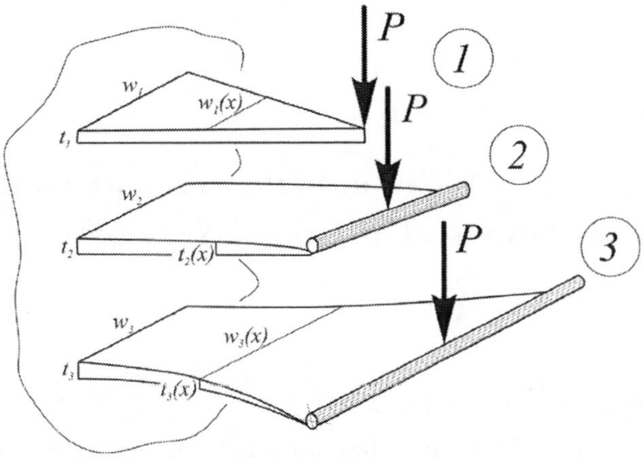

Figure 4. Equistrong beams for the case of loading by end force: 1 – triangle beam of uniform thickness; 2 – parabolic beam of uniform width; 3 – constant-area beam with the uniform cross section area.

The last-named beam type is the most convenient for composites as it allows saving the initial number of uncut fibres, which is of critical importance for strength implementation. However, in addition, the beam width increases unrestrictedly and the significant fibre misalignment occurs. Consequently, to apply to nature for experience and to observe how tree crown increases its flexibility to withstand wind loads is reasonable.

So, simultaneous branching and shaping have the most positive effects for composite leaf-spring design.

Acknowledgments

This work was financially supported by the Russian Foundation for Basic Research, grant No. 18-08-00372 A.

References

1] Polilov A N and Tatus' N A 2018 Evaluation of fiber's misorientation effect on compliance and load carry capacity of shaped composite beams *New Materials 3 IOP Publishing IOP Conf. Series: Materials Science and Engineering* 347. doi:10.1088/1757-899X/347/1/012021

[2] Polilov A N and Tatus' N A 2017 Designing branching or shaped composite elements by analogy with the structure of treetops *Journal of Machinery Manufacture and Reliability* **46** 385–93

[3] Malakhov A V and Polilov A N 2016 Design of composite structures reinforced curvilinear fibres *Composites: Part A* **87** 23-8

IOP Conf. Series: Materials Science and Engineering **489** (2019) 012024 doi:10.1088/1757-899X/489/1/012024

Method for determining the composite and real hardness of hardened surfaces considering the effect of retained stresses in the coating

M S Pugachev and N A Voronin

Mechanical Engineering Research Institute of the Russian Academy of Sciences, 4 Maly Kharitonyevsky Pereulok, Moscow, 101990, Russia

pugachevmax@mail.ru

Abstract. A computational and experimental method for determining the real hardness of the coating material and the effective (composite) hardness of the surface of engineering products hardened with hard protective coatings is developed. This method can be used to control the quality and reproduction of the technology of constructing parts with protective coatings.

1. Introduction

Hardness value of the surface layers is of a significant interest to predict the performance and tribotechnical properties of materials used in friction units. Hardness of the materials is determined by the method of indentation in the surface of the test material. Brinnel hardness (BH) is widely used for metal surfaces of the low and medium hardness and for hard and superhard compact materials, as well as coatings, including thin one – Vickers hardness (VH) [1].

Method for determining the Vickers hardness (VH) is to indent a four-sided diamond pyramid with a square base into the surface of the test material. Berkovich triangular pyramid of is widely used.

Nowadays the method of evaluation of mechanical characteristics of materials by the kinetic (instrumental) diagram "load-introduction" is widely used. The method consists in continuous indentation of the indenter into the surface of the sample with recording of the diagram of indenter introduction during loading and unloading [2]. Most of the research is conducted using diamond pyramids such as Vickers or Berkovich. The loading branch is used to determine the plastic characteristics of the material, in particular hardness.

Hardness (microhardness) of the coated surfaces is determined by the size of the imprint left by the diamond indenter after its indentation with a given load or by the depth of penetration under normal load.

If it is necessary to obtain information about the hardness of the coating material only, the measurement procedure is performed when small loads are applied to the indenter, which ensure the introduction of an indenter into the coating surface to a depth of not more than a tenth of the coating thickness. In the technique of measuring the hardness of thin coatings on the basis of numerous experimental data it is assumed that the values of the measured hardness corresponds (or approaches) to the value of the hardness of the coating material only when the thumb rule is fulfilled, which requires that the depth of the imprint under the diamond prism will be 0.08 – 0.1 of the coating thickness. At large depths of penetration the calculated microhardness of the surface is the value of the composite (effective) one considering the hardness not only of the coating, but also of the base material.

The technology of producing thin coatings and surface modified layers is continuously developing. Technologies based on the effect of highly concentrated flows of energy and matter on the surface are widely used in industry over the last years. In some cases, if not most, the surface layers may differ significantly from the base material (substrate) in physical and mechanical characteristics. During operation of products with thin coatings the retained stresses affect the performance inevitably arising in the process of obtaining the coating. They make their own sometimes significant contribution to the formation of the stress field in the surface volume of the coated product material. The depth, type and magnitude of retained stresses depend on the type and mode of processing. The strength of the surface layer will be largely determined by the presence and type of retained stresses. If the deformation of the material in contact is elastic, the stress at any point is a superposition of the retained microstresses and stresses generated by the external load. It is necessary to avoid the occurrence of retained stresses in the parts of the same sign as the voltage from the operating loads.

The problem of assessing the quality of the coating after its manufacture is inextricably associated with the problem of assessing the reproduction of the process. In modern technologies based on the impact of high-energy corpuscular fluxes on the surface, the synthesis of chemical compounds in the form of a thin coating occurs on the working surface of the hardened product under extremely non-equilibrium thermodynamic conditions, as a rule in a volume isolated from the external environment (reactor) at a reduced working gas pressure or in vacuum. Processing in most cases is multiprocessing. The products (machine parts or tools) are located in the volume of the reactor (chamber) at different distances and not with the same orientation with respect to the generator of the corpuscular flux. Tools affect the temperature distribution in the chamber during the processing of products. All this results in a fairly high percentage of defective processed products, even for the waste of the production process of hardening (e.g., carbide profile cutting of plates, drills, rings, mechanical seals, plungers, fuel pumps). It is important to have indicators that assess the quality of coated products and ways to quantify such indicators.

The objective of the study is to develop a method for assessing the real hardness of the coating and the composite surface hardness of the coated product, which takes into account the contribution of the magnitude and type of retained stresses in the coating.

2. Experimental details

The following types of coatings are used as the targets of research: diamond-like coating (DLC) applied to the substrate of 12X18H10T steel, titanium nitride coating on the substrate of D16T aluminum alloy, aluminum nitride coating on the substrate of 12X18H10T steel. All coatings are applied using ion-plasma method.

Coating thickness was measured by S3P profilometer (Germany) by scanning the step obtained as a result of local screening of the treated surface with a metal screen.

The hardness of the coated surface was measured by PMT3 microhardness tester with an indenter in the form of Vickers pyramid.

3. Experimental procedure

There are usually high retained stresses of different types on the surface of the samples with the test coatings. To remove these stresses the authors propose a method that is performed as follows:

- coating thickness h on the surface of the product is determined;
- based on the geometry of the pyramidal indenter the size of the diagonal d_h of indenter is calculated at a distance s equal to the coating thickness from the tip of the indenter;
- two slots identical in width and length are cut in the coating perpendicular to the surface with a depth that equals not less than the coating thickness figure 1, in the local place of the test surface of the coated product at a distance l from each other not less than the triple value of the calculated diagonal d_h, but not more than four times of its value;

- the coated product is placed in microhardness tester;
- a local place is combined with slots relative to the tops of the pyramidal indenter of microhardness tester in such a way that the indenter tip gets into the center of the local area under study between the two slots on further loading of the indenter;
- procedure of indentation in the test area of the surface is carried out by the method of the restored print by step increase in the load;
- load values P_i are recorded at each step of loading fixing the average value of the diagonal of the imprint d_i (according to the results of the measurement of two diagonals of the imprint); Step loading keep to the diagonal of the imprint value equal to or close to d_h;
- depth of penetration s_i is calculated at each step by formulas (1) and (2):

- for a quadrangular pyramid with a square base (Vickers pyramid):

$$s_i = \frac{d_i}{7} \quad (1)$$

- for a triangular pyramid with a base in the form of an equilateral triangle (Berkovich pyramid):

$$s_i = \frac{d_i}{6.43} \quad (2)$$

- the current values HV_i of the surface layer material are calculated using the known formulas [2]:

- for a quadrangular pyramid with a square base (Vickers pyramid):

$$HV_i = \frac{P_i}{26.43 s_i^2} \quad (3)$$

- for a triangular pyramid with a base in the form of an equilateral triangle (Berkovich pyramid):

$$HV_i = \frac{P_i}{26.44 s_i^2} \quad (4)$$

- the curve is built on the base obtained from experiments for penetration into the surface with a coating of array data on hardness and the corresponding penetration depth $HV_i = \varphi(s_i)$ in coordinates $HV_i - s_i$ or curve $HV_i = \varphi(s_i/h)$ in coordinates $HV_i - (s_i/h)$;
- within the penetration depth values of the obtained diagram from ~ 0,1 h up to ~ 1 h the curve $HV_i = \varphi(s_i)$ is the curve of change of dependence of composite hardness of the surface with the coating considering the absence of retained stresses in the coating;
- within the penetration depth values (0,08 – 0,1) h hardness value on the curve $HV_i = \varphi(s_i)$ is the evidence of real hardness of the coating considering the absence of retained stresses in the coating.

The method provides the use of parallel slots as alternative options, the use of rectilinear slots converging at an angle to each other or curved, including annular one, figure 1.

4. Verification of experimental procedure

All the samples studied in this work the coating thickness is ~ 10 μm and the initial microhardness varies in the coating depth from the maximum value on the surface to the minimum one closer to the substrate figure 2, 3, 4 curve 1.

Since drilling we cut through only the coating, the release from retained stresses occurs in it; therefore, the maximum depth of the penetration of Vickers indenter should not exceed the coating thickness. We will determine DLC coating, using a known ratio between the height and diagonal of the Vickers pyramid, we find that the value of the maximum depth of the indenter penetration, i.e. 10 μm, corresponds to the diagonal of the imprint of 70 μm and multiply it by 4, to obtain the size l of the distance between the slots, which will be 280 μm.

We drill the coating 3 see (figure 1) up to the substrate (base) with a core drill with a diamond abrasive at the end with an external diameter of 1 mm, moreover, an internal diameter is approximately 0.8 mm, figure 1. To obtain the required diameter of the inner platform 6 of the coating remaining after drilling, it is necessary to move the drill along one axis 4 in parallel, changing the

IOP Conf. Series: Materials Science and Engineering **489** (2019) 012024 doi:10.1088/1757-899X/489/1/012024

distance between the drilling centers. You can just change this distance arbitrarily and find a suitable size of the platform 5 or 6 after drilling. To obtain an inner platform 6 coated with a transverse axis size of 280 microns it is necessary that the distance between the drilling centers will be approximately 520 microns (depends on the cartridge beat with a core drill). After drilling we place the sample in the hardness tester and choose the place of hardness measurement, the center of the coated area 6. We measure the hardness at a depth of 0.9 to 1 coating thickness to determine the composite hardness of the coating corresponding to the obtained size of the platform 6, i.e. 280 microns. Its own platform 5 or 6 with the calculated size between the slots is created for each diagonal of the imprint and accordingly the penetration depth of the indenter. After drilling it is necessary to make sure that the coated area 6 does not have cracks and damages on the surface.

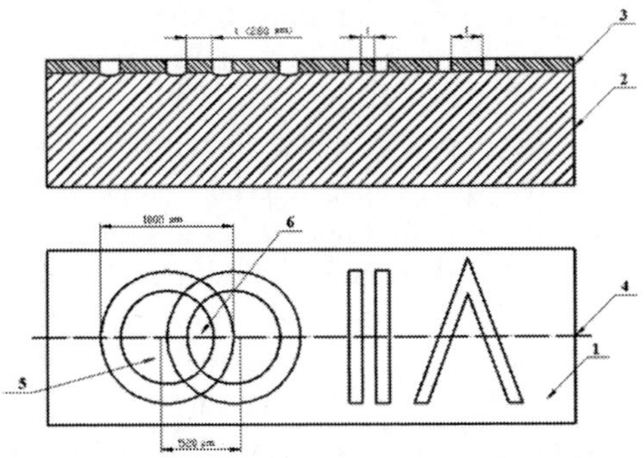

Figure 1. Drawing of types of slots of the surface layer obtaining different sizes of platforms of coating. 1 - sample (product), 2 – substrate, 3 - thin coating, 4 – displacement axis of the core drill, 5 – coating platform with a large size, 6 - coating platform with a smaller size.

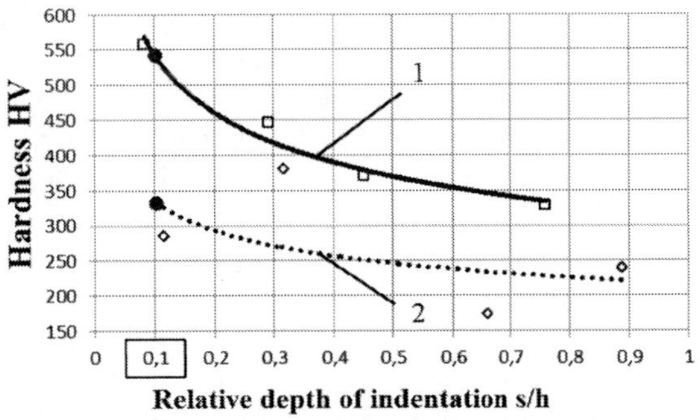

Figure 2. Hardness dependence of DLC coating on the relative depth of indentation. 1 - initial, 2 - after the release of the coating from retained stresses, • - values of the real hardness of coating.

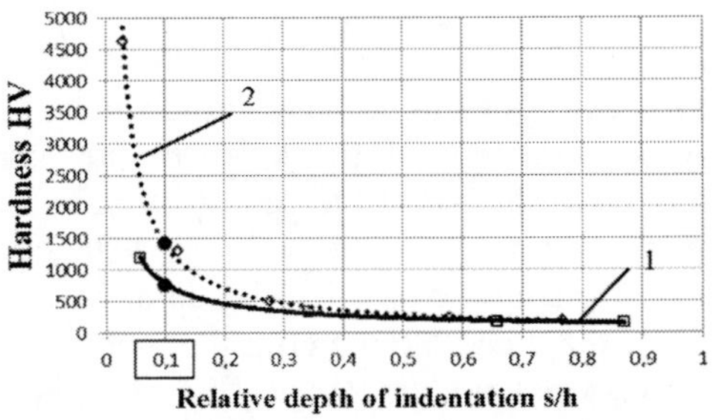

Figure 3. Hardness dependence of TiN coating on the relative depth of indentation. 1 - initial, 2 - after the release of the coating from retained stresses, • - values of the real hardness of the coating.

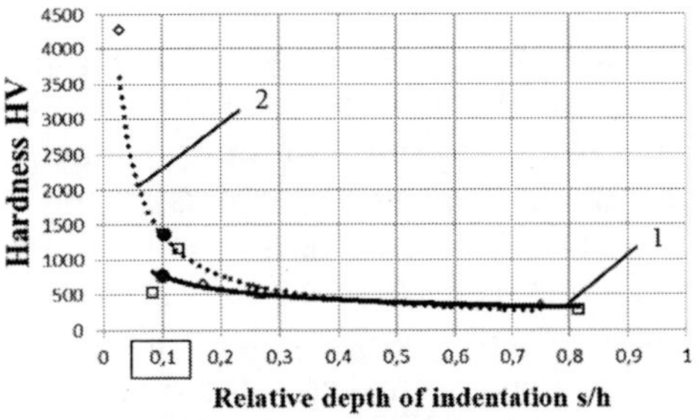

Figure 4. Hardness dependence of TiAlN coating on the relative penetration depth of the indenter. 1 - initial, 2 - after the release of the coating from retained stresses, • - values of the real hardness of the coating.

5. Theoretical background

According to the proposed method, the study area is allocated on the surface of the coated product see figure 1, the local area of coating 6, where two slots identical in width and length with a depth equal to the minimum coating thickness are formed perpendicular to the surface of the coating. Slots can be obtained mechanically, through electrospark or ultrasonic methods, chemical or physical etching. The width of the slots is not subject to strict requirements. They must create safely two free side surfaces in the coating that are not in contact with each other after their formation. An important condition for the formation of the slots is the choice of the method of their manufacture and processing modes, providing the minimum possible force effect and the minimum depth of plastic deformation of the surface on the side walls of the slots.

Slot length is free, but not less than 4 diagonals of the maximum indenter imprint in the test surface during a specific experiment for penetration. It was established experimentally that slot length is from 5 diagonals of the imprint and more reliably ensures the reproduction of the results of the experiment for one point of indentation. When conducting a study in a local place with a multiple number of imprints, the slot length should be increased by a multiple of the number of experimental imprints.

The most important and essential parameter is the distance between parallel slots. Distance value between the slots must ensure removal of retained stresses from previous operations (coating application) and create the conditions for conducting the test without any additional retained stresses from the indentation process.

It is known that the ratio between the maximum penetration depth of the indenter s in the elastic drive (typically s =5-50 nm) and the equivalent radius R of the tip of the real indenter (usually R >100-200 nm) is such that the elastoplastic transition occurs under conditions under which the indentation is actually carried out by a quasi-sphere radius slightly larger than the diagonal of the imprint. The stress state is characterized by the presence of an area of almost pure hydrostatic compression adjacent to the contact surface and propagating to a depth of several tenths of the imprint diagonal [3, 4]. Therefore, the best conditions for the initiation of plastic deformation on indentation exist at a depth = 0.5 of the radius of the quasi-sphere from the contact surface. This depth of the initiation point for surfaces with thin coatings is already in the substrate material [5]. This suggests that the deformation of the surface of the coated product is carried out due to the elastic deformation of the coating and the elastoplastic deformation of the substrate, if loading is increased. Consequently, plastic deformation in the coating material during indentation to a depth of more than 0.1 of the coating thickness up to the depth equal to the coating thickness does not occur and deformation of the coating does not create additional retained stresses, which could distort the values of the measured hardness.

Further increase in the indentation force results in the development of elastoplastic deformation of the substrate material and the increase in the size of the locally deformed coating area to comparable with the diagonal of the imprint. Therefore, if the imprint with a diagonal of the imprint d is positioned so that the diagonal of the imprint is perpendicular to the longitudinal direction of the future slot, then the distance between the imprint and the side wall of the slot should be about the size of the diagonal of the imprint. From the symmetry of the local place of study it follows that the distance l between the side walls of the slots should be approximately equal to 3 d at the central, symmetrical location of the indenter imprint relative to the slots. Moreover, a significant increase in the distance $l > 3d$ causes the increase in the measurement error, since there are conditions when the deformation of the coating begins to experience a hindered deformation from the available material that complicates its deformation. A smaller value of the distance $l < 3d$ is also not preferable, but its effect is less significant, since it is a question of reducing the elastic deformations of the coating and they are much less than the plastic deformations of the substrate. Theoretically it is preferably to place the slots at a distance from each other on the value (3-3,5) d. Experience has shown that the upper limit of the distance range l is not more than $4\ d$ functional in this method.

6. Measurement results

Figure 2, 3 and 4 show the measurement results of the real hardness of the coating and the composite hardness of the surface coating for all samples with coatings under study.

IOP Conf. Series: Materials Science and Engineering **489** (2019) 012024 doi:10.1088/1757-899X/489/1/012024

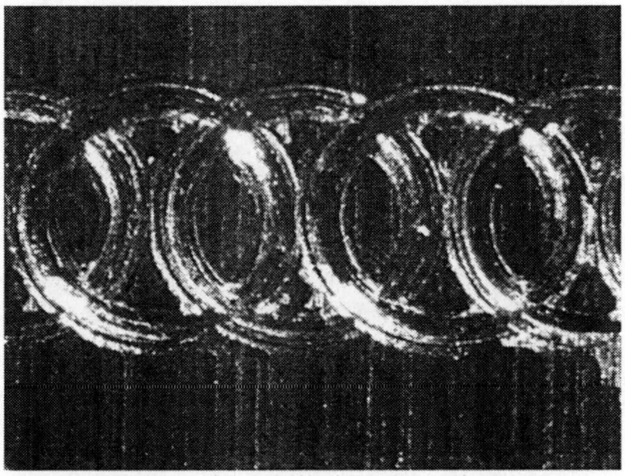

Figure 5. View of the drilled DLC coating x50.

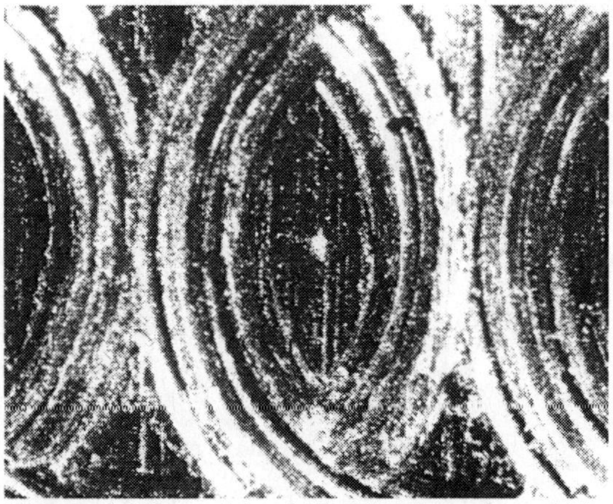

Figure 6. View of the drilled DLC coating x100.

This method, according to the above dependencies, allows comparing the coatings among themselves by the type of the available retained stresses. Figure 2 shows that microhardness falls at all indentation depths after drilling, which indicates the presence of compression stresses in the initial coating. Figure 3 shows on the contrary that tensile stresses are present on the surface of the initial coating, since microhardness increases after drilling and this is true only up to the middle of the coating thickness. When the depth of indenter penetration is more than 5 μm, the microhardness does not change almost, which suggests the absence of retained stresses at a depth below the middle of the coating. Figure 4 shows that the initial coating contains also tensile stresses, but over the entire depth.

These dependences confirm the well-known statement that the greatest retained stresses affect the coating surface; this is evidenced by the large differences between the hardness of the initial hardness and after drilling at the low load and small differences at the high load.

7. Conclusions

Method for determining the real hardness of the coating material and the composite hardness of the surface of engineering products hardened with hard protective coatings is developed.

This method of determining the hardness of hardened surfaces can be used to control the quality and reproduction of the technology of construction parts with protective coatings, including friction units. The proposed method for determining the real hardness of the coating and the effective surface hardness of the coated product is applied for the invention.

Acknowledgments

The work was financially supported by the Ministry of education and science, Agreement No. 14.607.21.0166, the project RFMEFI60717X0166.

References

[1] *Measurement of microhardness by indentation of diamond tips* 1993 Ru Standard No GOST 9450-76 (ST SEV 1195-78) Moscow: Izdatel'stvo standartov [in Russian language]

[2] ISO 14577-1:2015 2015 *Metallic Materials—Instrumented Indentation Test for Hardness and Materials Parameters Part I: Test method*

[3] Johnson K L 1985 *Contac Mechanics* Cambridge University press Cambridge

[4] Golovin Yu I 2008 Nanoindentation and Mechanical Properties of Solids in Submicrovolumes, Thin Near-Surface Layers, and Films: a Review *Phys. Solid State* **50** 2205–36

[5] Voronin N A 2010 Composite, effective, and real hardnesses of thin coatings *J. Frict. Wear* **31** 241–52

IOP Conf. Series: Materials Science and Engineering **489** (2019) 012025 doi:10.1088/1757-899X/489/1/012025

Stress-strains state calculation of a rod at uniaxial tension with non-local effects

I Y Savelyeva, A V Cherednichenko and A P Shukhtin
Bauman Moscow State Technical University, Russian Federation
shukhtin.andrey@mail.ru

Abstract. A numerical calculation of a uniaxial extension of a rod is carried out in terms of the nonlocal integral theory of elasticity using the finite element method for various parameters of the material, taking into account its microstructure, and contributions of nonlocal effects. The dependencies of the strain energy are presented with increasing number of finite elements, which confirm the convergence and correctness of the mathematical model and method.

1. Introduction

The creation of new structurally sensitive materials based on nanotechnology is an important direction in the development of modern materials science. Consolidated structurally sensitive materials are usually obtained by compacting nanopowders, deposition on the substrate, crystallization of amorphous alloys and other methods. Such materials have unique physicomechanical properties that allow them to be used effectively in structures exposed to high-intensity external influences [1, 2].

An important step in the creation and usage of new structurally sensitive materials is the construction of mathematical models that allow describing their behavior in a wide range of changes in external influences.

Currently, there are at least three main approaches to the theoretical modeling of materials with a complex internal structure: 1) quantum mechanical approaches, molecular dynamics methods and Monte Carlo models [3, 4]; 2) generalized continuum mechanics approaches that include theories of effective properties [5 — 8], mixture models [9, 10], models based on averaging procedures [11], Cosserat-type micropolar media [4, 5] and statistical models [12]; 3) hybrid molecular-mechanical approach combining molecular dynamics methods and macroscale models [3, 13].

It is known that the usage of methods of continuum mechanics for materials with a microstructure is limited to scale and boundary effects. Direct application of continuum mechanics methods for materials modified by micro- and nanostructured inclusions is incorrect [5, 14 — 16]. For these reasons, there is an interesting theory where, on the one hand, is taken into account the presence of the microstructure, and on the other hand the equations have the form of the usual equations of continuum mechanics, integro-differential in the general case, for solutions which are applied the methods of continuum mechanics. Approaches of classical continuum mechanics media with micro- and nanostructure is called the method of continuous approximation [14, 15]. The field of science in which the behavior of materials with micro- and nanostructures is studied using the method of continuous approximation is sometimes called

IOP Conf. Series: Materials Science and Engineering **489** (2019) 012025 doi:10.1088/1757-899X/489/1/012025

generalized continuum mechanics. The key points in this method are the establishment of connections between the characteristics of the micro (nano-) level and the macro level, as well as taking into account the effects of spatial and temporal non-locality of the media.

2. Mathematical model

Let us consider a mathematical model of a uniaxial tension of a rod (fig. 1) with a length L, a cross-sectional area S and a concentrated force on the P right-hand side.

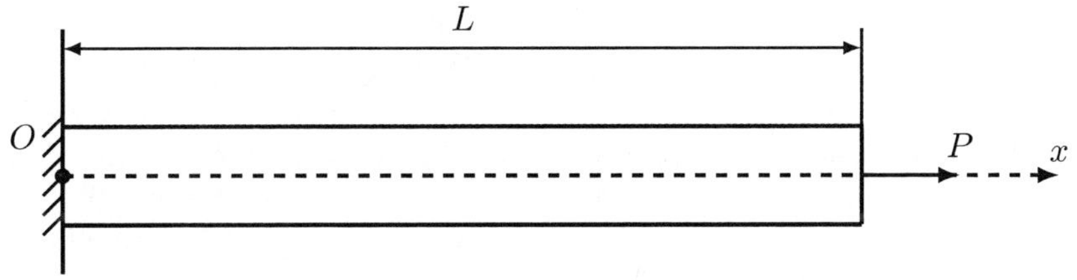

Figure 1. Uniaxial tension of a rod.

The formulation of the boundary value problem of the mechanics of a deformed body in terms of small deformations has the form [11]:

$$\frac{\partial \sigma(x)}{\partial x} + b(x) = 0, \ u(0) = 0, \ \sigma(L) = \frac{P}{S} \tag{1}$$

where $b(x)$ is the force distributed over the volume, $\sigma(x)$ and $u(x)$ are the stress and displacement along the axis Ox. However, according to the non-local theory of elasticity [12], the relation between stresses and strains can be represented as

$$\sigma(x) = p_1 E\varepsilon(x) + p_2 \int\limits_0^L \varphi(|x' - x|) E\varepsilon(x') dx', \tag{2}$$

where $\varepsilon(x)$ — the strains along the axis Ox, E — the Young modulus, p_1 and p_2 are the proportions of the influence of local and non-local effects such as $p_1 + p_2 = 1$; $\varphi(|x' - x|)$ is an influence function that determines spatial nonlocality, while

$$\int\limits_{-\infty}^{+\infty} \varphi(|x' - x|) dx' = 1.$$

The influence function was used by A.C. Eringen [4, 5, 14, 15] to solve problems in the theory of elasticity and based on the idea that long-range forces that responsible for the non-local deformation of the material at a given point of the space x are adequately described using the decreasing distance function $\varphi(|x' - x|)$, while increasing $|x' - x|$.

As shown in [18], problem (1) can be reformulated into the Fredholm integral equation of the second kind with respect to strains, which does not have a strict analytical solution in the general case.

3. Nonlocal FEM-formulation

From the varational approach, the solution of problem (1) is the equivalent to minimizing a functional, which, in its physical sense, is the total potential energy of a loaded rod [11]:

$$\Pi = \frac{1}{2}\int\limits_V \varepsilon\sigma dv - \int\limits_V ubdv - \int\limits_S uPds, \tag{3}$$

where V is the volume of the rod and S is the cross-sectional area of the right side. Using the discretization procedure, this problem can be solved in terms of isoparametric finite elements in displacements [13], then the original functional (3) can be rewritten in view of the relations in the form of FEM

$$\Pi[\mathbf{u}] = \frac{1}{2}p_1\sum_{n=1}^{N_e}\mathbf{u}_n^T\mathbf{K}_n^{loc}\mathbf{u}_n + \frac{1}{2}p_2\sum_{n=1}^{N_e}\sum_{m=1}^{N_e}\mathbf{u}_n^T\mathbf{K}_{nm}^{nl}\mathbf{u}_m - \sum_{n=1}^{N_e}\mathbf{u}_n^T(\mathbf{f}_n^{(b)} + \mathbf{f}_n^{(P)}), \tag{4}$$

where N_e is the number of elements, $\mathbf{u}_n$ and $\mathbf{u}_m$ are the displacement field in the element with number n and m, $\mathbf{K}_n^{loc}$ is the stiffness matrix of the element with the number n in terms of the classical theory of elasticity, $\mathbf{K}_{nm}^{nl}$ is matrix of non-local effects [27] that represents the influence of the m-th element on the n-th one (fig. 2), $\mathbf{f}_n^{(b)}$ and $\mathbf{f}_n^{(P)}$ is vector-columns of forces distributed over the volume and distributed load over the surface of the element with the number n. The relations included in (2) can be represented as

$$\mathbf{K}_n^{loc} = \int\limits_{V_n}\mathbf{B}_n^T E\mathbf{B}_n S dx, \qquad \mathbf{K}_{nm}^{nl} = \int\limits_{V_n}\int\limits_{V_m}\varphi(|x'-x|)\mathbf{B}_n^T E\mathbf{B}_m S dx' dx, \tag{5}$$

$$\mathbf{f}_n^{(b)} = \int\limits_{V_n}\mathbf{N}_n^T bS dx, \qquad \mathbf{f}^{(P)} = \int\limits_S \mathbf{N}_n^T P dx, \tag{6}$$

where $\mathbf{N}_n$ is the matrix of shape functions on the element with number n, $\mathbf{B}_n$ and $\mathbf{B}_m$ are the matrices of connection strains and stresses of the elements with numbers n and m, containing derivatives of the shape functions, V_n is the length of the element with number n.

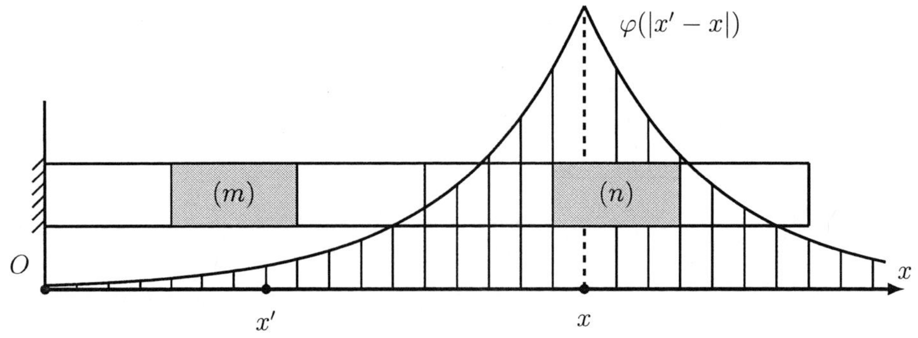

Figure 2. Influence of field of the strains at the element with a number m at the field of the strains at the element with number n.

A necessary condition for the extremum of the functional (4) is the fulfillment of the relation $\frac{\partial \Pi}{\partial \mathbf{U}} = 0$ [13], where $\mathbf{U}$ is the vector of nodal displacements on the entire rod. Using this, one can obtain a system of linear algebraic equations for displacements with help of (5) and (6)

$$\mathbf{KU} = \mathbf{F},$$

where $\mathbf{K} = \sum_{n=1}^{N_e} \left(p_1 \mathbf{K}_n^{loc} + p_2 \sum_{m=1}^{N_e} \mathbf{K}_{nm}^{nl} \right)$ and $\mathbf{F} = \sum_{n=1}^{N_e} \left(\mathbf{f}_n^{(b)} + \mathbf{f}_n^{(P)} \right)$. The global stiffness matrix $\mathbf{K}$ would be completely filled, whereas in terms of the classical theory of elasticity it would have a three-diagonal structure, and the global vector of the right side $\mathbf{F}$ would have one single nonzero component corresponding to the node number coinciding with the right side of the rod.

Like the matrix $\mathbf{K}_n^{loc}$, the matrix of nonlocal effects $\mathbf{K}_{nm}^{loc}$ is calculated with isoparametric relations taking into account

$$\mathbf{K}_{nm}^{nl} = \int_{-1}^{1} \int_{-1}^{1} \varphi(|x'(\eta) - x(\xi)|) \mathbf{B}_n^T E \mathbf{B}_m S \det \mathbf{J} d\xi d\eta,$$

where ξ and η is local coordinates, $\mathbf{J}$ is the Jacobi matrix of the derivatives connection in the local coordinate system with the derivatives in global coordinate system, while $\mathbf{B}_n = \mathbf{B}_n\left(x(\xi)\right)$ and $\mathbf{B}_m = \mathbf{B}_m\left(x'(\eta)\right)$.

Similarly with the classical theory of elasticity, as shown in [13], the strain energy (7) in terms of the non-local theory of elasticity is finite in its physical meaning

$$W = \frac{1}{2} \int_V \varepsilon \sigma dv. \tag{7}$$

At the same time, turning to a finite element formulation, the strain energy should tend to its limit as the number of elements increases and the number of their characteristic size decreases, that is $\lim_{h \to 0} W_h = W$, where W_h is strain energy, computing at the current amount of elements, which can be represented as

$$W_h = \frac{1}{2} \mathbf{u}^T \mathbf{K} \mathbf{u},$$

h is the maximum characteristic size of a finite element, which in the one-dimensional case is the length of the element.

IOP Conf. Series: Materials Science and Engineering **489** (2019) 012025 doi:10.1088/1757-899X/489/1/012025

4. Numerical results

The numerical experiment was carried out without influence of volume forces and with the following values of the material parameters: $L = 100$ mm, $S = 10$ mm^2, $E = 2.1 \cdot 10^6$ MPa and $P = 10$ N. The influence function selected for calculations was looked like

$$\varphi(|x' - x|) = \frac{1}{2a} \exp\left(\frac{|x' - x|}{a}\right),$$

where a is the characteristic of the material, which describes its microstructure. To calculate the matrices $\mathbf{K}_n^{loc}$ and $\mathbf{K}_n^{nl}$ numerical integration with Gauss quadratures over two points was used. Each calculation used linear finite elements of the same length in the amount of 100 elements.

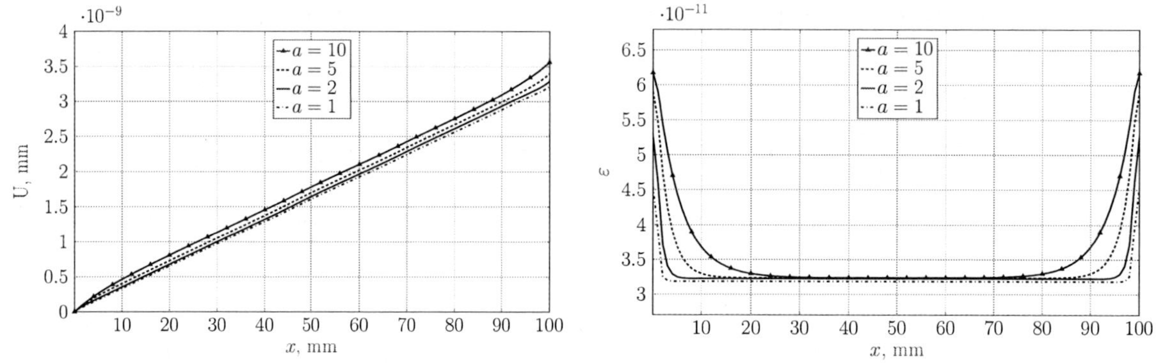

a) b)

Figure 3. Various nodal distributions depending on parameter a with non-local contribution parameter $p_1 = 0.25$: a) displacements; b) strains.

Figure 3 shows the dependencies of the distributions of displacements, strains, and stresses for various parameters a and the contribution parameter $p_1 = 0.25$. The dependencies of strains and stresses are of the greatest interest, since there is a sharp increase and change in gradients near the boundary of a rod.

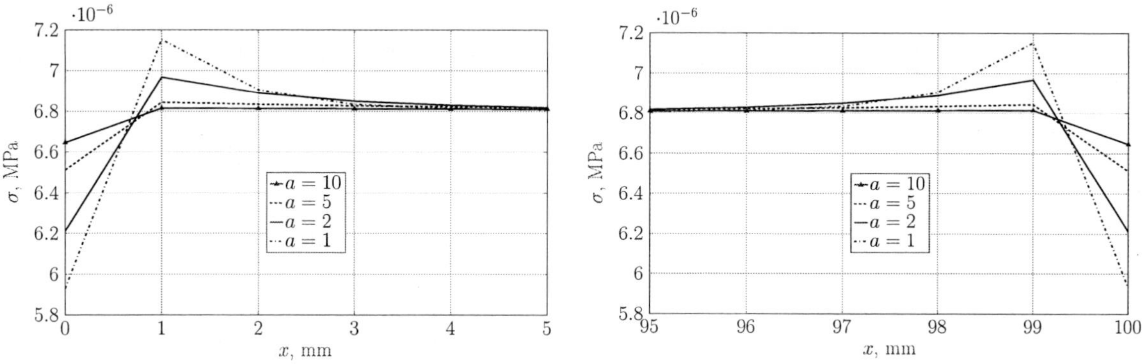

a) b)

Figure 4. Various nodal stress distributions depending on parameter a with non-local contribution parameter $p_1 = 0.25$: a) at the left side of a rod; b) at the right side of a rod.

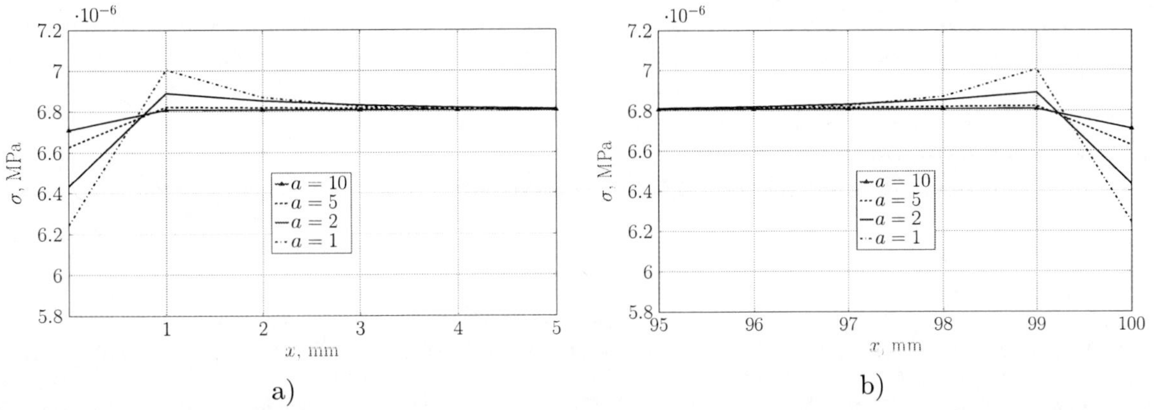

Figure 5. Various nodal stress distributions depending on parameter a with non-local contribution parameter $p_1 = 0.5$: a) at the left side of a rod; b) at the right side of a rod.

The plots of the stress distribution at different values of the parameters a and p_1 are divided into two parts (fig. 4 – 5), since they differ slightly at the points far from the borders, but a sharp drop in stress is noticeable near the boundaries; in this case, the larger value of the parameter a, the smaller drop in stresses near the rod boundaries, whereas for strains (fig. 3b) the opposite is true.

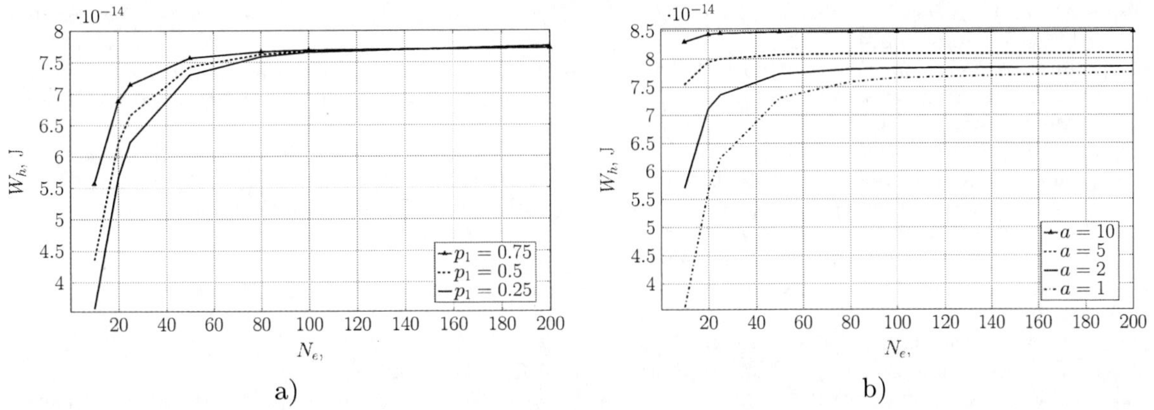

Figure 6. Strain energy by increasing elements amount: a) depending on parameter p_1, when $a = 1$; b) depending on parameter a, when $p_1 = 0.25$.

As can be seen from the plots of the strain energy depending on the number of elements for various values of parameter a and various values of parameter of non-local effects p_1 (fig. 6), there is a convergence of numerical results confirmed.

5. Conclusion

Calculations of the stress-strain state of the rod at uniaxial tension, with non-local effects taking into account, showed: 1) in terms of the non-local integral theory of elasticity, a numerical solution obtained by the finite element method tends to its limit; 2) by increasing the value of parameter a, which determines the microstructure of the rod's material, leads to increasing strains in the zone of growth near the sides of the rod with increasing the maximum values of the strains; 3) by increasing the value of parameter a, the maximum displacement of the rod grows slightly; 4) increasing the value of parameter a, leads to increasing strains drop in the region of the sides of the rod while decreasing the maximum stress values; 5) an increase in the share of the contribution of non-local effects does not have a significant effect on the strain energy of the entire rod.

6. Acknowledgements
The reaserch was supported by the Russian Federation for Basic Research project No. 18-38-20108.

References
[1] Andrievskiy R A and Ragulya A V 2005 *Nanostructual Materials* (Moscow: Academiya) p 192
[2] Alymov M I and Zelenskiy V A 2005 *Methods of Obtaining and Physico-Mechanical Properties of Bulk Nanocrystalline Materials* (Moscow: Mifi) p 51
[3] Gopalakrishan S and Narendar S 2013 *Wave Propagation in Nanostructures. Nonlocal Continuum Mechanics Formulations* (New York: Springer) p 365
[4] Ibragimov I M, Kovshov A N and Nazarov Yu F 2010 *Introduction to the Copmputer Modeling of nanosystems* (Saint-Petersburg: Lany) p 384
[5] Eringen A C 2002 *Nonlocal Continuum Field Theories* (New York: Springer-Verlag) p 393
[6] Krivcov A M 2007 *Deformation and Fracture of Solids with a Microstructure* (Moscow: FIZMATLIT) P 307
[7] Zarubin V S, Kuvyrkin G N and Savelyeva I Yu 2016 *Mech. of Comp. M.* **52** 143-52
[8] Zarubin V S, Kuvyrkin G N and Savelyeva I Yu 2016 *Comp. and N.* **7** 1-13
[9] Rushintskiy Ya Ya 1991 *Elements of Mixture Theory* (Kiev: Nauka dumka) p 160
[10] Bahvalov N S and Panasenko G P 1984 *Averaging of Processes in Periodic Media: Mathematical Problems in the Mechanics of Composite Materials* (Moscow: Nauka) p 352
[11] Sheremgor T D 1977 *Elasticity Theory of Peterogeneous Media* (Moscow: Nauka) p 400
[12] Karlicic D, Murmu T, Adhikari S and McCarthy M 2016 *Non-local Structual Mechanics* (Hoboken: John Willey & Sons) p 374
[13] Kunin I A *Elastic Media with Microstructure. Nonlocal Elasticity Theory* (Moscow: Nauka) p 416
[14] Onami M, Ivasimidzu S, Genka K, Sinodzava K and Tanaka K 1987 *Introduction to Micromechanics* (Moscow: Metallurgiya) p 280
[15] Zarubin V S and Kuvyrkin G N 2008 *Mathematical Model of Mechanic and Electrodynamic Continuum Media* (Moscow: Bauman Moscow State Technical University press) p 512
[16] Gusev A I 2005 *Nanomaterials, Nanostructures, Nanotechnologies* (Moscow: FIZMATLIT) p 416
[17] Koboyashi N 2005 *Introdution to Nanotechnology* (Moscow: BINOM) p 134
[18] Pool Jr Ch P and Owens F 2000 *Introduction to Nanotechnology* p 400
[19] Kelsall R W, Hamley I and Geoghegan M 2005 *Nanoscale Science and Technology* (Chichester: John Wiley & Sons) p 474
[20] Peddieson J, Buchanan G R and McNitt R 2003 *Int. J. Eng. Sc.* **41** 305-12
[21] Kuvyrkin G N and Savelyeva I Yu 2016 *J. of Mech. of Sol.* **51** 256-62
[22] Zienkiewicz O C and Taylor R L 2000 *The Finite Element Method. The Basis* vol 1 (Oxford: Butterworth-Heinemann) p 708
[23] Polizzotto C 2001 *Int. J. of Sol. and Str.* **38** 7359–80
[24] Di Paola M, Failla G, Pirrotta A, Sof A and Zingales M 2013 *J. of Ph. Tr A*
[25] Di Paola M, Failla G and Zingales M 2009 *J. of Elas.* **97** 103-30
[26] Di Paola M, Failla G and Zingales M 2010 *J. of Sol. and Str.* **47** 2347-58
[27] Pisano A A, Sofi A and Fuschi P *Int. J. of Sol. and Str.* **46** 3836-49

IOP Conf. Series: Materials Science and Engineering **489** (2019) 012026 doi:10.1088/1757-899X/489/1/012026

Research of the friction surfaces regular microgeometry parameters effect on the hydro-mechanical characteristics of the «piston-cylinder» tribounit

K Gavrilov, A Doikin, M Izzatulloev and S Surovtsev

Department of Automobile Transport, South Ural State University (national research university), 76 Lenin prospekt, Chelyabinsk 454080, Russia

gavrilovkv@susu.ru

Abstract. The article provides an overview of the friction surfaces microprofiling main types of the «piston-cylinder» tribounit. The calculation model is created and the program of the tribounit calculation analysis is developed. Calculations of the tribounit hydro-mechanical characteristics for different types of microgeometry are performed.

1. Introduction

At the new stage of development of tribology, the processes of friction and wear are presented as complex processes of physical and chemical mechanics, and it is impossible to describe them without certain simplifications. In this regard, various models are widely used, among which two directions are highlighted: a mathematical description of the microrelief of friction surfaces and a model of their contact interaction representation [1-6].

The interaction of rough surfaces, as well as surfaces with a technologically specified microrelief, determines many processes in electrical, heat and mechanical engineering, automotive industry, engine construction and other areas.

Micro-profiling is designed to reduce friction in tribounits, to increase their bearing capacity, reduce wear, improve the reliability and efficiency of friction units. Thus, the main types of microprofiling in the "piston-cylinder" connection are honing and texturing.

In this article we consider some types of regular microgeometry of friction surfaces (microgrooves) of tribounit "piston-cylinder" in the internal combustion engine (ICE) and their impact on the main hydromechanical characteristics (HMC): $\inf h_{\min}$ – minimum thickness of the lubricant layer, $\sup p_{\max}$ – maximum hydrodynamic pressure, Q^* – average lubricant consumption, $h^*_{\min}$ – average thickness of the lubricant layer, $p^*_{\max}$ – average hydrodynamic pressure, N^*_{fr} - average power of friction losses.

As the parameters of microgrooves were set their depth (h_k), the distance between them (b_k), and the inclination angle relative to the vertical axis of the piston (a_k), that allows to simulate the honing of the friction surface.

2. Method of calculation

The field of hydrodynamic pressures was determined from the Reynolds equation:

$$\frac{d}{d\varphi}\left[\frac{h^3}{12\overline{\mu}_E^*}\cdot\frac{d\overline{p}}{d\varphi}\right]+\frac{1}{a^2}\cdot\frac{d}{d\overline{z}}\left[\frac{h^3}{12\overline{\mu}_E^*}\cdot\frac{d\overline{p}}{d\overline{z}}\right]=\frac{\overline{w}_{21}}{2}\cdot\frac{d\overline{h}}{d\overline{z}}+\frac{d\overline{h}}{d\tau}.$$

Where $\overline{p}\psi^2/\mu_0\omega_0$ – dimensionless pressure in the lubricant layer, $\psi = h_0/R$ – the relative characteristic of the lubricant layer thickness; $\overline{h} = h/h_0$; $h = h_0 + h_k - e\cdot\cos\varphi + Z\cdot\mathrm{tg}\gamma\cdot\cos\varphi$; $\overline{\mu}_E^* = \mu_E^*/\mu_0$; $\overline{z} = z/R$; $-a \le z \le a$; $\varphi = x\cdot R$; $a = B/2R$; $\overline{w}_{21} = (w_2 - w_1)/\omega_0 R$; τ, $\overline{h}$, $\overline{\mu}_E^*$ – dimensionless time, the thickness and the effective viscosity of the lubricant, corresponding to the temperature T_E^*; B, R – the height and the radius of the piston skirt; μ_0, h_0, ω_0 – lubricant viscosity film thickness at the central position of the spike and shaft speed; w_1, w_2 – speed of the translational motion of the movable elements; w_{21} – dimensionless translational velocity of the piston movement; e_c – the offset of the center of mass of the piston from the cylinder axis; γ – the angle of the piston; φ – circumferential coordinate.

Calculation of the piston trajectory on the lubricant layer in the engine cylinder (figure 1) was made while the coordinate system was fixed on a motionless cylinder and the beginning of the movable coordinate system was located in the center of the piston mass.

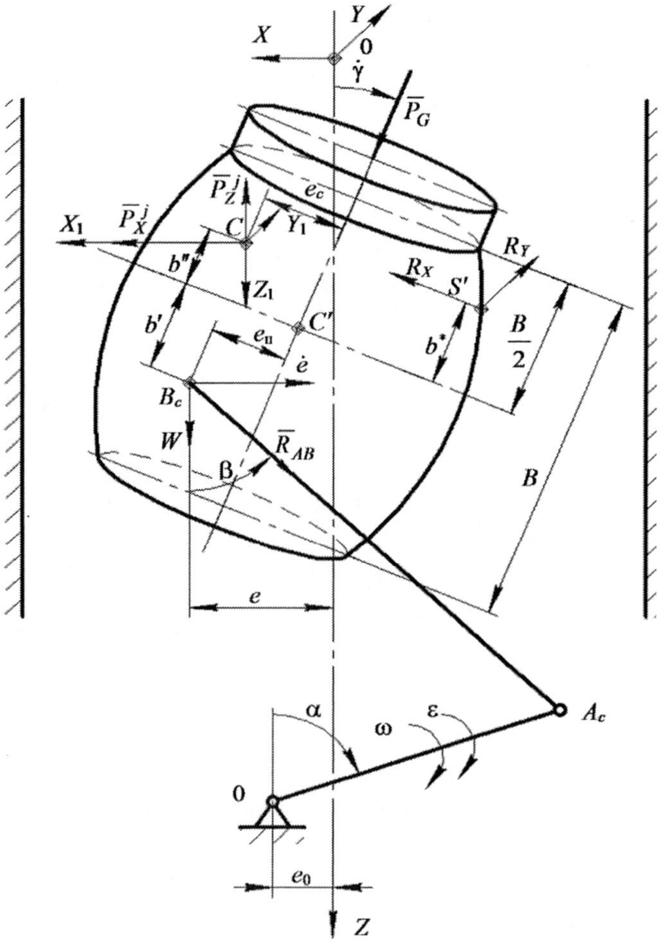

Figure 1. The dynamics of the piston on the lubricating layer in the diesel engine cylinder.

On the picture above P_G – pressure of gases; P_x^j, P_z^j – the inertial forces projection P_j of translational moving of the piston (attached in the center); e_c – the offset of the piston's center of mass; e_p – offsetting

IOP Conf. Series: Materials Science and Engineering **489** (2019) 012026 doi:10.1088/1757-899X/489/1/012026

of the piston; e_o – offsetting of the engine; R_x, R_y – projection of the lubricating layer reaction, α – angle of rotation of the crankshaft (ROC).

Complex movements of the piston consist of translational movements along and across the axis of the cylinder (respectively, with speeds $w = w_2$ and e), and of rotational movement around the axis of the piston finger with speed γ. Gravity and friction forces were not considered due to their smallness. Thus, the piston has the ability to move in the general case in the *XOZ*, *YOZ* subspaces. The acceleration of the point C along the *OZ*-axis was assumed to be equal to the acceleration of the piston translational motion determined in the kinematics of the plane-parallel motion of the crank mechanism [7].

The force R was considered as given force in a subspace parallel to the subspacie XOZ, the cylinder is non-rotatable ($w_1 = 0$).

3. Software complex

For the calculation analysis of the internal-combustion engine of the hydrodynamic tribounits, taking into account the described technique, program of tribological analysis "THE MICROGEOMETRY OF THE PISTON-CYLINDER TRIBOSYSTEM" was developed [8]. The initial data were: the indicator diagram of the working process in the cylinder, the macrogeometric parameters of tribounit and the parameters of regular microgeometry in the form of radial microgrooves, cross microgeometry with different inclination angles, and sinusoidal microgeometry.

The program provides the following functions: calculation of external forces acting on the piston on the basis of the hodographs' indicator diagram; determination of the field of hydrodynamic pressures in the lubricant layer, construction of the trajectory of the piston, calculation of tribological characteristics, determination of the friction mode and its duration by the angle of rotation of the crankshaft, as well as the optimization problem solution to determine the parameters of the macroprofile of the piston guide part.

4. The results of calculation

4.1. The adequacy check of the model

Using the developed program, comparative computational studies were performed on the example of a gasoline four-cylinder, in-line turbocharged engine. The initial parameters of the engine are presented in table 1.

Table 1. Initial data

Parameter	Value
The radius of the crank, m	0.045
The length of the connecting rod, m	0.135
The diameter of the piston, m	0.084
The area of the piston, m^2	0.0055
The mass of the piston, kg	0.434
The height of the piston skirt, m	0.0305
The nominal diameter clearance in «piston – cylinder» system, m	0.0000145

Figure 2 shows the lateral force acting on the piston, and figure 3 presents the results of the calculation of friction losses in the tribounit obtained by the authors of [9] and calculated in accordance with the presented technique.

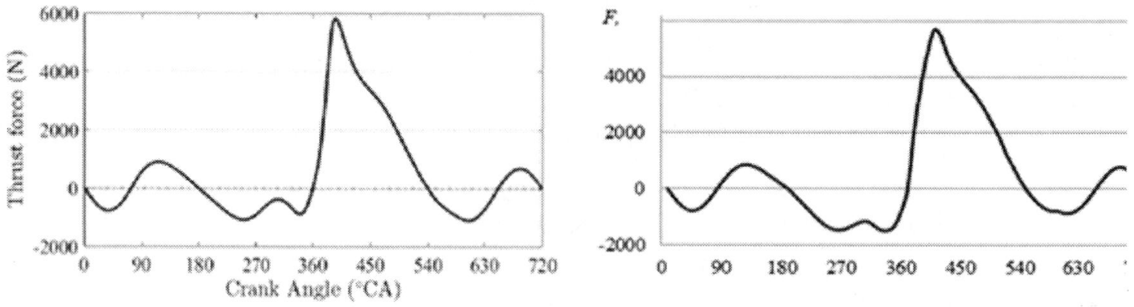

Figure 2. Lateral force acting on the piston from the ROC: on the left - results of work [9], on the right - developed method.

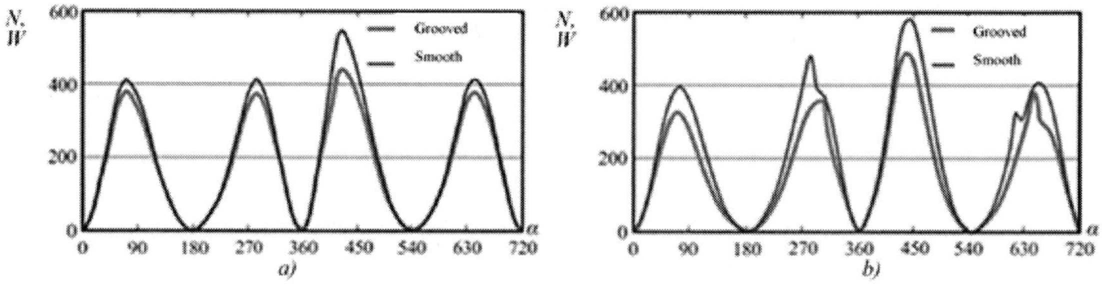

Figure 3. Friction losses at the angle of ROC: a) results of work [9], b) developed method.

The results of the calculation indicate the qualitative and quantitative coincidence of the calculated parameters. With this in mind, it can be concluded that the developed calculation method is adequate.

4.2. The calculation results of a diesel engine

The calculation of the hydromechanical characteristics of the coupling is performed on the example of diesel type ChN 13/15 (4 stroke supercharged, piston diameter 130 mm, piston stroke 150 mm), the initial data for which are shown in table 2. In addition, the indicator diagram and three-constant viscosity-temperature characteristic of motor oil SAE 5W-50 with characteristic values of dynamic viscosity coefficients at a certain temperature were used as initial data: $\mu_{40} = 0.074$ Pa·s, $\mu_{80} = 0.0237$ Pa·s, $\mu_{100} = 0.0154$ Pa·s.

Table 2. Initial data for calculation of diesel ChN 13/15

Parameter	Value
The radius of the crank, m	0.075
The length of the connecting rod, m	0.260
The diameter of the piston, m	0.130
The area of the piston, m^2	0.013
The mass of the piston, kg	5.035
The height of the piston skirt, m	0.084
The nominal diameter clearance in «piston – cylinder» system, m	0.00002

Table 3 presents calculations of HMC of diesel ChN 13/15.

Table 3. Results of HMC calculations

Types of microgeometry	inf h_{min}, μm	sup p_{max}, MPa	h^*_{min}, μm	p^*_{max}, MPa	Q^*, s^{-1}	N^*_{fr}, W
Without microgeometry	5.62	7.86	19.99	1.63	0.00169	298.99
Circular microgeometry	4.84	10.51	18.92	2.05	0.00178	262.15
Cross microgeometry	4.16	13.80	18.02	2.58	0.00186	236.09

The results indicate a decrease in the calculated friction losses in vehicles with regular microgeometry to 20% with non-critical changes in other HMC.

Graphs of the main HMC depending on the angle of rotation of the crankshaft for diesel are presented in figures 4-7.

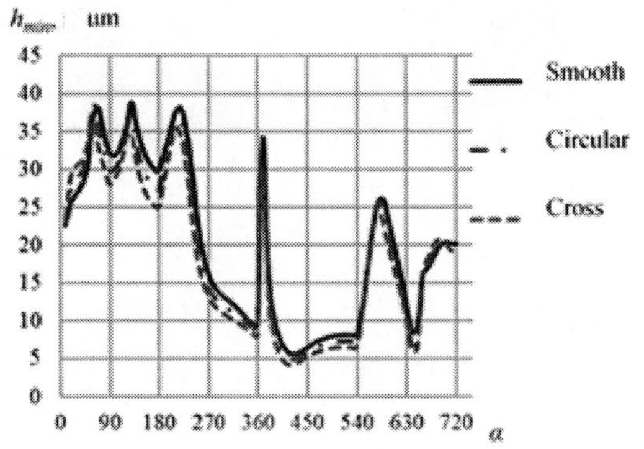

Figure 4. Dependence of lubricant layer thickness on the angle of ROC.

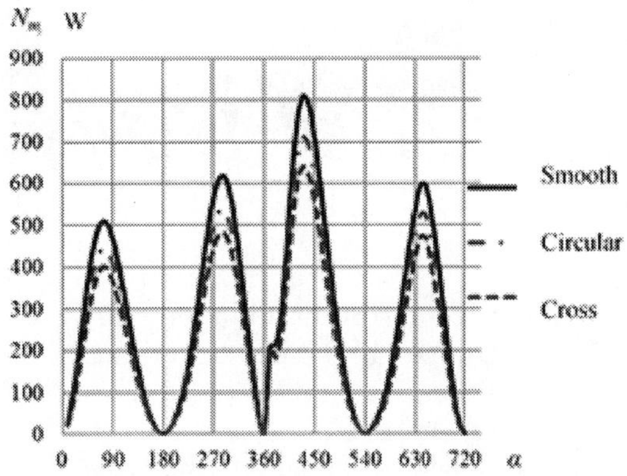

Figure 5. Dependence of the power losses due to friction on the angle of ROC.

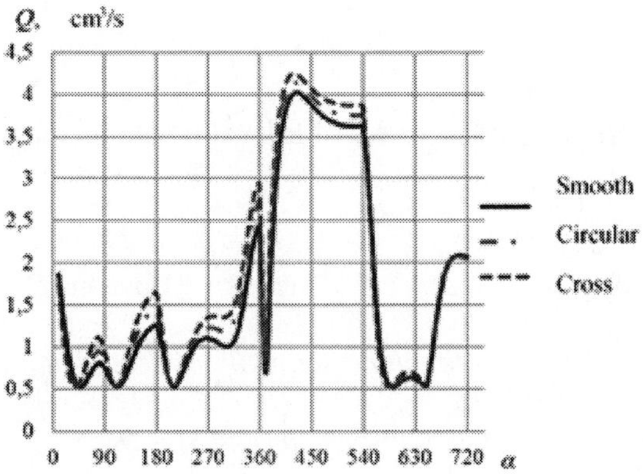

Figure 6. Dependence of the lubricant flow rate on the angle of ROC.

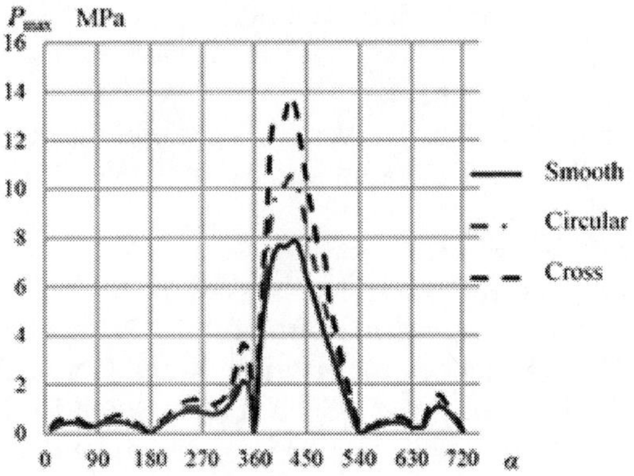

Figure 7. Dependence of the maximum hydrodynamic pressure on the angle of ROC.

From the graphs it can be seen that during the working cycle of a diesel engine there are no points in time where we can see the deterioration of such parameters as h_{min} and Q. Moreover, power of friction losses are reduced throughout the entire working cycle, and the maximum hydrodynamic pressure increases at the moments of maximum load, which provides an increase in the carrying capacity of the coupling and prevents contact of friction surfaces.

Figure 8 shows hydrodynamic pressure diagrams in the lubricant layer for a smooth and textured piston skirt (radial microgrooves) and we see the positive effect of regular microgeometry of the piston guide surface, which consists in increasing the bearing capacity of the tribounit for the angle of crankshaft rotation corresponding to the greatest load acting on the piston (α=390 deg. ROC).

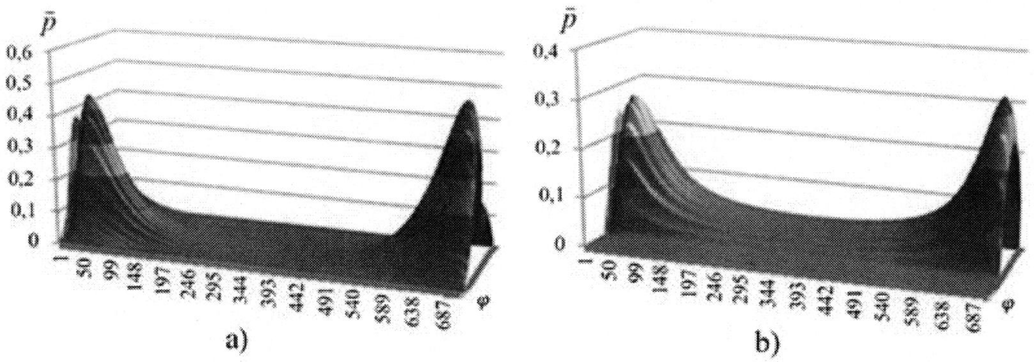

Figure 8. Hydrodynamic pressure in the lubricant layer for α=390 deg. ROC: a) radial microgrooves, b) without texturing.

5. Conclusion

The mathematical model and the program of the tribounit calculation analysis were developed. With their help, the study of influence of various kinds of regular microgeometry on the main HMC "piston-cylinder" tribounit was carried out. Its positive effect is shown, which consists in reducing friction losses (up to 20% depending on the type of microgeometry of the piston skirt surface), which are one of the main indicators of efficiency of the structure. As well as in increasing of the bearing capacity of the "piston-cylinder" tribounit at the moments of the maximum load on the piston and a possible transition to the mixed lubrication mode.

Acknowledgments

South Ural State University is grateful for financial support of the Ministry of Education and Science of the Russian Federation (grant No 9.7881.2017/8.9).

References

[1] Tigetov G and Goritskiy Y 2010 *Vestnik of National Research University "Moscow Power Engineering Institute"* **3** 3
[2] Izzatulloev M, Gavrilov K and Gordienko A 2017 *Proc. of the ninth sci. conf. of grad. stud. and doct. stud.* (Chelyabinsk: South Ural State University) 33
[3] Gavrilov K, Goritskiy Y, Migal I and Izzatulloev M 2017 *Trib. in Industry* **39(3)** 329
[4] Jeng Y, Lin Z and Shyu S 2004 *J. of Trib.* **126(3)** 620
[5] Novotny P 2017 *Trib. in Industry* **39(1)** 82
[6] Ohue Y and Tanaka H 2013 *J. Sci. Res.* **4** 379
[7] Rozhdestvenskii Y, Gavrilov K, Izzatulloev M and Gordienko A 2018 *Proc. of the* the tenth *sci. conf. of grad. stud. and doct. stud.* (Chelyabinsk: South Ural State University) 71
[8] Gavrilov K, Izzatulloev M and Sadreev T 2018 *Cert. of state reg. of the comp. prog. "MICROGEOMETRY OF THE PISTON-CYLINDER TRIBOSYSTEM"* № 2018663437

IOP Conf. Series: Materials Science and Engineering **489** (2019) 012026 doi:10.1088/1757-899X/489/1/012026

[9] Fang C, Meng X and Xie Y J. 2017 *Trib. Int.* **110** 42

IOP Conf. Series: Materials Science and Engineering **489** (2019) 012027 doi:10.1088/1757-899X/489/1/012027

Wear of foundation with a nonuniform coating by rough punch

K E Kazakov

Ishlinsky Institute for Problems in Mechanics of the Russian Academy of Sciences, Moscow, Russia

kazakov-ke@yandex.ru

Abstract. Plane problem of contact interaction for a viscoelastic foundation with multilayer coating and a rigid punch is considered. It is assumed that thin coating consists of several different elastic layers. The rigidity of the each layer depends on longitudial coordinate and can be described by own rapidly changing function. We obtain basic integral equation and additional conditions for this problem. The analytic solution of the problem is presented.

Introduction

In works [1] and [2] we solved some contact-wear problems. Paper [1] was dedicated to the obtaining of solutions for the plane contact-wear problem for flat punch and foundation with coating which has longitudially nonuniformity. A mixed integral equation with one complex function was obtained for this problem. Paper [2] considered an axisymmetric problem in the case where the punch had a complex shape. A mixed integral equation with two complex functions (punch shape and coating nonuniformity) was obtained for this problem.

In the current work we state and solve plane contact-wear problem for surface nonuniform forundation and rough punch. This plane problem has more degrees of freedom than axisymmetric problem [2]: punch can move and rotate. It has another kernels, operators, basis and etc. Some new effects can be detected in a plane problem.

1. Statement of the linear wear plane problem

Elastic layer of constant thickness lies on an underlying undeformable foundation (figure 1). Another elastic layer of constant thickness lies on the first layer. There are ideal contact between the layers and between lower layer and underlying foundation. We assume that upper layer is nonuniform and it is soft, i.e. its variable rigidity does not exceed the rigidity of lower layer [3]. We assume that rigidity of upper layer depends on longitudinal coordinate, i.e. $R = R(x)$.

A rigid punch indents into such foundation along the axis Oz. Moreover the punch move along the axis perpendicular to the plane Oxz with an average velocity V. The average modulus of the punch velocity is taken into account when considering the reciprocating motion. The layered foundation is worn due to described punch motion.

Further we will assume that the contact region is constant and bounded by the boundaries $x = -a$ and $x = a$, the characteristic dimension $2a$ of the contact region is much greater than the upper layer thickness h_1, and there is a smooth contact between the punch and upper layer.

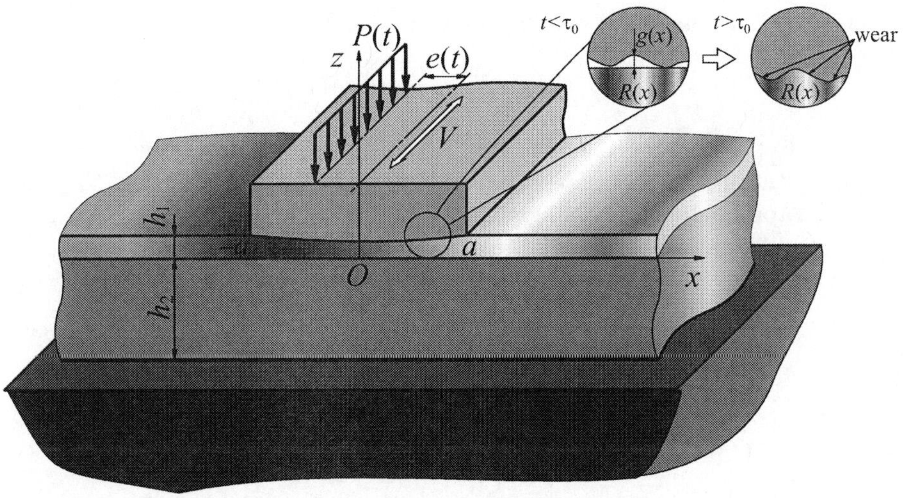

Figure 1. Contact-wear problem

Numerous experiments have shown that, in the case of linear wear, the velocity of body surface points is proportional to the normal load and inversely proportional to the material hardness [4–8]:

$$v_{\mathrm{w}}(x,t) = -\frac{k_{\mathrm{w}} V q(x,t)}{H(x)}$$

where $q(x,t)$ is normal contact pressure, $H(x)$ is hardness of coating (upper layer), k_{w} is constant and depends on several intensities of vibrations. Then the vertical displacement of the top face of coating due to wear has the form

$$u_{\mathrm{w}}(x,t) = -\int_{\tau_0}^{t} v_{\mathrm{w}}(x,\tau)\,d\tau = -\frac{k_{\mathrm{w}} V}{H(x)} \int_{\tau_0}^{t} q(x,\tau)\,d\tau. \tag{1}$$

Here τ_0 is time of start punching. The vertical displacement of the coating top face due to applied load $-q(x,t)$ has a form [9]

$$u_{\mathrm{q}}(x,t) = -\frac{q(x,t)h_1}{R(x)} - \frac{2(1-\nu_2^2)}{\pi E_2} \int_{-a}^{a} k_{\mathrm{pl}}\left(\frac{x-\xi}{h_2}\right) q(\xi,t)\,d\xi, \tag{2}$$

Here $R(x) = E_1(x)[1-\nu_1(x)]/[1-\nu_1(x)-2\nu_1^2(x)]$ (see, for example, [9]); $\nu_1(x)$, $E_1(x)$ and ν_2, E_2 are where the Poisson's ratios and the Young modulus of the upper and bottom layers, respectively; $k_{\mathrm{pl}}(s) = \int_0^{+\infty} L(u)\cos(su)/u\,du$ and is the plane contact problem kernel [10] ($L(u) = [2\kappa_2 \sinh(2u) - 4u]/[2\kappa_2 \cosh(2u) + \kappa_2^2 + 1 + 4u^2]$, $\kappa_2 = 3 - 4\nu_2$). There exist several cases when the hardness of the materials is proportional to their contact rigidity (see, for example, [11] and [12]), i.e.

$$H(x) = k_{\mathrm{H}} R(x). \tag{3}$$

So, total displacement is a sum of displacement $u_{\mathrm{w}}(x,t)$ and displacement $u_{\mathrm{q}}(x,t)$, i.e. (according to (1)–(3))

$$u(x,t) = -\frac{k_{\mathrm{w}} V}{H(x)} \int_{\tau_0}^{t} q(x,\tau)\,d\tau - \frac{k_{\mathrm{H}} q(x,t) h_1}{H(x)} - \frac{2(1-\nu_2^2)}{\pi E_2} \int_{-a}^{a} k_{\mathrm{pl}}\left(\frac{x-\xi}{h_2}\right) q(\xi,t)\,d\xi. \tag{4}$$

IOP Conf. Series: Materials Science and Engineering **489** (2019) 012027 doi:10.1088/1757-899X/489/1/012027

But it is easy to show that total displacement is equal to

$$u(x,t) = -\delta(t) - \alpha(t)x + g(x),$$

where $\delta(t)$ and α are the punch settlement and tilt angle, $g(x)$ is gap between coating and punch in nondeformable state ($g(x) \geq 0$, $\exists x_0 \in [-a,a] : g(x_0) = 0$). Note that if the stamp is rough, the function $g(x)$ rapidly changes. Using this relation and equation (4) we obtain main integral equation of our problem

$$\frac{k_{\rm w}V}{H(x)}\int_{\tau_0}^{t} q(x,\tau)\,d\tau + \frac{k_{\rm H}q(x,t)h_1}{H(x)} - \frac{2(1-\nu_2^2)}{\pi E_2}\int_{-a}^{a} k_{\rm pl}\left(\frac{x-\xi}{h_2}\right)q(\xi,t)\,d\xi = \delta(t)+\alpha(t)x-g(x), \quad (5)$$
$$x \in [-a,a], \quad t \geq \tau_0.$$

The resulting equation should be supplemented by equilibrium conditions of the punch:

$$\int_{-a}^{a} q(\xi,t)\,d\xi = P(t), \quad \int_{-a}^{a} \xi q(\xi,t)\,d\xi = P(t)e(t), \quad t \geq \tau_0, \quad (6)$$

where $P(t)$ is applied force and $e(t)$ is eccentricity.

We use following change of variables in equations (5) and (6)

$$\begin{gathered}
x^* = \frac{x}{a}, \quad \xi^* = \frac{\xi}{a}, \quad t^* = \frac{t}{\tau_0}, \quad \tau^* = \frac{\tau}{\tau_0}, \quad \lambda = \frac{H}{a}, \quad V^* = \frac{k_{\rm w}V\tau_0}{k_{\rm H}h_1} \\
\delta^*(t^*) = \frac{\delta(t)}{a}, \quad \alpha^*(t^*) = \alpha(t), \quad g^*(x^*) = \frac{g(x)}{a}, \quad m^*(x^*) = \frac{k_{\rm H}h_1E_2}{2a(1-\nu_2^2)H(x)}, \\
q^*(x^*,t^*) = \frac{2(1-\nu_2^2)q(x,t)}{E_2}, \quad P^*(t^*) = \frac{2(1-\nu_2^2)P(t)}{E_2 a}, \quad M^*(t^*) = \frac{2(1-\nu_2^2)P(t)e(t)}{E_2 a^2}, \\
\mathbf{F}^*f(x^*) = \int_{-1}^{1} k_{\rm pl}^*(x^*,\xi^*)f(\xi^*)\,d\xi^*, \quad k_{\rm pl}^*(x^*,\xi^*) = \frac{1}{\pi}k_{\rm pl}\left(\frac{x-\xi}{H}\right) = \frac{1}{\pi}k_{\rm pl}\left(\frac{x^*-\xi^*}{\lambda}\right).
\end{gathered} \quad (7)$$

Having omitted asterisks in the relations obtained, we arrive at the integral equation and additional condition in the form

$$\begin{gathered}
m(x)\left[q(x,t) + V\int_{1}^{t} q(x,\tau)\,d\tau\right] + \int_{-1}^{1} k_{\rm pl}(x,\xi)q(\xi,t)\xi\,d\xi = \delta(t) + \alpha(t)x - g(x), \\
\int_{-1}^{1} q(\xi,t)\,d\xi = P(t), \quad \int_{-1}^{1} \xi q(\xi,t)\,d\xi = P(t), \quad x \in [-1,1], \quad t \geq 1.
\end{gathered} \quad (8)$$

Note that functions $m(x)$ and $g(x)$ in (8) connect with hardness of upper layer and punch base form, respectively; these parameters can be described by a rapidly changing function. Hardness can be described even by discontinuous function.

So it is required to solve mixed integral equation (i.e. equations with two different integral operators, see [14]) with 2 rapidly changing functions.

Plane contact problem has four versions of mathematical statements. We construct only one solution for the case with known force $P(t)$ and moment $M(t)$, unknown settlement $\delta(t)$, tilt angle $\alpha(t)$, and contact pressures $q(x,t)$. In this case integral equation (8) will have unknown terms both on the left and on the right.

IOP Conf. Series: Materials Science and Engineering **489** (2019) 012027 doi:10.1088/1757-899X/489/1/012027

2. Special representation and basis

In main integral equation and additional condition (8), we make new change of variables

$$q(x,t) = \frac{Q(x,t)}{\sqrt{m(x)}} - (\mathbf{I}+\mathbf{V})\frac{g(x)}{m(x)}, \quad k_{\mathrm{pl}}(x,\xi) = k(x,\xi)\sqrt{m(x)}\sqrt{m(\xi)},$$
$$\mathbf{V}f(t) = -V\int_1^t e^{-V(t-\tau)} f(\tau)\,d\tau, \tag{9}$$

where $Q(x,t)$ is new unknown function, $\mathbf{I}$ is identity operator, $\mathbf{V}$ is Volterra operator with kernel $-Ve^{-V(t-\tau)}$ (note that $-Ve^{-V(t-\tau)}$ is kernel of resolvent for Volterra integral equation with kernel $-V$, see [15]). Then the (8) become

$$Q(x,t) + V\int_1^t Q(x,\tau)\,d\tau + \mathbf{F}Q(x,t) = \frac{\delta(t)}{\sqrt{m(x)}} + \frac{\alpha(t)x}{\sqrt{m(x)}} + \frac{\tilde{c}(t)\tilde{g}(x)}{\sqrt{m(x)}} \equiv f(x,t),$$
$$\int_{-1}^1 \frac{Q(\xi,t)}{\sqrt{m(\xi)}}\,d\xi = \tilde{P}(t), \quad \int_{-1}^1 \frac{\xi Q(\xi,t)}{\sqrt{m(\xi)}}\,d\xi = \tilde{M}(t), \quad x\in[-1,1], \quad t\geq 1, \tag{10}$$

where

$$\mathbf{F}f(x) = \int_{-1}^1 k(x,\xi)f(\xi)\,d\xi, \quad \tilde{g}(x) = \int_{-1}^1 \frac{k_{\mathrm{pl}}(x,\xi)g(\xi)}{m(\xi)}\,d\xi, \quad \tilde{c}(t) = (\mathbf{I}+\mathbf{V})1 = e^{-V(t-1)},$$
$$\tilde{P}(t) = P(t) + \tilde{c}(t)\int_{-1}^1 \frac{g(\xi)}{m(\xi)}\,d\xi, \quad \tilde{M}(t) = M(t) + \tilde{c}(t)\int_{-1}^1 \frac{\xi g(\xi)}{m(\xi)}\,d\xi. \tag{11}$$

Note, there is no function $m(x)$ in the left side of integral equation (10).

We will see the solution of our problem in the class of time-continuous functions in a Hilbert space $L_2[-1,1]$ (e.g., see [14]). First of all we should construct special basis. To this end we will orthonormal on $[-1,1]$ the linearly independent system of functions

$$\left\{\frac{1}{\sqrt{m(x)}}, \frac{x}{\sqrt{m(x)}}, \frac{x^2}{\sqrt{m(x)}}, \dots\right\}$$

by the formulas (we will use scalar product $(f_1(x), f_1(x)) = \int_{-1}^1 f_1(\xi)f_2(\xi)\,d\xi$)

$$J_m = \int_{-1}^1 \frac{\xi^m d\xi}{m(\xi)}, \quad d_{-1} = 1, \quad d_m = \begin{vmatrix} J_0 & \cdots & J_m \\ \vdots & \ddots & \vdots \\ J_m & \cdots & J_{2m} \end{vmatrix},$$
$$p_m^*(x) = \frac{1}{d_{m-1}d_m}\begin{vmatrix} J_0 & J_1 & \cdots & J_m \\ \vdots & \vdots & \ddots & \vdots \\ J_{m-1} & J_m & \cdots & J_{2m-1} \\ 1 & x & \cdots & x^m \end{vmatrix}, \quad p_m(x) = \frac{p_m^*(x)}{\sqrt{m(x)}}, \quad m = 0,1,2,\dots, \tag{12}$$

3. Projection method and solution of contact-wear problem

Divide the space $L_2[-1,1]$ into two parts (two subspaces): $L_2^{(1)}[-1,1]$ and $L_2^{(2)}[-1,1]$ whose bases are the functions $\{p_0(x), p_1(x)\}$ and $\{p_k(x)\}_{k=2,3,4,\dots}$, respectively.

Consequently [14], we introduce the operators of orthogonal projection $\mathbf{P}_1$ and $\mathbf{P}_2$ mapping the space $L_2[-1,1]$ onto $L_2^{(1)}[-1,1]$ and $L_2^{(2)}[-1,1]$, respectively, $\mathbf{P}_1 f(x) = \int_{-1}^1 [p_0(x)p_0(\xi) +$

$p_1(x)p_1(\xi)]f(\xi)\,d\xi$, $\mathbf{P}_2 = \mathbf{I} - \mathbf{P}_1$, where $\mathbf{I}$ is the identity operator. Then $\mathbf{P}_i f(x,t) = f_i(x,t)$ and $\mathbf{P}_i Q(x,t) = Q_i(x,t)$ $(i = 1, 2)$.

Then functions $Q(x,t)$ and $f(x,t)$ can be represented as sums of functions in the spaces $L_2^{(1)}[-1,1]$ and $L_2^{(2)}[-1,1]$; i.e.,

$$\begin{aligned}
&Q(x,t) = Q_1(x,t) + Q_2(x,t), \quad f(x,t) = f_1(x,t) + f_2(x,t),\\
&f_1(x,t) = \left[\sqrt{J_0}\delta(t) + \frac{J_1}{\sqrt{J_0}}\alpha(t) + g_0\tilde{c}(t)\right]p_0(x) + \left[\sqrt{\frac{J_0J_2 - J_1^2}{J_0}}\alpha(t) + g_1\tilde{c}(t)\right]p_1(x),\\
&f_2(x,t) = \frac{\tilde{c}(t)g_2(x)}{\sqrt{m(x)}},\\
&Q_1(x,t), f_1(x,t) \in L_2^{(1)}[-1,1], \quad Q_2(x,t), f_2(x,t) \in L_2^{(2)}[-1,1],
\end{aligned} \tag{13}$$

where, as follows from (12) and (13), the coefficient g_0 and the function $g_2(x)$ are determined by the expressions

$$g_0 = \sum_{l=0}^{\infty} K_{0l}\int_0^1 \frac{p_l(\xi)g(\xi)}{\sqrt{m(\xi)}}\,d\xi, \quad g_1 = \sum_{l=0}^{\infty} K_{1l}\int_0^1 \frac{p_l(\xi)g(\xi)}{\sqrt{m(\xi)}}\,d\xi, \quad \frac{g_2(x)}{\sqrt{m(x)}} = \frac{\tilde{g}(x)}{\sqrt{m(x)}} - g_0p_0(x). \tag{14}$$

Here K_{ml} are the coefficients of the expansion of the kernel $k(x,\xi)$ into a double series in the functions $p_k(x)$ $(k = 0, 1, 2, \ldots)$,

$$K_{ml} = \int_0^1\int_0^1 k(x,\xi)p_m(x)p_l(\xi)\,d\xi\,dx. \tag{15}$$

Note that $Q_1(x,t) \in L_2^{(1)}[-1,1]$ can be obtained from the supplementary condition (10),

$$Q_0(x,t) = z_0(t)p_0(x) + z_1(t)p_1(x), \quad z_0(t) = \frac{\tilde{P}(t)}{\sqrt{J_0}}, \quad z_1(t) = \frac{J_0\tilde{M}(t) - J_1\tilde{P}(t)}{\sqrt{J_0(J_0J_2 - J_1^2)}}, \tag{16}$$

and the right-hand side of the integral equation contains the known term $f_2(x,t) \in L_2^{(2)}[-1,1]$. In this case, the terms $Q_2(x,t) \in L_2^{(2)}[-1,1]$ and $f_1(x,t) \in L_2^{(1)}[-1,1]$ are unknown. These characteristic features permit classifying the problem thus obtained as a special case of the generalized projection problem, which was posed and solved in [14].

As in [14], we act by the operator $\mathbf{P}_2$ on the integral equation (10)) to obtain the equation

$$Q_2(x,t) + V\int_1^t Q_2(x,\tau)\,d\tau + \mathbf{P}_2\mathbf{F}Q_2(x,t) = -\mathbf{P}_2\mathbf{F}Q_1(x,t) + \frac{\tilde{c}(t)g_2(x)}{\sqrt{m(x)}} \tag{17}$$

with known right-hand side for the unknown function $Q_2(x,t)$. We seek the solution in the form

$$Q_2(x,t) = \sum_{k=1}^{\infty} z_k(t)\varphi_k(x), \tag{18}$$

where $z_k(t)$ is the desired function and the eigenfunctions $\varphi_k(x)$ of the operator $\mathbf{P}_2\mathbf{F}$ are determined by solving the spectral problem for the operator $\mathbf{P}_2\mathbf{F}$,

$$\mathbf{P}_2\mathbf{F}\varphi_k(x) = \gamma_k\varphi_k(x). \tag{19}$$

The eigenfunction system $\{\varphi_k(x)\}_{k=2,3,4,\dots}$ is related to the initial system $\{p_k(x)\}_{k=2,3,4,\dots}$ of basic functions in $L_2^{(2)}[-1,1]$ as

$$\varphi_k(x) = \sum_{l=2}^{\infty} \psi_{kl} p_l(x). \tag{20}$$

Therefore, solving the spectral problem is reduced to solving the system of algebraic equations

$$\sum_{l=2}^{\infty} K_{ml}\psi_{kl} = \gamma_k \psi_{km}, \quad k, m = 2, 3, \dots, \tag{21}$$

where K_{ml} are the coefficients of the expansion of the kernel $k(x, \xi)$ in a double series in the functions $p_k(x)$ $(k = 0, 1, 2, \dots)$ determined by formula (15).

We substitute the representation (18) into the integral equation (17) and use relations (11), (12), (19), and (20) to obtain the expression for the function $z_k(t)$,

$$z_k(t) = (\mathbf{I} + \mathbf{W}_k)\frac{g_k \tilde{c}(t) - z_0(t)K_k^{(0)} - z_1(t)K_k^{(1)}}{1 + \gamma_k},$$

where coefficients g_k and $K_k^{(0)}$ are calculated by the formulas

$$g_k = \sum_{m=2}^{\infty} \psi_{km} \sum_{l=0}^{\infty} K_{ml} \int_{-1}^{1} \frac{p_l(\xi)g(\xi)}{m(\xi)}\, d\xi, \quad K_k^{(0)} = \sum_{m=2}^{\infty} K_{0m}\psi_{km}, \quad K_k^{(1)} = \sum_{m=2}^{\infty} K_{1m}\psi_{km}, \tag{22}$$

and $\mathbf{W}_k$ is the Volterra operator whose kernel $R_k(t, \tau)$ is the resolvent of the kernel $-V/(1+\gamma_k)$:

$$R_k(t, \tau) = -\frac{V}{1 + \gamma_k} \exp\left[-\frac{V(t-\tau)}{1 + \gamma_k}\right]. \tag{23}$$

Then the function $z_k(t)$ can be represented as

$$\begin{aligned} z_k(t) = {} & \frac{g_k}{\gamma_k}\left\{e^{-V(t-1)} - \frac{1}{1+\gamma_k}\exp\left[-\frac{V(t-1)}{1+\gamma_k}\right]\right\} \\ & - \frac{K_k^{(0)}}{1+\gamma_k}\left\{z_0(t) - \frac{V}{1+\gamma_k}\int_1^t \exp\left[-\frac{V(t-\tau)}{1+\gamma_k}\right] z_0(\tau)\, d\tau\right\} \\ & - \frac{K_k^{(1)}}{1+\gamma_k}\left\{z_1(t) - \frac{V}{1+\gamma_k}\int_1^t \exp\left[-\frac{V(t-\tau)}{1+\gamma_k}\right] z_1(\tau)\, d\tau\right\}. \end{aligned} \tag{24}$$

As a result, by using (9), (13), (16), (18), and (20), we finally obtain the expression for the contact stresses

$$q(x, t) = \frac{1}{m(x)}\left[z_0(t)p_0^*(x) + z_1(t)p_1^*(x) + \sum_{k=1}^{\infty} z_k(t) \sum_{m=2}^{\infty} \psi_{km} p_m^*(x) - e^{-V(t-1)}g(x)\right],$$

where the functions $z_0(t)$, $z_1(t)$, and $z_k(t)$ are determined by relations (16) and (24), the coefficients ψ_{kl} can be obtained from (21), and $p_0^*(x)$, $p_1^*(x)$, $p_l^*(x)$ can be obtained from (12).

By applying the operator $\mathbf{P}_1$ to Eq. (10), we obtain the formulas for settlement and tilt angle

$$\delta(t) = \frac{1}{\sqrt{J_0}}\left\{z_0(t) + V\int_0^t z_0(\tau)\, d\tau - g_0 e^{-V(t-1)} + K_{00}z_0(t) + K_{01}z_1(t) + \sum_{k=2}^{\infty} K_k^{(0)} z_k(t)\right\} - \alpha(t)\frac{J_1}{J_0},$$

$$\alpha(t) = \sqrt{\frac{J_0}{J_0 J_2 - J_1^2}}\left\{z_1(t) + V\int_0^t z_1(\tau)\, d\tau - g_1 e^{-V(t-1)} + K_{10}z_0(t) + K_{11}z_1(t) + \sum_{k=2}^{\infty} K_k^{(1)} z_k(t)\right\},$$

IOP Conf. Series: Materials Science and Engineering **489** (2019) 012027 doi:10.1088/1757-899X/489/1/012027

Here the functions $z_0(t)$, $z_1(t)$, and $z_k(t)$ are determined from (16) and (24), the coefficients g_0, g_1, K_{00}, K_{01}, K_{10}, K_{11}, $K_k^{(0)}$, and $K_k^{(1)}$ are determined from (14), (15), and (22), and the integrals J_0, J_1, J_2 is calculated by formulas (12) ($k = 2, 3, 4, \ldots$).

Conclusions

The contact-wear plane problem for regular system of rigid punches with complex base forms and vicoelastic aging layer with nonuniform elastic coating are formulated and solved. The solution for one problem version is constructed in analytical form. Solutions for other versions can be constructed in a similar way. Expressions for contact pressures have a structure in which the functions are associated with the coating contact rigidity and the punch base forms are distinguished by separate terms and factors. This analytical solution allows one to carefully analyze the influence of contact characteristics on the stress-strain state of the foundation and positions of bodies. Moreover, such solution representation allows one to perform efficient calculations of linear wear of foundations in cases when the coating elastic properties and punch base form described by complex rapidly changing functions.

Acknowledgments

The paper is financially supported by the Ministry of Science and Higher Education (State Registration Number AAAA-A17-117021310381-8) and, in part, by the Russian Foundation for Basic Research (under grant No. 18-51-05012-Arm_a).

References

[1] Manzhirov A V and Kazakov K E 2017 Contact problem with wear for a foundation with a surface nonuniform coating *Doklady Physics* **62** (7) 344–9
[2] Manzhirov A V and Kazakov K E 2018 Axisymmetric problem of fretting wear for a foundation with a nonuniform coating and rough punch *AIP Conference Proceedings* **1959** 070023
[3] Alexandrov V M and Mkitaryan S M 1979 *Contact Problems for Bodies with Thin Coatings and Interlayers* (Moscow: Nauka) [in Russian]
[4] Pronikov A S 1957 *Wear and Durability of Machines* (Moscow: Mashgiz) [in Russian]
[5] Khrushchev M M and Babichev M A 1970 *Abrasive Wear* (Moscow: Nauka) [in Russian]
[6] Collins J 1993 *Failure of Materials in Mechanical Design: Analysis, Prediction, Prevention* (New York: Wiley)
[7] Goryacheva I G and Dobychin M N 1988 *Contact Problems in Tribology* (Moscow: Mashinostroenie) [in Russian]
[8] Soldatenkov I A 2010 *Wear Contact Problem with Applications to Engineering Calculation of Wear* (Moscow: Fizmatkniga) [in Russian]
[9] Alexandrov V M and Kovalenko E V 1986 *Problems in Mechanics of Continuous Media with Mixed Boundary Conditions* (Moscow: Nauka) [in Russian]
[10] Vorovich I I, Alexandrov V M, and Babeshko V A 1974 *Nonclassical Mixed Problems in the Theory of Elasticity* (Moscow: Nauka) [in Russian]
[11] Archard J F 1953 Contact and rubbing of flat surfaces *Journal of Applied Physics* **24** (8) 981–8
[12] Schalliamach A 1954 On the abrasion of rubber *Proceedings of the Physical Society. Section B* **67** (12) 883–91
[13] Manzhirov A V and Kazakov K E 2018 Modeling the contact interaction between a nonuniform foundation and a rough punch *Mathematical Models and Computer Simulations* **10** (3) 314–321
[14] Manzhirov A V 2016 A mixed integral equation of mechanics and a generalized projection method of its solution *Doklady Physics* **61** (10) 489–93
[15] Polyanin A D and Manzhirov A V 2008 *Handbook of Integral Equations, 2nd edition* (Boca Raton: Chapman & Hall/ CRC)

Effect of fillers on the coefficient of friction of a polyurethane sealing element for a coiled tubing sealing device

A Matsko and V Kopchenkov

North Caucasus Federal University, Stavropol, RF

anton_matsko@mail.ru

Abstract. The results of studies of wear resistance and friction coefficient of polyurethane grade PT-74 filled with solid lubricant powders are presented. The abrasion patterns of the obtained polyurethanes are presented. The type and quantity of the filler providing smaller wear of a sealing element are defined. The regression analysis revealed the physical and mechanical properties of the material that have the greatest effect on wear.

1. Introduction

Every year, the amount of the well workover, including usage of coiled tubing (CT) is growing. Due to environmental friendliness and cost-effectiveness coiled tubing is widely used, as well as the ability to work during a depression on a productive formation which excludes its pollution. The last one involves sealing at the space between the tubing and CT at the wellhead. It provides tightness with a sealing element (SE), made of an elastic material such as polyurethane – due to a higher (in reference to oil-gasoline-resistant rubber) endurance. Moreover, the wellhead sealer is an important component of a coiled tubing installation which guarantee safe and high-quality performance for work in the well.

In the process of trips, the SE exposed to intensive wear, which is associated with considerable friction forces at the contact of the SE-CT. For example, the lifetime of the sealing element in the wells of the Valanginian stage of the Yamburg field lasts for 1-2 trips.

Increased attrition (coefficient of attrition) not only accelerates the wear of the SE, but also reduces the range of loads applied on the CT during reciprocation, complicates the control of the load on the face and causes different sorts of problems [1]. The number of problems has been identified due to statistical researches of the operating experience of CI at the Yamburg field. The analysis and classification of complications created by the SE of the sealer gave us an opportunity to identify their main types and reasons.

1. The CT sticking in blowout preventer (BOP) due to hydrate formation. The reason is the wear of the SE and the formation of a gap that initiates the hydration mode of the well, rising of local thermobaric conditions which is a good opportunity for the hydrate formation. Existence in the stagnant zone of various inclusions which are good centers of crystallization.

2. The sticking of CT in the stripper due to the adhesion of the SE to the CT. The reason is increasing of trips speed (above 15 m/min), heating of the surface layer of the SE in the contact zone and, as a reason, temperature destruction of polyurethane. After a sharp stop, the destructured layer sticks to the CT.

3. Depressurization of the wellhead due to insufficient lubrication, high wear rate associated with high coefficients of friction in the SE-CT pair.

4. Hanging CT. The reason is insufficient lubrication, high wear rate associated with the high coefficient of friction in a pair of SE – CT.

In the down process to reduce the friction force the lubrication applied on the surface of CT and in the process of lifting - lubrication is carried out by the well environment [2]. Despite the presence of a lubricant, it cannot always provide sufficient efficiency under the existing conditions of operation of the seals. Actually, the lubricant is supplied to the pipe in a dusty atmosphere; the friction unit operates at high loads and lower temperatures, which makes some difficulties in lubrication of the friction unit. Pipe lifting during gas well development is accompanied by well development, therefore the lubricant is practically missing out.

The aim of the work is study the influence fillers on the friction coefficient of a polyurethane sealing element for a coiled tubing sealing device.

2. Materials and methods

A new method is proposed to reduce the wear of the seal of the sealer, due to the low efficiency of the measures implemented. This is a reduction in the friction coefficient for the SE material due to the introduction of a solid lubricant into this material. For the research, mixtures of fillers, such as graphite, molybdenum disulfide and PTFE (Polytetrafluoroethylene) or fluoroplastic, were prepared. The choice is made on the basis of the analysis of the efficiency of the used solid lubricant fillers, as well as on the basis of the price and availability of raw materials on the market. As a matrix used polyurethane PT-74 which is a domestic analogue of polyurethane brand *Adipren L100. Adipren L 100* is a material used for seals in the oil and gas industry. It is widely used for the manufacture of SE sealants coiled tubing units.

The introduction of solid lubricants usually weakens the base material, as they have low adhesion and do not provide a strong chemical bond with the matrix polymer. Consequently, the weakening of the material may reduce the durability. Therefore, an important problem is the selection of the type of filler and its percentage.

The following physical and mechanical characteristics were determined for the manufactured samples: conditional strength, critical deformation, conditional modulus of elasticity, Young's modulus, residual deformation, and hardness. The tests were carried out in the laboratory at a temperature of 25°C using a set of test equipment, including the *Instron 3365* test facility. Conditions for sample preparation and testing were in accordance with the corresponding GOST (State standards).

The experiment planning matrix was prepared for the experiments. Abrasion tests were carried out in the laboratory of the North Caucasus Federal University.

Typical laboratory installation MI-2 took for testing polymer samples for wear. It allows determining the force and coefficient of friction in sliding friction and the amount of the wear test samples.

Figure 1. Design of the test device

To reproduce the operating conditions of the sealing elements, the typical laboratory installation MI-2, figure 1, for testing polymer samples for wear was taken as the base, which allows determining the force and coefficient of friction during sliding friction and the amount of abrasion of the samples under study.

The principle of operation of the machine MI-2 for testing elastomers for abrasion is to determine the wear resistance of the material when samples slid on a rotating disk at a speed of 0.32 m/s in a controlled downforce. The installation design includes an electric motor (1), a presser disc with two samples (2), a control plate in the form of a wire disc (3), a lever with sensors of friction force (4), nozzles for blowing the contact zone (5). Temperature sensors and load device are not shown.

In order to better approximate the operating conditions, the installation was improved. The metal mesh wire was used as an abrasive surface to account for the nature of the surface of the CT [3]. To determine the coefficient of friction and temperature in the contact zone in real time with the display on the computer in the form of graphs, special sensors were installed and the software was developed.

An optical microscope Levenkhuk D70L was used to study the surface of worn samples and determine the type and mechanism of wear, and the VL-224 analytical balance with an accuracy of 0.0001g was used to measure material wear.

3. Results and discussion

The results of studies of the friction coefficient of samples on a metal mesh wire are shown in the graph, figure 2. From figure 2 it follows that filling polyurethane with components of all types of solid lubricant reduces the friction coefficient compared to the unfilled matrix, and the samples filled with fluoroplastic have the lowest values.

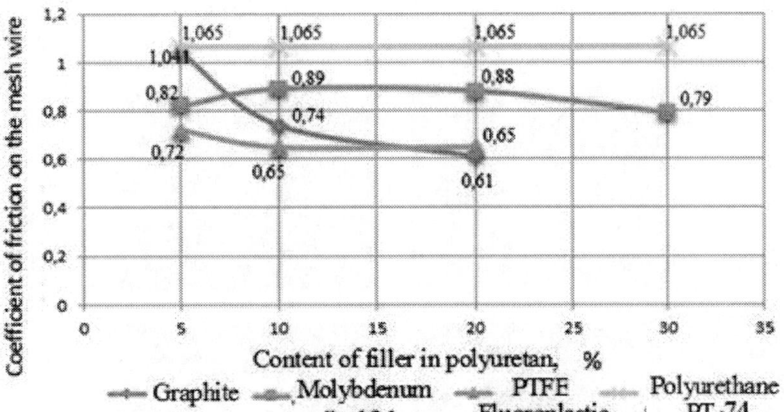

Figure 2. The dependence of the friction coefficient on the filler content

The results of the study of abrasion of the samples on the metal wire mesh are shown in figure 3. As can be seen from figure 3, an increase in the percentage of solid lubricants in polyurethane leads to an increase in the abrasion of the samples

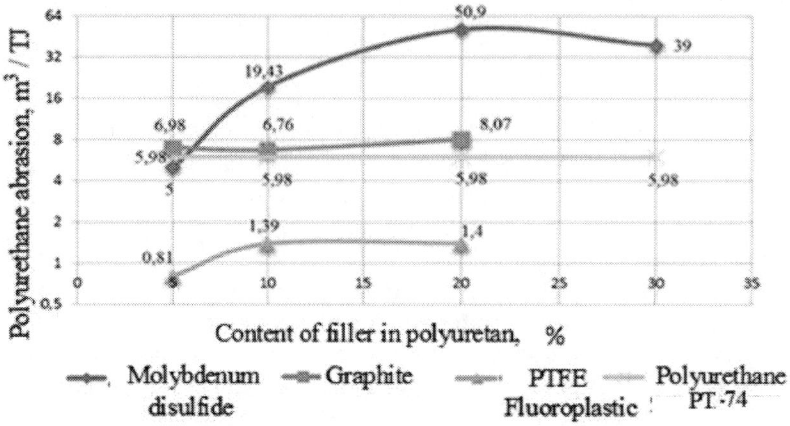

Figure 3. The dependence of the polyurethane wear on the filler content

This tendency is preserved for filling with fluoroplastic, however, the abrasion rate is less than that of unfilled polyurethane PT-74. An increase in abrasion is associated with a decrease in the strength properties of samples filled with solid lubricants.

The study of the surface of the samples allows us to associate the nature of wear with the wear rate. Figure 4 shows the wear surfaces obtained after the abrasion of the samples at the test installation MI-2. Pure polyurethane PT-74 and polyurethane filled with graphite and molybdenum disulfide have a frictional wear mechanism. Proof of this is the presence of transverse ridges (waves) [4]. At the same time, the less intense fatigue wear mechanism is implemented when filling with (PTFE) fluoroplastic, figure 4(d). Proof of this is the presence of a relatively smooth surface

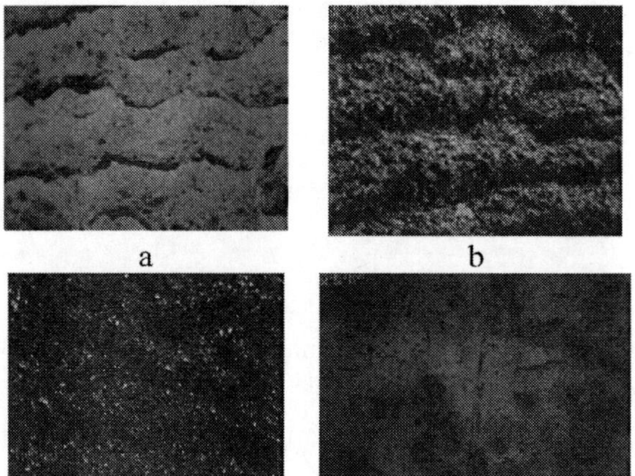

Figure 4. Abrasion pictures on the grid of filled samples of polyurethane PT-74 (x100):
(a) without filling, (b) filling graphite, (c) filling molybdenum disulfide, (d) filling fluoroplastic.

Thus, when polyurethane is filled with solid lubricants, the friction coefficient decreases, but at the same time, the strength properties are reduced. The test results shown in Table establish the connection between the polyurethane filling and the change in the friction coefficient f on the mesh wire and the specific power wear Y measured in $m^3 \cdot TJ^{-1}$ (where TJ is terajoule) with the physical and mechanical properties of the composites such as conditional strength T (MPa), critical deformation at break K_d (%), conditional modulus of elasticity E (MPa), Young's shear modulus G (MPa), residual deformation under compression ε (%).

Table 1. Physical and mechanical properties and wear of the samples

	Filling Material	%	Physical and mechanical properties GOST 270-75					f	Y, m³/TJ
			T, MPa	K_d, %	E, MPa	G, MPa	ε, %		
1	Graphite	5	21.96	821	4.9	16.7	18	1.04	6.98
2	Graphite	10	13.3	492	5.7	20.4	10	0.74	6.76
3	Graphite	20	11.7	732	6.6	26.6	10	0.61	8.07
4	MoS_2	5	22.14	1116	4.2	12.3	20	0.82	5
5	MoS_2	10	20.2	1163	4.5	18.1	20	0.89	19.43
6	MoS_2	20	12.1	1039	4.3	14.1	50	0.88	50.9
7	Fluoroplastic	5	16.71	622	5.2	22.7	14	0.72	0.81
8	Fluoroplastic	10	15.1	588	5.6	26.6	20	0.65	1.39
9	Fluoroplastic	20	12	552	6.1	29.7	16	0.65	1.4

To identify the degree of influence of individual physic-mechanical properties on abrasion, a regression analysis was performed. The analysis is made on the basis of data about the relationship of physical and mechanical properties of filled polyurethanes with the value of wear, table.

The influence factors are the physic-mechanical properties of polyurethane, which are regulated by the introduction of fillers. The response is the specific power wear of polyurethane, which is determined using the abrasion machine MI-2.

As a result of the regression analysis, using the *Statistica* program, a regression equation (1) for specific power wear is obtained:

$$Y = -27{,}29 + 0{,}0316 \cdot K_d + 89{,}52 \cdot f - 3{,}5 \cdot T \qquad (1)$$

As follows from the equation, the coefficient of friction f and the conventional strength of polyurethane have a determining and competing effect on the abrasion of composites. Reducing the strength properties reduces and even eliminates all the advantages obtained from reducing the coefficient of friction.

4. Conclusions

As a result of research, experimental dependences of the influence of such solid lubricants as graphite, molybdenum disulfide and fluoroplastic on the physic-mechanical properties of polyurethane-based composites, the coefficient of friction and wear were obtained.

(1) It has been established that the fluoroplastic filling has the greatest effect on the reduction of the friction coefficient. Moreover, the increase in the content of the filler over 10-12% does not influence the magnitude of the friction coefficient and wear.

(2) It is proved that the main effect on reducing wear has a decrease in the friction coefficient f and the conditional strength of polyurethane.

(3) Shows the limits of filling, allowing increasing the service life of seals for sealers. When exceeding 10% of the filling, the strength decreases and, as a result, the wear increases.

References

[1] Livescu S, Craig S H and Watkins T 2014 *SPE Annual Technical Conf. and Exh.* (Amsterdam) SPE-170635-MS (http://dx.doi.org/10.2118/170635-MS)

[2] Livescu S and Craig S H 2015 Increasing Lubricity of Downhole Fluids for Coiled-Tubing Operations. *SPE J.*20 (2) SPE-168298-PA (http://dx.doi.org/10.2118/168298-PA)

[3] Kopchenkov V G and Matsko AV 2015 *Sci. Innovations. Technologies* **4** 91.

[4] Kopchenkov V G 2004 *Problems of mechanical Engineering and reliability of machines* **6** 51.

IOP Conf. Series: Materials Science and Engineering **489** (2019) 012029 doi:10.1088/1757-899X/489/1/012029

Empirical correlation of heat generation in ball bearings depending on the operational conditions in the supports of aero-engine rotor

N I Petrov and Yu L Lavrentyev

CIAM, Moscow, Russia

yllavrentev@ciam.ru

Abstract. Modern gas turbine bearings operate at high rotational speeds, with a speed index of up to $3 \cdot 10^6$ mm·rpm. The methods currently used to calculate heat generation in aero-engine split-inner-ring ball bearings do not always give correct results, because they were developed based on tests at relatively slow speeds. To solve this problem, a number of split inner-ring ball bearings with three-point contact bearings have been tested at various levels of load, rotational speed and oil flow rate. The following bearings are tested: 126114 (d = 70 mm), 126126 (d = 130 mm) and 126130 (d = 150 mm). Tests results allowed authors to develop an empirical correlation for heat generation in split inner-ring ball bearings. Comparison with experimental data, presented in the domestic and foreign literature, shows that the coefficients in the empirical formula are selected correctly and accurately describe the thermal state of the bearing.

1. Introduction.

Modern gas turbine engines' bearings operate at high rotational speeds, reaching the speed parameter $d_m \cdot n = 2.7...3 \cdot 10^6$ mm·rpm. To support the heavy thrust loads in aero-engines, split inner-ring ball bearings are used. The inner ring of these bearings consists of two axial halves. High speeds and load leads to a significant heat generation in ball bearings. As such, correct estimation of heat generation and determination of required amount of oil for heat removal from bearings are vital problems solved at the development stage of a gas turbine engine.

Heat generation in a bearing is related to the friction that occurs between bearing elements. For an oil-lubricated bearing, additional heat is generated due to the balls passing through the lubricant in the bearing's free space. Calculating heat generation in a bearing by accurately accounting all the factors of a complex process of friction is very difficult. In this case, the calculations should take into account the actual distribution of loads among the rolling elements, estimation of elastohydrodynamic lubrication impact and the actual distribution of speeds over the contact area of each rolling element [1, 2].

In this regard, for estimating heat generation in practice, empirical correlations are used, obtained by generalizing a significant amount of experimental data. In article [3], the authors present a review of domestic (such as CIAM [4, 5] and KAI [6]) and foreign (such as FAG [7], SKF [8], MTU [9]) empirical methods for estimating heat generation in split inner-ring ball bearings. A comparison of the calculated values of heat generation, obtained by various methods, with experimental data was carried out, which showed that for aircraft bearings operated at high rotational speeds, the CIAM [5] and MTU [9] methods show the best convergence with the experiment.

The purpose of this study was to develop an empirical correlation for heat generation in split inner-ring ball bearings for high rotational speeds.

2. Bearings testing

In recent years, CIAM has been conducted a number of researches on heat generation in split inner-ring ball bearings at various levels of load, rotational speeds and oil flow rates. The objects tested in the present study are split inner-ring ball bearings with three-point contacts, used in aircraft engines. The rings and rolling elements of the bearings are made of bearing heat-resistant steel EI347Sh (similar to bearing steel M50). The cage is made from bronze with an anti-friction coating. The following bearings are tested:

- 126114 (70x110x18 mm);
- 126126 (130x200x30 mm);
- 126130 (150x225x35 mm).

The bearings test was carried out on a bearing test rig T14-15/2 (figure 1), intended to test heavily loaded rolling bearings installed in the supports of the rotors of gas turbine engines.

Figure 1. Bearing test rig T14–15/2.

Monitoring the test modes, evaluating the technical condition of the examined bearings and the state of the equipment of the test bench was carried out by registering the following parameters:

- rotational speed (n);
- applied axial (F_a) and radial (F_r) loads;
- temperatures of outer (T1-T3) and inner (T4-T6) ring of bearing;
- oil temperature, inlet (T_{oil_in}) and outlet (T_{oil_out});
- oil flow rate (V);
- vibrations.

For lubrication and cooling, synthetic IPM-10 oil was used, with a viscosity of 3.47 cSt at 100°C. Filtration of the circulating oil was carried out with a 10-micron filter.

The tests were carried out at different oil flow rates through the bearing (from 2 to 10 $l \cdot min^{-1}$). The amount of oil was determined by the number and diameter of the nozzles, as well as by the oil pressure in the oil system. The installation of the nozzles is shown in Figure 2. To ensure a flow rate in range from 2 to 5 $l \cdot min^{-1}$, one nozzle (N1) was installed; for a flow rate in range from 5 to 7 $l \cdot min^{-1}$, two nozzles (N1 and N2) were installed; and for a flow rate up to 10 $l \cdot min^{-1}$, three nozzles (N1, N2 and N3) were installed.

To determine the temperature state of the bearing, thermocouples were installed on the inner and outer rings of the bearing. Thermocouples were also installed to measure the oil temperature at the inlet and outlet of the bearing. The layout of the thermocouples is shown in figure 2. A photograph of the bearing with the installed thermocouples is shown in figure 3.

IOP Conf. Series: Materials Science and Engineering **489** (2019) 012029 doi:10.1088/1757-899X/489/1/012029

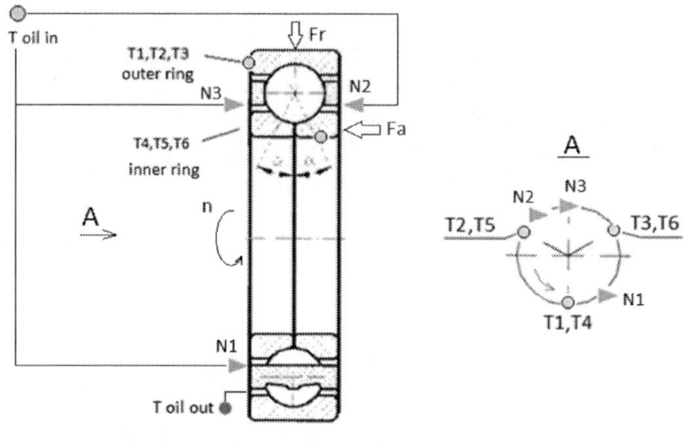

Figure 2. Layout of the thermocouple (T1–T6) and oil supply nozzles (N1–N3).

Figure 3. Installing the outer ring of bearing 126130 (d = 150 mm) in the housing with the thermocouples on the outer ring.

3. Empirical correlation of heat generation in the ball bearing.

To plan an experimental study of heat generation in split inner-ring ball bearings it is necessary to choose the sufficient number of experiments to obtain an empirical relationship, by which it is possible to determine the heat generation in a ball bearing under different operational conditions. An analysis of theoretical and experimental researches of ball bearings [5, 6, 9], carried out earlier, showed that heat generation from friction in a ball bearing (Q) depends on the bore diameter of the bearing (d), rotational speed (n), axial load (F_a), oil flow rate through the bearing (V), and viscosity or oil temperature (T).

The goal of the study was to obtain an empirical relationship $Q = f(d, n, F_a, V, T)$, which connects the heat generation in a ball bearing with the magnitude of these factors, kW:

$$Q = 10^{x0} \cdot d^{x1} \cdot n^{x2} \cdot F_a^{x3} \cdot V^{x4} \cdot T^{x5}, \qquad (1)$$

where $x0$, $x1$, $x2$, $x3$, $x4$ and $x5$ are empirically determined coefficients.

After taking the logarithm of expression (1), the heat generation value can be represented as a linear dependence on the parameters *d, n*, F_a, *V* and *T*:

$$\log_{10}(Q) = \log_{10}(10^{x0} \cdot d^{x1} \cdot n^{x2} \cdot F_a^{x3} \cdot V^{x4} \cdot T^{x5}), \tag{2}$$

$$\log_{10}(Q) = x0 + x1 \cdot \log_{10} d + x2 \cdot \log_{10} n + x3 \cdot \log_{10} F_a + x4 \cdot \log_{10} V + x5 \cdot \log_{10} T, \tag{3}$$

This method of obtaining empirical correlation is described in [9]. The authors of the article use in the formula for heat generation determination the oil inlet temperature instead of oil viscosity. Such an approach is convenient for engineering calculations because it connects the level of heat generation with the known operating conditions of the bearing. In addition, the difference between [4], [5] and [9] is that they investigate different ranges of variable parameters. Heat generation studies have been carried out with changes in the following independent factors:

- bore diameter (*d*), from 70 to 150 mm;
- rotational speed (*n*), from 1500 to 25000 rpm;
- axial load (F_a), from 2.9 to 34.3 kN;
- oil flow rate (*V*), from 2 to 10 $l \cdot min^{-1}$;
- oil inlet temperature (*T*), from 30 to 115°C.

To develop the empirical dependence of heat generation, the range of the bearing's operating modes included 156 experimental points. For each of the indicated points, the experimental value of heat generation was calculated. Heat generation in a ball bearing was determined by the increase in the heat of the oil from the time it entered to the time it was removed from the bearing, without taking into account heat transfer through the parts associated with the bearing, as follows:

$$Q_{exp} = Q_{oil} = \frac{C_p \cdot \rho \cdot V}{60} \cdot (T_{oil\_out} - T_{oil\_in}), \tag{4}$$

where C_p is the heat capacity of the oil in $J \cdot (kg \cdot °C)^{-1}$; ρ is the density of the oil in $g \cdot sm^{-3}$; V is the oil flow rate in $l \cdot min^{-1}$; and T_{oil_out} and T_{oil_in} are the oil temperatures at the outlet and inlet of the bearing, respectively, in °C.

4. Results and discussion.

The experimental values of heat generation for the bearings with bore diameters of d = 70 mm, d = 130 mm and d = 150 mm are shown in figure 4.

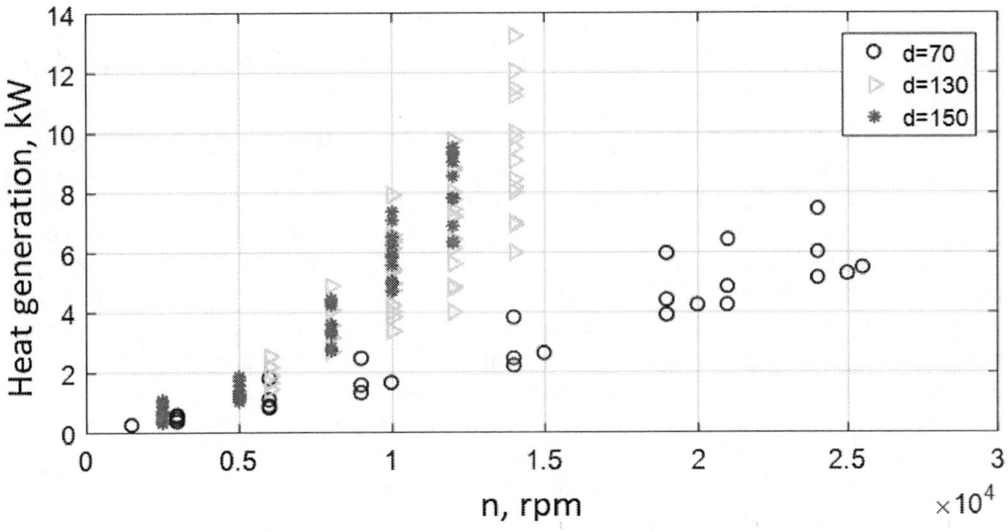

Figure 4. Heat generation in bearings and the experiment results.

IOP Conf. Series: Materials Science and Engineering **489** (2019) 012029 doi:10.1088/1757-899X/489/1/012029

The results of the experiment were processed by formula (3) using the least squares method, which allows the regression coefficients determination. These coefficients ensure the minimum of the sum of squares of the experimental data deviations from the values calculated by the usual linear regression equation (that is, the minimum of the following expression):

$$S = \left(\frac{1}{n} \sum_{i=1}^{n} (Q_{calc\_i} - Q_{exp\_i})^2 \right)^{1/2}, \qquad (5)$$

where Q_{calc_i} and Q_{exp_i} are calculated and experimental values of heat generation, and n is the number of experimental points . According to the regression analysis results, the coefficients $x0$, $x1$, $x2$, $x3$, $x4$ and $x5$ are determined, representing the degrees of the independent parameters d, n, F_a, V and T in formula (1). For each point, the heat generation is calculated from the obtained empirical dependence (formula [1]).

Figure 5 presents a comparison of the obtained calculated values with experimental data. It can be seen from the figure that the calculated values correctly describe the thermal state of the bearings with a diameter of 70 to 150 mm within the studied modes. Some variation may be associated with different ambient temperatures during research. Depending on the air temperature, the intensity of the heat transfer from the body of the test unit to the environment changes.

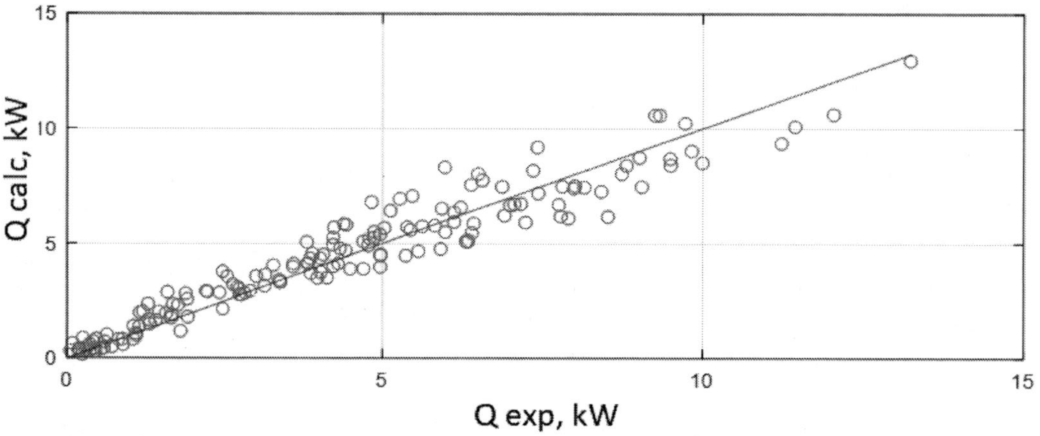

Figure 5. Comparison of calculated values of heat generation with experimental data.

It is worth noting that in order to ensure the performance of the bearing, the temperatures of the bearing parts and oil have primary importance. Heat-resistant bearing steels are designed for operation up to temperatures of 425…500°C. The limiting factor is the maximum allowable temperature of the oil. At high temperatures, the decomposition of oil additives occurs, and the oil begins to coke. In this regard, an important task at the design stage of the oil cooling system is to determine the required oil flow rate to ensure the oil temperature at the outlet of the bearing is below the acceptable level.

According to the obtained empirical dependence (1), the oil temperature at the bearing outlet was calculated (according to formula [4]). Figure 6 compares the calculated and actual oil temperatures at the bearing outlet. The results of the comparison show that, for 94% of the experimental points, the estimated oil outlet temperature differs from the actual temperature by no more than 10°C.

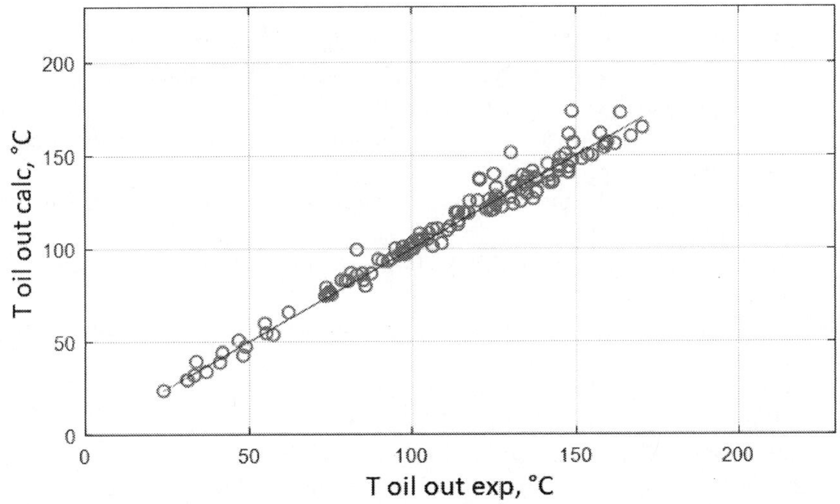

Figure 6. Comparison of the calculated and actual temperatures of the outlet oil.

To verify the correctness of the obtained coefficients from the empirical formula, the calculated values were compared with the experimental data presented in the literature by researchers from KAI (Russia) [6], MTU (Germany) [10] and the National Aeronautics and Space Administration (NASA; USA) [11]. Figure 7 shows graphs comparing the experimental and calculated heat generation values. It can be seen from the figure that the obtained empirical correlation correctly describes the bearing thermal state for these tests. Some deviations may be due to the difference between the viscosity of the oil used in the experiments and the viscosity of the IPM-10 oil.

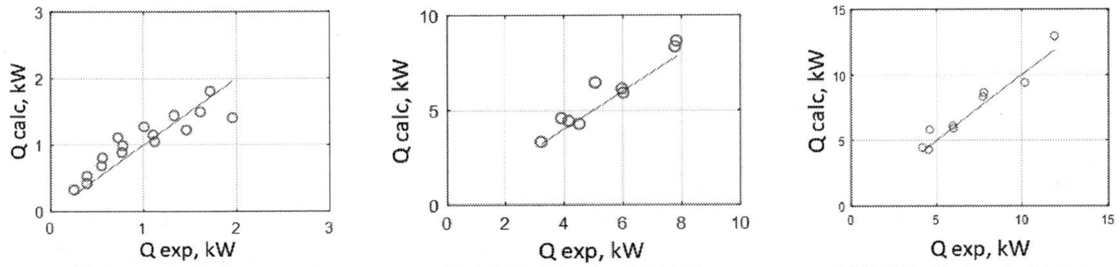

a) KAI, Demidovich V M [6] b) MTU, Flouros M [10] c) NASA, Parker R J [11]

Figure 7. Comparison the experimental data presented in the literature, with the calculated data obtained from the developed empirical correlation.

5. Conclusion.

The authors have developed an empirical relationship for determining heat generation in split inner-ring bearings at a wide range of rotational speeds, loads, oil flow rates and bore diameters, which correctly describes the experimental results in the specified region of the defining parameters:

- bore diameter (d), from 70 to 150 mm;
- rotational speed (n), from 1500 to 25000 rpm;
- axial load (F_a), from 2.9 to 34.3 kN;
- oil flow rate (V), from 2 to 10 $l \cdot min^{-1}$;
- oil inlet temperature (T), from 30 to 115°C.

A comparison with the experimental data presented in the literature shows that the coefficients in the empirical formula are selected correctly and accurately describe the thermal state of the bearing.

IOP Conf. Series: Materials Science and Engineering **489** (2019) 012029 doi:10.1088/1757-899X/489/1/012029

References

[1] Harris T A. and Kotzalas M N 2014 Advanced Concepts of Bearing Technology
[2] Gupta P K 2014 ADORE Advanced Dynamics Of Rolling Elements
[3] Petrov N I and Lavrentyev Yu L 2018 Comparison of Heat Generation Estimation Methods for Angular Contact Ball Bearings *Vestnik of Samara University. Aerospace and Mechanical Engineering* **17** pp 154-63
[4] Birger I A, Shorr B F and Iosilevich G B 1979 Strength Analysis of Machine Elements (Moscow: Mashinostroenie) pp 152-56
[5] Aksyonov N K, Petrov N I and Strukov A A 2008 *XIII Int. Gas Turbine Engines Congress* pp 69-72
[6] Demidovich V M 1978 Study of the Thermal Condition of Gas Turbine Bearings (Moscow: Mashinostroenie)
[7] FAG 2012 Rolling Bearing Lubrication WL 81 115/4 EA
[8] SKF 2016 Rolling bearings PUB BU/P1 10000/3 EN
[9] Flouros M 2006 *Aerospace Science and Technology* **10** pp 611-17
[10] Flouros M 2004 Oil Pumping in High Speed and High Loaded Ball Bearings. *Proc. of ASME Turbo Expo,* GT2004-53406
[11] Parker R J 1984 Comparison of Predicted and Experimental Thermal Performance of Angular Contact Ball Bearing *NASA Technical Paper* 2275

IOP Conf. Series: Materials Science and Engineering **489** (2019) 012030 doi:10.1088/1757-899X/489/1/012030

Modelling of quasistatic rolling of pneumatic tyre in Abaqus program complex

M A Shilov [1,2], P V Korolev [1,2], L B Maslov[1,2] and E S Shirokova[2]

[1] Department of theoretical and applied mechanics, Faculty of Electromechanics, Ivanovo State Power Engineering University, 153003, Ivanovo, Russia
[2] Institute of Chemistry and Ecology, Vyatka State University, 610000, Kirov, Russia

Mshilov@yandex.ru

Abstract. In the last ten years enterprises producing tyres for cars have had a tendency to the full automation that includes not only development of an independent line on tyre production, but also digital design based on the creation of a whole set of pneumatic tyres computer models with a series of calculations of their main characteristics at the car movement. Special attention is paid to the influence of the protector geometry, as well as the structure and properties of the materials of the tyre basic elements on its stress-strain state in the course of stationary rolling. To define these characteristics within the conditions of quasistatic rolling, the Abaqus program complex is used. The finite element modelling of a car tyre is carried out in the three-dimensional statement. The analysis of inflation and pressing is made both in the static state and during dynamic rolling. The results of modelling are epures of contact pressure distribution on the tyre surface.

1. Introduction

Tyres play an important role in running of vehicles. The controllability of cars and the safety of road users largely depend on the tyres quality and operating parameters. The tyre worn-out state has a considerable impact on the reliability of the tyre-road adhesion.

The tyre wear is a complex process that depends on various factors, such as material, design, bad alignment of wheels etc. “Bald” tyres increase fuel consumption and noises of the car at its high speed, they cannot provide a smooth ride, etc. All of this raises the emergency risk. One of the possible solutions of the problem is the development of new composite materials that have all the characteristics necessary for trouble-free operation. However, due to the growing number of new composite materials (including rubber-based) the benchmark tests requirements are considerably raised. One of the important mechanical characteristics of tyre operation is fatigue endurance. Though its value can be estimated by physical tests, they can last for months [1]. The finite element modelling and visualization program complexes are a convenient and effective tool for a theoretical research of tyres characteristics as well as tyres digital design.

The object of the present research is a radial pneumatic tyre, where the framework threads are located in the direction of the tire profile radius from one board to the other so that the threads in all the layers are parallel to each other. Only the breaker has a diagonal design. Such an arrangement of the framework threads does not allow the rubber to stretch strongly in the crosswise direction; the breaker holds the framework threads from the longitudinal movement. At the above-described arrangement of the framework threads the emerging tensions are approximately twice as low as those at the diagonal arrangement. This gives an opportunity to reduce the number of the cord layers; hence the weight of radial tyres is less than that of diagonal ones. Since the framework of radial tyres is less thick it is more elastic and has smaller internal friction. Consequently, there is a smaller amount of heat during its work which allows to increase the thickness of the tyre tread and the depth of its drawing thus extending the service life of the tyres.

The breaker, on the contrary, is very rigid and almost inextensible in the radial direction. Radial tyres have a greater stability of the form of the spot of contact with the road paving and create a lower rolling resistance thus providing a lower fuel consumption [2].

The present work sets the task to carry out finite and element modelling of a car tyre in three-dimensional statement, to investigate the tyre mechanical behavior at inflation and pressing in the static state as well as during stationary rolling.

2. Materials and methods

Due to the peculiarities of the material of the research object, the program complex of finite and element modelling Simulia Abaqus made by the Dassault Systems was chosen as the instrument of modelling. The main reasons for choosing this product are ample opportunities of the nonlinear analysis, the possibility to use an independent library of materials and elements and also special opportunities to solve the problem of tyre stationary rolling in the quasistatic statement.

The example of the tyre studied in the present work as well as its basic elements are given in figure 1.

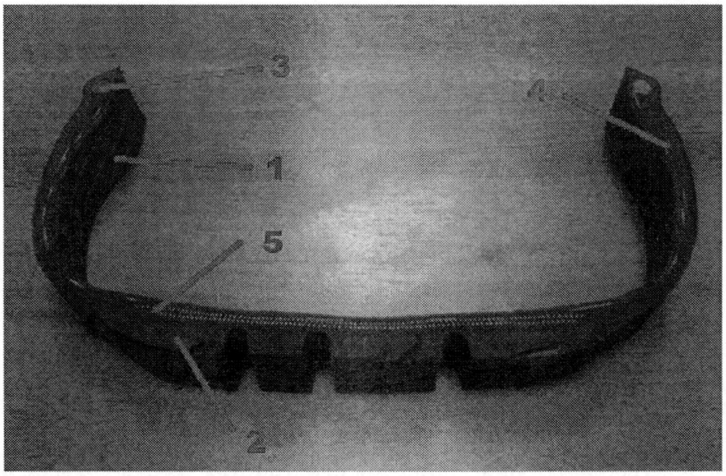

Figure 1. Tyre structure: 1) innerliner; 2) tyre tread; 3) steel board; 4) framework; 5) breaker

The following assumptions were accepted for the development of the finite element model of the radial pneumatic tyre. The tyre radius is 305 mm, the width is 187 mm, the profile height is116 mm. The value of the wheel load was varied discretely: 300, 350 and 400 kg respectively. The load was applied from top downward to the tyre axis of symmetry. The tyre consists of five separate parts: a rubber tyre tread, metal rings, a nylon two-layer cord, a two-layer breaker and an innerliner. The properties of the materials according to the chosen rheological model are presented in Table 1.

The equation of the material of Neo-Hookean model is presented in (1):

IOP Conf. Series: Materials Science and Engineering **489** (2019) 012030 doi:10.1088/1757-899X/489/1/012030

$$U = C_{10}(\bar{I}_1 - 3) + \frac{1}{D_1}(J^{el} - 1)^2 \quad (1)$$

where U is density of deformation energy per unit of volume; C and D are the material parameters depending on the temperature; I is the first invariant of the deformation tensor.

Table 1. Properties of the pneumatic tyre components

Structural elements of the pneumatic tyre	Density, model of material, dimensions	value of the parameters
framework	density, kg/m^3	1500
	Marlow hyper elastic model	
innerliner	density, kg/m^3;	1100
	Neo-Hookean model coefficients:	
	C_{10}, MPa,	0,6
	D_1, MPa^{-1}	0,03
tyre tread	density, kg/m^3;	1100
	Neo-Hookean model coefficients:	
	C_{10}, MPa,	0,5
	D_1, MPa^{-1}	0,04
metal rim	density, kg/m^3	7500
	Young modulus, MPa	207000
	Poisson coefficient	0,3
metal-clad breaker threads	density, kg/m^3	7500
	Young modulus, MPa	207000
	Poisson coefficient	0,3

The nylon threads of the framework cord are perpendicular to each other in the two composite reinforcing tyre layers: along and across the rotation axes of the tyre. The metal threads of the breaker are laid diagonally at the angle of 65 and -65 degrees to the rotation axis of the tyre in the rubber filler. As a whole it forms metal cord composite material.

The properties of the nylon threads and the rubber were set by the hyper elastic model of the material. The materials of the rings of the rim and the metal threads of the breaker layers were defined by the linear function. The determination of the supporting cords properties was simplified due to the assumption about the independence of supporting cords geometry on the material surrounding them. Thus, the creation of the grid for the cords was carried out without regard to the layers of the rubber components of the tyre. For this purpose, the program complex ABAQUS offers a possibility to use superficial elements describing the supporting cords properties. The superficial elements are used to define the cords geometry only. They are "cut into" the three-dimensional Solid elements and applied in order to model the rubber components of the tyre by introduction of the kinematic connections between the knots of the grid of the rubber material matrix and the knots of the superficial elements of the supporting cords.

Using the basic data we constructed a finite element model of the pneumatic tyre in the axisymmetric and three-dimensional statements (fig. 2) in order to study the influence of the axial vertical load value on the contact spot form and the stress-strain state of the tread material.

IOP Conf. Series: Materials Science and Engineering **489** (2019) 012030 doi:10.1088/1757-899X/489/1/012030

Figure 2. Geometrical model of the complicated tyre tread

Figure 3 presents the dependence $\sigma = \sigma(\varepsilon)$ of normal tension on relative lengthening. The maximum lengthening was 50%.

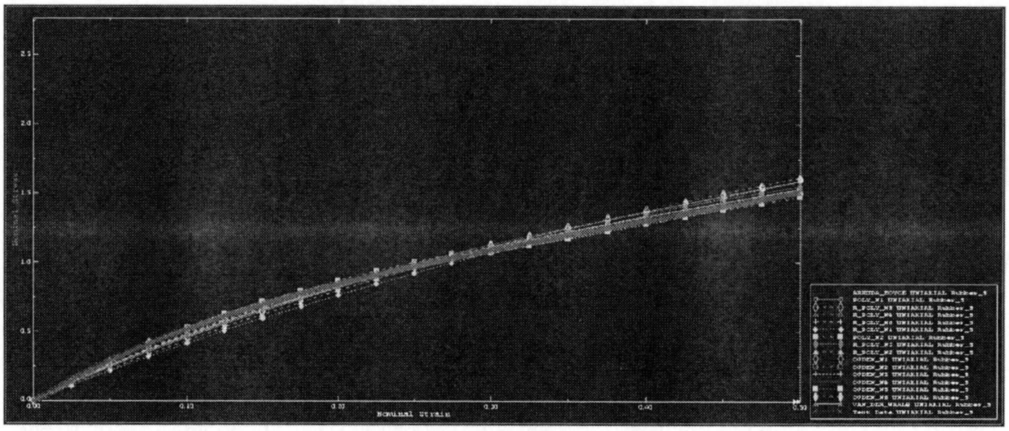

Figure 3. Dependence of normal tension on relative lengthening $\sigma = \sigma(\varepsilon)$. $\varepsilon = 50\%$

Comparing the experimental results with the data obtained from the models we should mention that at such lengthening all the material models work identically. However, some of the material models stop working when we increase relative lengthening (fig. 4). While modelling the pneumatic tyre three models, namely Mooney-Rivlin, Neo-Hookean and Yeoh models, describe the stress-strain state of the tread material most adequately.

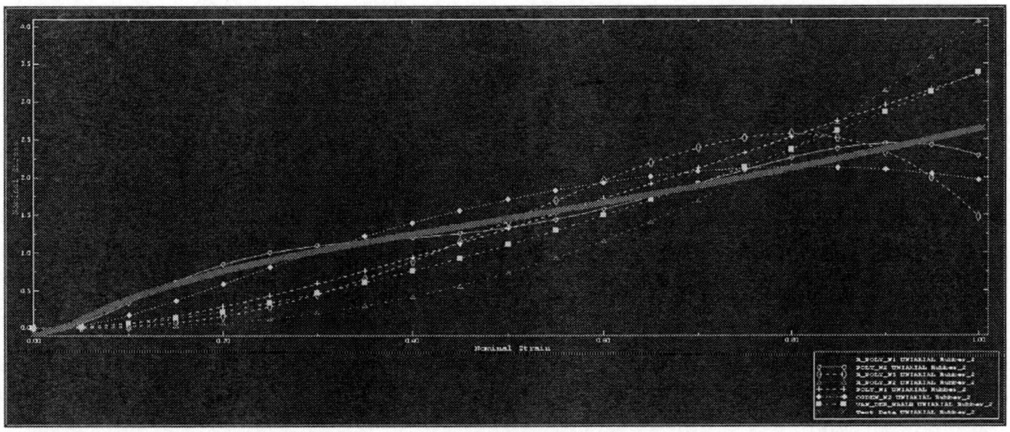

Figure 4. Dependence of normal tension on relative lengthening $\sigma = \sigma(\varepsilon)$. ε = 100%

3. Numerical approaches to the problem solution

The numerical experiment represented dynamic rolling of the model of the pneumatic tyre with a simplified drawing of the tread on the rigid level surface. All the model parts were joined into a single entity by imposing additional connections.

Modelling of the tyre static load was conducted in three stages:

1 stage – tyre inflation up to the operating pressure;

2 stage – initial pressing of the tyre to the rigid level surface described by the analytic plane. The initial pressing was set up by an artificial vertical shift of the analytic plane against the tire to provide tyre–surface interaction in the contact zone;

3 stage – after a steady tyre–surface contact is achieved, release of the tyre from the movement set at the second stage and application of the weight load to the tyre center.

A universal direct approach is creation of the tyre model in the three-dimensional statement ready to all types of computing experiments and adapted for a rapid change of boundary and initial conditions, geometry and properties. The limitation of this method is a considerable time span necessary to solve the dynamic problems of the tyre rolling, which is connected with a large number of the model degrees of freedom, as well as with the need to use a short lead time and a long period of time before the system gets into the stationary mode. Such method to solve the problems of stationary rolling is the most widespread in the tyre industry to assess the efficiency of the product. Unfortunately, it cannot be used for a digital design of tyres. So we developed a method to solve the problem of rolling at the constant speed based on the decomposition of the stress-strain state and the use of the independent Lagrange-Euler approach well-known in the continuum mechanics.

This method of calculation of the tyre stationary rolling was realized in the Abaqus Standard and consists of several stages:

1 stage – creation of the model of cross section of the pneumatic tyre with regard to its internal structure and solution of the problem of inflation in the axisymmetric statement;

2 stage – transformation of the constructed section into the three-dimensional model with the help of the Abaqus Standard. The Abaqus program complex makes it possible to pull the model from the section around the circumference with the preset number and type of final elements and to impose kinematic conditions to ensure the tyre surface pressing.

3 stage – using the elaborated three-dimensional finite element model of the tyre, imposing of power conditions in the form of weight and calculation of the stress-strain state of the tyre under the influence of weight;

4 stage – calculation of the dependence of the tyre stress-strain state in the conditions of the established rolling at the preset load on the car weight.

The established rolling of the tyre was studied at the wheel speed equal to 80 km/h and the coefficient of friction between the road and the tire equal to 0.5.

4. Results and their discussion

Computer modelling was carried out in two stages: static and dynamic. The static test modelling was carried out using two techniques with the invariable sizes and properties of the materials. The pressure in the tire was equal to 0.245 MPas, the normal load was equal to 3.3 kN. The results of the calculations are given in table 2.

Table 2. Calculation results for the tyre contact spot

Problem definition / model		Contact pressure, MPa	Print sizes, mm×mm
Static pressing of the tyre to the rigid surface	A full three-dimensional model of the tyre	0,72	141×132
	A simplified model without a rim	0,57	127×125
Stationary rolling of the tyre on the rigid surface	A full three-dimensional model of the tyre	0,77	126×123

The analysis of the results of the experiment showed the existence of an insignificant deviation of the contact pressure value for the tyre model in the full three-dimensional statement and that in the axisymmetric statement. It allows to state the expediency of the use of all the three ways of making a wheel. However, due to a great time expenditure on the numerical calculation of the tyre model in the three-dimensional statement the priority is given to the tyre model in the axisymmetric statement. But the use of the simplified model does not allow to take into account the geometrical features of the tread (fig. 4). That is why to create a complicated asymmetrical geometry of the tread it is necessary to use the tyre model in the three-dimensional statement.

The analysis of the results showed an increase of the contact pressure in the contact zone which can be explained by the increase of the tyre volume and the centrifugal effect. Fig. 5 shows the forms of the print of the tyre in contact with the road depending on the size of the imposed load and the material model.

5. Conclusion

Thus, the present research includes:

- numerical modelling of tyres with account of nonlinear analysis;
- calculation of tyres behavior at static and dynamic loads with account of the material models;
- creation of our own models of materials;
- modelling of a complex structure of the tyre;
- optimization of the tyre structure, for example, of the tread drawing or the cord characteristics.

The analysis of the results of the pneumatic tyre modelling in different statements and with variation of external and internal conditions showed that there are several solutions to the given task: in the axisymmetric statement and in the full three-dimensional statement. To describe the material properties we chose Yeoh material model as the one that describes the stress-strain state of the material most adequately. It is possible to use the received results in future in calculations of the tread rubber wear [3].

IOP Conf. Series: Materials Science and Engineering **489** (2019) 012030 doi:10.1088/1757-899X/489/1/012030

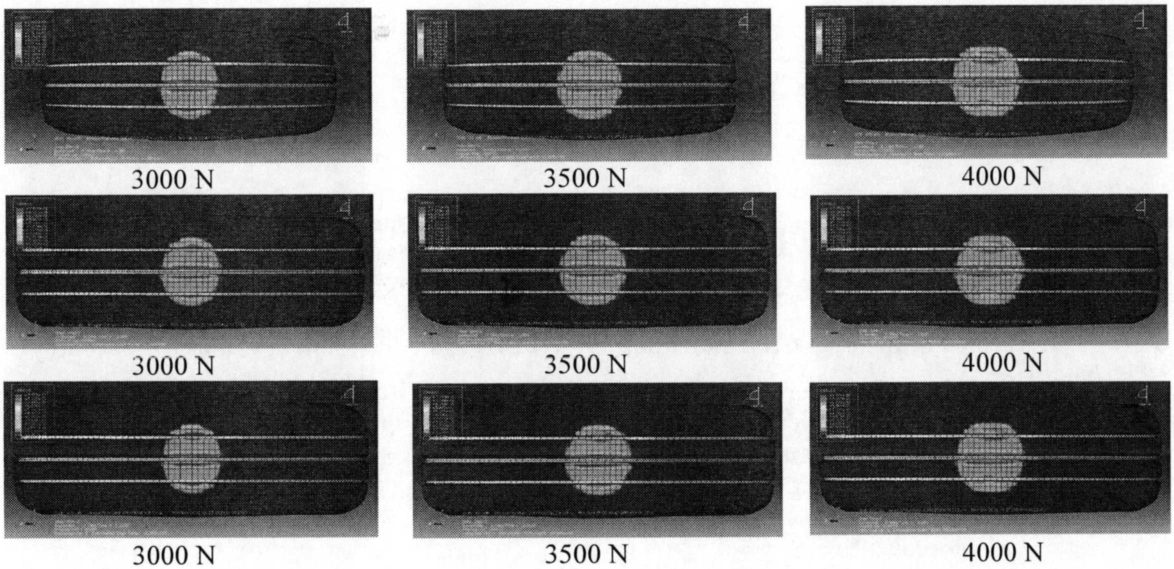

Figure 5. The form of the print of the tyre in contact with the road depending on the load and the material model (from top to down Mooney-Rivlin model, Neo-Hookean model, Yeoh model)

References

[1] Tolomeo D et al. *Cyclic Properties of Superelastic Nitinol: Design Implications SMST-2000: International Conference on Shape Memory and Superelastic Technologies* USA (California: Pacific Grove) P 471

[2] Ivanov A 2005 *Bases of Car Design Textbook for higher education institutions* Moscow LLC «*Za rulyom*»

[3] Shilov Mikhail A Maslov Leonid B Korolev Pavel V 2018 *Liq. Cryst. and Their Appl* **18** 73–8

IOP Conf. Series: Materials Science and Engineering **489** (2019) 012031 doi:10.1088/1757-899X/489/1/012031

Methodology development of the materials erosion tests

N I Smirnov[1], A N Yagovkina[1], M V Prozhega[1], N N Smirnov[1] and I A Safonov[2,3]

[1]Mechanical Engineering Research Institute of the Russian Academy of Sciences, 4 Maly Kharitonyevsky pereulok, Moscow, Russia
[2]National University of Science and Technology MISIS RF, Leninskiy prospekt 4, Moscow, Russia
[3]State Research Center JSC "RPA "CNIITMASH", Sharikopodshipnikovskaya st. 4, Moscow, Russia

prmaksim@gmail.com

Abstract. The research paper describes a centrifugal stand and a methodology of materials testing for erosive wear in a simulative medium, which contains water, abrasive particles, and corrosive-active elements, which are analogous to the formation fluid of oil wells. There are test results in water with abrasive of four types powder materials based on iron and copper, alloyed with N1, Cr, Mo. The experiment parameters: the fluid jet speed is 11...16 $m \cdot s^{-1}$, the impact angle is 45°...90°, silica sand and corundum particles of different dispersion F100, F40, F24. A linear dependence of the relative wear rate on the experiment time at various impact was revealed. The wear of powder materials in liquid with corundum F100 is approximately 5 times more than with similar fraction silica. The material PK90N4MG2KD15 has the greatest erosion wear resistance under the influence of F100 fine particles.

1. Introduction

Erosion-corrosive damage of details is often found in oil production equipment, pulp pipe, mine pumps, hydraulic turbines, and other types of machine-building installations, which leads to great economic losses in case of their failure and repair. In submersible multistage centrifugal pumps for oil producing, erosion is manifested in the form of "washouts" of the flow-through part of the pump stages and components of gas separators (figure 1). Powder materials are widely used in the pump stage manufacturing. However, their erosion resistance is not regulated by the design and technological documentation, and there are no methodologies of the corresponding tests.

The research papers of many authors have been devoted to the erosion processes study. A fundamental review of the methods and design models is given in the article [1]. Erosive wear models of plastic and brittle materials were described in the research papers [2-4]. In research paper [5], mechanistic models were considered for predicting the erosive wear of pipelines and it is based on computational fluid dynamics. On the basis of physical models, stands have been created to determine the wear resistance of materials, to experimentally estimate the rate of erosion, and to study the influence of various factors on it.

IOP Conf. Series: Materials Science and Engineering **489** (2019) 012031 doi:10.1088/1757-899X/489/1/012031

Figure 1. Erosion-corrosion of multistage centrifugal pumps component.

Figure 2. Erosion-corrosion of gas separator component.

Stands differ in the method of abrasive particles acceleration to the specimen surface: due to the pressure of the working medium (gas jet) [6] or a centrifugal accelerator [7]. As a working medium, both gaseous and liquids or suspensions mediums are used. Stands with the use of a gas jet, as a rule, are single-positioned (one specimen); stands with the use of a liquid are mainly multi-positioned. In various types of stands it is necessary to solve a number of technical problems. For example, in research paper [8], a developed centrifugal accelerator with eight specimens, simultaneously installed at different angles, was presented. However, the circulation of the same mixture in a closed loop during the experiment leads to the loss of cutting properties by the abrasive particles, the wear of the service pump and the decrease in the accuracy of the experiment. Similar problems may arise in the test procedure under a sliding drop of a jet of liquid with abrasive particles on specimens [9]. Most of the known devices are intended for erosion wear testing mainly in an abrasive-containing environment.

The purpose of this work is to develop a stand for erosion wear, test methodologies, as well as conducting of experiments with powdered materials in an abrasive-containing medium.

2. Experimental details

2.1. Experimental setup

Based on the study of international experience in the Mechanical Engineering Research Institute of the Russian Academy of Sciences a rotating test machine for erosion-corrosion measurement has developed [10], which has several advantages. A special feature of the rotating test machine is the single use of abrasive. This is achieved by separating the abrasive from the liquid in a sand feeder. Thus, the service pump that is pumping the liquid is not subject to wear. The materials used in the rotating test machine and the equipment allow experiments with a corrosive liquid at temperatures up to *80° C*.

The installation is based on the principle of a rotating nozzles with a back-to-back arrangement of fluid motion and consists of a plate with a test chamber, an adjustable-speed electric motor, feeder sand, a mixer, a rotor, an auxiliary pump, a gravity separator, flow meter, measurement and control systems. Prior to each test, the separator tank is filled with working liquid. With the help of an auxiliary pump for chemical liquid, the liquid is delivered into the mixer, where it mixes with the abrasive. The fluid jet speed in the specimen is determined indirectly through flow rate measurement using flow meter. In test chamber 1 there is a rotating shaft 2 on which rotor 3 is installed with two (or four) channels and hard alloy nozzles 4, figure 2. Specimens 5 are installed in specimen holder made of fluorine plastic on the rotor opposite the nozzles under different angles to the fluid jet. The chamber is closed with a transparent cover 6. Seals 7 prevent liquid from flowing at the points of contact with the rotating rotor. When the rotor rotates, the liquid that is coming from the mixer, under the action of centrifugal forces, is accelerated in the rotor channels and through the nozzles at a certain angle on the specimens, producing erosive destruction of their surface. Then the liquid flows from the chamber through the pipe into the gravity separator, where the abrasive particles are deposited.

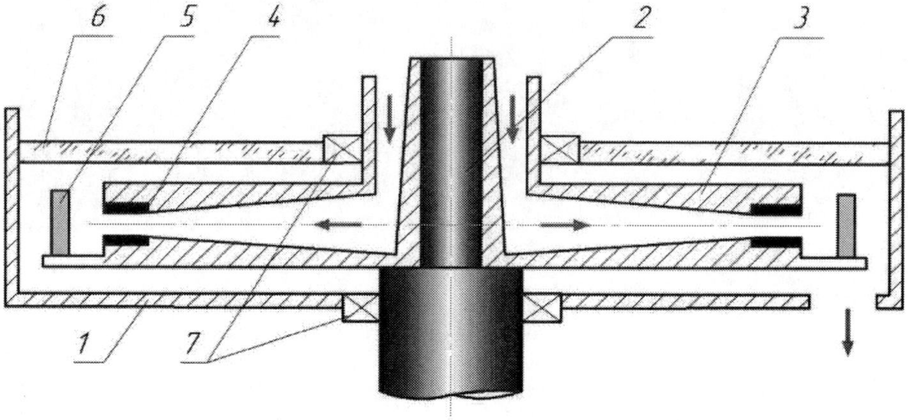

Figure 3. Test chamber scheme: 1 – test chamber; 2 – shaft; 3 – rotor; 4 – nozzle; 5 – specimen; 6 – cover; 7 – seals

The distance from the specimen to the nozzle can vary; the specimens installation angle relative to the flow (the angle between the specimen surface and the nozzle axis) is adjustable from 30° to 90°. At the same time, two (or four) specimens are installed at a certain angle and at a certain distance from the nozzle. The abrasive dispenser allows you to adjust the flow of solid particles into the mixer, thereby achieving a change in the concentration of particles in the model liquid. During the experiment, the rotational speed of the rotor and the fluid jet speed in each specimen remain constant.

After the experiment, the specimens are weighed and determine the weight-loss. As a criterion of materials wear resistance during erosion, the relative wear rate I_r was adopted [4]:

$$I_r = \frac{\Delta m}{M_a}$$

where Δm – the specimen weight-loss, *g*; M_a - abrasive mass in the experiment, *g*.
The test machine allows you to analyze the influence of various factors on the erosion: abrasive concentration, particle shape, velocity and temperature of the liquid, fluid properties, impact angle, material properties of the specimen, etc.

2.2. Experiment parameters

Water with an abrasive was used as a model liquid. The concentration of solid particles varied from 0.6 to *1.1* $g \cdot l^{-1}$, the rotational speed of the rotor was changed from *950* to *1500 rpm*, the flow rate of the liquid was from *1* to *1.5* $m^3 \cdot h^{-1}$. The experiments were carried out at two values of the impact angle, a = *45°* and a = *90°*, fluid jet speed in the specimen was *11* and *16* $m \cdot s^{-1}$, and at room temperature. The repeat number of experiments under the same conditions is not less than three.

One of the research paper objectives is the timing of experiment in testing powder materials. With a long duration of the experiment, the material costs of abrasive materials and electricity increase. With a short experiment time, significant wear values can be obtained by increasing the abrasive concentration. However, at the same time, the results can be distorted due to collisions of particles with each other and a large amount of equipment wear. Also the invariance condition of the specimen wear mechanism is important during conducting the experiment. For example, if in the process of experiment, microcutting at some stage passes into fatigue failure, then when materials with different properties are tested, the ratio of their relative wear rates may change. For this reason, a comparison of the wear resistance of these materials will be incorrect.

2.3. Material and specimen preparation

Testing specimens for wear were made by "Keramet" Company in the form of rectangular plates of $25\times 15\times 3$ *mm* made of four types iron-based powder materials, table 1. The materials were obtained by method of powder metallurgy: pressing a mixture of powders and burning in dissociated ammonia atmosphere with simultaneous infiltration with copper. The roughness of the specimens is $R_a=0.14$ *μm*. Table 1 presents the results of measuring the microhardness of materials.

The microhardness was determined using a PMT-3 microhardness tester at a load of *0.98 N*. The specimens were pressed into a conductive epoxy resin at 180° C using a specialized *CitoPress* hot press. The preparation of thin sections was carried out using *Struers Rotopol 21* machine. During polishing, paper-based circles of silicon carbide were used, the surface was polished using diamond paste and *OP-8* oxide suspension based on SiO_2.

The microstructure of the specimen surface was studied using a scanning electron microscope *Hitachi S*-3400N at accelerating voltage of 15 *kV*. The microscope is equipped with an X-Ray energy dispersive spectrometer *NORAN* for conducting local X-ray spectral analysis. The weight of the specimens was measured on the "Ohaus" weighting system with a measurement accuracy of 0.0001 *g*. The profile of the wear area was determined using a *Model 202* profile meter (Caliber, Russia).

The powder steels under study have either a ferritic-pearlitic structure (see figures 4,6,7) or pearlite (figure 5) with copper inclusions and alloying depending on the composition Ni, Cr, Mo and other elements.

Table 1 Chemical composition and microhardness of powder steels

	Material brand	Composition, % (by weight)							Microhardness (HV)
		C	Cu	Cr	Ni	Mo	S	Mn	
1	PK70D15	0.61…0.9	13…17	–	–	–	–	–	346.3±3.1
2	PK90H4MG2KD15	0.81…1.2	13…17	–	3…5	0.3…0.7	0.15…0.45	1…3	504.9±4.8
3	PK10H16N9D20	< 0.3	18…22	15…17	8…10	–	–	–	269.6±2.4
4	PK10H7N4D20	< 0.3	18…22	6…8	3…5	–	–	–	410.2±2.8

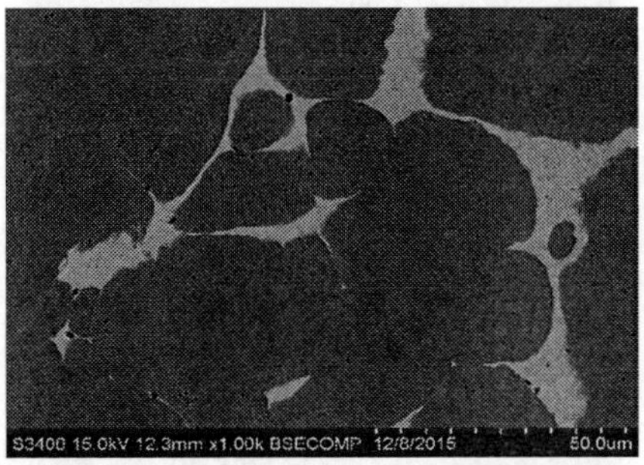

Figure 4. Structure of PK70D15 powder steel.

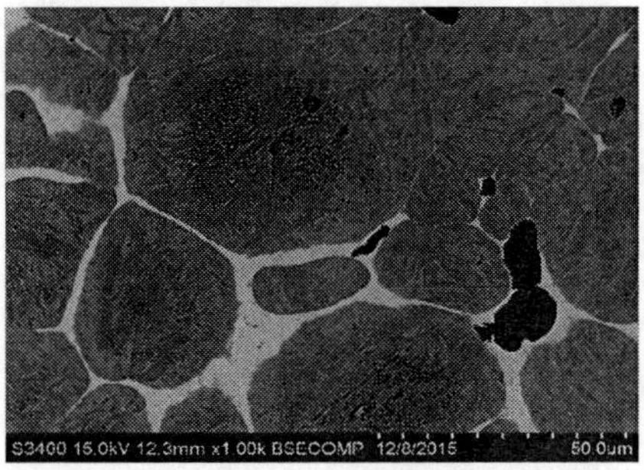

Figure 5. Structure of PK90N4MG2KD15 powder steel.

IOP Conf. Series: Materials Science and Engineering **489** (2019) 012031 doi:10.1088/1757-899X/489/1/012031

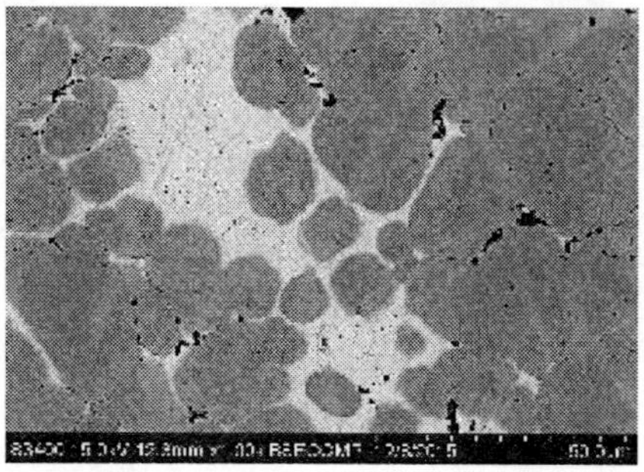

Figure 6. Structure of PK10H16N9D20 powder steel.

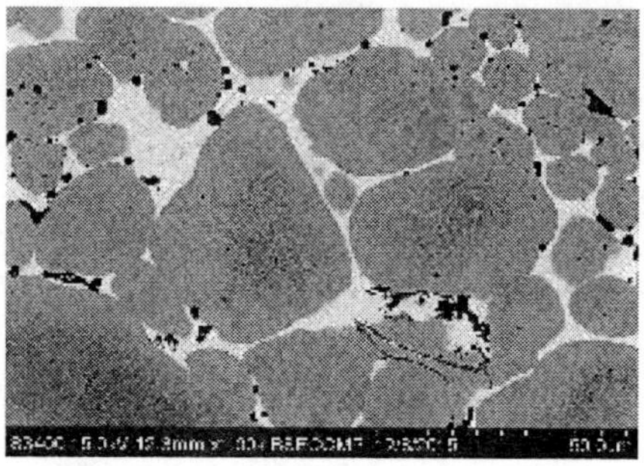

Figure 7. Structure of PK10H7N4D20 powder steels.

2.4. Abrasive particles size and shape

The grain-size composition and the average particle size of the abrasive were measured using *Analysette 22 MicroTec Plus FRITSCH* Company's microanalyzer for laser beam diffraction, scanning a suspension of the analyzed powder in water.

An abrasive represents silica sand or corundum particles of different dispersion. The shape of silica particles is round (see figure 8), corundum is acute-angled (see figures 9, 10, 11). Silica sand of this shape and size is found in the formation water of oil wells. Corundum is used by many researchers and has good wear properties.

Figure 8. Silica sand form.

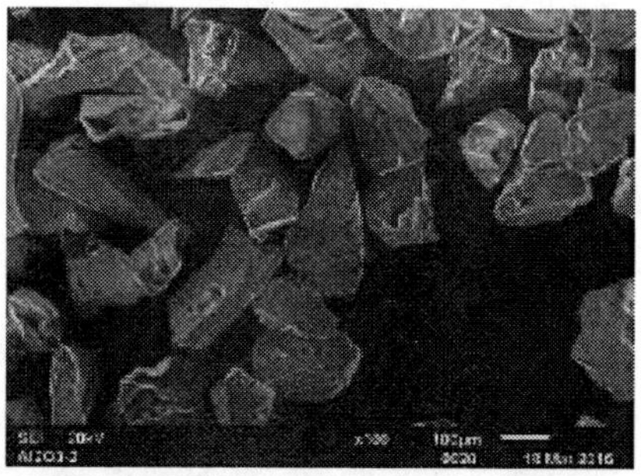

Figure 9. Corundum F100 form.

Figure 10. Corundum F40 form.

Figure 11. Corundum F24 form.

The grain-size composition of the silica sand powder includes *10%* of the fine fraction with a particle size up to *100 μm*, about *80%* of the particles are from *100* to *280 μm*, and the remaining *10%* are represented by a larger fraction – up to *420 μm*. The same trend is observed in powders of fine *F100* corundum. The main fraction of the powder (*83%*) has a size from *100* to *280 μm*. The average particle size of silica sand and corundum *F100* is D_m *=180 μm*. *F40* corundum powder consists of two fractions of *320...580 μm* and 580...880 *μm*, the amount of which differs slightly. The average particle size is D_m=530 *μm*. The particle size of *F24* corundum is in the range from *600 μm* to *1.1 mm*.

3. Results

For methodology development, we analyzed the kinetics of erosion, on the basis of which the experiment time was chosen. The specimen made of PK70D15 material was tested in abrasive liquid at a jet speed of $16\ m{\cdot}s^{-1}$ and an impact angle of *45°* and *90°*. After each experiment with the duration of 160 *min*, the specimen was weighed, the profile diagram of the erosion area was recorded, and the relative wear rate was determined and set again in the same way, table 2. The recorder profiles are presented in the figures 12 and 13. The numbers 1, 2, 3, 4 at figures 12 and 13 correspond to test number in table 2.

The results of experiments showed that the kinetic of erosion is linear, i.e. with an increase in the duration of the experiment, the value of relative wear rate I_r practically does not change. With such duration of experiment, it is possible to obtain significant wear values for both the least wear resistant and more wear resistant ones with acceptable measured values.

Table 2 Results of wear kinetics tests

Test number	Test duration *T, min*	$I_r{\cdot}10^6,\ g{\cdot}g^{-1}$	
		a=45°	*a= 90°*
1	160	2.13	2.27
2	160	1.93	1.93
3	160	2.13	2.00
4	160	2.26	2.26

The average value of the relative wear rate under the specified conditions is $2.11{\cdot}10^{-6}\ g{\cdot}g^{-1}$, the variation of results is within *7%*.

The profile measuring of the worn surface showed that the wear profile depends on the impact angle of the specimen. In both cases, at *a=90°* and *a=45°*, the wear is unequal. The maximum depth of the worn surface at *a=90°*, figure 12, is 1.5 times greater than at *a=45°*, figure 13. A similar relationship is observed when using corundum of the same fraction (figure 9). It should be noted that the intensity of wear in these cases is approximately the same.

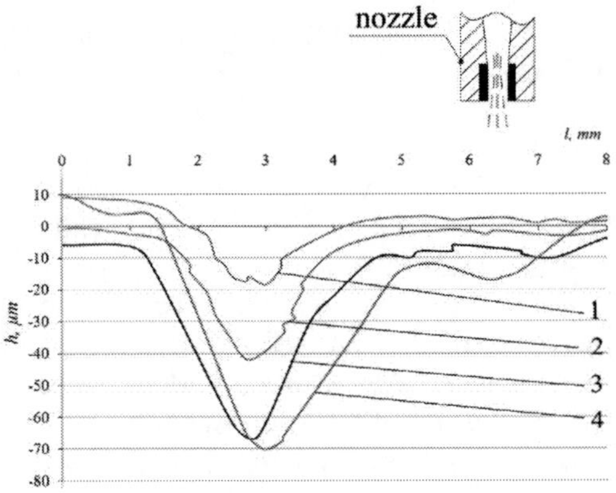

Figure 12. Wear track of surface in moving direction (test with silica sand, impact angle *a=90°*), *h* – wear, *l* – length

IOP Conf. Series: Materials Science and Engineering **489** (2019) 012031 doi:10.1088/1757-899X/489/1/012031

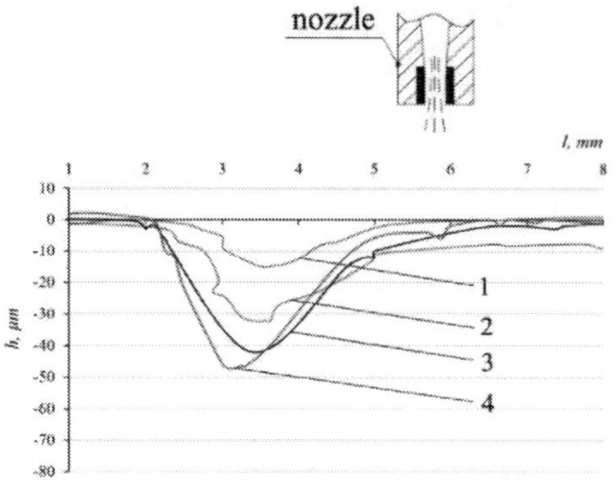

Figure 13. Wear track of surface in moving direction (test with silica sand, impact angle *a=45°)*

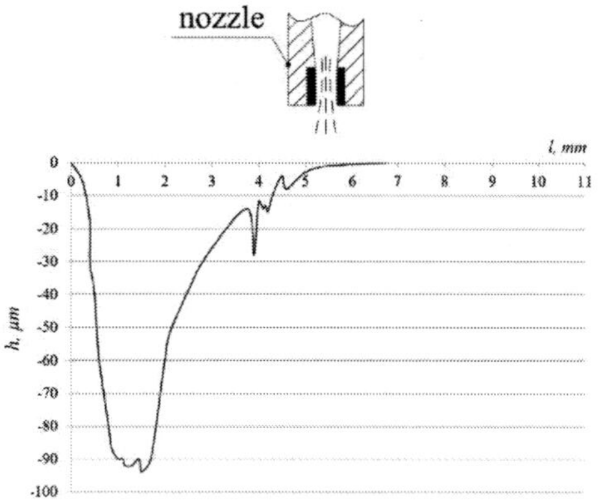

Figure 14. Wear track of surface in moving direction (test with corundum F100, impact angle *a=90°*)

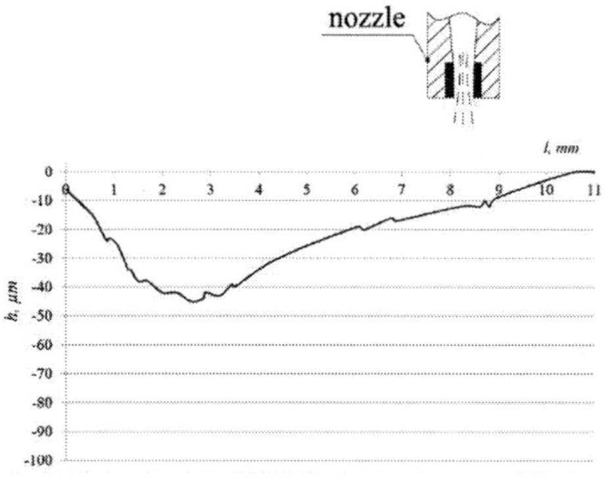

Figure 15. Wear track of surface in moving direction (test with corundum F100, impact angle *a=45°*)

The region of maximum specimen wear is displaced relative to the center of the nozzle due to flow divergence in the centrifugal accelerator as the rotor rotates. Figures 12-15 show schematically the location of the nozzle relative to the wear area of the specimen surface.

Figures 16 and 17 presents the testing results of four types powder materials according to the developed methodology. It is revealed that the type of abrasive influences strongly the wear resistance of materials. The relative wear rate of specimens when tested in water with F100 corundum is about 5 times greater (depending on impact angle) than in water with silica sand of a similar fraction.

The relative wear rate of powder materials increases with size increasing of the abrasive. The greatest wear is observed when tests with *F24* corundum is conducted, the average particle size of which is $D_m = 900\ \mu m$. For example, for material No. 1 (PK70D15), the relative wear rate when tested with *F24* abrasive is 1.4 times higher than when tested with *F40* abrasive ($D_m = 530\ \mu m$), and 1.8 times higher than the results with tests with F100 abrasive ($D_m = 180\ \mu m$).

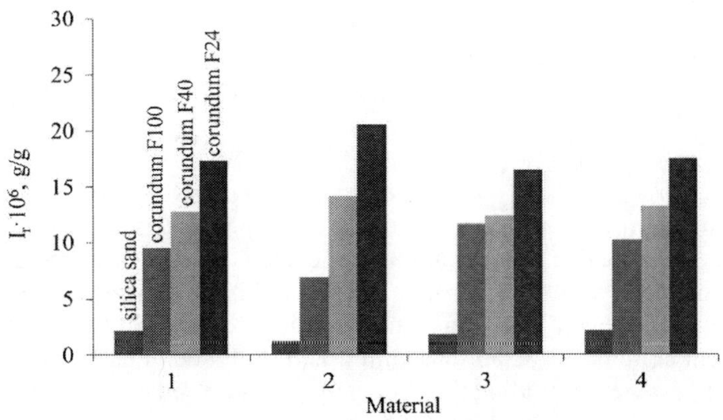

Figure 16. The relative wear rate of the study materials (see table 1) at the jet speed 16 $m \cdot s^{-1}$ and impact angle a = 90°

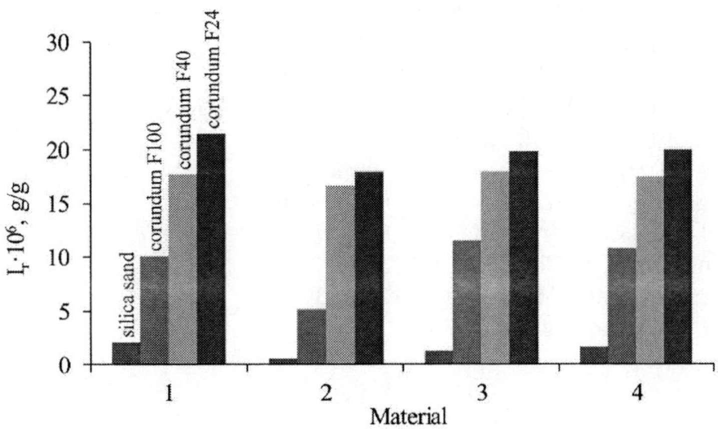

Figure 17. The relative wear rate of the study materials (see table 1) at the jet speed *16 $m \cdot s^{-1}$* and impact angle *a=45°*

As shown by the test results, the difference in the wear resistance of specimen No. 2 (PK90N4MG2KD15), which has the greatest hardness among the tested materials (see table 1), is noticeable only when exposed to small abrasive particles of *F100* dimension. When large particles of *F40* and *F24* size are applied, the difference decreases and is within the range of the measured values. However, there is a noticeable difference in the intensity of wear associated with the installation angle of the specimens. When exposed to large particles, *F40* and *F24*, the relative wear rate is somewhat lower at an impact angle of *90°*.

4. Conclusion

Based on the results obtained, the following main conclusions can be drawn.

The most significant parameter affecting the erosion wear resistance of powder materials at a given jet speed is the abrasive type. The wear of powder materials in water with *F100* corundum is ~ 5 times greater than with silica sand of a similar fraction.

The size of the abrasive also affects the relative wear rate of powder materials. With an increase in the size of corundum from *F100* to *F24*, the wear rate of material PK70D15 increased 1.8 times.

IOP Conf. Series: Materials Science and Engineering **489** (2019) 012031 doi:10.1088/1757-899X/489/1/012031

The hardness of the powder materials under the research has little effect on their erosion wear resistance to wear under given conditions. The material PK90N4MG2KD15 has the greatest wear resistance under the conditions of exposure to abrasive particles of *F100* dimension.
When testing for erosion wear, it is necessary to take into account the unevenness of specimen surface wear.

5. Summary
Erosion-corrosion damage to the multistage centrifugal submersible pumps installations leads to large economic losses in the repair of equipment, and the mechanism of such damage is not well known. A methodology of erosion testing on the centrifugal type stand basis has been developed, which makes it possible to analyze the features of wear and rank materials for durability. The test results of powder materials with different content of nickel, chromium and molybdenum can be used in the design and manufacture of pumping stages, providing the necessary resource and cost savings. The further direction of test method improvement is the development of a test procedure in a corrosive, abrasive-containing medium.

References

[1] Meng H C and Ludema K C 1995 *Wear* **181–183** 443
[2] Finnie I 1972 Wear **19** 81
[3] Hutchings I M 1981 Wear **70** 269
[4] Kragelskiy I V, Dobychin M N and Kombalov V S 1977 *Basics of calculations for friction and wear* (Moscow: Mashinostroenie) p 526
[5] Kozyrev S P 1971 *Hydroabrasive metal wear at cavitation* (Moscow: Mashinostroenie) p 240
[6] Yang K, Rong J, Liu C et al. 2016 *Tribology International* **93** 29
[7] Söderberg S, Hogmark S, Engman U and Swahn H 1981 *Tribology International* **14(b)** 333
[8] Azimian M and Bart H J 2015 *Wear* **328–329** 249
[9] Santa J F, Baena J C and Toro A 2007 *Wear* **263** 258
[10] Patent 2444719 Russian Federation 2012

IOP Conf. Series: Materials Science and Engineering **489** (2019) 012032 doi:10.1088/1757-899X/489/1/012032

Effect of laser hardening modes on the hardening zone geometric parameters and tribological properties of 40Cr steel

D A Starostin[1], V P Biryukov[2], D V Klevetov[1] and V F Yudkin[2]

[1]All-Russian Scientific Research Institute Signal, 57, Krupskaya Street, Kovrov, Vladimir region, 601903, Russia

[2]Mechanical Engineering Research Institute named by A.A. Blagonravov, Russian Academy of Sciences, 4 Maly Kharitonyevsky Pereulok, Moscow, 101990, Russia

Laser-52@yandex.ru

Abstract. In the work experiments on the beam of a multichannel CO_2-laser (ALTCU-3) with a unique layout within 25-130 mm using a full factorial experiment, we constructed the surface to depth and width of the zones of hardening through varying the speed of the beam in the range of 10-20 mm/s, a power of 1-2 kW and a beam diameter of 4-14 mm. Comparative wear tests showed that with the increase in the hardening area to 50% of the surface friction, the wear resistance increases by 3.2 times in comparison with the normalized steel 40Cr. It is shown that the value of the average microhardness decreases with increasing power and reducing the processing speed by 300 MPa, and the depth of the quenching zone increases up to 2.5 times.

1. Introduction

The reliability parameters of the system general depend on the reliability of its individual elements. According to preliminary studies, it was found that the quality of the surface layer in the friction pairs of hydraulic units affects the duration and intensity of their running-in process. One of the formation directions of the surface layer during the finishing treatment is laser heat treatment, which allows the hardened surface layer with certain properties [1]. In this case, laser treatment is usually not subjected to the entire surface of friction, and its individual sections. The result is a surface of consistently alternating "soft" (the original structure of cast iron) and solid areas of the laser impact zone. It has been experimentally established [1] that laser treatment not only increases the wear resistance of piston rings (PR) by 1.5-2.4 times, but also reduces the wear of serial cylinder sleeves. The alternation of wear-resistant laser tracks having a structure of high hardness bleached cast iron of, and cast iron areas of with graphite inclusions on the PR working has a positive effect on the performance of the friction pair. Based on field tests [2] in marine conditions of diesel engines 6 CRN 36/45 found that the laser-treated PRs with an operating time of 7.5 thousand are highly completive with a chrome piston rings ("RUMO") JSC in respect to wear-resistance and as compared to "surface analysis" (Zavod Nizhegorodskii Teplokhod) exceeds 1.8-2.2 times.

However, the main problem of laser processing industrial application is the need to obtain the dependence of the hardened layer parameters on the processing modes for a particular type of laser.

So in [3], for experimental studies, structural steels 65Mn, 30CrNiSiA, 45 were chosen. Processing was carried out on a specialized laser technological complex ALTCU-5 on the basis of a multichannel

CO_2 -laser with a unique arrangement of tubes. Mode selection was carried out on the variation in optimal treatment intervals: power 1000 – 4600 W, the moving speed of the laser beam 10 to 30 mm. However, the presented results of hardening are given in tabular form are particular, and do not show the General dependence of the laser hardening zone parameters necessary as a tool for determining the laser processing modes for real production.

Effects of laser treatment with diode laser on the material structure with the conclusion of the necessary dependencies of the quenching zone parameters on the processing modes [4]. The results are obtained by processing a rectangular unfocused laser beam and by defocusing the beam by 25 mm. The hardness of the hardened layer is reduced by half, and can, not be applied to the surface of the dies of bending dies, threads and other complex shape parts.

For 40CrNi2MnA steel, the technology of laser hardening by fiber laser using IPG 2D scanners has been developed, which allows hardening zones with a width of 15-50 mm and a depth of 0.2-2.5 mm per one robot am pass at transverse vibrations up to 100 Hz [5].

A special place in the laser thermal hardening is the processing of multi-channel CO_2-laser, which has its own characteristics. At the same time, its operation involves a number of difficulties associated with the uniformity question of the rays, focusing and their arrangement. At the same time, for the industrial use of laser thermal hardening technology is relevant research on materials that have found wider application in engineering, including in hydraulics, in particular steel 40Cr. The sizes of the laser quenching zones are obtained by regression equations depending on the processing modes for the CO_2-laser and the parameters of the material surface layer for the "Kometa-M" installation [6]. However, as noted above, such studies need to be carried out for each specific type of equipment.

The purpose of this work is to establish the influence of laser processing modes of multi-channel CO_2-laser with a unique layout on the changes in the depth and width of the quenching zones and the quenching area influence on the wear resistance of hardened samples and subsequent performance for steel 40Cr.

2. Materials and equipment

Laser hardening was performed on an automated laser technological complex [7-9], on samples of steel grade 40Cr with dimensions 10×20×80 mm. To increase the surface absorption capacity before laser treatment, the samples were phosphating. Metallographic studies were carried out using a digital microscope AM413ML and metallographic microscope Altami MET 1C, microhardness measurements on PMT-3M at a load of 0.98 N.

In the first series of experiments investigated the power density effect of laser radiation on the depth H, width B, and the microhardness HV of the hardened tracks with a radiation power of P 1000 W and 2000 W and the beam speed V 10 $mm \cdot s^{-1}$ and 20 $mm \cdot s^{-1}$ for the defocusing of the laser beam Z is from 25 mm to 130 mm. the Diagram of the multichannel laser treatment CO_2 - laser is presented in figure 1.

In the second series of experiments, the influence of processing modes on the parameters of hardened tracks was determined using the full factorial experiment (FFE) method [10]. Radiation power P (1000 -2000 W), processing speed V (10-20 $mm \cdot s^{-1}$), and laser beam diameter d (4-6 mm) were chosen as experimental factors. The depth H and width B of the laser quenching zones were considered as the system responses for the construction of mathematical models. The sections were made according to the standard procedure and the depth and width of the hardened zones were measured three times. The calculation determined all possible interactions of factors. Since FFE 2^3 was performed, the number of experiments was 8.

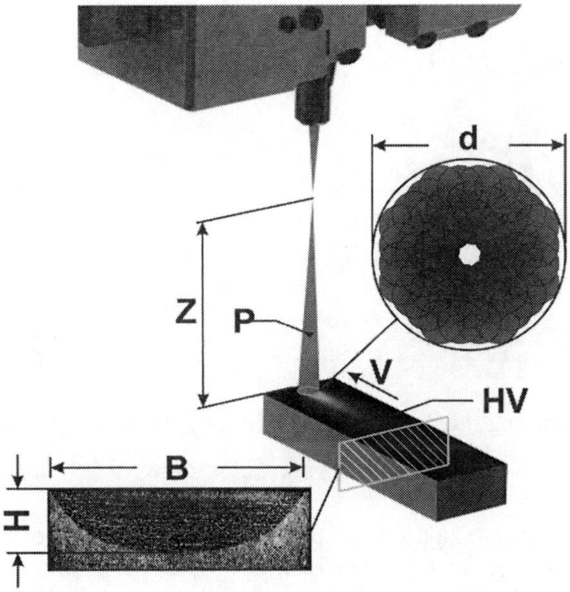

Figure 1. Scheme of laser heat treatment by multichannel CO_2 – laser

In the third series of experiments, the samples were strengthened with a different step of laying tracks to obtain a quenching area S 25, 50, 75, 100% and with overlapping tracks 10-30%, figure 2. Tests on wear resistance and scoring resistance was carried out on the machine friction MTU-01 according to the diagram plane (sample-ring, counter-sample, 40Cr steel, HRC 48-52) at a load of 2 MPa. As a lubricant used transmission oil Tszp-8.

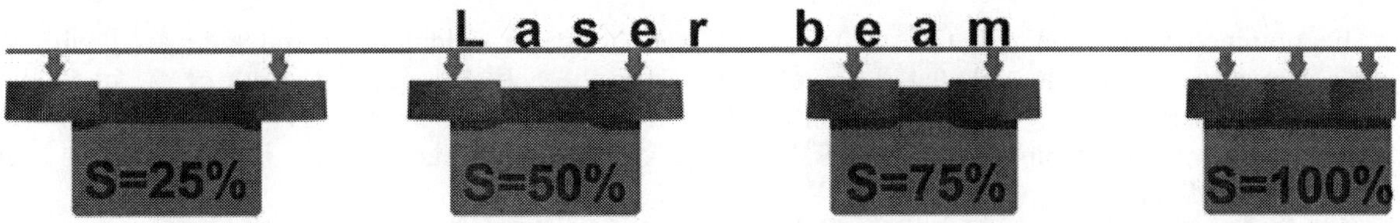

Figure 2. Scheme of laser heat treatment with different quenching area

3. Results of experiments and calculations

According to the measurements results of the hardened zones parameters in the first series of experiments, graphs of the depth and width of the quenching zones on the change in the distance to the focal plane or the radiation power density are shown in Figure 3. The left and right part of the graph in Figure 1a can be represented as linear dependencies with a limit of their definition. For example, for the left part of the graph of the limit values of the laser spot diameters on the surface of part 4 to 6 mm, and for the right – 10-12 mm. Then the solution of the linear regression equations will have adequate values.

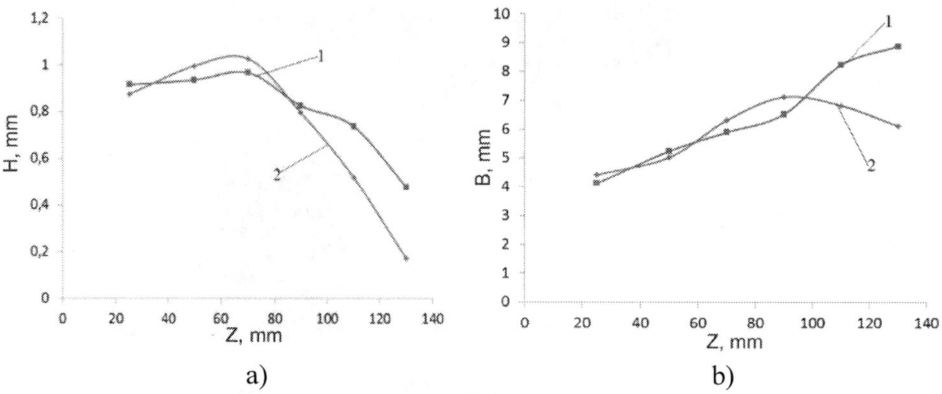

Figure 3. Changing the depth (a) and width (b) of the laser quenching zones:

1-P = 1 kW, V=10 mm·s^{-1}; 2-P = 2 kW, V=20 mm·s^{-1}

As a result of the regression analysis, a system of regression equations for the geometric parameters of the hardened zones, depending on the processing modes in physical quantities, is obtained to determine the width of the quenching zone in (1) and to determine the depth of the quenching zone in (2).

$$H = 0{,}95 + 0{,}179 \cdot \left(\frac{P-1500}{500}\right) - 0{,}174 \cdot \left(\frac{V-15}{5}\right) + 0{,}035 \cdot (d-5) - 0{,}019 \cdot \left(\frac{P-1500}{500}\right) \cdot \left(\frac{V-15}{5}\right) + 0{,}01 \cdot \left(\frac{P-1500}{500}\right) \cdot (d-5) + 0{,}015 \cdot \left(\frac{V-15}{5}\right) \cdot (d-5) \quad (1)$$

$$B = 4{,}682 + 0{,}361 \cdot \left(\frac{P-1500}{500}\right) - 0{,}341 \cdot \left(\frac{V-15}{5}\right) + 0{,}445 \cdot (d-5) + 0{,}0173 \cdot \left(\frac{P-1500}{500}\right) \cdot \left(\frac{V-15}{5}\right) - 0{,}042 \cdot \left(\frac{P-1500}{500}\right) \cdot (d-5) - 0{,}087 \cdot \left(\frac{V-15}{5}\right) (d-5) \quad (2)$$

The obtained regression equations (1) and (2) were used for calculations and compared with the experimental results. The calculated values differ from the actual depth and width of the quenching zones by no more than 5%. Regression models of dependencies of type H (P,V), B (P,V) were introduced into the table editor of MS Excel and comparative surfaces for these functions were constructed, figure 4.

Based on the graphs (figure 4), the influence on the geometric parameters of the quenching zones has the power, speed and beam diameter. With increasing power, all other things being equal, the depth parameter of the quenching zone changes to a greater extent, as well as with a change in the processing speed. As the beam diameter increases, so does the depth and width of the quenching zones.

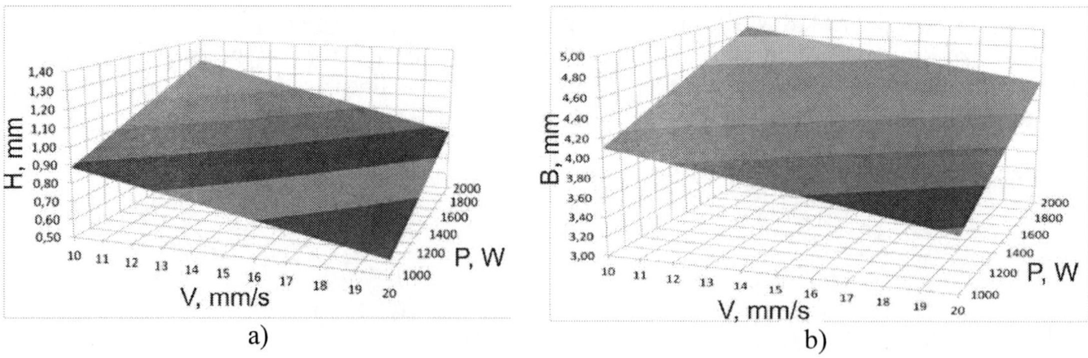

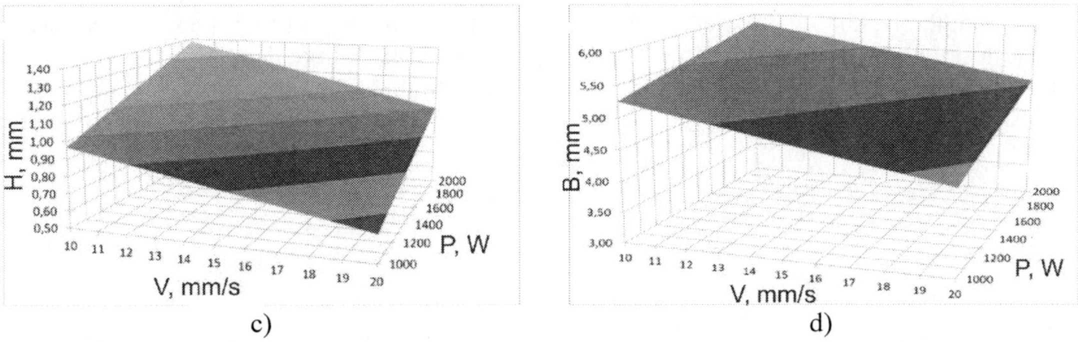

Figure 4. Graphs of the depth (a, b) and width (c, d) of the laser quenching zone depending on P and V processing at d = 4 and 6 mm, respectively

The microhardness of the hardened zones varied 6180–7160 MPa depending on the processing modes. The regression analysis is conducted out and the equations system of average microhardness regression of the hardened zones depending on the processing modes in natural quantities is obtained (3).

$$HV(P, V, d) = -0{,}2208 \cdot P + 12{,}9219 \cdot V + 0{,}1237 \cdot d + 6262{,}0182 \qquad (3)$$

Comparative surfaces of this function were constructed using the regression model of type HV (P, V) (figure 5). Based on the graphs, the value of the average microhardness decreases with increasing power and reducing the processing speed by 300 MPa, which occurs due to the increase in the depth of the quenching zone up to 2.5 times.

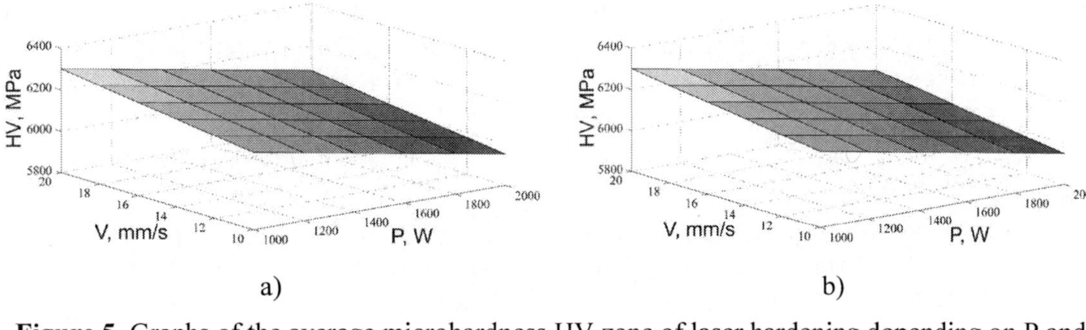

Figure 5. Graphs of the average microhardness HV zone of laser hardening depending on P and V processing at d = 4 mm (a) and d = 6 mm (b)

At the same time, the study revealed the optimal radiation parameters (P=2 kW, V=20 $mm \cdot s^{-1}$, d=10mm) at which the values of width B = 7.12 mm, depth H = 0.81 mm, and the average microhardness 6700 MPa.

The maximum wear resistance is obtained at 100% of the laser quenching area (see figure 6). It exceeds the wear resistance of the sample base material by 3.5 times. However, for parts operating under sliding friction conditions with a load not exceeding 3 MPa and under satisfactory lubrication conditions, it is advisable to use a technology with laser hardening of 50-60% of the friction surface. In this case, the wear resistance is increased by 3.2 times compared to the non-hardened sample, and the cost of laser treatment is two times lower.

From the presented research results it follows that the smallest losses of laser radiation energy are achieved when the beam is defocused up to 60 mm. Further increase in the beam diameter leads to a significant loss of energy along the hardening path edges sometimes more than 50%, but in some cases this is permissible when hardening without melting, for example, when processing the surface of the punches.

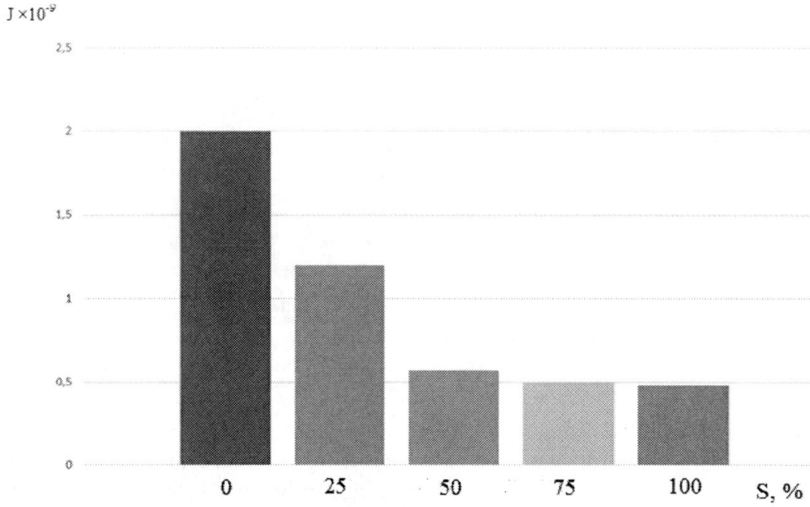

Figure 6. Wear intensity diagram depending on the area of the hardened surface

4. Conclusion

The dependences of the depth and width of the hardened zones on the beam defocus of 25-130 mm for the speed and power of the laser beam of 10-20 mm·s^{-1} and 1000-2000 W, respectively, are obtained.
With the help of regression analysis, the surfaces demonstrating the dependence of the hardening zones parameters on the power and velocity of the laser beam are constructed.
Wear resistance of samples at 50% of the hardened layer area is 3.2 times higher than the base material, normalized steel 40Cr.

References

[1] Kazakov S S and Goyim V V 2015 *Journal Karelian scientific* **11** 120-3
[2] Kazakov S S and Matveev Y I 2011 *Bulletin of NGIEI* **2** 55-60
[3] Arakelyan S M, Estonin G A, Skryabin I O, Abrahan S I and Novikov O A 2016 *Journal Modern high technologies* **5** 9-13
[4] Moradi M and Karami Moghadam M 2019 *Journal Optics and Laser Technology* **111** 554-70
[5] Biryukov V P, Fishkov A A, Tatarkin D Yu and Hreptovich E. V. 2017 *Journal Photonics* **3** 28-34
[6] Biryukov V P 2017 *Journal Photonics* **2** 22-32
[7] Yegorov A P, Gavrilova V S and Voronov S A 2015 *Eighth all-Russian conf. of young scientists and specialists "Future of machine-building in Russia"*: collection of reports of the Russian machine-builders Moscow MGTU N E Bauman pp 449-51
[8] Klevetov V D, Starostin D A and Egorov A P 2018 *Technology. Security. Management: materials of VIII all-Russian scientific and technical conference* FGBOU VO "KGTU named after V A Degtyarev" (Kovrov, Russia) pp 21-30
[9] Yugov V I 2012 *Journal Photonics* **4** 12-20
[10] Evdokimov Yu A, Kolesnikov V I and Teterin P I 1980 *Planning and analysis of experiments in*

IOP Conf. Series: Materials Science and Engineering **489** (2019) 012032 doi:10.1088/1757-899X/489/1/012032

solving problems of friction and wear (Moscow: Nauka) in Russian

IOP Conf. Series: Materials Science and Engineering **489** (2019) 012033 doi:10.1088/1757-899X/489/1/012033

Numerical analysis of nonlinear forced vibrations of a cylindrical shell with combinational internal resonance in a fractional viscoelastic medium

B Ajarmah and M V Shitikova

Voronezh State Technical University, Research Center on Dynamics of Solids and Structures, 20-letija Oktjabrja Street 84, 394006 Voronezh, Russia

basem@pass.ps mvs@vgasu.vrn.ru

Abstract. Nonlinear force driven vibrations of thin cylindrical shells in a viscoelastic medium, the damping properties of which are described by the fractional derivative Kelvin–Voigt model, are studied. A comparative analysis has been carried out for numerical results of both free and forced shell's oscillations using two different numerical methods for the case of the combination internal resonance.

1. Introduction

The analysis of forced nonlinear vibrations of thin cylindrical shell is the important area of applied mechanics, since shells are widely used as structural elements in many fields of industry and technology [1]. It will suffice to mention the state-of-the-art articles [2-3] and the monograph [4]. involving the extensive review of literature in the field of internal resonances in different mechanical systems. Various types of the internal resonance: one-to-one, two-to-one, three-to-one, as well as a variety of combinational resonances, when three and more natural modes interact, have been discussed. The enumerated internal resonances were investigated in various mechanical systems with multiple degree-of-freedom, as well as in strings, beams, plates, and shells.

Free damped nonlinear vibrations of cylindrical shells in a fractional derivative medium have been studied in [5], wherein the procedure resulting in decoupling linear parts of nonlinear equations has been proposed with the further utilization of the method of multiple scales for solving nonlinear governing equations of motion, in so doing the amplitude functions are expanded into power series in terms of the small parameter and depend on different time scales.

It has been shown that the phenomenon of the internal resonance between vibrational subsystems of the cylindrical shell under consideration could be very critical, since in the circular cylindrical shell of such a type the two-to-one [5], one-to-one, three-to-one [6] internal resonances, as well as combinational internal resonances [7] could occur, which are governed by the order of smallness of viscosity. All possible cases of the internal resonance have been recently revealed in [7], which belong to the resonances of the constructive type, since all of them depend on the geometrical dimensions of the shell under consideration and its mechanical characteristics, that is why such resonances could not be ignored and eliminated for a particularly designed shell. It has been shown that the energy exchange could occur between two or three subsystems at a time: normal vibrations of the shell, its torsional vibrations and shear vibrations along the shell axis. Such an energy exchange, if it takes place for a rather long time, could result in crack formation in the shell, and finally to its failure. The energy exchange has been illustrated pictorially by the phase portraits, wherein the phase trajectories of the phase fluid motion are depicted [5-7].

In the present paper, the approach proposed in [5] has been extended to the force driven vibrations of a thin cylindrical shell under the combinational internal resonance with the force frequency approximately equal to one of natural frequencies.

2. Problem formulation

Let us consider the dynamic behavior of a simply-supported thin cylindrical shell with geometrical nonlinearities, vibrations of which in a viscoelastic fractional derivative medium are described in the general coordinates by the following three differential equations written in the dimensionless form (free damped nonlinear vibrations described by the partial nonlinear differential equations presented in [5] are supplemented herein by the orthogonal harmonic force applied to the thin cylindrical shell middle surface at the point with coordinates x_0,φ_0,r):

$$u_{xx} + \frac{1-\nu}{2}\beta_1^2 u_{\varphi\varphi} + \frac{1+\nu}{2}\beta_1 v_{x\varphi} - \nu\beta_1 w_x + w_x\left(w_{xx} + \frac{1-\nu}{2}\beta_1^2 w_{\varphi\varphi}\right) + \frac{1+\nu}{2}\beta_1^2 w_\varphi w_{x\varphi} = \ddot{u} + æ_1\left(\frac{d}{dt}\right)^\gamma u, \tag{1}$$

$$\beta_1^2 v_{\varphi\varphi} + \frac{1-\nu}{2} v_{xx} + \frac{1+\nu}{2}\beta_1 u_{x\varphi} - \beta_1^2 w_\varphi + \beta_1 w_\varphi\left(\beta_1^2 w_{\varphi\varphi} + \frac{1-\nu}{2} w_{xx}\right) + \frac{1+\nu}{2}\beta_1 w_x w_{x\varphi} = \ddot{v} + æ_2\left(\frac{d}{dt}\right)^\gamma v, \tag{2}$$

$$\begin{aligned}
&\frac{\beta_2^2}{12}(w_{xxxx} + 2\beta_1^2 w_{xx\varphi\varphi} + \beta_1^2 w_{\varphi\varphi\varphi\varphi}) - \nu\beta_1 u_x - \beta_1^2 v_\varphi + \beta_1^2 w \\
&+\frac{1}{2}\nu\beta_1 (w_x)^2 + \frac{1}{2}\beta_1^3 (w_\varphi)^2 - w_x\left(u_{xx} + \frac{1-\nu}{2}\beta_1^2 u_{\varphi\varphi} + \frac{1+\nu}{2}\beta_1 v_{x\varphi}\right) \\
&-\beta_1 w_\varphi\left(\beta_1^2 v_{\varphi\varphi} + \frac{1-\nu}{2} v_{xx} + \frac{1+\nu}{2}\beta_1 u_{x\varphi}\right) - w_{xx}\left(u_x + \nu\beta_1 v_\varphi - \nu\beta_1 w\right) \\
&-\beta_1^2 w_{\varphi\varphi}\left(\nu u_x + \beta_1 v_\varphi - \beta_1 w\right) - (1-\nu)\beta_1 w_{x\varphi}\left(\beta_1 u_\varphi + v_x\right) \\
&= -\ddot{w} - æ_3\left(\tfrac{d}{dt}\right)^\gamma w - F_0\delta(x-x_0)\delta(\varphi-\varphi_0)\cos(\Omega_F t),
\end{aligned} \tag{3}$$

subjected to the initial

$$u|_{t=0} = v|_{t=0} = w|_{t=0} = 0, \quad \dot{u}|_{t=0} = \dot{v}|_{t=0} = \dot{w}|_{t=0} = 0 \tag{4}$$

as well as the boundary conditions

$$w|_{x=0} = w|_{x=l} = 0, \quad v|_{x=0} = v|_{x=l} = 0, \quad u_x|_{x=0} = u_x|_{x=l} = 0, \quad w_{xx}|_{x=0} = w_{xx}|_{x=l} = 0, \tag{5}$$

where the x-axis is directed along the axis of the cylinder, r and φ is the polar radius and angle, respectively, $u = u(x,\varphi,t)$, $v = v(x,\varphi,t)$, and $w = w(x,\varphi,t)$ are the displacements of points located in the shell's middle surface in three mutually orthogonal directions x,φ,r, $\beta_1 = l/R$, $\beta_2 = h/l$ are parameters defining the dimensions of the shell , h is the thickness, l is the length, R is the radius, t is time, an over dot denotes the time-derivative, lower indices label the derivatives with respect to the corresponding coordinates, $F_0\cos(\Omega_F t)$ is the harmonic force with the amplitude F0 = const and frequency ΩF, δ is the Dirac delta function, æi are damping coefficients, $\left(\frac{d}{dt}\right)^\gamma$ is the fractional order of the operator of differentiation which is equivalent to the Riemann-Liouville fractional derivative of the γ-order [8]

$$D^\gamma F = \frac{d}{dt}\int_0^t \frac{F(t-t')}{\Gamma(1-\gamma)}dt' . \tag{6}$$

To solve the partial differential equations (1)-(3) subjected to the initial (4) and boundary (5) conditions, we will use the approach suggested in [5], which involves the decoupling of linear parts of equations (1)-(3) with further utilization of the generalized method of multiple time scales [9] to expand the amplitude functions into power series in terms of the small parameter. Assuming thereafter that the vibrational process occurs in such a way that only three natural modes corresponding to the generalized displacements X_{imn} $(i=1,2,3)$ predominate and are coupled by the combinational resonance, resulting in the energy transfer between the coupled modes. Thus, equations (1)-(3) are reduced to the following set of equations:

$$\ddot{X}_{1mn} + æ_1 D^\gamma X_{1mn} + \Omega_{1mn}^2 X_{1mn} = -\sum_{i=1}^{3} F_{imn}L_{imn}^{I} - \tilde{F}\cos(\Omega_F t)L_{3mn}^{I}, \tag{7}$$

$$\ddot{X}_{2mn} + æ_2 D^\gamma X_{2mn} + \Omega_{2mn}^2 X_{2mn} = -\sum_{i=1}^{3} F_{imn}L_{imn}^{II} - \tilde{F}\cos(\Omega_F t)L_{3mn}^{II}, \tag{8}$$

$$\ddot{X}_{3mn} + æ_3 D^\gamma X_{3mn} + \Omega_{3mn}^2 X_{3mn} = -\sum_{i=1}^{3} F_{imn}L_{imn}^{III} - \tilde{F}\cos(\Omega_F t)L_{3mn}^{III}, \tag{9}$$

where m and n are integers, $\tilde{F} = F_0\cos(\Omega_F t)\sin(\pi m x_0)\sin(\pi n \varphi_0)$, Ω_{1mn}, Ω_{2mn} and Ω_{3mn} are the eigenvalues of the matrix S_{ij}^{mn} and L_{imn}^{I}, L_{imn}^{II} and L_{imn}^{III} are eigenvectors of the same matrix S_{ij}^{mn} with the elements

$$S_{ij}^{mn} = \begin{bmatrix} S_{11}^{mn} & S_{12}^{mn} & S_{13}^{mn} \\ S_{21}^{mn} & S_{22}^{mn} & S_{23}^{mn} \\ S_{31}^{mn} & S_{32}^{mn} & S_{33}^{mn} \end{bmatrix} = \begin{bmatrix} \left(\pi^2 m^2 + \frac{1-\nu}{2}\beta_1^2 n^2\right) & \frac{1+\nu}{2}\beta_1 \pi m n & \nu\beta_1\pi m \\ \frac{1+\nu}{2}\beta_1\pi m n & \left(\frac{1-\nu}{2}\pi^2 m^2 + \beta_1^2 n^2\right) & \beta_1^2 n \\ \nu\beta_1\pi m & \beta_1^2 n & \frac{\beta_2^2}{12}(\pi^2 m^2 + \beta_1^2 n^2)^2 + \beta_1 \end{bmatrix} \tag{10}$$

3. Methods of solution

Utilizing the procedure described in detail in [7] and assuming that $\tilde{F} = \varepsilon^2 f$, we will solve equations (7)-(9) using two different methods for the case of the combinational internal resonance $\Omega_1 = \Omega_2 + \Omega_3$ accompanied by the primary resonance, i.e. when the frequency of the external force is close to the natural frequency of the vertical mode $\Omega_F = \Omega_3 + \varepsilon\sigma$, where σ is the detuning parameter.

The first method resides in the discretization of all derivatives in equations (7)-(10). For this purpose we will follow the procedures suggested by Diethelm in [10,11].

Let us introduce the following notation:

$$\begin{gathered} Y_1 = X_1, \quad Y_2 = D^\gamma X_1 = D^\gamma Y_1, \quad Y_3 = DX_1 = DY_1, \\ \ddot{X}_1 = DDX_1 = DY_3 = -\sum_{i=1}^{3} F_{1mn}L_{i\,mn}^{I} - æ_1 Y_2 - \Omega_1^2 Y_1 - \tilde{F}\cos(\Omega_F t)L_{3mn}^{I}. \end{gathered} \tag{11}$$

The first time-derivative could be discretized using the trapezoidal rule

$$DY_1 = Y_3 \to Y_{1\,i} = Y_{1\,i-1} + \frac{1}{2}h\left(Y_{3\,i} + Y_{3\,i-1}\right), \tag{12}$$

therefore

$$Y_{i\,i} - \frac{1}{2}hY_{3\,i} = Y_{1\,i-1} + \frac{1}{2}h\left(Y_{3\,i-1}\right) = s_{2\,i-1}. \tag{13}$$

The fractional derivative could be discretized using the Diethelm's method [10,11] according to the definition

$$D^{\gamma}Y_1 = \frac{1}{{}^{\alpha}\chi_i}\left(\sum_{k=0}^{i} {}^{\gamma}\omega_{k,i}Y_{1\ i-k} + \frac{Y_{1\ 0}}{\gamma}\right) = Y_2\,, \tag{14}$$

whence it follows

$${}^{\gamma}\chi_i Y_2 - {}^{\gamma}\omega_{0,i}Y_{1\ i} = \left(\sum_{k=1}^{i} {}^{\gamma}\omega_{k,i}Y_{1\ i-k} + \frac{Y_{1\ 0}}{\gamma}\right) \tag{15}$$

Assuming that $s_{1\ i-1} = \left(\sum_{k=1}^{i} {}^{\gamma}\omega_{k,i}Y_{1\ i-k} + \frac{Y_{1\ 0}}{\gamma}\right)$ and applying the described above procedure for other two equations, equations (7)-(9) could be represented in the following matrix form:

$$\begin{bmatrix} -{}^{\gamma}\omega_{0,i} & {}^{\gamma}\chi_i & 0 & 0 & 0 & 0 & 0 & 0 & 0 \\ 1 & 0 & -\frac{h}{2} & 0 & 0 & 0 & 0 & 0 & 0 \\ \frac{h}{2}\Omega_1^2 & \frac{h}{2}æ_1 & 1 & 0 & 0 & 0 & 0 & 0 & 0 \\ 0 & 0 & 0 & -{}^{\gamma}\omega_{0,i} & {}^{\gamma}\chi_i & 0 & 0 & 0 & 0 \\ 0 & 0 & 0 & 1 & 0 & -\frac{h}{2} & 0 & 0 & 0 \\ 0 & 0 & 0 & \frac{h}{2}\Omega_2^2 & \frac{h}{2}æ_2 & 1 & 0 & 0 & 0 \\ 0 & 0 & 0 & 0 & 0 & 0 & -{}^{\gamma}\omega_{0,i} & {}^{\gamma}\chi_i & 0 \\ 0 & 0 & 0 & 0 & 0 & 0 & 1 & 0 & -\frac{h}{2} \\ 0 & 0 & 0 & 0 & 0 & 0 & \frac{h}{2}\Omega_3^2 & \frac{h}{2}æ_2 & 1 \end{bmatrix} * \begin{bmatrix} Y_{1i} \\ Y_{2i} \\ Y_{3i} \\ Y_{4i} \\ Y_{5i} \\ Y_{6i} \\ Y_{7i} \\ Y_{8i} \\ Y_{9i} \end{bmatrix} = \begin{bmatrix} s_{1\,i-1} \\ s_{2\,i-1} \\ s_{3\,i-1} \\ s_{4\,i-1} \\ s_{5\,i-1} \\ s_{6\,i-1} \\ s_{7\,i-1} \\ s_{8\,i-1} \\ s_{9\,i-1} \end{bmatrix} \tag{16}$$

It is quite straightforward to solve (16) for **Y** column by finding the inverse of the left matrix and multiplying it by **s** (the right column which represents the already known values or initial conditions from the previous iteration step).

The second method is based on the generalized multiple time scales approach proposed by Rossikhin and Shitikova [9], resulting in a set of six nonlinear differential equations in terms of amplitudes a_i and phases φ_i

$$\begin{aligned} &(a_1^2\dot{)} + s_1a_1^2 = \Omega_1^{-1}a_{23}^{I}a_1a_2a_3\sin\delta, && \dot{\varphi}_1 - \tfrac{1}{2}\sigma_1 - \tfrac{1}{2}\tfrac{a_{23}^{I}}{\Omega_1}\tfrac{a_2a_3}{a_1}\cos\delta = 0, \\ &(a_2^2\dot{)} + s_2a_2^2 = -\Omega_2^{-1}a_{13}^{II}a_1a_2a_3\sin\delta, && \dot{\varphi}_2 - \tfrac{1}{2}\sigma_2 - \tfrac{1}{2}\tfrac{a_{13}^{II}}{\Omega_2}\tfrac{a_1a_3}{a_2}\cos\delta = 0, \\ &(a_3^2\dot{)} + s_3a_3^2 = -\Omega_3^{-1}a_1a_2a_3\sin\delta - \Omega_3^{-1}fa_3\sin\varphi_3, && \dot{\varphi}_3 - \tfrac{1}{2}\sigma_3 - \tfrac{1}{2}\tfrac{a_{12}^{III}}{\Omega_3}\tfrac{a_1a_2}{a_3}\cos\delta - \tfrac{1}{2(\Omega_3a_3)}f\cos\varphi_3 = 0, \end{aligned} \tag{17}$$

where $\delta = \varphi_2 + \varphi_3 - \varphi_1$ is the phase difference, and all coefficients are given in [7].

Equations (17) could be solved numerically using the Runge-Kutta fourth-order algorithm via MatLab.

4. Numerical results

Reference to figure 1 shows the relation between β_2 and β_1 representing shell's parameters under the combinational internal resonance $\Omega_1=\Omega_2+\Omega_3$. Figure 2 represent the interdependency between the shell's parameters and the natural frequencies satisfying the condition of the combinational resonance under investigation. From Figure 2(A) it is seen that the shell parameters β_2 and β_1 under the combinational internal resonance are exponentially interrelated, while from Figures 2(B) and 2(C) it is evident that magnitudes of β_2 are linearly related to every frequency of combinational internal resonance in the case of ($\Omega_1=\Omega_2+\Omega_3$), and β_1 are exponentially related to them.

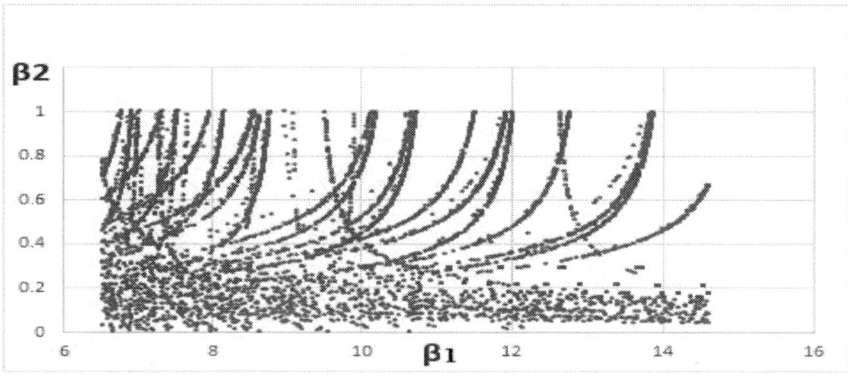

Figure 1 Shell's parameters under the condition of the combinational internal resonance ($\Omega_1=\Omega_2+\Omega_3$)

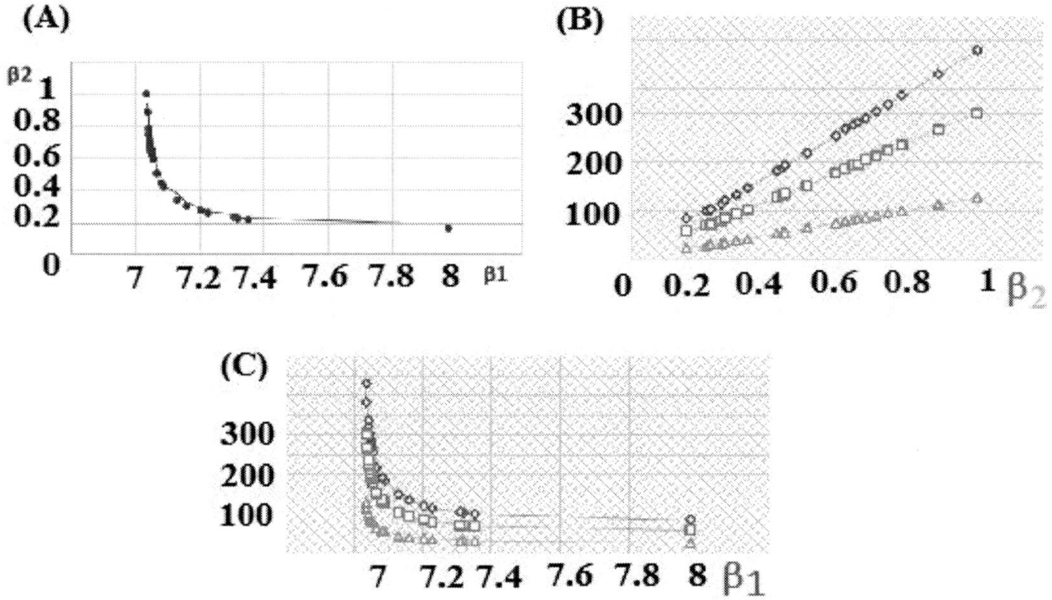

Figure 2 Interdependency between parameters β_1, β_2 and frequencies Ω_1, Ω_2, Ω_3: (A) β_1 - β_2 dependence at $m_1=n_1=m_2=m_3=5$, $n_2=4$, $n_3=2$, (B) β_2-dependence of frequencies -◇- Ω_1, -□- Ω_2, -▫- Ω_3, (C) β_1-dependence of frequencies

The numerical solution of (16) using the multi-step method has been carried out for the dimensionless parameters given in figure 1, and the results are shown in figure 3 which represents the time-dependence estimation of the generalized displacements X_1, X_2, and X_3 during forced vibrations at the fractional parameter $\gamma=0.25$.

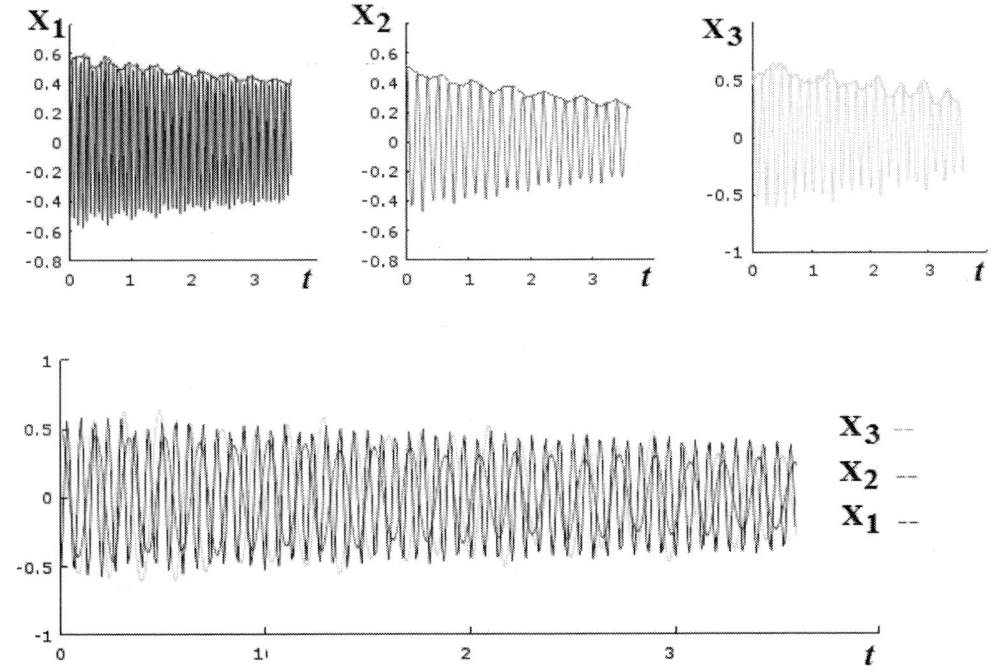

Figure 3 Time-dependence of the generalized displacements X_1, X_2, and X_3 calculated via the Diethelm's multi-term method

The solution of equations (17) has been carried out numerically using the Runge-Kutta forth-order algorithm, and the results are shown for forced and free vibrations in Figures 4 and 5, respectively. The magnitudes of the fractional parameter γ are shown by figures near the corresponding curves. Figures 4 and 5 show the extensive energy exchange between the coupled modes of vibrations, in so doing the external force increases the amplitudes of vibrations. With the increase of the fractional parameter from zero (undamped vibrations) to unit (conventional Kelvin-Voigt model) attenuation of vibrations increases.

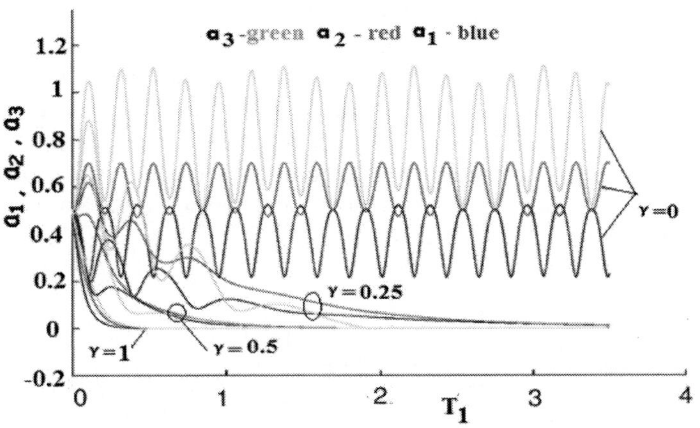

Figure 4: Amplitudes vs T_1: forced vibrations analyzed by the Rossikhin-Shitikova generalized method of multiple-time scales

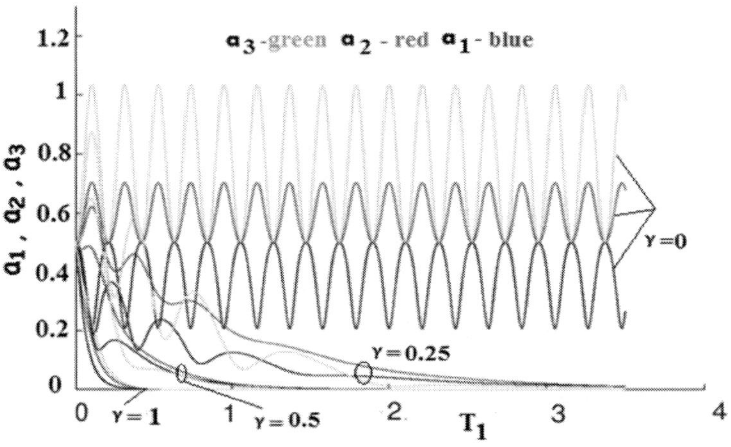

Figure 5 Amplitudes vs T_1: free vibrations analyzed by calculated by the Rossikhin-Shitikova generalized method of multiple-time scales

Conclusion

In the present paper, nonlinear force driven vibrations of thin cylindrical shell in a viscoelastic medium have been studied, when the motion of the cylindrical shell is described by a set of three coupled nonlinear differential equations subjected to the conditions of the combinational internal resonance, resulting in the interaction three orthogonal modes, corresponding to the mutually orthogonal displacements, using two different numerical methods. To describe the nonlinear damped vibrations of the thin shell under consideration, the fractional derivative Kelvin-Voigt model is used, because its prediction is in a good compliance with experimental data. The nonlinear set of resolving equations has been obtained in terms of the generalized displacements and in terms of the amplitudes and phases. The two systems have been solved numerically by two different methods.

Within the first method, the generalized displacements of a coupled set of nonlinear ordinary differential equations are estimated using numerical solution of nonlinear multi-term fractional differential equations by the procedure based on the reduction of the problem to a system of fractional differential equations. According to the second method, the amplitudes and phases of nonlinear vibrations are estimated from the governing nonlinear differential equations describing amplitude-and-phase modulations for the case of the combinational internal resonance. A good agreement between the results obtained by the two methods has been found.

Relationships (16) and (17) allow one to trace the external force influence on vibrational process, resulting in the energy exchange between the coupled modes. The interrelation between the parameters β_2 and β_1, which define the geometrical parameters of the shell, and the shell's natural frequencies Ω_1, Ω_2, and Ω_3, which could fall within the condition of the internal combinational resonance $\Omega_1=\Omega_2+\Omega_3$, has been revealed in the form of linear and exponential relations. It has been shown that with the increase of the fractional parameter from zero (undamped vibrations) to unit (conventional Kelvin-Voigt model) attenuation of vibrations increases, and the external force influence becomes less effective as the fractional parameter γ approaches to 1.

Acknowledgments

This research was made possible by the Grant No. 9.5138.2017/BP as a Government task from the Ministry of Education and Science of the Russian Federation.

References

[1] Alijani F, Amabili M 2014 Non-linear vibrations of shells: A literature review from 2003 to 2013 *Int. J. Non-Linear Mech.* **58** 233–57

[2] Rossikhin Y A, Shitikova M V 2018 Soliton-like solutions in the problems of vibrations of nonlinear mechanical systems: Survey *Nonlinear Systems - Modeling, Estimation, and Stability* (London: InTechOpen) Chapter 4 pp 65-87

[3] Nayfeh A, Balachandran S 1989 Modal interactions in dynamical and structural systems. Appl. Mech. Rev. **42** 175-201

[4] Nayfeh A 2000 *Nonlinear Interactions: Analytical, Computational, and Experimental Methods* (New York: John Wiley & Sons)

[5] Rossikhin Y A, Shitikova M V 2018 2015 A new approach for studying nonlinear dynamic response of a thin fractionally damped cylindrical shell with internal resonances of the order of ε *Adv. Struct. Mater.* **45** 301-21

[6] Rossikhin Y A, Shitikova M V 2015 Nonlinear dynamic response of a fractionally damped cylindrical shell with a three-to-one internal resonance. *Appl. Math. Comp.* **257** 498-525

[7] Rossikhin Y A, Shitikova M V 2015 Analysis of non-linear vibrations of a fractionally damped cylindrical shell under the conditions of combinational internal resonance. *Computational Problems in Science and Engineering* **343** 59-107

[8] Rossikhin Y A, Shitikova M V 2009 Application of fractional calculus for dynamic problems of solid mechanics: Novel trends and recent results *Appl. Mech. Rev.* **63**(1) 010801

[9] Rossikhin Y A, Shitikova M V, and Shcheglova TA 2009 Forced vibrations of a nonlinear oscillator with weak fractional damping *J. Mech. Mat. Struct.* **4** (9) 1619- 36

[10] Baleanu D, Diethelm K, Scalas E, Trujillo J J 2012 *Fractional Calculus* (Singapore: World Scientific)

[11] Diethelm K 1997 An algorithm for the numerical solution of differential equations of fractional order *Elect. Trans. Num. Anal.* **5** 1-6

IOP Conf. Series: Materials Science and Engineering **489** (2019) 012034 doi:10.1088/1757-899X/489/1/012034

Linear orbital stability analysis of the pendulum-type motions of a Kovalevskaya top with a suspension point vibrating horizontally

M V Belichenko

Moscow Aviation Institute (National Research University), Moscow, Russia

tuzemec1@rambler.ru

Abstract. The motion of a heavy rigid body is considered, with a point (suspension point) performing high-frequency horizontal vibration. The center of mass of the body is on the principal axis of inertia for this point. Within the framework of an approximate autonomous system, pendulum-type motions of the body are studied, they occur in a vertical plane containing the axis of vibration or in a plane perpendicular to it. For a body with a mass geometry corresponding to S. V. Kovalevskaya case, a linear orbital stability analysis of these motions is performed.

1. Introduction

The presence of high-frequency vibrations of the rigid body suspension has a significant influence on its motion. Vibrations can lead to the emergence of new types of body motions impossible for a body with a fixed point, or to a change in the stability of known particular cases of motion.

At the beginning of the 20th century, the phenomenon of dynamic stabilization of the inverted position of a mathematical pendulum due to vertical vibrations of the suspension point was first described [1]. Later, the mathematical pendulum and the pendulum systems were studied.

Recently, the influence of high-frequency vibrations on the dynamics of a heavy rigid body with a more complicated mass geometry, as well as more complex vibrations of its suspension point, have been investigated. In [2], an approximate autonomous system of Euler-Poisson type equations is obtained, which describes the motion of a heavy rigid body with an arbitrary mass geometry, with suspension point performing arbitrary periodic or conditionally periodic fast vibrations of small amplitude in three-dimensional space. Within the framework of this approximate system, a number of problems on dynamics of a body with a vibrating suspension has been solved. In particular, the orbital stability of the pendulum-type motions of the Lagrange top with a vibrating suspension has been studied [3].

Pendulum-type motions of a rigid body with a fixed point were first described in [4]. The orbital stability of the pendulum-type motions of a Kovalevskaya top with a fixed point was studied in [5] and [6].

This paper studies the influence of high-frequency horizontal periodic vibrations on the existence and stability of the pendulum-type motion of a rigid body with the center of mass on the principal axes of inertia for the suspension point. The problem of linear orbital stability of the pendulum-type motions for the Kovalevskaya top in the presence of such vibrations is considered in detail.

2. Statement of the problem. Equations of motion

We consider the motion of a heavy rigid body in a uniform gravitational field, assuming that one of the points of the body O (the suspension point) makes a periodic, with frequency Ω, motion along the horizontal line according to the law $O'O = \xi(t)$ relative to a fixed point O'. The average value of $\langle \xi(t) \rangle$ for the period is assumed to be zero.

We introduce a coordinate system $OXYZ$ moving translationally, with the axis OY directed vertically upwards, and OX along the direction of vibrations. The second coordinate system $Oxyz$ is rigidly attached to the body, with axes directed along the body principal axes of inertia for the point O. Let the center of mass G of the body lie on the principal axis Oz, and $OG = z_G$. The principal moments of inertia for the point O are A, B and C, and the mass of the body is m. The orientation of the system $Oxyz$ relative to $OXYZ$ is defined by Euler angles ψ, θ and φ.

Let the maximum deviation h_* of the suspension point from the point O' is small compared with the reduced body length $l = B/(mz_G)$, and the vibration frequency Ω of this point is large relative to the characteristic frequency $\omega_* = \sqrt{g/l}$. We introduce a small parameter $\varepsilon = \sqrt{h_*/l}$ and assume that $h_*\Omega$ □ 1.

We describe the motion of the body using the canonical Hamiltonian equations, choosing the Euler angles ψ, θ and φ for the generalized coordinates. Using the methods of the perturbation theory, we can obtain a near-identity canonical change of variables that reduces the Hamiltonian function of the system to a form, in which the main (approximate) part is autonomous.

We introduce the dimensionless parameters, pulses and time by the formulas

$$\alpha = \frac{\langle (\xi')^2 \rangle m^2 l^2}{A^2 \omega_*^2}, \quad \mu = \frac{B}{A}, \eta = \frac{B}{C}, P_\psi = B\omega_* P_1, \; P_\theta = B\omega_* P_2, \; P_\varphi = B\omega_* P_3, \; \tau = \omega_* t$$

Leaving the original notation for the Euler angles, we can write the approximate Hamiltonian in the form (here $\hat{\Pi}_v$ is the vibration potential [2])

$$\hat{H} = \frac{\eta P_3^2}{2} + \frac{(\sin^2\varphi + \mu\cos^2\varphi) P_2^2}{2} + \frac{(\mu - 1)\sin\varphi\cos\varphi (P_1 - P_3\cos\theta) P_2}{\sin\theta} + \frac{(\cos^2\varphi + \mu\sin^2\varphi)(P_1 - P_3\cos\theta)^2}{2\sin^2\theta} - \cos\psi\sin\theta + \hat{\Pi}_v \tag{1}$$

$$\hat{\Pi}_v = \frac{\alpha}{2}\left(\mu(\cos\psi\sin\varphi + \sin\psi\cos\varphi\cos\theta)^2 + (\cos\psi\cos\varphi - \sin\psi\sin\varphi\cos\theta)^2\right) \tag{2}$$

Further research is carried out in the framework of the approximate system with Hamiltonian (1), (2). The errors of the solutions of this system with respect to ones of the complete system as well as the time interval, on which the approximate solutions are studied, are given in [2].

The system with the Hamiltonian (1), (2) admits particular solutions that correspond to pendulum-like body motions, where one of its principal axes of inertia not containing the center of mass G, takes a fixed horizontal position, and the radius vector OG moves in the vertical plane containing the axis of vibration (plane OXY) or perpendicular to it (plane OYZ).

In the first case $\theta = \pi/2, \varphi = 0, \; P_2 = 0, P_3 = 0$, and the change in the quantities ψ and P_1 is described by canonical equations with the Hamiltonian function

$$H_0 = \frac{P_1^2}{2} - \cos\psi + \frac{\alpha}{2}\cos^2\psi \tag{3}$$

To describe motions in the plane perpendicular to the axis of vibrations, we redirect the axes of the coordinate system $OXYZ$ so that the vibrations of the suspension point occur along OZ. Part of the Hamiltonian (1) without the last term remains unchanged, and the vibration potential takes the form

$$\hat{\Pi}_v = \frac{\alpha}{2}\left(\mu\cos^2\varphi + \sin^2\varphi\right)\sin^2\theta \tag{4}$$

The motions at hand are given by the relations $\theta = \pi/2, \varphi = \pi/2$, $P_2 = 0, P_3 = 0$, while the quantities ψ and P_1 satisfy the canonical equations of the mathematical pendulum, defined by the Hamiltonian

$$H_{01} = \frac{\mu P_1^2}{2} - \cos\psi \tag{5}$$

Note that in the case of the absence of vibrations of the body suspension point, the pendulum-like motions of the radius vector center OG can occur in any fixed vertical plane. However, there is a cyclic coordinate in the system if we choice the other coordinates axes and these motions are unstable with respect to spatial perturbations.

Further, we will assume that the body mass geometry corresponds to S. V. Kovalevskaya case, which is realized for $\mu = 2, \eta = 1$.

The aim of the work is to describe and study the linear orbital stability of the pendulum-like motions of the Kovalevskaya top with a horizontally vibrating suspension point that correspond to the model systems with Hamiltonians (3) and (5).

3. Pendulum-type motions

Let us consider equilibrium positions of the model systems with Hamiltonians (3) and (5). The system with Hamiltonian (3) with $\alpha < 1$ has a lower stable equilibrium at $\psi = 0$ and upper unstable equilibrium at $\psi = \pi$. If $\alpha > 1$, then there are also stable inclined equilibrium positions at $\psi_* = \pm\arccos(\alpha^{-1})$, and the lower and upper equilibria are unstable. The motions of the system with Hamiltonian (5) do not depend on the vibration intensity, and there are stable lower and unstable upper equilibria.

The model systems under study have the first integrals (energy integrals) of the form $H_0 = h = \text{const}$, $H_{01} = h_1 = \text{const}$. Depending on the value of the energy constant, the pendulum-like motions are oscillations around stable equilibrium positions, rotations, or asymptotic motions. Figure 1 a and 1 b show bifurcation diagrams on the parameter plane α, h and α, h_1 for systems with Hamiltonians (3) and (5), respectively.

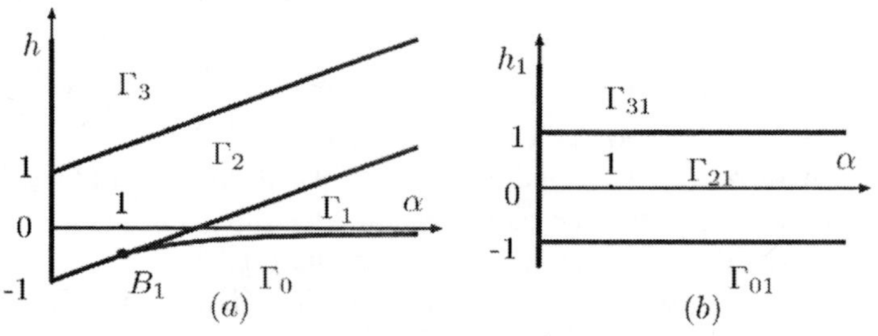

Figure 1. Bifurcation diagrams.

In the diagram in figure 1 a the boundaries of regions with different types of motions are straight lines $h=\pm 1+\alpha/2$ and part of the hyperbola $2\alpha h=-1, \alpha>1$. The regions in the diagram in figure 1 b separated by straight lines $h_1=\pm 1$. In the regions Γ_0 и Γ_{01} motion is impossible. In the region Γ_1, the radius vector of the center of mass of the body oscillates around the inclined equilibrium. The regions Γ_2, Γ_{21} correspond to the oscillations of this vector near the lower position, and the regions Γ_3 and Γ_{31}, to the rotations.

Let us integrate the equations of motion of the system with Hamiltonian (3) in each of the regions in figure 1 a. We solve the energy integral $H_0=h$ with respect to P_1 and, given that $\psi'=P_1$, we write the differential equation of the pendulum-type motion in the form

$$\frac{d\psi}{d\tau}=\pm\left(2h+2\cos\psi-\alpha\cos^2\psi\right)^{1/2}$$

Assuming that $u=\cos\psi$ ($|u|\le 1$), we rewrite this equation as

$$\tau=\pm\int_{u_0}^{u}\frac{du}{\sqrt{f(u)}},\quad f(u)=-\alpha\left(1-u^2\right)(u-u_1)(u-u_2) \tag{6}$$

Here $u_1=\left(1-\sqrt{1+2\alpha h}\right)\alpha^{-1}$, $u_2=\left(1+\sqrt{1+2\alpha h}\right)\alpha^{-1}$.

The integral on the right side of (6) is expressed in terms of elliptic functions. Further, we use the notation sn for elliptic sine, $\mathrm{F}(u,k)$ for elliptic integral of the first kind, $\mathrm{K}(k)$ for complete elliptic integral of the first kind, and k for module of the elliptic integral.

If the values of the parameters α and h correspond to the region Γ_1, then the roots of the polynomial $f(u)$ from (6) satisfy the condition $-1<u_1<u_2<1$. For oscillations in the vicinity of the inclined position, condition $u_1\le u\le u_2$ satisfies. Assuming that $u|_{\tau=0}=u_2$ and using some formulas from [7], we obtain

$$\tau=-\frac{\mathrm{F}(u_*,k_1)}{\xi_1}, u_*=\arcsin\left(\frac{(1-u_1)(u_2-u)}{(1-u)(u_2-u_1)}\right)^{1/2}, \xi_1=\frac{1}{2}\left(\alpha(1-u_1)(u_2+1)\right)^{1/2}, k_1=\left(\frac{2(u_2-u_1)}{(1-u_1)(u_2+1)}\right)^{1/2}$$

Reversing the function $\tau(u)$, we have

$$u=\frac{(u_2-u_1)\mathrm{sn}^2(\xi_1\tau,k_1)-u_2(1-u_1)}{(u_2-u_1)\mathrm{sn}^2(\xi_1\tau,k_1)-1+u_1}$$

The resulting solution is periodic, with frequency $\omega_1=\pi\xi_1/(2K(k_1))$.

The motions in the region Γ_2 are oscillations that cover the lower and inclined equilibrium positions (if they exist), and $-1<u_1\le u\le 1<u_2$. In this case, the solution with the initial condition $u|_{\tau=0}=1$ is

$$u=\frac{u_2(1-u_1)\mathrm{sn}^2(\xi_2\tau,k_2)+u_1-u_2}{(1-u_1)\mathrm{sn}^2(\xi_2\tau,k_2)+u_1-u_2},\quad \xi_2=\frac{1}{2}\left(2\alpha(u_2-u_1)\right)^{1/2},\quad k_2=\left(\frac{(1-u_1)(u_2+1)}{2(u_2-u_1)}\right)^{1/2}$$

and its frequency is $\omega_2=\pi\xi_2/2K(k_2)$.

For the rotations in the region Γ_3 we have $u_1<-1\le u\le 1<u_2$. The solution with the initial condition $u|_{\tau=0}=1$ is represented as

IOP Conf. Series: Materials Science and Engineering **489** (2019) 012034 doi:10.1088/1757-899X/489/1/012034

$$u = \frac{2u_2\mathrm{sn}^2(\xi_3\tau, k_3) - u_2 - 1}{2\mathrm{sn}^2(\xi_3\tau, k_3) - u_2 - 1}, \quad \xi_3 = \frac{1}{2}\left(\alpha(1-u_1)(u_2+1)\right)^{1/2}, \quad k_3 = \left(\frac{2(u_2-u_1)}{(1-u_1)(u_2+1)}\right)^{1/2}$$

and its frequency is $\omega_3 = \pi\xi_3 / 2K(k_3)$.

At the boundary $h = 1 + \alpha / 2$, the function $f(u)$ has a double root $u = -1$. Integrating, we find

$$u = \frac{\mu^2 - (4\alpha+6)\mu + 1}{-\mu^2 - (4\alpha+2)\mu - 1}, \quad \mu = \exp\left(-2\tau\sqrt{1+\alpha}\right), \quad u\big|_{\tau=0} = 1$$

This expression tends with time to the point $u = -1$ corresponding to the upper equilibrium position.

At the boundary of the regions Γ_1 and Γ_2, the function $f(u)$ has a double root $u = 1$. Integration yields

$$u = \frac{\mu^2 - (4\alpha-6)\mu + 1}{\mu^2 + (4\alpha-2)\mu + 1}, \quad \mu = \exp\left(2\tau\sqrt{\alpha-1}\right), \quad u\big|_{\tau=0} = u_1$$

In this solution, the radius vector of the center of mass asymptotically approaches the lower equilibrium position.

The motion of the system with Hamiltonian (5) is described by the equation of the ordinary mathematical pendulum. The frequencies of the motions in the regions Γ_{21}, Γ_{31} are respectively

$$\omega_{21} = \frac{\pi}{2\sqrt{2}\,\mathrm{K}(k_{21})}, \quad \omega_{31} = \frac{\pi k_{21}}{\sqrt{2}\,\mathrm{K}(k_{21}^{-1})}, \quad k_{21} = \left(\frac{h_1+1}{2}\right)^{1/2}$$

4. The study of orbital stability

To study the orbital stability of the motions considered, we introduce action-angle variables I, w in the regions Γ_1, Γ_2 $_3\Gamma$ $_{2}\Gamma$ $_{31}, \Gamma$ of oscillations and rotations of the model systems. The corresponding particular solutions of the complete three-degree-of-freedom systems, in which the "pendulum part" is written in variables I, w, will be taken as unperturbed motions.

In the system with Hamiltonian (1), we introduce perturbations by the formulas $r = I - I_0, p_2 = P_2, p_3 = P_3, q_2 = \theta - \pi/2, q_3 = \varphi - \varphi_0$, where I_0 and φ_0 are values of the variable action and the angle φ on the unperturbed motion. We consider the quadratic part, in the variables $r^{1/2}, p_2, q_2, p_3, q_3$, of the perturbed Hamiltonian, representing it in the form

$$H = H_2 + \ldots, \quad H_2 = \omega_0 r + H_{2*}(p_2, q_2, p_3, q_3, w) \tag{7}$$

Here ω_0 is a frequency of unperturbed motion. If pendulum-like motions of the system with Hamiltonian (3) are taken for the unperturbed motion, then the function H_{2*} has the form

$$H_{2*} = p_2^2 + P_1 q_3 p_2 + \frac{\cos\psi + 2\alpha\sin^2\psi + P_1^2}{2} q_2^2 + P_1 p_3 q_2 - \frac{\alpha\sin 2\psi}{2} q_3 q_2 + \frac{p_3^2}{2} + \frac{P_1^2 + \alpha\cos^2\psi}{2} q_3^2 \tag{8}$$

If the unperturbed motion is described by a system with Hamiltonian (5), then we have

$$H_{2*} = \frac{p_2^2}{2} - P_1 q_3 p_2 + \frac{p_3^2}{2} + 2P_1 q_2 p_3 + \frac{2P_1^2 - \alpha + \cos\psi}{2} q_2^2 + \frac{\alpha - P_1^2}{2} q_3^2 \tag{9}$$

IOP Conf. Series: Materials Science and Engineering **489** (2019) 012034 doi:10.1088/1757-899X/489/1/012034

In expressions (8) and (9), the functions $\psi(w)$ and $P_1(w)$ are calculated for the unperturbed motion.

Further, we will consider the linearized equations of perturbed motion described by the Hamiltonian H_2 from (7) with regard to relations (8) and (9). At the energy level $H_2 = 0$ that corresponds to the unperturbed motion we realize the isoenergetic reduction and, taking the quantity w as a new independent variable, we consider a non-autonomous reduced two-degree-of-freedom system with the Hamiltonian $K = H_{2*}\omega_0^{-1}$.

The stability criteria for the trivial equilibrium of a system with the Hamiltonian H_2 from (7) and the reduced system coincide [8]. The characteristic equation of the latter is

$$\rho^4 - a_1\rho^3 + a_2\rho^2 - a_1\rho + 1 = 0 \tag{10}$$

Here a_1 is the trace of the monodromy matrix $\mathbf{X}(2\pi)$, and a_2 is the sum of its main second-order minors. Stability conditions are given by inequalities

$$-2 < a_2 < 6, \quad 4(a_2 - 2) < a_1^2 < (a_2 + 2)^2 / 4$$

5. Generating points

At energy levels of unperturbed motion corresponding to stable equilibrium positions, the system with the Hamiltonian K is autonomous. The eigenfrequencies $\Omega_{i,j}$ (i is a region number, $j = 1,2$) of the equations of small linear oscillations in the regions $\Gamma_1, \Gamma_2, \Gamma_{21}$ have the form

$$\Omega_{1,1} = \frac{\left(4\alpha^2 - 1 + 2\sqrt{1 - 16\alpha^2 + 16\alpha^4}\right)^{1/2}}{\sqrt{\alpha^2 - 1}}, \quad \Omega_{2,1} = \frac{\sqrt{2}}{\sqrt{1-\alpha}}, \quad \Omega_{21,1} = \sqrt{2\alpha},$$

$$\Omega_{1,2} = \frac{\left(4\alpha^2 - 1 - 2\sqrt{1 - 16\alpha^2 + 16\alpha^4}\right)^{1/2}}{\sqrt{\alpha^2 - 1}}, \quad \Omega_{2,2} = \frac{\sqrt{\alpha}}{\sqrt{1-\alpha}}, \quad \Omega_{21,2} = \sqrt{2(1-\alpha)}$$

The points of the resonances of the first and second orders are generating points for parametric resonance regions and single resonant curves on the parameter plane α, h or α, h_1. At resonances of the first order, the relations $\Omega_{i,j} = N$ are satisfied, and at resonances of the second order, the relations $\Omega_{i,j} = 2N - 1$, $\Omega_{i,1} - \Omega_{i,2} = N$, $\Omega_{i,1} + \Omega_{i,2} = N$. Here N is an integer number. Let $\alpha_{i,k}$ be resonant values of the parameter α characterizing intensity of vibrations, where i is region number, and k resonant relation.

Consider the generating points of the resonances of the first and second orders for pendulum- type motions of the system with Hamiltonian (3). For all types of resonances at hand, the sequences of values of $\alpha_{i,k}$ monotonically tend to $\alpha = 1$ as the value N increases. Further, we will limit ourselves to the following points:

$$\alpha_{1,\Omega_1=2,3,4} = 1/2,\ 7/9,\ 7/8; \qquad \alpha_{1,\Omega_1=3/2,5/2,7/2} = 1/9,\ 17/25,\ 41/49;$$

$$\alpha_{1,\Omega_2=1,2} = 1/2,\ 4/5; \qquad \alpha_{1,\Omega_2=1/2,3/2,5/2} = 1/5,\ 9/13,\ 25/29;$$

$$\alpha_{1,\Omega_1+\Omega_2=2,3,4,5} = 0.16808,\ 1/2,\ 0.68575,\ 0.78804; \qquad \alpha_{1,\Omega_1-\Omega_2=1} = 1/2.$$

$$\alpha_{2,\Omega_1=1} = 1.80892; \qquad \alpha_{1,\Omega_1=1/2,3/2} = 3.05607,\ 1.44949;$$

$$\alpha_{2,\Omega_2=5} = 1.43404;\ \alpha_{2,\Omega_2=9/2} = 1.81363;\ \alpha_{1,\Omega_1+\Omega_2=5,6,7} = 2.28032,\ 1.57899,\ 1.35168$$

For pendulum motions of the system with Hamiltonian (5), at the lower boundary of the region there is a finite number of resonance points:

IOP Conf. Series: Materials Science and Engineering **489** (2019) 012034 doi:10.1088/1757-899X/489/1/012034

$\alpha_{21,\Omega_1=1}=1/2;$ $\alpha_{21,\Omega_1=1/2}=7/8;$ $\alpha_{21,\Omega_2=1}=1/2;$ $\alpha_{21,\Omega_2=1/2}=1/8;$

$\alpha_{21,\Omega_1+\Omega_2=2}=1/2;$ $\alpha_{21,\Omega_1-\Omega_2=1}=0.06698;$ $\alpha_{21,\Omega_2-\Omega_1=1}=0.93301$

Note that in the regions Γ_2 and Γ_{21} at $\alpha=1/2$ the cases of multiple resonances $\Omega_{2,1}=2,\Omega_{2,2}=1$ and $\Omega_{21,1}=1,\Omega_{21,2}=1$ occur.

The study showed that for the pendulum-type motions of the first model system, the generating points for parametric resonance regions are only points $\alpha=1/2,\ 7/9,\ 0.78804,\ 4/5,\ 7/8$ on the lower boundary of the region Γ_2, and points $\alpha=1.80892,\ 1.57899,\ 1.43404$ on the lower boundary of the region Γ_1. For pendulum motions of the second model system, at the lower boundary of the region Γ_{21}, only one point $\alpha=1/2$ of multiple resonance generates a parametric resonance regions.

From each of the remaining generating points for both systems, the single resonant curve is arisen, and there are no instability regions.

6. Stability diagrams

To solve the orbital stability problem of the pendulum-type motions considered, one needs to integrate the equations of motion of a non-autonomous system with the Hamiltonian K and calculate the coefficients of the characteristic equation (10). The integration of the equations of motion has been carried out numerically. For the system corresponding to the model system with Hamiltonian (3), the range of parameters $\alpha\in(0;5]$ and $h\in[-1;3.5]$ is considered, and for the system corresponding to the model system with Hamiltonian (5), the region $\alpha\in(0;5]$ and $h_1\in[-1;2]$. The results of the orbital stability study of the pendulum-type motions defined by Hamiltonians (3) and (5) are shown in figure 2 a and 2 b, respectively.

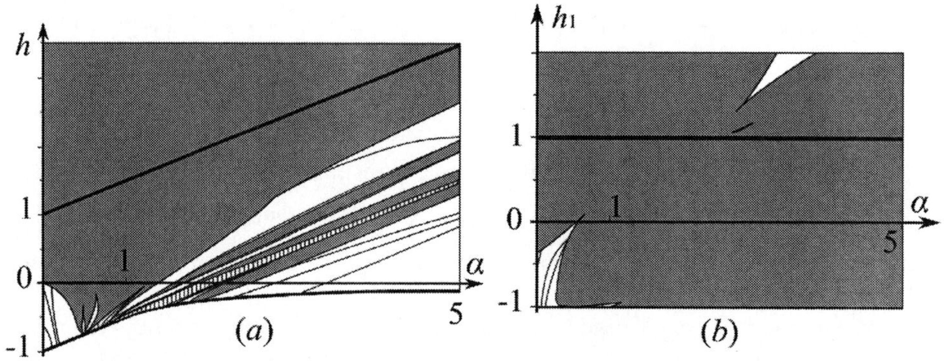

Figure 2. Stability diagrams.

Here, thin lines are the curves of resonances of the first and second orders. The instability regions are indicated in gray, the linear orbital stability regions are not painted over.

We note that in the vicinity of the point $\alpha=1$ on the lower boundary curves in the regions Γ_1 and Γ_2, there is a countable set of points generating parametric resonance regions and single resonant curves. These regions and curves are located in the vicinity of the ray $h=\alpha/2-1,\alpha>1$ corresponding to the asymptotic motion of the radius vector of the body mass center. The domain, in which they are located in figure 2 b, is shaded. It is bounded by sections of two resonant curves obtained.

6.1. Orbital stability of pendulum-type motions of the system corresponding to model Hamiltonian (3)
For oscillations in the region Γ_1, there is a large instability region near the oscillation energy level, corresponding to the asymptotic motion. It is generated by three points indicated in the previous section, and on its upper border has two corner points with coordinates $(2.71815; 0.28091)$ and $(2.99157; 0.43130)$. The stability domains for oscillations near the inclined equilibrium position are found up to the upper boundary of the region Γ_1. With an increase in the intensity of vibrations, the stability regions are expanding.

For oscillations in the region Γ_2, three significant instability regions can be distinguished. One of these regions is adjacent to the upper boundary of the oscillation region and originates at the generating points $\alpha = 1/2$ and $\alpha = 7/9$. Parts of its borders have corner points $(0;0)$, $(0.49252;-0.34305)$, $(0.64130;-0.17655)$, $(1.22538;-0.05679)$, and $(2.77955;1.25049)$. Two instability regions generated by dots $\alpha = 0.78804$ and $\alpha = 4/5$, with an increase in the intensity of vibrations, go along the lower boundary of the region. The instability region arising from the point $\alpha = 7/8$ is small and merges into the lower one of the large instability regions.

The significant stability regions for oscillations near the lower position are available at low and high vibration intensities. With an increase in the intensity of vibrations, the stability regions are expanding. For the rotation, of the stability regions are not detected.

6.2. Orbital stability of pendulum-type motions of the system corresponding to model Hamiltonian (5)
For oscillations around the lower equilibrium position, there is an instability region that occupies almost the entire region Γ_{21}. It is born from the point of multiple resonance $\alpha = 1/2$ and has angular points $(0;-0.37233)$, $(0.66882;0.43844)$, $(1.15900;-0.37413)$ and $(1;-0.5)$. Two stability regions are adjacent to the lower boundary of the region Γ_{21} at $\alpha < 1$. A significant stability region exists only for small values of the vibration intensity of the suspension point.

In the rotation region for the range of parameters considered, two stability regions are found. One of them, with a corner point $(2.72745;2.68839)$, is expanding with increasing level of rotation energy, and the other one is enclosed between the corner points $(2.83023;2.53740)$, $(2.67780;2.40480)$ and $(2.70412;2.41939)$.

Note that in the limiting case where the suspension point of the body is fixed the results on the orbital stability of the pendulum-type motions of the Kovalevskaya top obtained in the paper are consistent with the results obtained in [5, 6].

Acknowledgments
This work was carried out within the framework of the state assignment (project No. 3.3858.2017/4.6).

References
[1] Stephenson A 1908 *Mem. and Proc. of the Manchester Literary and Phil. Soc.* **52 Pt.2 (8)** 1-10
[2] Markeyev A P 2011 *J. of App. Maths and Mechs* **75 (2)** 132-9
[3] Belichenko M V 2018 *Russ. J. of Nonlinear Dynamics* **14 (2)** 243-63
[4] Mlodzeevsky B K 1894 *Proc. of the Department of physical Sciences of Society of Devotees of Natural Science, Anthropology, and Ethnography* **7 (1)** 46-8
[5] Markeyev A P 2001 *J. of App. Maths and Mechs* **65 (1)** 47-54
[6] Markeyev A P, Medvedev S V, Chehovskaya T N 2003 *Mechanics of solids* **1** 3-9
[7] Byrd P F, Friedman M D 1971 *Handbook of Elliptic Integrals for Engineers and Scientists.* (Verlag, New York, Heidelberg, Berlin: Springer)
[8] Markeyev A P 1978 *Libration points in celestial mechanics and cosmodynamics* (Moscow:

IOP Conf. Series: Materials Science and Engineering **489** (2019) 012034 doi:10.1088/1757-899X/489/1/012034

Nauka)

IOP Conf. Series: Materials Science and Engineering **489** (2019) 012035 doi:10.1088/1757-899X/489/1/012035

Experimental study on active damping of compressor blade forced vibrations using piezoelements

A D Bortnikov

Central Institute of Aviation Motors (CIAM), 111116 Moscow, 2 Aviamotornaya Street, Russia

bortnikovad@gmail.com

Abstract. This paper is dedicated to active damping of gas turbine engine blades forced vibrations using the capabilities of piezoelements. Experimental investigations are carried out with a non-rotating compressor blade, rigidly clamped on a shaker. Vibration damping was carried out by using of feedback based on a piezoelement pair fixed on the blade surface and used as a sensor and an actuator. A significant decrease of resonant stresses in the blade is shown.

1. Introduction

The problems of vibration durability of gas turbine engines blades are well studied [1-2], however, the blade failures due to increased alternating stresses continues to occur, significantly complicating the operation and reducing the economic performance of engines.

It is necessary to find newer methods of damping to ensure low- and high-cycle fatigue, especially at resonant and near-resonant modes of details operation. Besides the passive methods consisting in energy dissipation of diverging vibrations by various types of friction forces [2] the attention of researchers is aimed at finding active methods that would prevent the appearance of increased stresses. One of such methods is using the electromechanical devices (piezoelements) based on the piezoelectric effect of certain materials.

The piezoelectric effect is that under the influence of applied voltage to a piezoelement (PE) its size changes and, conversely, the deformation of PE leads to the appearance of an electric charge on it.

It is possible to use this effect to make a PE, which is attached to a structure, deform in such a way as to compensate for the vibrations of this structure.

Many studies are dedicated to theoretical and practical problems of the active vibration control using PE [3-5]. Studies are mainly directed on researching the influence of PE location, parameters and configurations of damping circuit, control laws to the efficiency of vibrations damping for various objects.

In the area of the active damping of forced vibrations of gas turbine engine parts, there are works [6-10] showing the effectiveness of PE-based damping methods.

In this work, the experiments were carried out on a full-scale blade using bimorph PE combined into a feedback system with the possibility of phase shift and gain of the control signal. Also, when using PE as an additional source of excitation, the new data, which may be useful in fatigue tests, were obtained.

IOP Conf. Series: Materials Science and Engineering **489** (2019) 012035 doi:10.1088/1757-899X/489/1/012035

2. Experimental setup

Active damping of a blade vibrations with PE is meant as the organization of an electromechanical system (blade → PE-sensor → signals processing unit, amplifier → PE-actuator) so that vibration impact of PE-actuator glued on the blade weakens the resulting values of the blade resonant stresses and displacement amplitudes. Figure 1 shows the experimental setup scheme and its basic elements.

In the works in which experimental studies of active piezo-damping are carried out, the experimental setup schemes contain the following main elements: an object of damping, PE-sensor, signals processing and formation unit, PE-actuator [3-5].

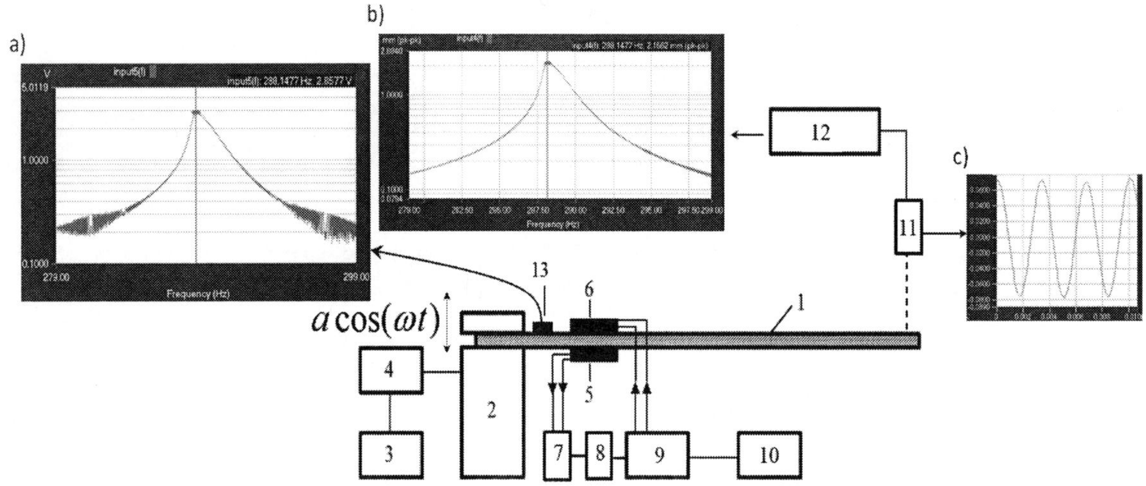

Figure 1. Experimental setup scheme. 1 – compressor blade; 2 – shaker; 3 – shaker frequency generator; 4 – shaker amplifier; 5 – piezo-sensor; 6 – piezo-actuator; 7 – active impedance; 8 – frequency filter; 9 – voltage amplifier; 10 – voltage amplifier power supply; 11 – laser vibrometer; 12 – recording equipment; 13 – strain sensor. a) sensor strain indications; b) resonant curve of blade tip displacement amplitude (peak-to-peak); c) displacement of blade tip.

In this work tests were carried out using electrodynamic shaker with a digital control system. The laser vibrometer was used to register the blade tip displacement under forced vibrations, and a sensor strain – to record the alternating mechanical stresses amplitudes.

The amplitude amplifier has the ability to increase the amplitude of the input voltage signal in required frequency range without changing of the phase between input and output signal. A frequency filter was used to control the phase of input signal from piezoelectric element. The filter has the ability to set phase between input and output signal either 0 ° or 180 °.

The PE-sensor wires are connected to the active resistance, which, at some value of resistance and due to the electrical properties of PE, allows the phase of the voltage to be shifted by -90 °.

During active vibration damping, a signal from the PE-sensor passed through the active resistance is fed to frequency filter input. Further, the signal is processed and the signal from frequency filter output, where it can change phase by 180 °, is fed to amplifier input, where it can be amplified in the range from 1 to 90. The amplified signal is fed to PE-actuator pasted on the other side of the blade, operating at phase shift of -90 ° in the active vibration damper mode, and when the phase shift of +90 ° – in the «anti-damping» mode.

The PE-actuator vibrating under the action of the control signal is the same source of vibration excitation for the blade as the shaker. Therefore, for effective action on the blade at resonance, it is necessary to set the control signal phase that differs from the signal phase of PE-sensor by ±90 °. It is obviously, the vibrations of PE-actuator and blade occur with the frequency of excitation.

IOP Conf. Series: Materials Science and Engineering **489** (2019) 012035 doi:10.1088/1757-899X/489/1/012035

Figure 2 shows the investigated compressor blade with two glued PE. In tests bimorph type PE with dimensions of 40.0 × 12.0 × 0.6 mm are used.

The PE placements were selected on the basis of a preliminary calculation of natural frequencies and vibration modes so, that these are the areas of increased deformations for the blade first bending mode of vibration. PE on the blade pressure side is a sensor, PE on the suction side is an actuator.

Figure 3 shows the investigated compressor blade mounted on the shaker rig with two glued PE connected to each other by the electrical network described above. Using the shaker the forced vibrations were excited in the blade, rigidly fixed on it, with passage through the natural frequency of the first bending mode of the blade vibrations at a constant rate of increase of the exciting frequency.

Figure 2. Compressor blade with strain sensor and PE, view from the pressure (left) and suction (right) side.

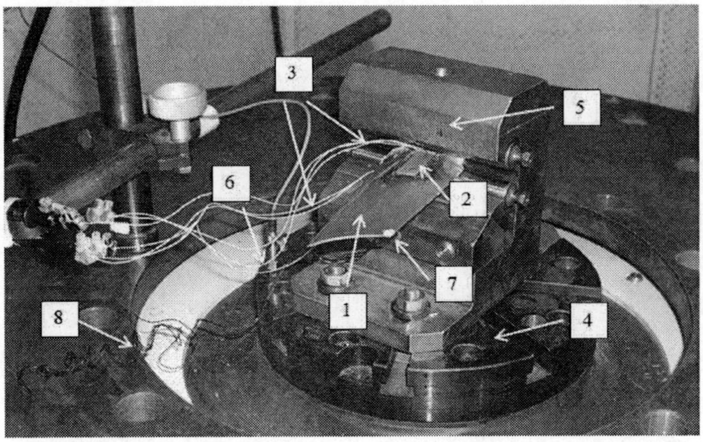

Figure 3. Compressor blade with PE on shaker. 1 – compressor blade; 2 – PE on suction side; 3 – PE wires; 4 – shaker table; 5 – shaker rig; 6 – accelerometer; 7 – laser vibrometer reflector (for measuring blade tip displacement); 8 – strain sensor wires.

When processing the tests results, special attention was paid to the accuracy of measurements of the blade tip displacements range and alternating mechanical stresses at the resonant frequency of the blade without PE, with switched on/off PE, as well as the reliability of measuring the resonant frequency of the system for different experimental conditions with varying excitation frequency.

3. Test results

Figure 4 shows the obtained resonant curves of displacement ($2A$) of the blade tip and alternating mechanical stress (σ_a) amplitudes at the input edge near the compressor blade root for different cases: without PE, with switched off PE, for active damping and «anti-damping» with different values of a feedback coefficient k and constant value of vibration acceleration of the shaker a = 4.9 m/s^2 and a = 9.8 m/s^2. The value k is defined as the ratio of the PE-actuator voltage amplitude and voltage amplitude received from PE-sensor. Active «anti-damping» is carried out by the joint unidirectional action of the exciting load and the PE-actuator, which leads to increase displacements range and alternating mechanical stresses in the blade.

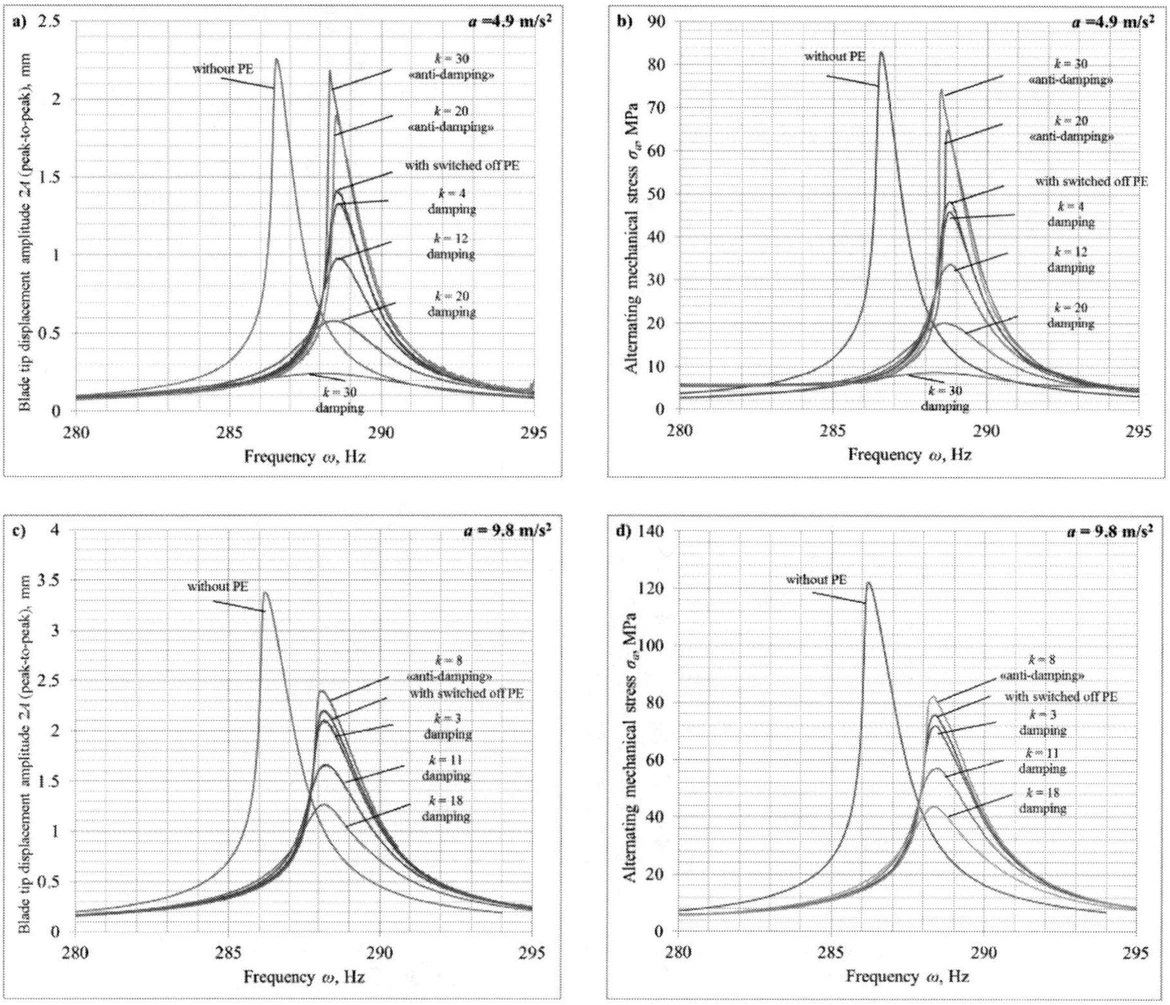

Figure 4. Experimental resonance curves of compressor blades without PE, with switched off PE, with active damping and «anti-damping».

As can be seen from figure 4, with using PE and vibration acceleration of the shaker a = 4.9 m/s^2 and k = 30 (damping mode) the amplitude range of resonant displacements of the blade tip decrease from 2.26 mm to 0.25 mm and the resonant mechanical stresses from 83.0 MPa to 8.8 MPa, in comparison to a blade without PE. With vibration acceleration of the shaker a = 9.8 m/s^2 and k = 18 the amplitude range of resonant displacements of the blade tip decrease from 3.38 mm to 1.26 mm, and the resonant mechanical stresses from 122.1 MPa to 43.8 MPa.

Such a decrease depends on two factors:

- additional rigidity in a structure and features of PE mounting;
- active damping directly.

Moreover, for a = 4.9 m/s^2, the resonant mechanical stresses decreases by ~ 40 MPa, and for a = 9.8 m/s^2 – by ~ 33 MPa only due to active damping.

In the «anti-damping» mode, the increase of registered values was recorded.

4. Conclusions

Based on the experiment results, the fundamentally possible efficiency of forced vibrations damping of gas turbine engine compressor blade using capabilities of piezoelements is confirmed.

With the vibration acceleration a = 4.9 m/s^2 and k = 30, the resonant stresses are reduced by ~ 85 % when the blade oscillates on first bending mode, in comparison to a blade without PE for initial stress amplitudes of 83 MPa.

As the level of exciting load increases, the efficiency of active vibration damping decreases, which can be explained by the limited ability of its PE to withstand higher loads. Another type of PE, more powerful, could probably be more effective.

After gluing PE on the blade, the frequency of resonant vibrations increases and the amplitude of the blade tip displacement and resonant mechanical stresses decreases due to the introduction of additional rigidity into the structure and mounting PE features.

In the active «anti-damping» mode the increase of recorded parameters confirms the correctness of chosen vibration control scheme using PE. The new data on active «anti-damping» mode, which may be useful in fatigue tests of various objects, were obtained.

Further studies will focus on the issues of integrating PE into a structure of engine elements (stator and rotor blades, casings, rotor bearings).

Acknowledgement

The work was carried out in the CIAM's department «Dynamics, Strength, Reliability» with the guidance of Professor, Doctor of Technical Sciences B.F. Shorr.

References

[1] Ayrapetov E L et al 1980 *Kolebaniya mashin, konstruktsiy i ikh elementov (Vibratsii v tekhnike: Spravochnik v 6-ti t* ed V N Chelomey) vol 3 ed F M Diementberg and K S Kolesnikov (Moscow: Mashinostroenie) p 544 (in Russian)
[2] Bauer V O et al 1981 *Dinamika aviatsionnykh gazoturbinnykh dvigateley* ed I A Birger and B F Shorr (Moscow: Mashinostroenie) p 232 (in Russian)
[3] Akopyan V A, Rozhkov E V, Soloviev A N and Shevtsov S N 2007 *Nano- i Mikrosistemnaya Tekhnika* **10** pp 36–61 (in Russian)
[4] Kim S J and Jones J D 1995 *J. Intelligent Material Systems and Structures* vol **6** pp 610–23
[5] Gundage V V and Sonawane P R 2016 *Int. J. Research in Engineering and Technology* vol **2** pp 56–65
[6] Duffy K P, Choi B B and Provenza A J 2012 *Proc. ASME Turbo Expo 2012* (Copenhagen, Denmark) p 11
[7] Schwarzendahl S M, Szwedowicz J and Neubauer M 2012 *Proc. ASME Turbo Expo 2012* (Copenhagen, Denmark) p 10
[8] Kelley C R and Kauffman J L 2017 *J. Intelligent Material Systems and Structures* vol **28 (16)**

pp 2275–85
[9] Kohler R, Kaletsch C, Rinderknecht S and Marszolek M 2011 *20th Int. Symp. Air Breathing Engines* (Gothenburg) vol 3 pp 1675–85
[10] Jiuzhou L, Lin L, Pengcheng D and Chao L 2016 *Proc. ASME Turbo Expo 2016: Turbomachinery Technical Conf. and Exposition* (Seoul) p 12

IOP Conf. Series: Materials Science and Engineering **489** (2019) 012036 doi:10.1088/1757-899X/489/1/012036

Nonlinear dynamics of multi-edge trepanning vibration drilling

A M Gouskov[1,2] and **Dinh Duc Tung**[1]

[1] Bauman Moscow State Technical University, Moscow, Russia
[2] Mechanical Engineering Research Institute of the Russian Academy of Sciences (IMASH RAN), Moscow, Russia

gouskov_am@mail.ru

Abstract. In this paper, the dynamics of multi-edge trepanning vibration drilling is considered as multi-edges cutting with regenerative mechanism of oscillations excitation. The equations of new surface formation, the equations of motion for a system with one degree of freedom, and the fractional-rational cutting law are used for the numerical simulation of the dynamics of the cutting process. It presents the results of modeling the variation in chip thickness and shape of the chips.

1. Introduction

One of the main methods for receiving large diameter holes is trepanning drilling. Because, the energy costs in this method are significantly lower than in other methods. One of the main problems encountered during drilling is the removal of chips from the cutting zone and the supply of coolant to this zone due to the drain chips. For solution of this problem, vibration is used in trepanning drilling, which, under certain conditions, allows to obtain segmented (discontinuous) chips. As in other processing methods with the rotation of a tool or workpiece (turning, milling, etc), in trepanning drilling, the regenerative mechanism is the most powerful, leading to the appearance of self-oscillations of tool and the appearance of intermittent cutting [1, 2, 3, 6].

In this paper, we studied the dynamics of vibration trepanning drilling with a multi-edge trepanning drill. Consider a tool in the form of the main body, which appears to be a hollow cylinder (shell), on the end surface of which n cutting elements are mounted (figure 1).

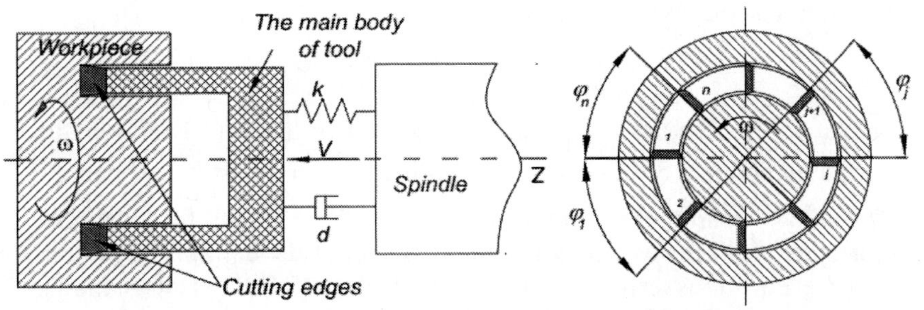

Figure 1. Schematic of multi-edges trepanning vibration drilling

2. The complete model of nonlinear dynamics

The study of the dynamics of this process is performed through a complete model of nonlinear dynamics, which can be described by a system consisting of three groups of equations: equations of new surfaces generation; cutting law and equations of motion of the technological system [5, 8, 9, 11].

Equations of new surfaces generation. In multi-edge trepanning drilling, *j*-th cutting edge processes the surface, that was formed by the previous *(j-1)*-th cutting edge in time $\left(t-t_{j-1}\right)$. The time t_{j-1} is the delay between cutting edges (*(j-1)*-th and *j*-th), that equal to the time the part rotates at angle between the cutting edges $t_{j-1}=\varphi_{j-1}/\omega$. Equations of new surfaces generation are written as [4, 6, 7, 8, 11, 13, 14]:

$$\begin{cases} D_j(t)=Vt-u(t)-L_{j-1}\left(t-t_{j-1}\right)+A-H_j, \\ h_j(t)=\max\left[0, D_j(t)\right], \\ L_j(t)=L_{j-1}\left(t-t_{j-1}\right)+h_j(t), \end{cases} \tag{1}$$

where $L_j(t)$- is the axial distances from the free end of the part to the surface under *j*-th cutting edge and; $L_{j-1}\left(t-t_{j-1}\right)$- is the axial distances from the free end of the part to the surface under *(j-1)*-th cutting edge; $D_j(t)$- is the axial distance from the *j*-th cutting edge to the surface being processed $L_{j-1}\left(t-t_{j-1}\right)$; $u(t)$ - axial deflection (vibration) of the tool; $h_j(t)$ - uncut chip thickness for the *j*-th cutting edge; $A=l-Z_0$- constant distance, l – length of the part being processed, Z_0 - initial axial position of the first cutting edge (nominal set value); H_{0j} - initial axial offset of the *j*-th cutting edge (nominal set value) from the first one, axial for which A is defied, t – current time.

Cutting forces. $F_j\ \left(j=\overline{1,n}\right)$ in the axial direction for each *j*-th cutting edge are described by the cutting law model in the form of a fractional rational function [4, 6, 10, 12]:

$$F_j(t)=K_0 h_j(t)\frac{c+rh_j(t)}{c+h_j(t)}, \tag{2}$$

where: $K_0=\gamma\sigma_L B$- static cutting stiffness, σ_L - characteristic stress value for given material, B - chip width; γ, r – non-dimensional coefficients, determined experimentally for given process conditions at hand, c – characteristic linear size of the cutting process.

Equation of motion. In the case of rigid attachment of the cutting edges with the main body of the tool, all cutting edges perform the same vibrations of the tool. We thus have a dynamical system with one degree of freedom (DOF) with the axial vibration of tool. Equation of motion for the axial vibration of the tool has the following form:

$$m\ddot{u}=-d\dot{u}-ku+F; \quad F=\sum_{j=1}^{n}F_j, \tag{3}$$

with m- mass of tool, d и k- linear damping and stiffness coefficients of fastening of tool with spindle, F_j- axial component of the cutting force, that acts on the *j*-th cutting edge.

Equations (1)-(3) constitutes a complete mathematical model for dynamics of the multi-edge vibration trepanning drilling, which is a system of delay differential algebraic equations (DDAE). Solutions of these equations can be carried out by the DDE-BIFTOOL module in the MATLAB system using the ε-embedding method [4, 6] for transition from system DDAE to system of delay differential equation (DDE). Thus, equations (1) have the form:

$$\begin{cases} D_j(t)=Vt-u(t)-L_{j-1}(t-t_{j-1})+A-H_j, \\ h_j(t)=\max\left[0,D_j(t)\right], \\ \varepsilon L_j'(t)=-L_j(t)+L_{j-1}(t-t_{j-1})+h_j(t),\ \ \varepsilon<<1 \end{cases} \quad (4)$$

We present equations (2) - (4) to non-dimensional form as the linear scale X_* feed per revolution of part h_0, as the time scale $T_*=2\pi\sqrt{m/k}$, T_* - free oscillation period the tool and as the scale of cutting forces $F_*=K_0 h_0$.

$$\zeta=d/2\sqrt{mk},\ \kappa=K_0/k,\ \Pi_j=F_j/K_0h_0,\ \rho=T/T_*=2\pi/\omega T_*,$$

$$\{\xi_j,\eta_j,\Lambda_j,\Delta_j,\mathrm{A},\mathrm{H}_j\}=\frac{1}{h_0}\{u_j,h_j,L_j,D_j,A,H_j\}.$$

Here ρ - is the ratio of the natural frequency of the tool to the frequency of rotation of the part (the parameter is a dimensionless cutting speed), the parameter κ - is the relative static stiffness of cutting. Then, equations (2) - (4) in dimensionless form take the form:

$$\begin{cases} \Delta_j(\tau)=\tau/\rho-\xi(\tau)-\Lambda_{j-1}(\tau-\tau_{j-1})+\mathrm{A}-\mathrm{H}_j,\ \ \sum_{j=1}^{n}\tau_j=\rho, \\ \eta_j(\tau)=\max(0,\Delta_j(\tau)),\ \ \Pi_j=\eta_j(\eta_*+r\eta_j)/(\eta_*+\eta_j) \\ \varepsilon\Lambda_j(\tau)=-\Lambda_j(\tau)+\Lambda_{j-1}(\tau-\tau_{j-1})+\eta_j(\tau), \\ \xi''=-4\pi\zeta\xi'-4\pi^2\xi+4\pi^2\kappa\Pi,\ \ \Pi=\sum_{j=1}^{n}\Pi_j \end{cases} \quad (5)$$

3. Numerical simulation of dynamics

The numerical solution is carried out with the case of three-edge trepanning drilling with unsymmetrical angular disposition of three cutting edges $(\varphi_1=180^\circ;\varphi_2=\varphi_3=90^\circ)$ and without axial offsets $(\mathrm{H}_j=0)$ for machining of hole with diameter of 150 mm in the part made of titanium alloy Ti-6Al-4V with characteristic stress $\sigma_L=1000\,\mathrm{MPa}$ [15]. In this case, the trepanning drill with mass of 12,6 kg and dimensionless cutting coefficient of cutting edges $\gamma=8{,}8456$ [16] is used; the chip width is set to B = 10 mm. Therefore, static cutting stiffness can be calculated by $K_0=\gamma\sigma_L B=8{,}8456.10^7$ N/m. Numerical calculations, based on the system (5) with small parameter $\varepsilon=10^{-4}$, have enabled the computation of axial vibration of the tool ξ (figure 2), thickness of chips and shape of chips (figure 3) as a function of dimensionless time τ/ρ for the following set of system properties: $\rho=10{,}4;\zeta=0{,}055;\ r=0{,}65;\kappa=0{,}15;\eta_*=0{,}15$.

In figure 2, it can be seen that with the trepanning drilling, the phenomenon of self-oscillation of the tool occurs under the interaction between the three cutting edges. At first, the amplitude of oscillation of the tool varies greatly and after some rotations of the rotation of part is established (stability).

In figure 3 shows the development process of vibrations (figure 3, a) and the occurrence of segmented chips (figure 3, b) - this is characteristic of the subcritical Poincare – Andronov – Hopf bifurcation [2, 7]. In virtue of the unsymmetrical arrangement of the cutting edges (φ_i) without axial offsets (H_j), the shape of the chip sections and their length are different, which can be used to optimize the removal of chips from the cutting zone.

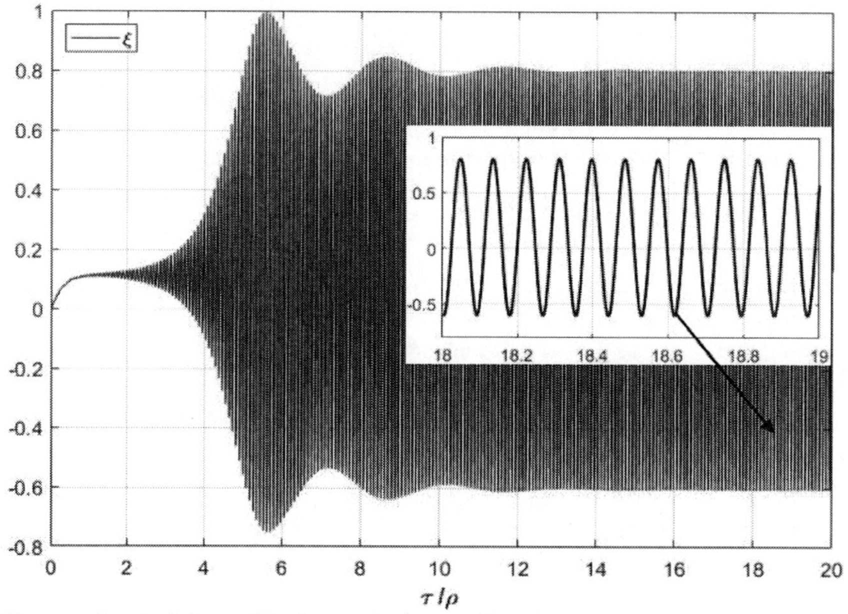

Figure 2. Axial oscillation of the tool with unsymmetrical angular disposition of three cutting edges without axial offsets

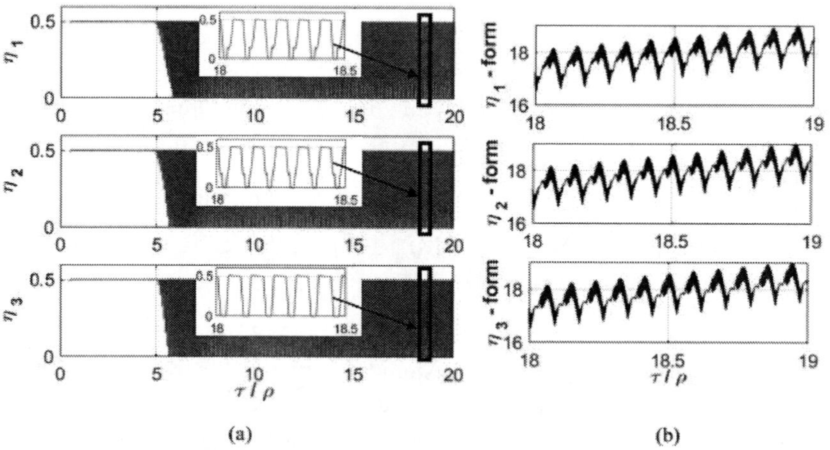

Figure 3. Thickness (a) and shape (b) of the chips removed by each cutting edge in the unsymmetrical arrangement of the three cutting edges without axial offsets

Conclusion

Maintaining of vibrations in the form of self-oscillation with three-edge trepanning drilling allows to obtain discontinuous chips for effective removal from the cutting zone.

Acknowledgments

The work has been realized under the support of the Russian Scientific Foundation, project № 18-19-00708

References

IOP Conf. Series: Materials Science and Engineering **489** (2019) 012036 doi:10.1088/1757-899X/489/1/012036

[1] Dombovari Z and Stepan G 2013 Dynamics of drill bits with cutting edges of varying parameters *Proc. ASME 2013 Int. Design Engineering Technical Conf. and Computers and Information in Engineering Conf.* vol 7B (Portland, Oregon, USA) p 7.

[2] Tlusty J and Spacek L 1954 *Self-excited vibrations on machine tools* (Nakl. CSAV, Prague. in Czech).

[3] Tobias S 1965 *Machine-tool Vibration* (Blackie, Glasgow).

[4] Gouskov A, Guskov M, Dinh Duc Tung and Panovko G 2018 *Vibroengineering Procedia* **17** pp 124-29.

[5] Kondratenko K, Gouskov A, Guskov M, Lorong Ph and Panovko G 2014 *Vibration Engineering and Technology of Machinery* **23** pp 929-37.

[6] Gouskov A and Dinh Duc Tung 2017 Features of the nonlinear dynamics of multi-cutter turning *Proc. 2nd All-Russian Scientific and Technical Conf. "Mechanics and Mathematical Modeling in engineering"* (Moscow: BMSTU) pp 107-11. (in Russian)

[7] Gouskov A 2000 Nonlinear Dynamic of vibration drilling. Role of equations of new surfaces formation *4th Int. Congress "Design - Technological Informatics"* (Moscow: MSTU STANKIN) pp 123-30. (in Russian)

[8] Gouskov A, Voronov S, Paris H and Batzer S 2001 Cylindrical Workpiece Turning Using Multiple Cutting Tool *Proc. of the Design Technical Conf. and Computers and Information Engineering Conf.* (Pittsburgh, Pannsylvania).

[9] Dombovari Z, Barton D, Wilson R, Stepan G 2011 *Int. J. of Nonlinear Mechanics* **46** pp 330-38.

[10] Brissaud D, Gouskov A, Guibert N and Rech J 2008 *CIRP Annals – Manufacturing Technology* **57** pp 385-88.

[11] Gerasimenko A, Guskov A, Gouskov A, Lorong P and Panovko G 2016 *Vibroengineering Procedia* **8** pp 179-84.

[12] Gouskov A, Guskov M, Lorong Ph and Panovko G 2017 *Int. J. of Machining and Machinability of Materials* **19** pp 17-39.

[13] Gouskov A, Voronov S, Paris H and Batzer S 2002 *Communications in Nonlinear Science and Numerical Simulation* **7** pp 207-21.

[14] Lamikiz A 2005 *Machining Science and Technology* **9** pp 411-36.

[15] Isik B and Kently A 2014 *Archives of Metallurgy and Materials* **9** pp 467-71.

[16] Brissaud D, Gouskov A, Paris H and Tichkiewitch S 2009 *Asian Int. J. of Science and Technology – Production and Manufacturing* **1** pp 267-75.

IOP Conf. Series: Materials Science and Engineering **489** (2019) 012037 doi:10.1088/1757-899X/489/1/012037

Influence of fractional calculus model parameters on nonlinear forced vibrations of suspension bridges

A L Katembo and M V Shitikova

Research Center on Dynamics of Solids and Structures, Voronezh State Technical University, Voronezh 394006, Russia

katembo2020@gmail.com

Abstract. Nonlinear forced vibrations of suspension bridges, when the frequency of an external force is approaching one of the natural frequencies of the suspension system, which, in its turn, undergoes the conditions of the one-to-one internal resonance, are investigated. The generalized method of multiple time scales is used as the method of solution. The damping features are described by the fractional derivative, which is interpreted as the fractional power of the differentiation operator. An approximate analytical solution has been obtained by the method of successive approximations.

1. Introduction

The experimental data obtained in [1] during ambient vibrations studies of the Golden Gate Bridge show that different vibrational modes possess different amplitude damping factors, and the order of smallness of these coefficients tells about low damping capacity of suspension combined systems, resulting in prolonged energy transfer from one partial subsystem to another.

Free damped vibrations of suspension bridges in the cases of the one-to-one and two-to-one internal resonance have been investigated in [2] when damping features of the system are prescribed by the first derivative of the displacement with respect to time. It has been shown that for the both types of the internal resonance the damping coefficients do not depend on the natural frequency of vibrations. However, this result is in conflict with the experimental data. To lead the theoretical investigations in line with the experiment, fractional derivatives were introduced in [3] for describing the processes of internal friction proceeding in suspension combined systems at free vibrations. The suspension bridge model put forward allows one to obtain the damping coefficient dependent on the natural frequency of vibrations. This model has been generalized further in [4] by using two different fractional parameters for analyzing vertical and torsional modes.

In the present paper, the model analyzed in [4] is used for investigating nonlinear forced vibrations of suspension bridges, when the frequency of an external force is approaching one of the natural frequencies of the suspension system, which undergoes the conditions of the one-to-one internal resonance.

2. Problem Formulation

The suspension bridge scheme presents a bisymmetrical thin-walled stiffening girder, which is connected with two suspended cables by virtue of vertical suspensions. The cables are thrown over the pylons and are tensioned by anchor mechanisms. The suspensions are considered as inextensible and uniformly distributed along the stiffening girder. The cables are parabolic, and the contour of the girder's cross-section is undeformable. It is assumed that the girder's contour translates as a rigid body vertically (in the y-axis direction) on the value of $\eta(z,t)$ and rotates with respect to the girder's axis (the z-axis) through the angle of $\varphi(z,t)$. The origin of the frame of references is in the center of gravity of the cross-section.

It is known for suspension bridges [2,5] that some natural modes belonging to different types of vibrations could be coupled with each other, i.e., the excitation of one natural mode gives rise to another one. Two modes interact more often than not, although the possibility for interaction of a greater number of modes is not ruled out.

If only two modes predominate in the vibrational process, namely: the vertical n-th mode with linear natural frequency ω_{0n} and the torsional m-th mode with the natural frequency Ω_{0m}, then the functions $\eta(z,t)$ and $\varphi(z,t)$ could be approximately defined as

$$\eta(z,t) \approx \upsilon_n(z) x_{1n}(t), \qquad \varphi(z,t) \approx \Theta_m(z) x_{2m}(t), \tag{1}$$

where x_{1n} and x_{2m} are the generalized displacements, and $\upsilon_n(z)$ and $\Theta_m(z)$ are the natural modes of the two interacting modes of vibrations.

The resolving set of equations describing forced vibrations is written in a dimensionless form as

$$\begin{gathered}\ddot{x}_{1n} + \omega_{0n}^2 x_{1n} + \beta D_+^{\gamma_1} x_1 + a_{11}^n x_{1n}^2 + a_{22}^{nm} x_{2m}^2 + \left(b_{11}^n x_{1n}^2 + b_{22}^{nm} x_{2n}^2\right) x_{1n} = \hat{F}\cos(\omega_1 t), \\ \ddot{x}_{2n} + \Omega_{0m}^2 x_{2n} + \beta D_+^{\gamma_2} x_2 + a_{12}^{nm} x_{1n} x_{2m} + \left(c_{11}^{nm} x_{1n}^2 + c_{22}^m x_{2m}^2\right) x_{2m} = 0,\end{gathered} \tag{2}$$

where the coefficients a_{ij}, b_{ij} and c_{ij} (i=1,2; j=2) are defined in [5] (subsequently the indices n and m will be omitted for ease of presentation), $\hat{F} = const$ is the amplitude of the external harmonic force, the terms $\beta D_+^{\gamma_1} x_1$ and $\beta D_+^{\gamma_2} x_2$ describe inelastic reaction of the system, β is the viscosity coefficient, $D_+^{\gamma} x$ $(\gamma = \gamma_1 \text{ or } \gamma_2)$ is the fractional derivative which is interpreted as the fractional power of the differentiation operator [3,4] with the fractional parameter γ (order of the fractional derivative).

$$D_+^{\gamma} x = \left(\frac{d}{dt}\right)^{\gamma} \qquad (0 < \gamma \le 1). \tag{3}$$

Let us consider the case of the one-to-one internal resonance, as well as suppose that the frequency of the external force is close to the natural frequencies of the interacting modes, i.e.,

$$\omega_0 = \Omega_0 + \varepsilon^2 \sigma, \qquad \omega_1 = \omega_0 - \varepsilon^2 \sigma_1, \tag{4}$$

where ε is a small parameter that is of the same order of magnitudes as the amplitudes of vibrations, and σ and σ_1 are detuning parameters.

3. Method for determining the governing equations

An approximate solution of (2) for small but finite amplitudes weakly varying with time can be represented by an expansion in terms of different time scales in the form

$$\begin{gathered}x_1(t) = \varepsilon x_{11}(T_0, T_2, ...) + \varepsilon^2 x_{12}(T_0, T_2, ...) + \varepsilon^3 x_{13}(T_0, T_2, ...) + ... \\ x_2(t) = \varepsilon x_{21}(T_0, T_2, ...) + \varepsilon^2 x_{22}(T_0, T_2, ...) + \varepsilon^3 x_{23}(T_0, T_2, ...) + ...\end{gathered} \tag{5}$$

Here, $T_0 = t$ is a fast scale characterizing motions with the natural frequencies ω_0 and Ω_0, and $T_2 = \varepsilon^2 t$ is a slow scale, characterizing the modulations of the amplitudes and phases. Moreover, we let $\beta = \varepsilon^2 \mu$, and $\hat{F} = \varepsilon^3 F$, where $\mu = \text{const}$ and F=const.

Considering that $\dfrac{d}{dt} = D_0 + \varepsilon^2 D_2 + ...$, $\dfrac{d^2}{dt^2} = D_0^2 + 2\varepsilon^2 D_0 D_2 + ...$, where $D_n = \partial / T_n$, and substituting (4) into (2), after equating the coefficients at like powers of ε to zero, one obtains

- To order ε: $D_0^2 x_{11} + \omega_0^2 x_{11} = 0, \qquad D_0^2 x_{21} + \Omega_0^2 x_{21} = 0,$ (6)

- To order ε^2: $D_0^2 x_{12} + \omega_0^2 x_{12} = -a_{11}x_{11}^2 - a_{22}x_{21}^2, \qquad D_0^2 x_{22} + \Omega_0^2 x_{22} = -a_{12}x_{11}x_{21},$ (7)
- To order ε^3:

$$D_0^2 x_{13} + \omega_0^2 x_{13} = -2D_0 D_2 x_{11} - 2a_{11}x_{11}x_{12} - 2a_{22}x_{21}x_{22} - b_{11}x_{11}^3 - b_{22}x_{21}^2 x_{11}$$
$$-\mu D_0^{\gamma_1} x_{13} + F\cos(\omega_1 T_0), \tag{8}$$
$$D_0^2 x_{23} + \Omega_0^2 x_{23} = -2D_0 D_2 x_{21} - a_{12}(x_{11}x_{22} + x_{12}x_{21}) - c_{22}x_{21}^3 - c_{11}x_{11}^2 x_{21} - \mu D_0^{\gamma_2} x_{23},$$

During the deduction of (6)-(8) it was assumed, following [3], that

$$D_+^\gamma = \left(\frac{d}{dt}\right)^\gamma = \left(D_0 + \varepsilon^2 D_2 + \ldots\right)^\gamma = D_0^\gamma + \varepsilon^2 \gamma D_0^{\gamma-1} D_2 + \ldots, \tag{9}$$

where D_0^γ is obtained from (3) by replacing t with T_0.

The solution of equations (6) will be sought in the form

$$x_{11} = A_1(T_2)\exp(\mathrm{i}\omega_0 T_0) + \overline{A_1}\exp(-\mathrm{i}\omega_0 T_0), \qquad x_{21} = A_2(T_2)\exp(\mathrm{i}\Omega_0 T_0) + \overline{A_2}\exp(-\mathrm{i}\Omega_0 T_0), \tag{10}$$

where A_1 and A_2 are unknown complex functions, and $\overline{A_1}$ and $\overline{A_2}$ are the complex conjugates of A_1 and A_2, respectively. Substituting (10) in (7) yields

$$x_{12} = \frac{1}{3\omega_0^2} a_{11}A_1^2 \exp(2i\omega_0 T_0) + \frac{a_{22}A_2^2 \exp(2i\Omega_0 T_0)}{4\Omega_0^2 - \omega_0} - \left(a_{11}A_1\overline{A_1} + a_{22}A_2\overline{A_2}\right)\omega_0^{-2} + cc,$$
$$x_{22} = \frac{a_{12}}{\omega_0(2\Omega_0 + \omega_0)} A_1 A_2 \exp[iT_0(\Omega_0 + \omega_0)] - \frac{a_{12}}{\omega_0(2\Omega_0 - \omega_0)} A_1\overline{A_2}\exp[iT_0(\Omega_0 - \omega_0)] + cc, \tag{11}$$

where cc is complex conjugate part to the preceding terms.

Substituting (11) in equations (8) with due account for the conditions (4) and $D_0^\gamma e^{aT_0} = a^\gamma e^{aT_0}$, and eliminating the terms that produce secular terms, we obtain the solvability conditions

$$-iD_2 A_1 - \frac{1}{2}\mu\omega_0^{-1}(\mathrm{i}\omega_0)^{\gamma_1} A_1 - \lambda_1 A_1^2 \overline{A_1} - \lambda_2 A_1 A_2 \overline{A_2} + \frac{1}{4}\Gamma_1 \overline{A_1} A_2^2 \exp(-2i\sigma T_2)$$
$$+\frac{1}{4} F\exp(-i\sigma_1 T_2)\omega_0^{-1} = 0, \tag{12}$$
$$-iD_2 A_2 - \frac{1}{2}\mu\Omega_0^{-1}(i\Omega_0)^{\gamma_2} A_2 - \lambda_3 A_1 \overline{A_1} A_2 - \lambda_4 A_2^2 \overline{A_2} + \frac{1}{4}\Gamma_2 A_1^2 \overline{A_2}\exp(2i\sigma T_2) = 0,$$

where coefficients λ_i and Γ_j $(i = 1,2,3,4; j = 1,2)$ are presented in [3, 5]. Substitution of functions A_1 and A_2 for $A_1 = A_1 \exp(-i\sigma_1 T_2)$ and $A_2 = A_2 \exp\left[i(\sigma - \sigma_1)T_2\right]$, respectively, in equations (12) leads to eliminating $\exp(\pm 2i\sigma T_2)$ and $\exp(\pm i\sigma_1 T_2)$.

Representing the functions A_1 and A_2 in (12) in their polar forms, i.e., $A_1 = a_1\exp(i\varphi_1)$ and $A_2 = a_2\exp(i\varphi_2)$, as a result we obtain the modulation equations

$$\dot{a}_1 + \frac{1}{2}\mu\omega_0^{\gamma_1-1}\sin\left(\frac{1}{2}\pi\gamma_1\right)a_1 - \frac{1}{4}\Gamma_1 a_1 a_2^2 \sin\delta + \frac{1}{4}F\omega_0^{-1}\sin\varphi_1 = 0,$$
$$\dot{a}_2 + \frac{1}{2}\mu\Omega_0^{\gamma_2-1}\sin\left(\frac{1}{2}\pi\gamma_2\right)a_2 - \frac{1}{4}\Gamma_2 a_1^2 a_2 \sin\delta = 0,$$
$$\dot{\varphi}_1 - \frac{1}{2}\mu\omega_0^{\gamma_1-1}\cos\left(\frac{1}{2}\pi\gamma_1\right) - \sigma_1 - \lambda_1 a_1^2 - \lambda_2 a_2^2 + \frac{1}{4}\Gamma_1 a_2^2\cos\delta + \frac{1}{4}F\omega_0^{-1}a_1^{-1}\cos\varphi_1 = 0, \tag{13}$$
$$\dot{\varphi}_2 - \frac{1}{2}\mu\Omega_0^{\gamma_2-1}\cos\left(\frac{1}{2}\pi\gamma_2\right) - (\sigma_1 - \sigma) - \lambda_3 a_1^3 - \lambda_4 a_2^2 + \frac{1}{4}\Gamma_2 a_1^2\cos\delta = 0,$$

where a dot denotes differentiation with respect to T_2, and $\delta = 2(\varphi_2 - \varphi_1)$ is the phase difference.

The solution of the set of equations (13), when $F = 0$, has been considered for $\gamma_1 \neq \gamma_2$ and $\gamma_1 = \gamma_2$ in [4] and [3], respectively. For the case $\gamma_1 = \gamma_2$, the analytical solution has been obtained in [6].

An approximate solution of the set of nonlinear equations (13) will be found in the next section by the method of successive approximations.

4. Method of solution

As the initial approximation, let us consider the solution of the homogeneous part of the system of equations (13):

$$\begin{aligned}
&\dot{a}_1 + \frac{1}{2}\mu\omega_0^{\gamma_1-1}\sin\left(\frac{1}{2}\pi\gamma_1\right)a_1 = 0,\\
&\dot{a}_2 + \frac{1}{2}\mu\Omega_0^{\gamma_2-1}\sin\left(\frac{1}{2}\pi\gamma_2\right)a_2 = 0,\\
&\dot{\varphi}_1 - \frac{1}{2}\mu\omega_0^{\gamma_1-1}\cos\left(\frac{1}{2}\pi\gamma_1\right) - \sigma_1 = 0,\\
&\dot{\varphi}_2 - \frac{1}{2}\mu\Omega_0^{\gamma_2-1}\cos\left(\frac{1}{2}\pi\gamma_2\right) - (\sigma_1 - \sigma) = 0,
\end{aligned} \tag{14}$$

which has the form

$$a_1 = a_{10}e^{-S_1T_2}, \quad a_2 = a_{20}e^{-S_2T_2}, \quad \varphi_1 = S_3T_2 + \varphi_{10}, \quad \varphi_2 = S_4T_2 + \varphi_{20}, \tag{15}$$

where a_{i0} and φ_{i0} (i=1,2) are, respectively, the initial values of amplitudes and phases to be found from the initial conditions, $\delta_0 = 2(\varphi_{20} - \varphi_{10})$ is the initial phase difference, and

$$\begin{aligned}
&S_1 = \frac{1}{2}\mu\omega_0^{\gamma_1-1}\sin\left(\frac{1}{2}\pi\gamma_1\right), \qquad S_2 = \frac{1}{2}\mu\Omega_0^{\gamma_2-1}\sin\left(\frac{1}{2}\pi\gamma_2\right),\\
&S_3 = \frac{1}{2}\mu\omega_0^{\gamma_1-1}\cos\left(\frac{1}{2}\pi\gamma_1\right) + \sigma_1, \quad S_4 = \frac{1}{2}\mu\Omega_0^{\gamma_2-1}\cos\left(\frac{1}{2}\pi\gamma_2\right) + (\sigma_1 - \sigma).
\end{aligned} \tag{16}$$

Now substituting (15) in equations (13) yields

$$\begin{aligned}
&\dot{a}_1 + S_1a_1 = \frac{1}{4}\Gamma_1 a_{10}e^{-(S_1+2S_2)T_2}a_{20}^2\sin(\Sigma T_2 + \delta_0) - \frac{1}{4}F\omega_0^{-1}\sin(S_3T_2 + \varphi_{10}),\\
&\dot{a}_2 + S_2a_2 = -\frac{1}{4}\Gamma_2 a_{10}^2 e^{-(2S_1+S_2)T_2}a_{20}\sin(\Sigma T_2 + \delta_0),\\
&\dot{\varphi}_1 - S_3 = \lambda_1 a_{10}^2 e^{-2S_1T_2} + \lambda_2 a_{20}^2 e^{-2S_2T_2} - \frac{1}{4}\Gamma_1 a_{20}^2 e^{-2S_2T_2}\cos(\Sigma T_2 + \delta_0) - \frac{1}{4}F\omega_0^{-1}a_{10}^{-1}e^{S_1T_2}\cos(S_3T_2 + \varphi_{10}),\\
&\dot{\varphi}_2 - S_4 = \lambda_3 a_{10}^2 e^{-2S_1T_2} + \lambda_4 a_{20}^2 e^{-2S_2T_2} - \frac{1}{4}\Gamma_2 a_{10}^2 e^{-2S_1T_2}\cos(\Sigma T_2 + \delta_0),
\end{aligned} \tag{17}$$

where $\Sigma = 2(S_4 - S_3)$.

To solve the first two equations in (17), we will use the method of variation of arbitrary functions, and assume the proposed solution in the form

$$a_1(T_2) = C_1(T_2)e^{-S_1T_2}, \qquad a_2(T_2) = C_2(T_2)e^{-S_2T_2}, \tag{18}$$

where $C_1(T_2)$ and $C_2(T_2)$ are arbitrary functions to be found.
Substituting the proposed solution (18) in equations (17) yields

$$\dot{C}_1(T_2)=\frac{1}{4}\Gamma_1 a_{10}a_{20}^2 e^{-2S_2T_2}\sin(\Sigma T_2+\delta_0)-\frac{1}{4}F\omega_0^{-1}e^{S_1T_2}\sin(S_3T_2+\varphi_{10}),$$
$$\dot{C}_2(T_2)=-\frac{1}{4}\Gamma_2 a_{10}^2 a_{20}e^{-2S_1T_2}\sin(\Sigma T_2+\delta_0). \tag{19}$$

Integrating equations (19), we have

$$C_1(T_2)=-\frac{1}{4}\Gamma_1 a_{10}a_{20}^2\frac{2S_2\sin(\Sigma T_2+\delta_0)+\Sigma\cos(\Sigma T_2+\delta_0)}{4S_2^{\ 2}+\Sigma^2}e^{-2S_2T_2}$$
$$-\frac{F}{4\omega_0}\frac{S_1\sin(S_3T_2+\varphi_{10})-S_3\cos(S_3T_2+\varphi_{10})}{S_1^2+S_3^2}e^{S_1T_2}+C_{10}, \tag{20}$$
$$C_2(T_2)=\frac{1}{4}\Gamma_2 a_{10}^2 a_{20}\frac{2S_1\sin(\Sigma T_2+\delta_0)+\Sigma\cos(\Sigma T_2+\delta_0)}{4S_1^{\ 2}+\Sigma^2}e^{-2S_1T_2}+C_{20},$$

where C_{10} and C_{20} are constants of integration.

Considering relationships (20), the amplitude functions take the form

$$a_1=a_{10}e^{-S_1T_2}-\frac{1}{4}\Gamma_1 a_{10}a_{20}^2\frac{2S_2\sin(\Sigma T_2+\delta_0)+\Sigma\cos(\Sigma T_2+\delta_0)}{4S_2^{\ 2}+\Sigma^2}e^{-(S_1+2S_2)T_2}$$
$$-\frac{F}{4\omega_0}\frac{S_1\sin(S_3T_2+\varphi_{10})-S_3\cos(S_3T_2+\varphi_{10})}{S_1^2+S_3^2}+C_{10}e^{-S_1T_2}, \tag{21}$$
$$a_2=a_{20}e^{-S_2T_2}+\frac{1}{4}\Gamma_2 a_{10}^2 a_{20}\frac{2S_1\sin(\Sigma T_2+\delta_0)+\Sigma\cos(\Sigma T_2+\delta_0)}{4S_1^{\ 2}+\Sigma^2}e^{-(2S_1+S_2)T_2}+C_{20}e^{-S_2T_2}.$$

Integrating the third and fourth equations in (17), we obtain the T_2-functions of the phases of vibration

$$\varphi_1=S_3T_2+\varphi_{10}-\frac{\lambda_1 a_{10}^2}{2S_1}e^{-2S_1T_2}-\frac{\lambda_2 a_{20}^2}{2S_2}e^{-2S_2T_2}$$
$$+\frac{1}{4}\Gamma_1 a_{20}^2\frac{2S_2\cos(\Sigma T_2+\delta_0)+\Sigma\sin(\Sigma T_2+\delta_0)}{4S_2^{\ 2}+\Sigma^2}e^{-2S_2T_2}$$
$$-\frac{1}{4}\frac{Fa_{10}^{-1}}{\omega_0}\frac{S_1\cos(S_3T_2+\varphi_{10})+S_3\sin(S_3T_2+\varphi_{10})}{S_1^2+S_3^2}e^{S_1T_2}+C_{30}, \tag{22}$$
$$\varphi_2=S_4T_2+\varphi_{20}-\frac{\lambda_3 a_{10}^2}{2S_1}e^{-2S_1T_2}-\frac{\lambda_4 a_{20}^2}{2S_2}e^{-2S_2T_2}$$
$$+\frac{1}{4}\Gamma_2 a_{10}^2\frac{2S_1\cos(\Sigma T_2+\delta_0)+\Sigma\sin(\Sigma T_2+\delta_0)}{4S_1^{\ 2}+\Sigma^2}e^{-2S_1T_2}+C_{40},$$

where C_{30} and C_{40} are constants of integration to be determined from the initial conditions.

Since the general solution of the system under consideration is the sum of the particular solution of the inhomogeneous set of equations and the general solution of the corresponding homogeneous system, then the arbitrary constants could be chosen in such a way that the initial conditions of all successive approximations would be zero. Thus, for the first approximation the constants to be found take the form

$$
\begin{aligned}
C_{10} &= \frac{1}{4}\Gamma_1 a_{10} a_{20}^2 \frac{2S_2 \sin\delta_0 + \Sigma\cos\delta_0}{4S_2^{\,2} + \Sigma^2} + \frac{F}{4\omega_0}\frac{S_1 \sin\varphi_{10} - S_3\cos\varphi_{10}}{S_1^2 + S_3^2}, \\
C_{20} &= -\frac{1}{4}\Gamma_2 a_{10}^2 a_{20} \frac{2S_1 \sin\delta_0 + \Sigma\cos\delta_0}{4S_1^{\,2} + \Sigma^2}, \\
C_{30} &= \frac{\lambda_1 a_{10}^2}{2S_1} + \frac{\lambda_2 a_{20}^2}{2S_2} - \frac{1}{4}\Gamma_1 a_{20}^2 \frac{2S_2 \cos\delta_0 + \Sigma\sin\delta_0}{4S_2^{\,2} + \Sigma^2} + \frac{1}{4}\frac{F a_{10}^{-1}}{\omega_0}\frac{S_1\cos\varphi_{10} + S_3 \sin\varphi_{10}}{S_1^2 + S_3^2}, \\
C_{40} &= \frac{\lambda_3 a_{10}^2}{2S_1} + \frac{\lambda_4 a_{20}^2}{2S_2} - \frac{1}{4}\Gamma_2 a_{10}^2 \frac{2S_1 \cos\delta_0 + \Sigma\sin\delta_0}{4S_1^{\,2} + \Sigma^2}.
\end{aligned}
\tag{23}
$$

Substitution of the found constants of integration (23) in relationships (21) and (22) results in the approximate analytical solution of the formulated problem.

5. Conclusion

Nonlinear forced vibrations of suspension bridges, when the frequency of an external force is approaching one of the natural frequencies of the suspension system, which, in its turn, undergoes the conditions of the one-to-one internal resonance, are investigated.

The approximate analytical solution has been obtained for the case $\gamma_1 \neq \gamma_2$ by the method of successive approximations.

Acknowledgments

This research was made possible by the Grant No. 9.5138.2017/8.9 as a Government task from the Ministry of Education and Science of the Russian Federation.

References

[1] Abdel-Ghaffar A M and Scanlan R H 1985 Ambient vibration studies of Golden Gate Bridge. I: Suspended structure *ASCE J. Eng. Mech.* **111** 463-82

[2] Rossikhin Y A and Shitikova M V 1995 Effect of viscosity on the vibrational processes in a combined suspension system *Mech. Solids* **30** 157-66

[3] Rossikhin Y A and Shitikova M V 1998 Application of fractional calculus for analysis of nonlinear damped vibrations of suspension bridges *ASCE J. Eng. Mech.* **124** 1029-36

[4] Rossikhin Y A and Shitikova M V 2005 Fractional calculus model of nonlinear free vibrations of suspension bridges *Proc. of the 5th Int. Conf. on Structural Dynamics EURODYN2005* (Sept 4-7 2005, Paris, France) vol 3 (Rotterdam: Milpress) pp 2033-38

[5] Abdel-Ghaffar A M and Rubin L I 1983 Nonlinear free vibrations of suspension bridges: Theory and application *ASCE J. Eng. Mech.* **109** 313-45

[6] Rossikhin Y A and Shitikova M V 2013 Nonlinear dynamic response of a fractionally damped suspension bridge subjected to small external force *Int. J. Mech.* 7(3) 155-63

IOP Conf. Series: Materials Science and Engineering **489** (2019) 012038 doi:10.1088/1757-899X/489/1/012038

Impact response of a nonlinear viscoelastic auxetic plate

A I Krusser, Yu A Rossikhin and M V Shitikova

Research Centre on Dynamics of Solids and Structures, Voronezh State Technical University, Voronezh 304006, Russia

an.krusser@yandex.ru

Abstract. Impact induced geometrically non-linear vibrations of a viscoelastic plate are investigated for the case when the shear operator is governed by the fractional derivative Kelvin-Voigt model in conjunction with the time-independent coefficient of volume extension-compression. Such a model could describe the behavior of so-called auxetic materials with negative Poisson's ratios. The modified method of multiple time scales involving the expansion of the fractional derivative in terms of a small parameter has been utilized for solving nonlinear governing equations of motion. The asymptotic behavior of the roots of the characteristic equation for determining the frequencies of the system as a function of the retardation time is studied and three modes of vibration in the process of impact interaction are described.

1. Introduction

Nonlinear vibrations of plates caused by different dynamic loads are an important area of applied mechanics, since they can lead to dangerous non-linear phenomena. In the majority of articles on this topic, studies were carried out for a conventional isotropic plate, the material of which is described by the Hooke's law [1-3]. But recently widely spread a new class of materials called auxetics which exhibit negative Poisson's ratio (NPR). In contrast to conventional materials (like rubber, glass, metals, etc.) they expand transversely when pulled longitudinally and contract transversely when pushed longitudinally [4].

The work, wherein materials with $\nu < 0$ were mentioned for the first time is dated by 1944 [5]. It was reported that crystals of FeS_2 have $\nu \approx -0.14$, but more recent studies have not confirmed this information. In 1987, Wojciechowski [6] proposed the first model of a two-dimensional isotropic thermodynamically stable systems with $\nu < 0$, composed of non-spherical particles. Potential interest in these materials has increased after the publication by Lakes the results of tests on synthesized polyester foam with $\nu = -0.7$ [7]. These new types of materials were named "auxetics" by Evans [8,9].

Auxetics cause great interest among scientists due to their special physical and mechanical properties such as resistance to indentation, fracture toughness, shear resistance, viscoelastic properties, acoustic absorption, shape memory, variable permeability, etc. [10-15]. Today auxetic properties have been identified in various classes of materials: composites, porous and granular materials, including a variety of polymers and foams, crystals of different compounds, natural and biological objects [16-18]. Much attention is paid to the synthesis of man-made auxetic materials, manufactured by using the modern technologies. A few studies have examined the modeling of metamaterials with negative Poisson's ratio by creating various geometrical configurations [8,9,19-21].

Auxetics have potentially important applications in the nearest future and can be advantageously used in the development of hydrophones and other sensors, for fiber reinforcement in composites, shock and sound absorbers, fasteners and rivets, air filters, for thermal protection in aerospace, in manufacturing defense protective clothing, etc. [16,22].

During the past 20 years many researchers have studied the behavior of materials with negative Poisson's ratio in various dynamic problems. Argatov et al. [23] studied implications of the negative Poisson's ratio with regard to the quasi-static and dynamic indentation compliances for isotropic homogeneous materials. Aw et al.[23] considered the deformation of thin auxetic membrane structures under indentation. It is shown that negative Poisson's ratios have direct influence on the membrane deformation domain, including the force–displacement curve, the deflection profile and the contact area. Alderson and Coenen [25] studied the problem of the low velocity impact response of auxetic carbon fibre laminates. Allen et al. [26] considered an analogous problem for auxetic and conventional open-cell polyurethane foams. The auxetic samples displayed a six times reduction in peak acceleration, showing potential in impact protector devices such as shin or thigh protectors in sports equipment applications. Strek et al. [27,28] studied the dynamic response of sandwich panels with auxetic cores. Jin et al. [29] extended this problem for sandwich structures with graded auxetic honeycomb cores under blast loading. Results show that both the graded honeycomb cores and cross-arranged honeycomb cores can significantly improve the resistance ability of the sandwich structures under blast loading.

Lim [30,31] considered thin auxetic plates and shells and found that the bending moment under uniform loading of circular plates suggests that the optimal Poisson's ratio is −1/3 if the plate is simply-supported at the edge. Auxetic materials are not suitable for uniformly loaded and simply supported square plates, as moment minimization study requests an optimal Poisson's ratio of 0.115, but are highly suitable for central point loaded and simply supported square plates, as moment minimization study suggests an optimal Poisson's ratio of −1.

However, as far as the authors know, the majority of papers in the field delivered the results obtained by experimental methods for studying structures [20, 23-29, 32], while the theory of mathematical modeling of auxetic behavior is only under development [30,31,33].

In the present paper, which continues the research started in [34], the problem of the impact interaction of an elastic sphere and a viscoelastic plate is considered.

2. Problem formulation

To describe the damping features of the plate, the shear operator is preassigned in terms of the fractional derivative Kelvin-Voigt model

$$\tilde{\mu} = \mu_0[1 + (\tau_\sigma^\mu)^\gamma D_{0+}^\gamma], \tag{1}$$

where μ_0 is the relaxed shear modulus, τ_σ^μ is the retardation time, $\gamma (0 < \gamma \le 1)$ is the fractional parameter, D_{0+}^γ is the Riemann-Liouville fractional derivative

$$D_{0+}^\gamma x(t) = \frac{d}{dt}\int_0^t \frac{x(t')dt'}{\Gamma(1-\gamma)(t-t')^\gamma}, \tag{2}$$

$\Gamma(1-\gamma)$ is the Gamma function, and $x(t)$ is an arbitrary function.

The bulk operator is assumed to be time-independent $\tilde{K} = K_0 = \text{const}$, i.e.

$$\tilde{E} = \frac{9K_0\tilde{\mu}}{3K_0 + \tilde{\mu}}. \tag{3}$$

It has been shown that such a model could describe the behaviour of so-called auxetic materials with negative Poisson's ratios [35].

Let us write Poisson's operator $\tilde{\nu}$ in the form [35]

$$\tilde{\nu} = -1 + \frac{E_0}{2\mu_0} \ni_\gamma^* (t_\sigma^\gamma), \tag{4}$$

where $\ni_\gamma^* (t_\sigma^\gamma)$ is the dimensionless Rabotnov's fractional operator defined in [36] as follows

$$\ni_\gamma^* (t_\sigma^\gamma) = \frac{1}{1 + t_\sigma^\gamma D_{0+}^\gamma}. \tag{5}$$

Assume that an elastic or rigid sphere of mass M moves along the z-axis towards a simply supported thin rectangular nonlinear plate with thickness h and dimensions a and b along the x- and y-axes, as shown in Figure 1. Impact occurs at the moment $t = 0$ with the low velocity εV_0 at the point N with Cartesian coordinates x_0, y_0, where ε is a small dimensionless parameter. According to the Donnell-Mushtari nonlinear shallow shell theory, the equations of motion could be obtained in terms of lateral deflection w and Airy's stress function ϕ

$$\frac{\tilde{D}}{h}\left(\frac{\partial^4 w}{\partial x^4} + 2\frac{\partial^4 w}{\partial x^2 \partial y^2} + \frac{\partial^4 w}{\partial y^4} \right) = \frac{\partial^2 w}{\partial x^2}\frac{\partial^2 \phi}{\partial y^2} + \frac{\partial^2 w}{\partial y^2}\frac{\partial^2 \phi}{\partial x^2} - 2\frac{\partial^2 w}{\partial x \partial y}\frac{\partial^2 \phi}{\partial x \partial y} + \frac{F}{h} - \rho \ddot{w}, \tag{6}$$

$$\frac{1}{\tilde{E}}\left(\frac{\partial^4 \phi}{\partial x^4} + 2\frac{\partial^4 \phi}{\partial x^2 \partial y^2} + \frac{\partial^4 \phi}{\partial y^4} \right) = -\frac{\partial^2 w}{\partial x^2}\frac{\partial^2 w}{\partial y^2} + \left(\frac{\partial^2 w}{\partial x \partial y} \right)^2, \tag{7}$$

where $\tilde{D} = \dfrac{\tilde{E}h^3}{12(1-\tilde{\nu}^2)}$ is the cylindrical rigidity operator, ρ is the density, $\tilde{E}$ and $\tilde{\nu}$ are the Young's operator and Poisson's operator, respectively, $F = P(t)\delta(x - x_0)\delta(y - y_0)$ is the contact force, $P(t)$ is yet unknown function, δ is the Dirac delta function, and overdots denote time-derivatives.

Coupled vibrations of the impactor (sphere) and the target (nonlinear viscoelastic plate) are described by an infinite set of coupled nonlinear ordinary differential equations of the second order with respect to time for defining the generalized coordinates which has been obtained in [34]:

$$\begin{aligned} &\ddot{\xi}_{mn}(t) + \tilde{\Omega}_{mn}^2 \xi_{mn}(t) + \frac{32\pi^4 \tilde{E}}{a^6 b^6 \rho} \sum_r \sum_s \sum_i \sum_j \sum_k \sum_l \sum_p \sum_q B_{rsijmn} B_{pqklij} \xi_{rs}(t) \xi_{pq}(t) \xi_{kl}(t) \\ &+ \frac{4M}{ab\rho h} \sin\left(\frac{m\pi x_0}{a} \right) \sin\left(\frac{n\pi y_0}{b} \right) \sum_p \sum_q \ddot{\xi}_{pq}(t) \sin\left(\frac{p\pi x_0}{a} \right) \sin\left(\frac{q\pi y_0}{b} \right) = 0, \end{aligned} \tag{8}$$

where coefficients B_{pqklmn} are represented in [37], $\tilde{\Omega}_{mn}$ is the viscoelastic operator corresponding to the natural frequency of the mnth mode of the plate vibration defined as

$$\tilde{\Omega}_{mn}^2 = \Omega_{0mn}^2 \tilde{d}; \quad \tilde{d} = [1 + \eta t_\sigma^\gamma D^\gamma - \zeta \ni_\gamma^* (A t_\sigma^\gamma)], \tag{9}$$

with

$$\Omega_{0mn}^2 = \frac{\pi^4}{\rho}\frac{h^2}{48}\left(\frac{m^2}{a^2} + \frac{n^2}{b^2} \right)^2 (9\lambda_0 + 10\mu_0), \; \eta = \frac{12(\lambda_0 + \mu_0)}{(9\lambda_0 + 10\mu_0)}, \; \zeta = \frac{(3\lambda_0 + 2\mu_0)^2}{(9\lambda_0 + 10\mu_0)(\lambda_0 + \mu_0)}, \; A = \frac{4(\lambda_0 + \mu_0)}{(\lambda_0 + 2\mu_0)}.$$

The last term in each equation from (8) describes the influence of the coupled impact interaction of the target with the impactor of the mass M applied at the point with the coordinates x_0, y_0.

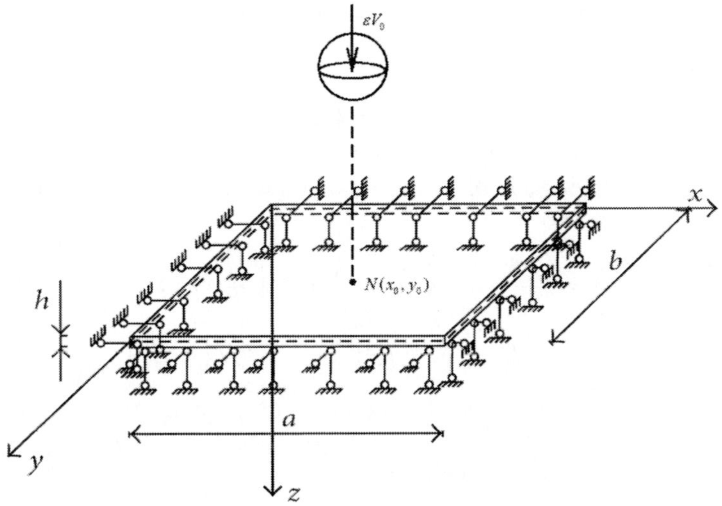

Figure 1. Geometry of a rectangular thin plate

It is known [38] that during nonstationary excitation of thin bodies not all possible modes of vibration would be excited. Moreover, the modes which are strongly coupled by any of the so-called internal resonance conditions are initiated and dominate in the process of vibration, in so doing the types of modes to be excited are dependent on the character of the external excitation.

Thus, in order to study the additional nonlinear phenomenon induced by the coupled impact interaction due to equation (8), we suppose that only two natural modes of vibrations are excited during the process of impact, namely, $\Omega_{\alpha\beta}$ and $\Omega_{\gamma\delta}$, in so doing each type of impact subjection should be considered separately.

3. Method of solution

The set of equations (8) is reduced to the following two nonlinear differential equations:

$$p_{11}\ddot{\xi}_{\alpha\beta} + p_{12}\ddot{\xi}_{\gamma\delta} + \tilde{\Omega}^2_{\alpha\beta}\xi_{\alpha\beta} + p_{16}\xi^3_{\alpha\beta} + p_{17}\xi_{\alpha\beta}\xi^2_{\gamma\delta} = 0, \tag{10}$$

$$p_{21}\ddot{\xi}_{\alpha\beta} + p_{22}\ddot{\xi}_{\gamma\delta} + \tilde{\Omega}^2_{\gamma\delta}\xi_{\gamma\delta} + p_{26}\xi^3_{\gamma\delta} + p_{27}\xi^2_{\alpha\beta}\xi_{\gamma\delta} = 0, \tag{11}$$

where coefficients p_{ij} $(i=1,2,\ j=3,4...7)$ are represented in [10] and p_{ij} $(i=1,2,\ j=3,4,5)=0$.

In order to solve a set of two nonlinear equations (10) and (11), ley us apply the method of multiple time scales [39] via the following expansions:

$$\xi_{ij}(t) = \varepsilon X^1_{ij}(T_0,T_1,T_2) + \varepsilon^2 X^2_{ij}(T_0,T_1,T_2) + \varepsilon^3 X^3_{ij}(T_0,T_1,T_2), \tag{12}$$

where $ij = \alpha\beta$ or $\gamma\delta$, $T_n = \varepsilon^n t$ are new independent variables, among them: $T_0 = t$ is a fast scale characterizing motions with the natural frequencies, and $T_1 = \varepsilon t$ and $T_2 = \varepsilon t^2$ are slow scales characterizing the modulation of the amplitudes and phases of the modes with nonlinearity.

Recall that the first and the second time derivatives are defined, respectively, as follows [39]

$$\frac{d}{dt} = D_0 + \varepsilon D_0 + \varepsilon^2 D_2 + ..., \tag{13}$$

$$\frac{d^2}{dt^2} = D_0^2 + 2\varepsilon D_0 D_1 + \varepsilon^2\left(D_1^2 + 2D_0D_2\right), \tag{14}$$

where $D_i^n = \partial^n / \partial T_i^n$ $(n=1,2,\ i=0,1)$.

Following Rossikhin and Shitikova [40], the fractional derivative is interpreted as the fractional power of the differential operator

$$D_+^\gamma = \left(\frac{d}{dt}\right)^\gamma = D_0^\gamma + \varepsilon\gamma D_0^{\gamma-1} D_1 + \ldots \tag{15}$$

Now let us expand the dimensionless Rabotnov's fractional operator (5) in a Taylor series. As a result we have

$$\begin{aligned} э_\gamma^*(\tau^\gamma) &= \frac{1}{1+\tau^\gamma D_{0+}^\gamma} = (1+\tau^\gamma D_{0+}^\gamma)^{-1} = \left[1+\tau^\gamma (D_0^\gamma + \varepsilon\gamma D_0^{\gamma-1} D_1)\right]^{-1} \\ &= (1+\tau^\gamma D_0^\gamma)^{-1} - \varepsilon(1+\tau^\gamma D_0^\gamma)^{-2}\tau^\gamma \gamma D_0^{\gamma-1} D_1 + \ldots \end{aligned} \tag{16}$$

Substituting relationships (12)-(16) in (10) and (11), after equating the coefficients at like powers of ε to zero, we are led to a set of recurrence equations at various orders:

to order ε

$$p_{11} D_0^2 X_1^1 + p_{12} D_0^2 X_2^1 + \tilde{\Omega}_1^2 X_1^1 = 0, \tag{17}$$
$$p_{21} D_0^2 X_1^1 + p_{22} D_0^2 X_2^1 + \tilde{\Omega}_2^2 X_2^1 = 0; \tag{18}$$

to order ε^2

$$p_{11} D_0^2 X_1^2 + p_{12} D_0^2 X_2^2 + \tilde{\Omega}_1^2 X_1^2 = -2p_{11} D_0 D_1 X_1^1 - 2p_{12} D_0 D_1 X_2^1 - \tilde{I}_{11}\gamma D_0^{\gamma-1} D_1 X_1^1, \tag{19}$$
$$p_{21} D_0^2 X_1^2 + p_{22} D_0^2 X_2^2 + \tilde{\Omega}_2^2 X_2^2 = -2p_{21} D_0 D_1 X_1^1 - 2p_{22} D_0 D_1 X_2^1 - \tilde{I}_{21}\gamma D_0^{\gamma-1} D_1 X_2^1; \tag{20}$$

to order ε^3

$$\begin{aligned} p_{11} D_0^2 X_1^3 + p_{12} D_0^2 X_2^3 + \tilde{\Omega}_1^2 X_1^3 &= -2p_{11} D_0 D_1 X_1^2 - 2p_{12} D_0 D_1 X_2^2 - p_{11}\left(D_1^2 + 2D_0 D_2\right) X_1^1 \\ &- p_{12}\left(D_1^2 + 2D_0 D_2\right) X_2^1 - \tilde{I}_{11}\gamma D_0^{\gamma-1} D_1 X_1^2 - p_{16}\left(X_1^1\right)^3 \tilde{E} - p_{17} X_1^1 \left(X_2^1\right)^2 \tilde{E}, \end{aligned} \tag{21}$$

$$\begin{aligned} p_{21} D_0^2 X_1^3 + p_{22} D_0^2 X_2^3 + \tilde{\Omega}_2^2 X_2^3 &= -2p_{21} D_0 D_1 X_1^2 - 2p_{22} D_0 D_1 X_2^2 - p_{21}\left(D_1^2 + 2D_0 D_2\right) X_1^1 \\ &- p_{22}\left(D_1^2 + 2D_0 D_2\right) X_2^1 - \tilde{I}_{21}\gamma D_0^{\gamma-1} D_1 X_2^2 - p_{26}\left(X_2^1\right)^3 \tilde{E} - p_{27} X_1^1 \left(X_2^1\right)^2 \tilde{E}, \end{aligned} \tag{22}$$

where for simplicity it is denoted $X_1^1 = X_{\alpha\beta}^1$, $X_2^1 = X_{\gamma\delta}^1$, $X_1^2 = X_{\alpha\beta}^2$, $X_2^2 = X_{\gamma\delta}^2$, $\tilde{\Omega}_1 = \tilde{\Omega}_{\alpha\beta}$, $\tilde{\Omega}_2 = \tilde{\Omega}_{\gamma\delta}$, and $\tilde{I}_{11(21)} = \Omega_{01(2)}^2 [\eta t_\sigma^\gamma - \zeta(1+At_\sigma^\gamma D_0^\gamma)^{-2} At_\sigma^\gamma]$.

The solution of the set of equations (17) and (18) we will seek in the form

$$X_1 = A_1(T_1, T_2) e^{\lambda_1 T_0} + A_2(T_1, T_2) e^{\lambda_2 T_0} + \text{cc}, \tag{23}$$
$$X_2 = \alpha_1 A_1(T_1, T_2) e^{\lambda_1 T_0} + \alpha_2 A_2(T_1, T_2) e^{\lambda_2 T_0} + \text{cc}, \tag{24}$$

where $\lambda_1, \lambda_2, \alpha_1, \alpha_2$ are yet uknown constants, *cc* is the complex conjugate part to the preceding terms.

Substituting equations (23) and (24) in (17) and (18) yields

$$\begin{aligned} &\{p_{11}\lambda_1^2 + \Omega^2{}_{01}[1+\eta t_\sigma^\gamma \lambda_1^\gamma - \zeta(1+At_\sigma^\gamma \lambda_1^\gamma)^{-1}] + \alpha_1 p_{12}\lambda_1^2\} A_1 e^{\lambda_1 T_0} \\ &+\{p_{11}\lambda_2^2 + \Omega^2{}_{01}[1+\eta t_\sigma^\gamma \lambda_2^\gamma - \zeta(1+At_\sigma^\gamma \lambda_2^\gamma)^{-1}] + \alpha_2 p_{12}\lambda_2^2\} A_2 e^{\lambda_2 T_0} = 0, \end{aligned} \tag{25}$$

$$\begin{aligned} &\{p_{21}\lambda_1^2 + \Omega^2{}_{02}\alpha_1[1+\eta t_\sigma^\gamma \lambda_1^\gamma - \zeta(1+At_\sigma^\gamma \lambda_1^\gamma)^{-1}] + \alpha_1 p_{22}\lambda_1^2\} A_1 e^{\lambda_1 T_0} \\ &+\{p_{21}\lambda_2^2 + \Omega^2{}_{02}\alpha_2[1+\eta t_\sigma^\gamma \lambda_2^\gamma - \zeta(1+At_\sigma^\gamma \lambda_2^\gamma)^{-1}] + \alpha_2 p_{22}\lambda_2^2\} A_2 e^{\lambda_2 T_0} = 0. \end{aligned} \tag{26}$$

In order to satisfy equations (25) and (26), it is a need to vanish to zero each bracket in these equations. As a result, from four different brackets we have

$$\alpha_i = -\frac{p_{11}\lambda_i^2 + \Omega^2{}_{01}[1+\eta t_\sigma^\gamma \lambda_i^\gamma - \zeta(1+At_\sigma^\gamma \lambda_i^\gamma)^{-1}]}{p_{12}\lambda_i^2}, \tag{27}$$

$$\alpha_i = -\frac{p_{22}\lambda_i^2}{p_{22}\lambda_i^2 + \Omega^2{}_{02}[1+\eta t_\sigma^\gamma \lambda_i^\gamma - \zeta(1+At_\sigma^\gamma \lambda_i^\gamma)^{-1}]}. \tag{28}$$

Since the left-hand side parts of relationships (27) and (28) are equal, then their right-hand side parts should be equal as well. Now equating the corresponding right-hand side parts, we are led to one and the same characteristic equation for determining the values λ_1 and λ_2

$$\begin{aligned}(p_{11}p_{22} - p_{12}^2)\lambda^4 + (p_{11}\Omega_{02}^2 + p_{22}\Omega_{01}^2)[1+\eta t_\sigma^\gamma \lambda^\gamma - \zeta(1+At_\sigma^\gamma \lambda^\gamma)^{-1}]\lambda^2 \\ + \Omega_{01}^2\Omega_{02}^2[1+\eta t_\sigma^\gamma \lambda^\gamma - \zeta(1+At_\sigma^\gamma \lambda^\gamma)^{-1}]^2 = 0.\end{aligned} \tag{29}$$

Let us study the asymptotic behavior of the roots of equation (29), supposing for this purpose that t_σ^γ is a small value. Then equation (29) at any $\gamma(0 \le \gamma \le 1)$ could be written in the form

$$(p_{11}p_{22} - p_{12}^2)\lambda^4 + (p_{11}\Omega_{02}^2 + p_{22}\Omega_{01}^2)(1-\zeta)\lambda^2 + \Omega_{01}^2\Omega_{02}^2(1-\zeta)^2 = 0, \tag{30}$$

where $p_{11}p_{22} - p_{12}^2 > 0$, and $1-\zeta = \dfrac{16\mu_0(\mu_0+\lambda_0)}{(9\lambda_0+10\mu_0)(\lambda_0+2\mu_0)} > 0$.

Solving equation (30), we find

$$\lambda_{1,2}^2 = \frac{(1-\zeta)}{2\left(p_{11}p_{22} - p_{12}^2\right)}\left[-\left(p_{22}\Omega_{01}^2 + p_{11}\Omega_{02}^2\right) \pm \sqrt{\left(p_{11}\Omega_{02}^2 - p_{22}\Omega_{01}^2\right)^2 + 4\Omega_{01}^2\Omega_{02}^2 p_{12}^2}\right] < 0 \tag{31}$$

From (31) it follows that λ_1 and λ_2 are purely imaginary values , i.e. $\lambda_{1,2} = i\omega_{1,2}$ and

$$e^{\lambda_{1,2}t} = e^{i\omega_{1,2}t}, \tag{32}$$

where ω_1 and ω_2 are real values. In this case, the plate perfomes undamped vibrations.

It is easy to show that at $t_\sigma^\gamma \to 0$ operator $\tilde{I}_{11(21)} = 0$ and $\tilde{\Omega}_{mn}^2 = \Omega_{mn}^2$, where Ω_{mn} is a constant value corresponding to the natural frequency of the *mn*th mode of the elastic plate.

Now let us suppose that t_σ^γ is a large value. Then equation (29) at any $\gamma(0 \le \gamma \le 1)$ could be written in the form

$$(p_{11}p_{22} - p_{12}^2)\chi^2 + (p_{11}\Omega_{02}^2 + p_{22}\Omega_{01}^2)\eta t_\sigma^\gamma \chi + \Omega_{01}^2\Omega_{02}^2\eta t_\sigma^{2\gamma} = 0, \tag{33}$$

where $\chi = \lambda^{2-\gamma}$.

Solving equation (33), we find

$$\chi_{1,2}^2 = \frac{\eta t_\sigma^\gamma}{2\left(p_{11}p_{22} - p_{12}^2\right)}\left[-\left(p_{22}\Omega_{01}^2 + p_{11}\Omega_{02}^2\right) \pm \sqrt{\left(p_{11}\Omega_{02}^2 - p_{22}\Omega_{01}^2\right)^2 + 4\Omega_{01}^2\Omega_{02}^2 p_{12}^2}\right] < 0 \tag{34}$$

Rewriting the solution of (34) in the form $\chi_{1,2} = -B_{1,2}$ when $B_{1,2} > 0$, we will obtain

$$\lambda_{1,2} = \chi_{1,2}^{\frac{1}{2-\gamma}} = (-B_{1,2})^{\frac{1}{2-\gamma}} = B_{1,2}^{\frac{1}{2-\gamma}} e^{i(2-\gamma)^{-1}\pi},$$

or

$$\lambda_{1,2} = B^{\frac{1}{2-\gamma}}[\cos(2-\gamma)^{-1}\pi + i\sin(2-\gamma)^{-1}\pi]. \tag{35}$$

Considering (35), relationships (32) are reduced to

$$e^{\lambda_{1,2}t} = e^{(\alpha_{1,2}+i\bar{\omega}_{1,2})t}, \tag{36}$$

where $\alpha_{1,2} = B_{1,2}^{\frac{1}{2-\gamma}}\cos(2-\gamma)^{-1}\pi$, and $\bar{\omega}_{1,2} = B_{1,2}^{\frac{1}{2-\gamma}}\sin(2-\gamma)^{-1}\pi$.

It follows from equation (36) that at $\gamma \neq 0$ and $\gamma \neq 1$ damped vibrations take place. At $\gamma = 0$, values $\alpha_{1,2} = 0$ and vibrations become undamped, while at $\gamma = 1$, values $\omega_{1,2} = 0$ and the plate possesses the aperiodic mode of vibration.

4. Conclusion

Large amplitude (geometrically non-linear) vibrations of viscoelastic rectangular plate were investigated for the case when the shear operator is governed by the fractional derivative Kelvin-Voigt model in conjunction with the time-independent coefficient of volume extension-compression, what is verified by experimental data. It has been shown that such a model could describe the behaviour of so-called auxetic materials with negative Poisson's ratios.

The governing set of nonlinear differential equations has been obtained using the modified method of multiple time scales. The asymptotic behavior of the roots of the characteristic equation for determining the frequencies of the system as a function of the retardation time was studied, and three types of vibration in the process of impact interaction were described. It has been shown that the characteristic equation (30) describes only one vibrational mode – undamped, while the characteristic equation (33) describes three vibrational modes at once: undamped, damped and aperiodic.

Acknowledgement

The research described in this publication has been supported by the Russian Science Foundation.

References

[1] Amabili M 2008 *Nonlinear Vibrations and Stability of Shells and Plates* (Cambridge: Cambridge University Press)
[2] Breslavsky I D, Amabili M and Legrand M 2014 *Int. J. Nonlinear Mech.* **58** 30–40
[3] Ribeiro P and Petyt M 2000 Non-linear free vibration of isotropic plates with internal resonance *Int. J. Nonlinear Mech.* **35** 263-78
[4] Liu Q 2006 *Literature review: Materials with negative Poisson's ratios and potential applications to aerospace and defence* (Victoria, Australia: DSTO Defence Science and Technology Organization)
[5] Love A E H 1944 *A treatise on the mathematical theory of elasticity* (New York: Dover)
[6] Wojciechowski K W 1987 *Molec. Phys.* **61** 1247-58
[7] Lakes R S 1993 *Adv. Mater.* **15** 293-6
[8] Evans K E 1991 *Endeavour* **4** 170-4
[9] Evans K E, Alderson K L 2000 *Engng. Sci. and Edu* 148-54
[10] Smith C W, Evans K E., Lehman F 1999 *Cell. Polym.* **18** 79-101
[11] Choi J B, Lakes R S 1992 *J. Mater. Sci.* **27** 5373-81
[12] Choi J B, Lakes R S 1996 *Int. J. Fracture* **80** 73-83
[13] Chen C P, Lakes R S 1993 *J. Mater. Sci.* **28** 4288-98
[14] Chen C P, Lakes R S 1996 *J. Eng. Mater. Technol.* **118** 285-8
[15] Scarpa F and Tomlin P J 2000 *Fatigue Fract. Engng. Mater. Struct.* **23** 717–20
[16] Alderson A, Alderson K L 2007 *Proc. IMechE* G **221** 565-75
[17] Ballato A, Fellow L 2010 *IEEE Trans Ultrason Ferroelectr Freq Control* **57** 7-15
[18] Goldstein R V, Gorodtsov V A and Lisovenko D S 2014 Young's modulus and Poisson's ratio for seven-constant tetragonal crystals and their nano/microtubes *Phys Mech* **17** 5-14
[19] Grima 2010 J N *Auxetic Metamaterials* (Strasbourg, France)
[20] Bertoldi K, Reis P M, Willshaw S and Mullin T 2010 *Adv. Mater.* **22** 361-6
[21] Saxena K K, Das R and Calius E P 2016 Three decades of auxetics research materials with negative Poisson's ratio: A review *Adv. Eng. Mat.* 101002
[22] Carneiro V H, Meireles J and Puga H 2013 *Materials Science-Poland* **31** 561-71
[23] Argatov I I, Guinovart-Díaz R, Sabina F J 2012 *Int. J.Eng. Sci.* **54** 42-57
[24] Aw J, Zhao H, Norbury A, et al. 2015 *Phys. Status Solidi* B **252** 1526–32

IOP Conf. Series: Materials Science and Engineering **489** (2019) 012038 doi:10.1088/1757-899X/489/1/012038

[25] Alderson K L, Coenen 2008 V L *Phys. stat. sol.* B **245** 489–96
[26] Allen T, Shepherd J, Hewage T A M, Senior T, Foster L and Alderson A 2015 *Phys. Status Solidi* B **252** 1631–39
[27] Strek T, Jopek H and Nienartowicz M 2015 *Phys. Status Solidi* B **252** 1540–50
[28] Strek T, Maruszewski B, Narojczyk J W and Wojciechowski K W 2008 *J Non Cryst Solids* **354** 4475–80
[29] Jin X, Wang Z, Ning J, Xiao G, Liu E and Shu X 2016 *Comp.* B **106** 206-17
[30] Lim T C 2015 *Auxetic Materials and Structures* (Springer Singapore)
[31] Lim TC 2013 *IMechE. J. Mat. Des. Appl.* **227** 111–23
[32] Nayfeh A H and Balachandran B 1990 *Int. J. Nonlinear Mech.* **25** 199–209
[33] Rossikhin Yu A and Shitikova M V 2017 Mathematical models of viscoelastic auxetics *Abstr. 3d Int. Conf. on Mechanics of Composites* (Bologna) p 29
[34] Rossikhin Yu A, Shitikova M V and Krusser A I 2017 Impact response of a nonlinear viscoelastic auxetic doubly curved shallow shell *Proc. 24th Int. Cong. on Sound and Vibration (London)* ed B Gibbs (London: Int. Institute of Acoustics and Vibration) pp 1-8
[35] Rossikhin Yu A, Shitikova M V and Krusser A I 2016 *Mech. Res. Commun.* **77** 44-9
[36] Rossikhin Yu A and Shitikova M V 2014 *Fract. Calc. Appl. Anal.* **17** 674-83
[37] Rossikhin Yu A, Shitikova M V and Salih Khalid J M 2015 *Int. J. Mech.* **9** 329-42
[38] Nayfeh A H 2000 *Nonlinear Interaction: Analytical, Computational and Experimental Methods* (New York: John Wiley & Sons)
[39] Nayfeh A H 2003 *Perturbation methods* (New York: John Wiley & Sons)
[40] Rossikhin Yu A and Shitikova M V 1998 *ASCE J. Eng. Mech.* **124** 1029-36

IOP Conf. Series: Materials Science and Engineering **489** (2019) 012039 doi:10.1088/1757-899X/489/1/012039

Modelling the granular medium dynamics on rough vibrating plane using method of large particles

I P Lyan[1] and G Y Panovko[1,2]

[1] Mechanical Engineering Research Institute of the Russian Academy of Sciences (IMASH RAN), Moscow, Russia
[2] Bauman Moscow State Technical University (BMSTU), Moscow, Russia

LyanIlyaIMASH@yandex.ru

Abstract. The article represents the results of mathematical modelling of behaviour of the granular medium on an oscillating rough surface. Granular medium under the action of vibration is considered as a non-Newtonian (dilatant) fluid. The system of differential equations of continuum mechanics is solved using the modified large particles method. An algorithm for solving the problem for granular media with different properties under external influences is proposed.

1. Introduction

In modern technological operations of processing granular medium various methods of vibration processing, such as transportation, mixing, compaction/loosening, separation and etc., are widely used. Vibration allows to significantly intensify the course of such processes, and in some cases it turns out to be the only possibility for technological process implementation [1, 2].

Vibration in a granular medium causes various effects, in particular, it can flow like a liquid [3]. This state of granular medium is called fluidization or pseudo fluidity. As the intensity of the vibration increases, the connection between individual particles of the medium can be broken and its behavior becomes similar to a gas (vibro-boiling). The properties of the granular medium should be taken into account to choose rational vibration parameters and design of the oscillating machine`s working body that leads to incremental growth of technological process efficiency.

For embodiment of vibration technological processes the oscillating machines with two and more self-synchronized inertial exciters are widely used. The effect of self-synchronization of exciter in such machines depends on a wide spectrum of parameters, including the position of a dynamic system`s center of mass. Moving granular medium on the tray of such a machine leads to a significant shift of the system`s entire center of mass coordinates due to heterogeneity of material distribution on the working body. This paper sets the goal to determine the displacement of the system`s center of mass in time during the movement of the granular medium on the rough vibrating surface (tray).

2. Setting the objectives and the loading scheme

The paper discusses the dynamics of granular medium (quartz sand) flowing down on the tray of the vibrating conveyor (figure 1). Granular medium under the action of gravity precipitates from hopper *1*, forming embankment *2*. Tray *3* of the vibrating conveyor is turned to an angle ϑ from the horizontal, commits unidirectional harmonic oscillations $\xi(t)$ to an angle α from the tray, which is considered to be absolutely rigid. In order to avoid falling cargo to the other side of the vibrating conveyor, there is a

IOP Conf. Series: Materials Science and Engineering **489** (2019) 012039 doi:10.1088/1757-899X/489/1/012039

limiter *4* on the left side. Supposed distribution of the granular medium on the tray under the action of vibration is shown as a dotted line *5*.

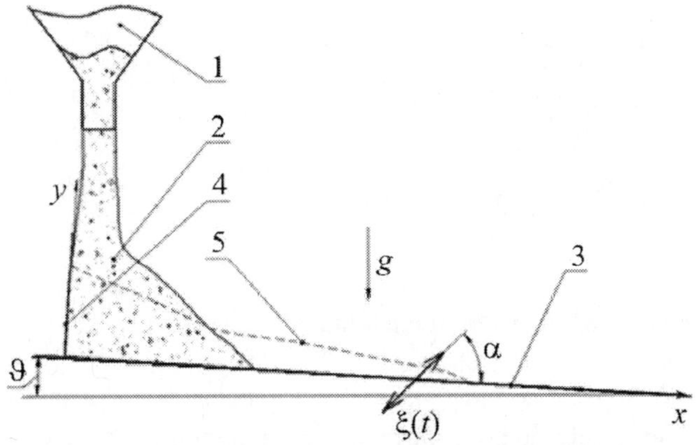

Figure 1. Loading scheme of the vibrating conveyor.

Description of granular medium behavior in a vibrating field is based on relevant computational models [3-9]. These models can be divided into two main types: discrete models of the granular medium regarded as set of material particles or solids with viscoelastic relations between them, and continual models using approaches of continuous medium mechanics in the whole volume of medium or in each of its elementary layer.

In this paper, the motion simulation of granular medium is based on the works [6-10] where moving granular medium under vibrational loads is assumed as a viscous dilatant fluid. Features of the flow of such continuous medium are laid in the rheological equation connecting strain rates with stresses. Together with the mass and momentum conservation equations they close the system of equations of continuum medium mechanics. Numerical solution of that system is based on the large particles method developed by O. M. Belotserkovsky and Yu. M. Davydov [10].

In a similar formulation, the problem has already been considered in [6]. However, analysis of these papers showed that the used finite difference numerical solution schemes of equations lead to the occurrence of asymmetry in characteristic medium movements, which, by nature, must possess symmetry.

3. Mathematical model

In the simulation of granular medium the following hypotheses and assumptions are accepted:

- the studying problem is considered to be plane in the gravity field;
- granular medium is considered to flow without separation from the tray surface;
- medium is considered to be single-phased (monogenic) and significant size of particle is considered to be many times smaller than calculation area.

Mathematical model of the studying process is described by the system consisting of three groups of equations: rheological equations connecting components of stress tensor with strain rates of granular medium, law of conservation of momentum and the law of conservation of mass of the system [6, 8, 10].

In this very work the rheological equations are obtained by using approaches described in [6]. In contrast to the work [6], the rates of angular deformations are taken into account in the form $\frac{1}{2}\left(\frac{\partial v_x}{\partial y}+\frac{\partial v_y}{\partial x}\right)$, which provides more correct description of the movement medium under the action of vibration. As a result, differential equations connecting strain rates and stresses in the projections on the axis x, y associating with the tray are written as:

$$\sigma_x = \alpha_0 + \alpha_1 \frac{\partial v_x}{\partial x} + \left(\alpha_4 + \alpha_5 \frac{\partial v_x}{\partial x}\right)\left[\frac{\partial v_x}{\partial x}\frac{\partial v_y}{\partial y} - \frac{1}{4}\left(\frac{\partial v_x}{\partial y} + \frac{\partial v_y}{\partial x}\right)^2\right] + \left(\alpha_2 + \alpha_3 \frac{\partial v_x}{\partial x}\right)\left(\frac{\partial v_x}{\partial x} + \frac{\partial v_y}{\partial y}\right);$$

$$\sigma_y = \alpha_0 + \alpha_1 \frac{\partial v_y}{\partial y} + \left(\alpha_4 + \alpha_5 \frac{\partial v_y}{\partial y}\right)\left[\frac{\partial v_x}{\partial x}\frac{\partial v_y}{\partial y} - \frac{1}{4}\left(\frac{\partial v_x}{\partial y} + \frac{\partial v_y}{\partial x}\right)^2\right] + \left(\alpha_2 + \alpha_3 \frac{\partial v_y}{\partial y}\right)\left(\frac{\partial v_x}{\partial x} + \frac{\partial v_y}{\partial y}\right); \quad (1)$$

$$\tau_{xy} = \frac{1}{2}\left\{\alpha_1 + \alpha_3\left(\frac{\partial v_x}{\partial x} + \frac{\partial v_y}{\partial y}\right) + \alpha_5\left[\frac{\partial v_x}{\partial x}\frac{\partial v_y}{\partial y} - \frac{1}{4}\left(\frac{\partial v_x}{\partial y} + \frac{\partial v_y}{\partial x}\right)^2\right]\right\}\left(\frac{\partial v_x}{\partial y} + \frac{\partial v_y}{\partial x}\right),$$

where $\sigma_x, \sigma_y, \tau_{xy}$ – components of the two-dimensional stress tensor; $\frac{\partial v_x}{\partial x}, \frac{\partial v_y}{\partial y}, \frac{1}{2}\left(\frac{\partial v_x}{\partial y} + \frac{\partial v_y}{\partial x}\right)$ – components of the strain rate tensor; $\alpha_i = \alpha_i(\upsilon)$, $i = 0,...,5$ – coefficients describing the dissipative properties of the medium, υ – volume concentration of the medium.

The coefficients $\alpha_i = \alpha_i(\upsilon)$, $i = 0,...,5$ are determined experimentally and they are functions of the volume concentration υ [6-8]. If the concentration of the medium exceeds some value of υ_d, then occurs hopping increase in a stress level in the medium at the same strain rates. To take into account this effect in the mathematical model, coefficients $\alpha_i = \alpha_i(\upsilon)$, $i = 0,...,5$ are defined as slopes (piece-liniar functions) that rapidly increases with concentration above υ_d.

The equations of conservation of momentum are:

$$\frac{\partial v_x}{\partial t} + v_x \frac{\partial v_x}{\partial x} + v_y \frac{\partial v_x}{\partial y} = \frac{1}{\upsilon\rho_0}\left(\frac{\partial \sigma_x}{\partial x} + \frac{\partial \tau_{xy}}{\partial y}\right) - g_x + \ddot{\xi}_x + R_x;$$
$$\frac{\partial v_y}{\partial t} + v_x \frac{\partial v_y}{\partial x} + v_y \frac{\partial v_y}{\partial y} = \frac{1}{\upsilon\rho_0}\left(\frac{\partial \tau_{xy}}{\partial x} + \frac{\partial \sigma_y}{\partial y}\right) - g_y + \ddot{\xi}_y + R_y, \quad (2)$$

where t - current time; ρ_0 - average density of solid particles; g_x, g_y – gravitational acceleration projections on the *x*, *y* axis; $\ddot{\xi}_x, \ddot{\xi}_y$ - projections on the *x*, *y* axes of external acceleration vibration exposure; R_x, R_y – projections on the x, y axis of aerodynamic drag forces.

The equations of conservation of momentum are:

$$\frac{\partial \upsilon}{\partial t} + \frac{\partial(\upsilon v_x)}{\partial x} + \frac{\partial(\upsilon v_y)}{\partial y} = 0.$$

As a result, there is a complete system of nonlinear partial differential equations with variable coefficients describing flow of a granular medium along a vibrating surface. The principal feature of this system of equations is that a free boundary of granular medium is unknown in advance. To solve such problems, large particles method can be used as well proven in accounting practice [6-10]. When using this method, the free boundary of medium is determined at each integration step.

4. Solution

The large particle method is a variation of the Harlow`s particles method in cells, which uses the Lagrange and Euler approaches in numerical solving of continuum mechanics problems [10]. This method combines the advantages of Lagrangian and Eulerian approaches. The calculation area is divided by a fixed (Eulerian) grid, but the continuous medium is treated as a discrete model, i.e. it is considered to be a set of particles of a fixed mass (Lagrangian grid of particles) that moves through the Eulerian grid. The motion of such system of particles during the time Δt is carried out first in the form of a change in the internal state, assuming that the grid is stationary - the Eulerian stage. Then the displacement of all particles without changing the internal state is considered - the Lagrangian stage. At the final stage, the grid is recalculated to its original state.

In accordance with the basic idea of the method of large particles, the computational area is divided into a rectangular grid with the cell size $\Delta x \times \Delta y$ (figure. 2 (a)). Each cell has four boundaries (figure. 2, (b)) on which normal and tangential stresses are determined, and the velocity vectors and the value of the volume concentration are calculated inside each cell.

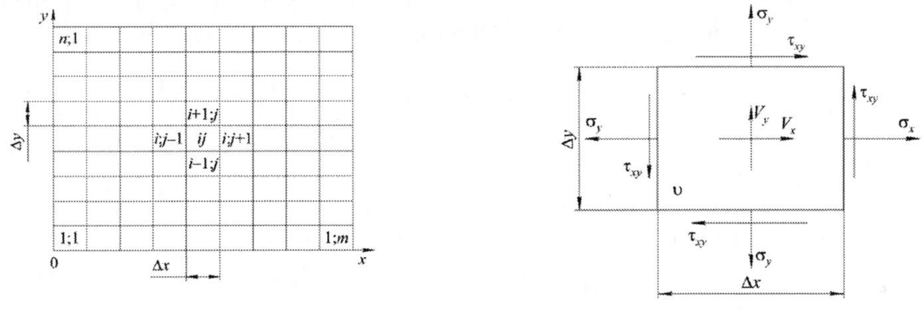

(a) (b)

Figure 2. Calculation area (a) and parameters of one cell (b) of large particles method.

4.1. Euler stage.

At the first (Eulerian) stage, it is assumed that the change of velocities, stresses and energy are calculated without mass transfer effects. For the subsequent numerical implementation, equations (1) are written in a finite-difference form at the current integration step:

$$\sigma_{x,i,j} = \alpha_0 + \alpha_1 V_{xx} + \left(\alpha_0' + \alpha_1' V_{xx}\right)\left(V_{xx} + V_{yy}\right) + \left(\alpha_0'' + \alpha_1'' V_{xx}\right)\left(V_{xx} V_{yy} - \frac{1}{4}\left[V_{xy} + V_{yx}\right]^2\right);$$

$$\sigma_{y,i,j} = \alpha_0 + \alpha_1 V_{yy} + \left(\alpha_0' + \alpha_1' V_{yy}\right)\left(V_{yy} + V_{xx}\right) + \left(\alpha_0'' + \alpha_1'' V_{yy}\right)\left(V_{yy} V_{xx} - \frac{1}{4}\left[V_{yx} + V_{xy}\right]^2\right); \qquad (3)$$

$$\tau_{xy,i,j} = \frac{1}{2}\left(V_{xy} + V_{yx}\right)\left(\alpha_1 + \alpha_1'\left[V_{xx} + V_{yy}\right] + \alpha_1''\left[V_{xx} V_{yy} - \frac{1}{4}\left\{V_{xy} + V_{yx}\right\}^2\right]\right),$$

$$V_{xx} = \frac{v_{x,i+1,j} - v_{x,i-1,j}}{2\Delta x}, \quad V_{yy} = \frac{v_{y,i+1,j} - v_{y,i-1,j}}{2\Delta y}, \quad V_{xy} = \frac{v_{x,i+1,j} - v_{x,i-1,j}}{2\Delta y}, \quad V_{yx} = \frac{v_{y,i+1,j} - v_{y,i-1,j}}{2\Delta x},$$

where i, j - row and column numbers of grid respectively in which the cell of the computational area is located.

The equations for finding the intermediate velocities for each time step in a finite-difference form are the following ones:

IOP Conf. Series: Materials Science and Engineering **489** (2019) 012039 doi:10.1088/1757-899X/489/1/012039

$$\tilde{v}_{x,i,j} = v_{x,i,j} + \frac{\Delta t}{2\upsilon\rho_0}\left(\frac{\sigma_{x,i+1,j} - \sigma_{x,i\text{-}1,j}}{2\Delta x} + \frac{\tau_{xy,i,j+1} - \tau_{xy,i,j\text{-}1}}{2\Delta y} + R_x\right) - g_x\Delta t + \ddot{\xi}_x\Delta t;$$
$$\tilde{v}_{y,i,j} = v_{y,i,j} + \frac{\Delta t}{2\upsilon\rho_0}\left(\frac{\tau_{xy,i+1,j} - \tau_{xy,i\text{-}1,j}}{2\Delta x} + \frac{\sigma_{y,i,j+1} - \sigma_{y,i,j\text{-}1}}{2\Delta y} + R_y\right) - g_y\Delta t + \ddot{\xi}_y\Delta t. \quad (4)$$

At this stage, according to the idea of the large particles method, the convective terms of equations (2) are omitted [6-8]. It should be emphasized that in equations (3) there is a division by concentration, so that in order to avoid the occurrence of singular terms in cells with zero concentrations, the velocities and stresses of the medium are considered to be equal zero.

4.2. Lagrange stage.

At this stage, the velocity of the element of the granular material volume on the cell boundaries at each integration step is averaged with the velocity in the adjacent cell on this boundary. For the cell with the number i, j, the velocity on the boundaries is determined by the following formula:

$$U_{k,i,j} = \frac{v_{q,i+s,j+s} + v_{q,i+s,j+s}}{2}; \quad (5)$$

here $k = \{1, 2, 3, 4\}$ - index of the cell boundary ($k = 1$ corresponds to the left boundary, $k = 2$ - the lower boundary, $k = 3$ - the right boundary, $k = 4$ - the upper boundary), $q = \{x , y\}$ - direction of the velocity, $s = \{- 1, 0, 1\}$ - coefficient determining the number of cell adjacent with the boundary k. For example, for cell i, j the flow velocity of material through the left boundary:

$$U_{1,i,j} = \left(v_{x,i,j} + v_{x,i,j-1}\right)/2.$$

Using the obtained values of the velocities on the cell boundaries the masses flowing through the border k of the cell during the time Δt is determined as:

$$\Delta M_{1,i,j} = \begin{cases} \Delta x U_{1,i,j}\upsilon_{i,j-1}\Delta t, & \text{если } U_1 > 0; \\ \Delta x U_{1,i,j}\upsilon_{i,j}\Delta t, & \text{если } U_1 \le 0; \end{cases} \qquad \Delta M_{3,i,j} = \begin{cases} \Delta x U_{3,i,j}\upsilon_{i,j}\Delta t, & \text{если } U_3 > 0; \\ \Delta x U_{3,i,j}\upsilon_{i,j+1}\Delta t, & \text{если } U_3 \le 0; \end{cases}$$
$$\Delta M_{2,i,j} = \begin{cases} \Delta x U_{2,i,j}\upsilon_{i-1,j}\Delta t, & \text{если } U_2 > 0; \\ \Delta x U_{2,i,j}\upsilon_{i,j}\Delta t, & \text{если } U_2 \le 0; \end{cases} \qquad \Delta M_{4,i,j} = \begin{cases} \Delta x U_{4,i,j}\upsilon_{i,j}\Delta t, & \text{если } U_4 > 0; \\ \Delta x U_{4,i,j}\upsilon_{i+1,j}\Delta t, & \text{если } U_4 \le 0. \end{cases} \quad (6)$$

Then the change in mass and the new volume concentration in the cell are:

$$\Delta M = \Delta M_{1,i,j} + \Delta M_{2,i,j} - \Delta M_{3,i,j} - \Delta M_{4,i,j};$$
$$\upsilon^{\{n+1\}}{}_{i,j} = \upsilon^{\{n\}}{}_{i,j} + \frac{\Delta M}{\Delta x \Delta y},$$

where {n} is current and {n+1} is next integration step.

4.3. Final stage.

To describe the cell state, the function $D_{k,i,j}$ is introduced, which is defined as following: $D_{k,i,j} = 1$ if the material flows into the cell (i, j) through the boundary k; $D_{k,i,j} = 0$ if the material flows from cell (i, j) through the border k.

Then the values of the cell velocities at the next integration step {n + 1} are determined by the formulas:

$$v_{x,i,j}^{\{n+1\}} = \{D_{1,i,j}^{\{n\}} \cdot \tilde{v}_{x,i,j+1}^{\{n\}} \cdot \left|\Delta M_{1,i,j}^{\{n\}}\right| + D_{2,i,j}^{\{n\}} \cdot \tilde{v}_{x,i-1,j}^{\{n\}} \cdot \left|\Delta M_{2,i,j}^{\{n\}}\right| + D_{3,i,j}^{\{n\}} \cdot \tilde{v}_{x,i,j-1}^{\{n\}} \cdot \left|\Delta M_{2,i,j}^{\{n\}}\right| +$$
$$+ D_{4,i,j}^{\{n\}} \cdot \tilde{v}_{x,i+1,j}^{\{n\}} \cdot \left|\Delta M_{4,i,j}^{\{n\}}\right| + \tilde{v}_{x,i,j}^{\{n\}} [\upsilon_{i,j}^{\{n\}} \Delta x \Delta y - \left(1 - D_{1,i,j}^{\{n\}}\right)\left|\Delta M_{1,i,j}^{\{n\}}\right| - \left(1 - D_{2,i,j}^{\{n\}}\right)\left|\Delta M_{2,i,j}^{\{n\}}\right| -$$
$$- \left(1 - D_{3,i,j}^{\{n\}}\right)\left|\Delta M_{3,i,j}^{\{n\}}\right| - \left(1 - D_{4,i,j}^{\{n\}}\right)\left|\Delta M_{4,i,j}^{\{n\}}\right|]\} / \left(\upsilon_{i,j}^{\{n+1\}} \Delta x \Delta y\right),$$

(7)

$$v_{y,i,j}^{\{n+1\}} = \{D_{1,i,j}^{\{n\}} \cdot \tilde{v}_{y,i,j+1}^{\{n\}} \cdot \left|\Delta M_{1,i,j}^{\{n\}}\right| + D_{2,i,j}^{\{n\}} \cdot \tilde{v}_{y,i-1,j}^{\{n\}} \cdot \left|\Delta M_{2,i,j}^{\{n\}}\right| + D_{3,i,j}^{\{n\}} \cdot \tilde{v}_{y,i,j-1}^{\{n\}} \cdot \left|\Delta M_{2,i,j}^{\{n\}}\right| +$$
$$+ D_{4,i,j}^{\{n\}} \cdot \tilde{v}_{y,i+1,j}^{\{n\}} \cdot \left|\Delta M_{4,i,j}^{\{n\}}\right| + \tilde{v}_{y,i,j}^{\{n\}} [\upsilon_{i,j}^{\{n\}} \Delta x \Delta y - \left(1 - D_{1,i,j}^{\{n\}}\right)\left|\Delta M_{1,i,j}^{\{n\}}\right| - \left(1 - D_{2,i,j}^{\{n\}}\right)\left|\Delta M_{2,i,j}^{\{n\}}\right| -$$
$$- \left(1 - D_{3,i,j}^{\{n\}}\right)\left|\Delta M_{3,i,j}^{\{n\}}\right| - \left(1 - D_{4,i,j}^{\{n\}}\right)\left|\Delta M_{4,i,j}^{\{n\}}\right|]\} / \left(\upsilon_{i,j}^{\{n+1\}} \Delta x \Delta y\right),$$

where $\Delta M_{k,i,j}^{\{n\}}$ play the role of weights of velocities. In contrast to [6], the values $\Delta M_{k,i,j}^{\{n\}}$ are taken in absolute value, which ensures the symmetry of the solutions obtained.

Additionally, the following condition is introduced: as soon as the concentration in the cell has reached the limit value, the velocity on the boundary through which the material flows into the cell (i, j) is zeroed:

$$\text{if } \begin{cases} \upsilon_{i,j} > \upsilon_d \\ D_{1,i,j} = 1 \end{cases} \rightarrow U_{1,i,j} = 0, \quad \text{if } \begin{cases} \upsilon_{i,j} > \upsilon_d \\ D_{3,i,j} = 1 \end{cases} \rightarrow U_{3,i,j} = 0,$$
$$\text{if } \begin{cases} \upsilon_{i,j+1} > \upsilon_d \\ D_{1,i,j} = 0 \end{cases} \rightarrow U_{1,i,j} = 0; \quad \text{if } \begin{cases} \upsilon_{i,j+1} > \upsilon_d \\ D_{3,i,j} = 0 \end{cases} \rightarrow U_{3,i,j} = 0;$$

(8)

$$\text{if } \begin{cases} \upsilon_{i,j} > \upsilon_d \\ D_{2,i,j} = 1 \end{cases} \rightarrow U_{2,i,j} = 0, \quad \text{if } \begin{cases} \upsilon_{i,j} > \upsilon_d \\ D_{4,i,j} = 1 \end{cases} \rightarrow U_{4,i,j} = 0,$$
$$\text{if } \begin{cases} \upsilon_{i,j+1} > \upsilon_d \\ D_{2,i,j} = 0 \end{cases} \rightarrow U_{2,i,j} = 0; \quad \text{if } \begin{cases} \upsilon_{i,j+1} > \upsilon_d \\ D_{4,i,j} = 0 \end{cases} \rightarrow U_{4,i,j} = 0.$$

Thus, the material in this cell does not flow into, but may leak. According to [6, 10], to obtain stable solutions, it is necessary that during the whole integration the following conditions are satisfied:

$$v_{q,i,j} \leq \frac{\Delta q}{\Delta t}, \quad \Delta q = \{\Delta x, \Delta y\}.$$

It should be highlighted that in formulas (3) - (8) the parameters of adjacent cells are used, which leads to the problem of their usage on the boundaries of the computational area. To overcome this problem, the computational area is surrounded by "fictitious" cells, which parameters are assumed to be equal (up to the signs of the speeds) to the parameters of adjacent cells of the computational area. To form the boundary conditions, the coefficients of "sticking" - β and "leaking" - γ [6, 8, 10] are introduced:

$$-1 \leq \beta \leq 1,$$
$$-1 \leq \gamma \leq 1.$$

For example, on the upper boundary of the computational area the boundary conditions for the integration domain of size N × M are written as following:

IOP Conf. Series: Materials Science and Engineering **489** (2019) 012039 doi:10.1088/1757-899X/489/1/012039

$$\upsilon_{i,M} = \upsilon_{i,M-1}, \qquad \sigma_{y,i,M} = \sigma_{y,i,M-1},$$
$$\sigma_{x,i,M} = \sigma_{x,i,M-1}, \quad v_{x,i,M} = \beta v_{x,i,M-1},$$
$$\tau_{xy,i,M} = \tau_{xy,i,M-1}, \quad v_{y,i,M} = \gamma v_{y,i,M-1}.$$

5. The results analysis

In the numerical implementation of the above algorithm, quartz sand was chosen as the medium continuously coming from the bunker to the inclined surface of the tray, harmonically vibrating in a direction parallel to its plane (see figure. 1). The upper boundary of the computational area coincides with the lower edge of the bunker. The movement of material through the bottom and left boundaries of the computational area is forbidden and the right and the top boundaries are free. It is assumed that the characteristic size of the cross section of the bunker's outlet is equal to four cells located in the direction of the x axis (figure. 3, (a)).

Numerical calculations were carried out with the values of the coefficients α_i, $i = 0,...,5$ for quartz sand [6]. Euler mesh parameters: $L = 5\,m$, $H = 0{,}5\ m$; $M = 20$, $N = 100$. Initial concentration of cells with material - $\upsilon_d = 0{,}53$. The plane is inclined at an angle $\beta = 3$ °.

The calculation results are shown in figure 3 in the form of several successive discrete in time distributions of the granular medium (figure 3 (a-e)) and the graphic illustrating how the centre of mass moves in the computational area is shown as well in figure 3 (f). Under the action of gravity the medium falls onto the surface of the tray, which is illustrated by filling the cells in the direction of the y axis, see figure 3 (a, b). The free fall of material from the feeder is indicated by dots in figure 3 (f): the center of mass moves strictly vertically. After the first layers of the medium touch the surface of the tray, its spreading begins immediately (see figure 3 (c)), so the centre of mass starts to move along the surface of the tray (crosses in figure 3 (f)). With the further outflow of the medium from the bunker, the spreading zone increases (see figure 3 (d, e)). During the formation of the embankment and movement of the medium along the vibrating tray, the center of mass shifts in the direction of the x axis and by the time the tray is completely filled, it is in a position close to the center of the computational area (circles in figure 3, (e)).

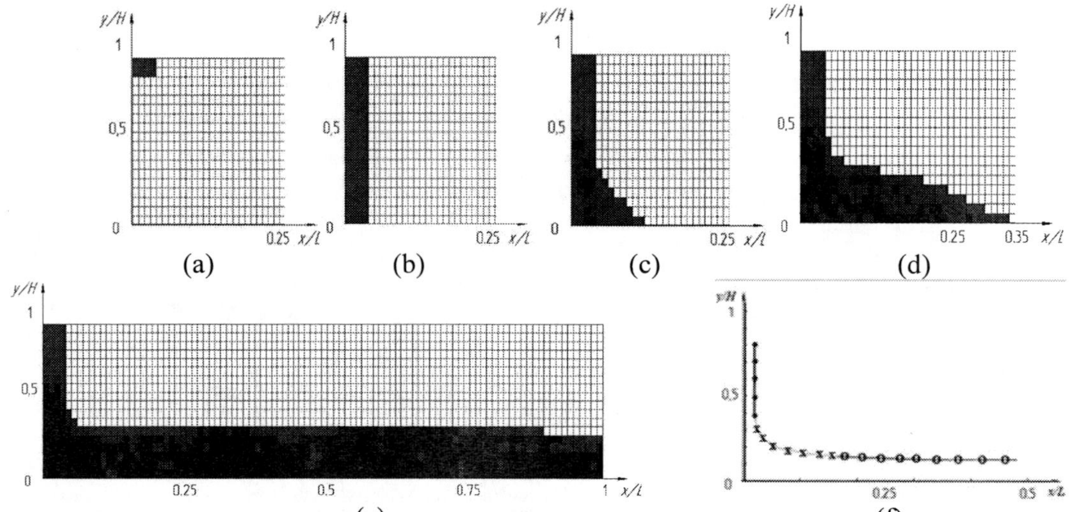

Figure 3. Calculational results. (a)-(e) - distributions of the granular medium, (f) – the trajectory of medium centre of mass.

IOP Conf. Series: Materials Science and Engineering **489** (2019) 012039 doi:10.1088/1757-899X/489/1/012039

6. Conclusion

As a result of the research a system of differential equations based on the concepts of continuum mechanics describing the granular medium dynamics on a rough vibrating surface was formed. To solve this system of equations a numerical algorithm based on the method of large particles has been developed and programmatically implemented. The proposed algorithm allows both global (displacement and velocity of the centre of mass, average transport speed, distribution and shape of the medium on the tray, depending on the vibration parameters), and local parameters (concentration and velocity of the medium in an arbitrary zone) to be obtained at an arbitrary time.

The results obtained during the numerical experiment correspond to the physical nature of this process. The proposed algorithm is universal and can be used to simulate the movement of granular medium under other conditions of vibration effects, in particular, with a non-uniform field of vibration impact.

Acknowlegments

The research was supported by Russian Science Foundation (project No. 18-19-00708).

References

[1] Spivakovsky A O and Goncharevich I F 1972 *Vibrating Conveyors, Feeders and Auxiliary Devices* (Moscow: Mashinostroenie)

[2] Gutman I 1968 *Industrial Uses of Mechanical Vibrations* (London: Business Books Ltd.)

[3] Blekhman I I and Janelidze G Yu 1964 *Vibratory Movement* (Moscow: Nauka)

[4] Jaeger H M, Nagel S R and Behringer R P 1996 Granular Solids, Liquids and Gases *Rev. of Mod. Phys.* **68** 1259–73

[5] Ogawa S 1978 Multitemperature Theory of Granular Materials *Proc. of the US-Japan Seminar on Continuum Mechanical and Statistical Approaches in the Mechanics of Granular Materials* 208-17

[6] Loktionova O G 2008 *Dynamics of Vibration Technological Processes and Machines for Processing Inhomogeneous Granular Media*: Diss. Doctoral (in Engineering) (Kursk)

[7] Yatsun S F 1997 Mathematical Modeling of Vibration Machines for Processing Granular Materials *Izvestia Kursk state tech. university* **1** 11-20

[8] Zhuravleva E V 2000 *Modeling the Dynamics of the Process of Vibrational Transportation of Granular Material*: Diss. Candidate of Eng. Sciences (Kursk)

[9] Klein G K 1977 *Structural Mechanics of Granular Media* (Moscow: Stroyizdat)

[10] Belotserkovsky O M 1994 *Numerical Modeling in Continuum Mechanics* (Moscow: Physmatlit)

IOP Conf. Series: Materials Science and Engineering **489** (2019) 012040 doi:10.1088/1757-899X/489/1/012040

Method of engine structural frame vibrations analysis during fan blade-out

V Y Myasnikov[1,2] and I I Ivanov[1,2]

[1] Dynamics and strength of aviation engines department, Central Institute of Aviation Motors, 111116 Aviamotornaya Street 2, Moscow, Russia
[2] Applied Mechanics Department, Bauman Moscow State Technical University, 105005 2nd Baumanskaya Street 5, Moscow, Russia

vumyasnikov@gmail.com

Abstract. The probability of fan blade destruction is extremely small but not equal to zero. The consequences of fan blade-out (FBO) can be catastrophic. The absence of dangerous consequences during FBO event should be confirmed by expensive tests. This causes urgency of developing accurate computational techniques for engine structural frame vibrations analysis. In this paper, we propose an approach to determining rotor-casing system dynamic response to unbalance associated with FBO. The problem is solved in a nonlinear non-stationary formulation. Rotor-casing contact interaction, blades crown stiffness and changes in rotational speed are taken into account.

1. Introduction

Fan blades of aviation gas turbine engines (GTE) are subjected to significant centrifugal loads. Their working conditions are further complicated by the possibility of damage due to ingress of foreign objects. These factors can lead to complete or partial destruction of the blade and to blade-out.

The probability of a fan blade-out (FBO) is extremely low, but not equal to zero. Regulatory industry documents [1, 2] oblige aviation GTE developers to experimentally confirm the absence of hazardous aftereffects for an aircraft in case of FBO. Blade-out engine test is an extremely expensive task. This determines the relevance of the engine structural frame dynamics analysis methods development in order to ensure successful testing at the first attempt.

The following factors can be identified as the main factors leading to an increase in loads on the elements of the engine structural frame components after FBO:

- The interaction of separated blade's fragments with a fan casing and succeed blades;
- The deflection of the overhung shaft part and the impact of the remaining fan impeller blades on the casing;
- Almost instantaneous unbalance occurrence in the low pressure (LP) rotor, many times higher than the nominal value;
- The passage of the unbalanced LP rotor through a critical speed when rotational frequency is slowed due to the fuel cut-off and intensive blades-casing contact interaction.

The first of the listed factors appears within the first 0.1 seconds after FBO, while the resonant mode passing can take place in 1-2 seconds. Different time scales make it possible to approximately consider these groups of factors separately, despite the nonlinearity of the system.

The tasks of the blade-out simulation and its interaction with the remaining blades and the fan casing are successfully solved by modern finite element analysis complexes [3, 4].The main interest in them is confirmation of the casing's ability to hold destroyed blade fragments.

The response of the rotor to an instantaneous unbalance and the passage of the unbalanced rotor through the resonant mode, with contact interaction being considered, have been studied for several decades and examined both in monographs and modern articles [5-7].

In this work, the focus is on the second group of factors. An approach is proposed to simulate the response of the aviation GTE rotor-casing system to an instantaneous unbalance associated with the FBO, taking into account possible contact interaction and changes in the rotor rotational frequency.

2. Rotor model development

The design model includes a fan casing and a shaft of a LP rotor which have typical structure for modern aviation GTE. The rotor is mounted on three elastic supports and carries a cantilever disk with a set of fan blades. 3D solid model of the design model is shown at figure 1.

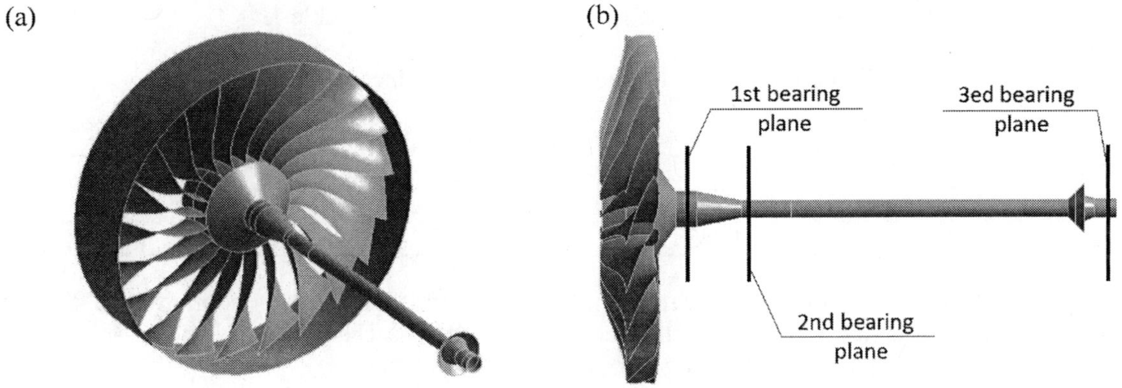

Figure 1. 3D solid model of rotor-casing system (a), rotor bearing planes (b)

The rotor model is built using the finite element (FE) method and subsequent static condensation according to the Guyana method [8]. The three-dimensional (3D) FE rotor model is created using four-node shell elements and has 111648 degrees of freedom. As condensation nodes, 15 nodes, that lie on the axis and do not belong to the three-dimensional FE model directly, are chosen. These nodes are connected to the nodes of the 3D model through multipoint constraints (MPC) of RBE2 and RBE3 types [9]. Each of the condensation nodes has 6 degrees of freedom. The reduced rotor model has 90 degrees of freedom. Fan blades and LP turbine structural components are not included in the model directly. The point inertial elements are used for their modeling. The rotor supports are modeled by binodal linear elastic elements. 3D FE model with MPC links and condensation nodes is shown at figure 2. MPC are shown by pink lines, condensation nodes are highlighted with green circles. Condensation nodes with point inertial elements are additionally highlighted with orange circles.

IOP Conf. Series: Materials Science and Engineering **489** (2019) 012040 doi:10.1088/1757-899X/489/1/012040

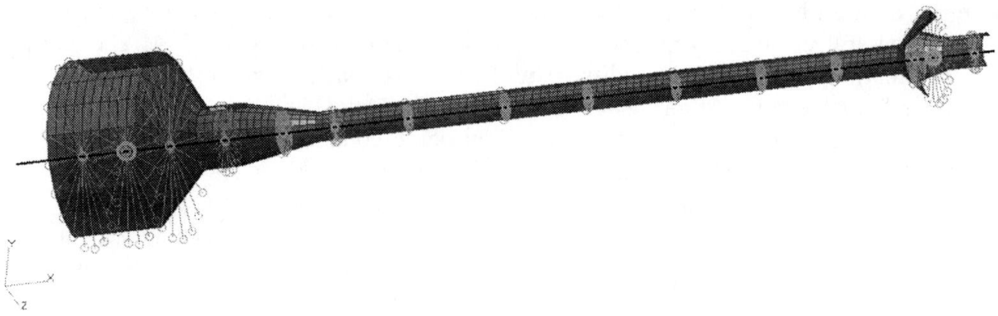

Figure 2. 3D FE rotor model, MPC and condensation nodes are highlighted.

The quality of the model obtained using the Guyana static condensation method is determined by the choice of condensation nodes and has to be evaluated. This estimation is made on the basis of calculating the rotor critical speeds using the full and reduced models, table 1.

Table 1. Rotor critical speeds (full and reduced models).

№	Mode shape	Critical speed, RPM		Error, %
		Full FE model	Reduced FE model	
1	1st bending	1514	1567	3.5
2	2nd bending	4349	4356	0.2

The error in the critical frequencies determined using the reduced model is no more than 4%. That indicates a fairly accurate reflection of the inertial and stiffness characteristics of the rotor by the reduced model and the possibility of using it for dynamics simulation.

3. Contact interaction modelling

The contact interaction of the rotor with the casing through the fan blades is taken into account using the penalty method. The normal component of the contact force F_n^{cont} was considered to be a nonlinear function of the difference between the radial displacement of the rotor node r_O, which lies in the plane of contact, and the initial value of the radial gap Δ. The tangential force component F_τ^{cont} was considered linear proportional to the normal component with the proportionality coefficient μ:

$$\delta = r_O - \Delta, \quad F_n^{cont} = \begin{cases} 0, \ \delta < 0 \\ F_n^{cont}(\delta), \ \delta \geq 0 \end{cases}$$
$$F_\tau^{cont} = \mu F_n^{cont} \tag{1}$$

The moment, created by the tangential component of the contact force relative to the axis of rotation, was taken into account:

$$M_x^{cont} = -\mathrm{sgn}\left(V_A^\tau\right) F_\tau^{cont}\left(r_{blade} + \Delta - r_O\right) \tag{2}$$

Here r_{blade} is the distance from the rotation axis to the most distant point at the periphery of the fan impeller, V_A^τ – the tangential speed component of this point. Contact forces are shown at figure 3, explanations of the speed V_A^τ calculation are given there, for simplicity the fan wheel is shown as a round disk.

IOP Conf. Series: Materials Science and Engineering **489** (2019) 012040 doi:10.1088/1757-899X/489/1/012040

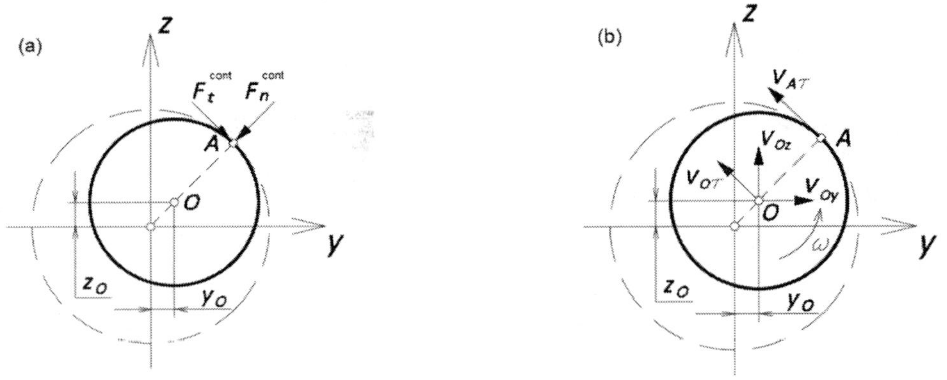

Figure 3. Contact forces (a) and the velocity of the fan impeller periphery point (b)

Velocity V_A^τ can be determined by:

$$V_A^\tau = V_O^\tau + \omega r_{blade}$$
$$V_A^\tau = -\dot{y}_O \frac{z_O}{r_O} + \dot{z}_O \frac{y_O}{r_O} + \omega r_{blade} \quad (3)$$

The contact forces projections on the axis of a fixed Cartesian coordinate system are defined as follows:

$$F_y^{cont} = -F_n^{cont}(\delta)\frac{y_O}{r_O} + \mathrm{sgn}(V_A^\tau)\mu F_n^{cont}(\delta)\frac{z_O}{r_O}$$
$$F_z^{cont} = -F_n^{cont}(\delta)\frac{z_O}{r_O} - \mathrm{sgn}(V_A^\tau)\mu F_n^{cont}(\delta)\frac{y_O}{r_O} \quad (4)$$

The fan blades compliance was taken into account by specifying the nonlinear elastic characteristic of the blades crown as a dependence $F_n^{cont}(\delta)$. The elastic characteristic of single blade was obtained from the solution of a contact problem. The blade was pressed against rigid casing. The radial displacement was set to the blade root section. The dependence of the contact force normal component projection on the radial direction on the displacement value is taken as the elastic characteristic. Fan blade deformed shape is shown at figure 4. The elastic characteristic of the blades crown is obtained under the assumption that it is a system of radially oriented springs, each of which has an elastic characteristic corresponding to one blade.

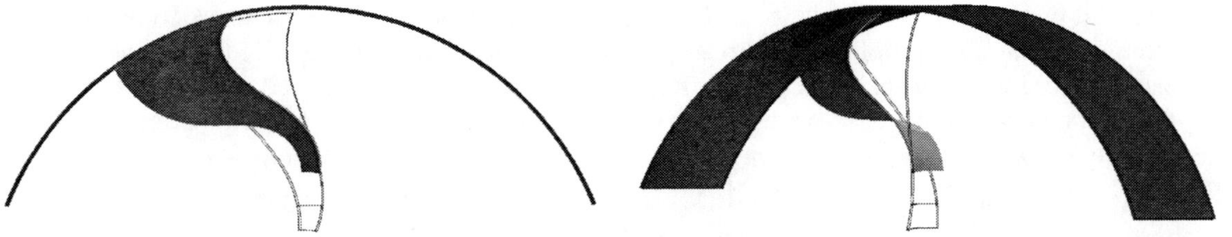

Figure 4. Fan blade deformed shape

Figure 5 shows the elastic characteristic of single blade and a crown consisting of 21 blades. Fractures on the blades crown elastic characteristics correspond to an increase number of the blades contacting with the casing.

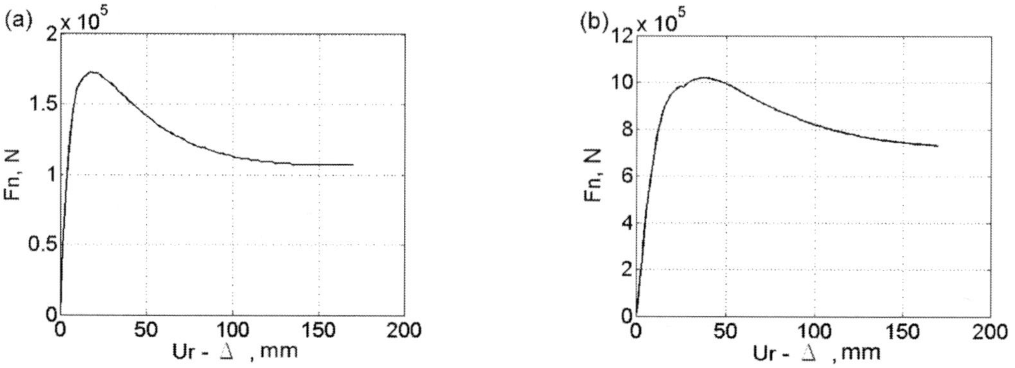

Figure 5. Elastic characteristics of one blade (a) and crown of 21 blades (b)

The elastic characteristic has an extremum associated with the type of deformed blade configuration and buckling of the blade. As the radial displacement of the root section increases, the contact zone of the blade with the casing shifts. This leads to an increase in the angle between the normal component of the reaction on the casing and the direction of displacement and to a decrease in the magnitude of the normal contact force projection on this displacement direction.

4. Rotor-casing FBO response

The problem of studying the response of the rotor-casing system to an instantaneous unbalance is solved in a non-linear non-stationary formulation based on the integration of the equations of motion in the time domain. The system of equations describing the rotor motion is supplemented by an equation describing the rotation around its own axis, as a rigid body:

$$\begin{cases} [M]\{\ddot{q}\}+[D]\{\dot{q}\}+\dot{\varphi}[G]\{\dot{q}\}+([K]+\ddot{\varphi}[C])\{q\}=\{f\} \\ J_p\ddot{\varphi}=M_x^{cont} \end{cases} \quad (5)$$

Equations (5) are integrated compatibly with the use of an implicit scheme according to the Newmark method. Thus, the contact forces influence on the change in rotor rotational speed is taken into account.

At each integration step, the contact conditions were checked. In the case of their accomplishment, the vector of nodal forces was supplemented by the contact forces. Otherwise, it was represented only by the load from the unbalance.

The stiffness and mass matrices of the system are obtained using MSC Nastran. To integrate the equations of motion in this formulation, a C ++ program has been developed. The program correctness has been tested for the case of taking into account only the normal component of the contact force and a constant rotor rotational speed. In this formulation, the problem can be solved using MSC Nastran.

The following parameters were used in simulation: unbalance characteristics (mass and eccentricity) – $m_{unb}=9$ kg, $e=621$ mm, which corresponds to the destruction of one and a half fan blades, rotor rotational speed $\omega=3800$ RPM, initial gap $\Delta=15$ mm. Damping matrix was considered proportional to the stiffness matrix, damping level is 7.5% of the critical at maximum rotation frequency. The time point $t=0$ corresponds to the unbalance occurrence. The results of the simulation using MSC Nastran and developed program are shown at figure 6.

IOP Conf. Series: Materials Science and Engineering **489** (2019) 012040 doi:10.1088/1757-899X/489/1/012040

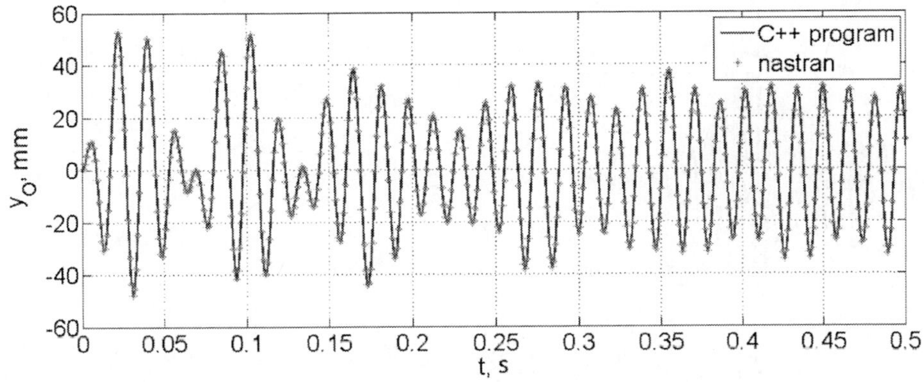

Figure 6. Rotor vertical displacements at the fan section

The obtained values are in good agreement, the error does not exceed 1%, which allows to make a conclusion about the correctness of the program.

The results of the calculation, taking into account both the normal and tangential components of the contact forces, as well as the moment and rotational speed variation for the case of constant fan blades stiffness are given at figure 7. The characteristics of the unbalance, damping and the initial clearance are assumed to be the same as in the first calculation, initial rotational speed $\omega = 3800$ RMP, the blades crown stiffness is assumed to be equal to $k = 3 \cdot 10^4 \ \dfrac{\text{N}}{\text{mm}}$, $F_n^{cont} = k\delta$.

(a) (b)

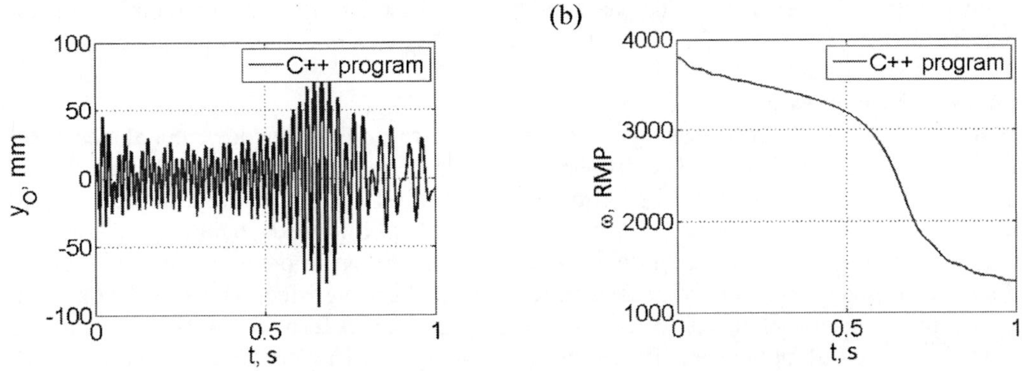

Figure 7. Rotor vertical displacements at the fan section (a) and calculated rotational velocity (b) for the case of constant blades crown stiffness

The largest rotor displacements in the fan section occur approximately 0.66 seconds later the unbalance appears and are achieved at the rotational frequency $\omega \approx 2380$ RPM. The highest displacement value is 94 mm. The difference in the frequency of achieving the largest displacements from the 1st critical speed value of the three-support rotor, given in Section 2, is due to the fact that when blade-casing contact interaction accurse the blades crown acts as an additional elastic support. This causes a change of the rotor 1st critical speed.

The most intensive change in the rotation frequency is realized, when the greatest displacements are achieved, which is associated with the adopted model of the contact forces formation and the constant rigidity of the blades crown in this design case.

The simulation results for the case of the blades crown nonlinear elastic characteristic are shown at figure 8. The elastic characteristic shown in figure 4 was used.

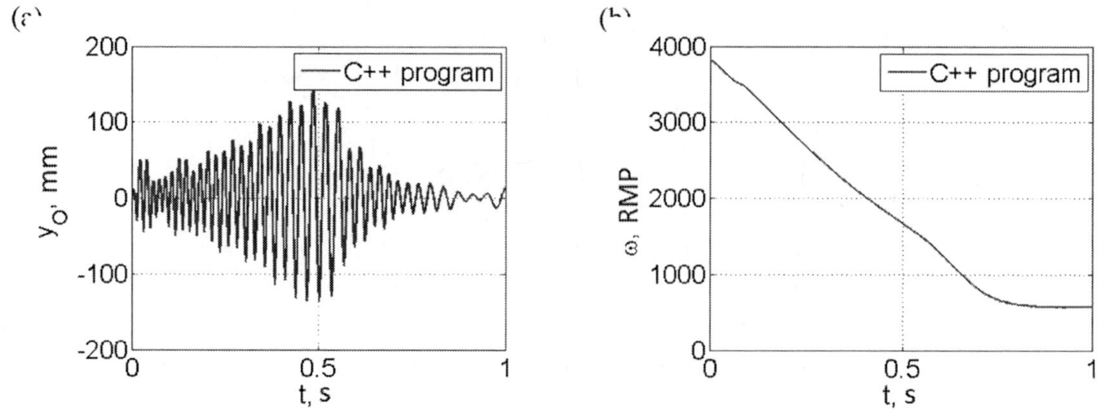

Figure 8. Rotor vertical displacements at the fan section (a) and calculated rotational velocity (b) for the case of a nonlinear elastic characteristic of the blades crown

In this design case, the largest displacements are observed approximately 0.49 seconds after the FBO. Their value is 147 mm. The frequency of reaching the greatest displacements is $\omega \approx 1718$ RPM. The change in rotational frequency happens to be almost linear. The absence of an increase in the intensity of deceleration during the passage of a resonance stems to the fact that the elastic characteristic of the blades has a maximum. The greatest value of contact forces, and therefore the braking torque $\left(M_x^{cont}\right)$, is limited in magnitude. Higher displacement values compared with the previous case are associated with differences in the elastic characteristic of the blades crown, as well as with a lower rate of the resonant mode passage. After $t = 0.86$ s rotation speed remains constant this corresponds to the rotor displacements in the fan section being less than the initial gap value.

5. Summary and conclusions

An approach to determining rotor-casing system dynamic response to sudden unbalance associated with FBO is proposed in this paper. The problem is solved in a non-linear non-stationary formulation based on the integration of supplemented equations of motion system in the time domain. Blades-casing contact interaction considering fan blades crown flexibility and changes in rotor rotational speed was taken into account. To integrate the equations of motion in presented formulation, a C ++ program has been developed. The program correctness has been verified. The simulation results for the cases of the constant blades stiffness and nonlinear blades crown elastic characteristic are presented. In case of constant blades stiffness the most intensive change in the rotor rotation frequency is realized, when the greatest displacements are achieved, which is associated with the adopted model of the contact forces formation. In case of nonlinear blades crown stiffness rotor deceleration curve is almost linear which stems to the fact that the elastic characteristic of the blades has a maximum. The greatest value of contact forces, and therefore the braking torque $\left(M_x^{cont}\right)$, is limited in magnitude.

The proposed approach has a practical value and can be applied to assess the impact of the engine structural frame parameters on the values of the forces realized after FBO. As part of further method improvement, it is planned to take into account the flexibility of the fan casing, presence of the abradable coating, as well as improvements in the blades crown elastic characteristic determination. The experience of the authors' calculations, as well as comparison with experimental data, indicate that the rigidity of the blades crown, obtained using the proposed approach, is found to be high estimated. Other sources of rotor deceleration besides contact forces are also planned to be considered.

References

[1] Federal Aviation Administration 1993 *Regulations F. A. Part 33 Airworthiness Standards:*

IOP Conf. Series: Materials Science and Engineering **489** (2019) 012040 doi:10.1088/1757-899X/489/1/012040

Aircraft Engines (Washington DC) p 59
[2] European Aviation Safety Agency 2015 *Certification Specifications and Acceptable Means of Compliance for Engines (CS-E) Amendment 4* p 211
[3] Shmotin Y, Gabov D, Ryabov A, Kukanov S and Rechkin V 2006 *42nd AIAA/ASME/SAE/ASEE Joint Propulsion Conference & Exhibit (July 9-12 Sacramento CA)*
[4] Carney K S, Lawrence C and Carney D V 2002 *Proc. 7th int. LS-DYNA users conf. (Livermore CA, USA)* pp 14–7
[5] Muszynska A 2005 *Rotordynamics,* ed L L Faulkner (Boca Raton Fl: CRC press) p 1128
[6] Wang C, Zhang D, Ma Y, Liang Z and Hong J 2016 *Mechanical Systems and Signal Processing* **76** 111–35
[7] Ivanov I I and Blinnik B S 2014 *29th Cong. of the Intern. Council of the Aeronautical Sciences (Sept. 7-12 St. Petersburg)*
[8] Bathe K J 2006 *Finite element procedures* (New Jersey: Prentice Hall) p 1037
[9] MYSTRAN Software Inc. 2011 *User Reference Manual For the MYSTRAN General Purpose Finite Element Structural Analysis Computer Program* p. 283

IOP Conf. Series: Materials Science and Engineering **489** (2019) 012041 doi:10.1088/1757-899X/489/1/012041

Multiple-degree-of-freedom rotor model with total unbalance and possible contact on stator

A Nikiforov

Mechanical Engineering Research Institute of the Russian Academy of Sciences, Moscow, Russian Federation

n.andre@mail.ru

Abstract. Mathematical model for the flexible rotor which has static and dynamic unbalance and enabled slip rolling on the compliant stator is developed. Also, rotation inertia, gyroscopic effect, bearings characteristics, internal hysteresis, and external viscous friction are included. The matrix approaches are used. The proposed model allows investigating the dynamic effect of total unbalance and stator displacements as an elastic and solid body on basis of the ordinary differential equations.

1. Introduction

The most of issues in engineering practice associated with the shaft or rotor vibration are due to the destructive effect of the unbalance and stator on the system. Therefore a large number of scientific papers are devoted to the mathematical modeling of rotor systems with the presence of unbalance or continuous rubbing on the stator. It is enough to refer, for example, to articles [1-8], as well as to the lists of references cited in them. However, the search for the simulation of dynamic unbalance (oblique disk seating on the shaft) or contact elasticity with the inertia of stator rotation very probably will result in failure.

The originality of this work is to build the unbalanced flexible rotor-elastic bearings-foundation system briefly called the rotor system and also the unbalanced flexible rotor-elastic bearings-rigid stator-elastic supports-foundation system which is simply called the rotor-stator system. If there is direct contact in the system between the flexible rotor and the compliant stator represented as an additional linear elastic coupling (spring between them) by the contact stiffness k_k then it is called the contact system, otherwise – contactless system.

2. Total unbalance simulation

So, let the center of rotor mass G not coincide with the geometric center of disk O and locate in a disk plane at the distance e from point O (Figure 1). In addition, let the disk is mounted on the shaft with a tilt, i.e. with an existence of angle e_1 between disk axis and axis of the undeformed shaft z or the same between vector $I_0\omega$ and tangent z_1 to axis of the bent shaft at point O. Finally, let a plane passing through the tangent z_1 and vector $OG=e$ be at an angle e_2 to vector $e_1\omega$ which is perpendicular to the plane passing through the tangent z_1 and the disk axis i.e. containing the angle e_1. The values e, e_1, e_2 characterize the total unbalance of the given rotor.

At the existence of linear and angular eccentricities, the motion equations of the light shaft with massive disk [9] should contain the accelerations and speeds of center of disk mass instead of the accelerations and speeds of geometric disk center:

$$
\begin{aligned}
&m\ddot{x}_G + k_{11}x + k_{12}\varphi_y = 0 \\
&m\ddot{y}_G + k_{11}y + k_{12}\varphi_x = 0 \\
&I\ddot{\varphi}_{yG} + I_0\omega\dot{\varphi}_{xG} + k_{21}x + k_{22}\varphi_y = 0 \\
&I\ddot{\varphi}_{xG} - I_0\omega\dot{\varphi}_{yG} + k_{21}y + k_{22}\varphi_x = 0
\end{aligned},
$$

where $x_G, y_G, \varphi_{xG}, \varphi_{yG}$ are coordinates of point G.

Internal friction is not taken into account because at rotation speed equal to whirl speed an elastic line of the bent rotor due to the unbalanced forces rotates around a support line so that the rotor does not rotate "inside" the elastic line, there are no changes in stress signs and, therefore, internal friction does not "work".

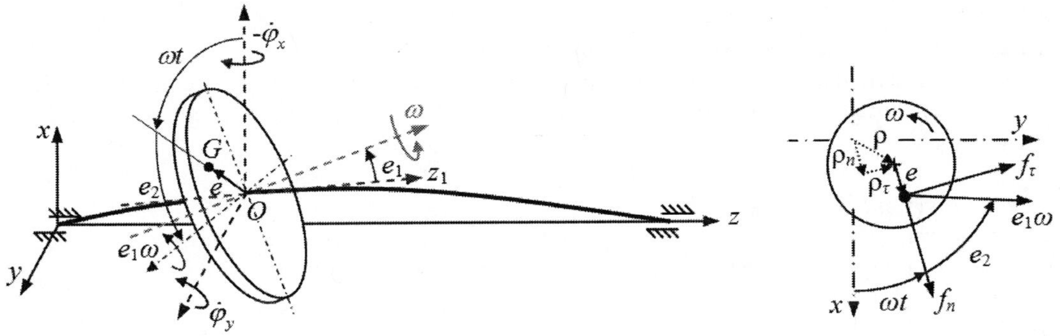

Figure 1. The motion of unbalanced rigid disk on the flexible shaft and unbalanced forces.

In view of geometric reasons (figure 1):

$$
\begin{aligned}
&x_G = x + e\cos\omega t \\
&y_G = y + e\sin\omega t \\
&\dot{\varphi}_{yG} = \dot{\varphi}_y + e_1\omega\cos(\pi/2 - \omega t - e_2) \\
&-\dot{\varphi}_{xG} = -\dot{\varphi}_x + e_1\omega\sin(\pi/2 - \omega t - e_2)
\end{aligned}.
$$

From here:

$$
\begin{aligned}
&\ddot{x}_G = \ddot{x} - e\omega^2\cos\omega t \quad && \dot{\varphi}_{yG} = \dot{\varphi}_y + e_1\omega\sin(\omega t + e_2) \quad && \ddot{\varphi}_{yG} = \ddot{\varphi}_y + e_1\omega^2\cos(\omega t + e_2) \\
&\ddot{y}_G = \ddot{y} - e\omega^2\sin\omega t, && \dot{\varphi}_{xG} = \dot{\varphi}_x - e_1\omega\cos(\omega t + e_2), && \ddot{\varphi}_{xG} = \ddot{\varphi}_x + e_1\omega^2\sin(\omega t + e_2).
\end{aligned}
$$

Corresponding replacements yields the following differential equations of dynamics for unbalanced single-disk flexible rotor:

$$
\begin{aligned}
&m\ddot{x} + k_{11}x + k_{12}\varphi_y = me\omega^2\cos\omega t \\
&m\ddot{y} + k_{11}y + k_{12}\varphi_x = me\omega^2\sin\omega t \\
&I\ddot{\varphi}_y + I_0\omega\dot{\varphi}_x + k_{21}x + k_{22}\varphi_y = (I_0 - I)e_1\omega^2\cos(\omega t + e_2) \\
&I\ddot{\varphi}_x - I_0\omega\dot{\varphi}_y + k_{21}y + k_{22}\varphi_x = (I_0 - I)e_1\omega^2\sin(\omega t + e_2).
\end{aligned}
$$

At "compressed" displacements:

$$
\begin{aligned}
&m\ddot{z} + k_{11}z + k_{12}\varphi_0 = me\omega^2 e^{i\omega t} \\
&I\ddot{\varphi}_0 - iI_0\omega\dot{\varphi}_0 + k_{21}z + k_{22}\varphi_0 = (I_0 - I)e_1(\cos e_2 + i\sin e_2)\omega^2 e^{i\omega t}
\end{aligned},
$$

IOP Conf. Series: Materials Science and Engineering **489** (2019) 012041 doi:10.1088/1757-899X/489/1/012041

where $z = x + iy$, $\varphi_0 = \varphi_y + i\varphi_x$.

At the entry of matrices and vectors:

$$\mathbf{M}\{\ddot{q}\} - i\omega\mathbf{G}\{\dot{q}\} + \mathbf{K}\{q\} = \omega^2 e^{i\omega t}\left(\{f_n\} + i\{f_\tau\}\right),$$

$$\text{where } \mathbf{M} = \begin{bmatrix} m & 0 \\ 0 & I \end{bmatrix}, \mathbf{G} = \begin{bmatrix} 0 & 0 \\ 0 & I_0 \end{bmatrix}, \mathbf{K} = \begin{bmatrix} k_{11} & k_{12} \\ k_{21} & k_{22} \end{bmatrix}, \{q\} = \begin{Bmatrix} z \\ \varphi_0 \end{Bmatrix},$$

$\{f_n\} = \begin{Bmatrix} me \\ (I_0 - I)e_1 \cos e_2 \end{Bmatrix}$, $\{f_\tau\} = \begin{Bmatrix} 0 \\ (I_0 - I)e_1 \sin e_2 \end{Bmatrix}$ are forces of unbalance in normal and tangential directions with respect to e (Figure 1).

The worst case is e_2=0 i.e. actions of the linear and angular unbalances coincide in phase, and, as a result, deflection of the flexible rotor will be greatest. The most favorable case is e_2=180°.

3. Flexible rotor-compliant stator system simulation

To research the dynamics of a rotor in contact with a stator, it is necessary to construct a vibrational model having two or more degrees of freedom i.e. containing at least several inertial, elastic and damping elements.

3.1. The rotor system or the rigid stator case

Any axisymmetric rotor system is easy and advantageous (from the viewpoint of performing sufficiently accurate calculations) to represent as a combination of the following three finite elements (figure 2).

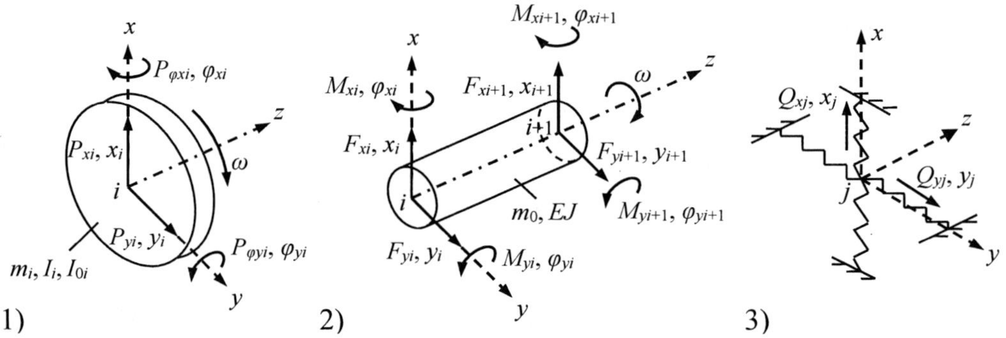

Figure 2. Finite elements of rotor: rigid dick (1), beam element (2), linearized short bearing (3) including generalized forces and displacements.

1. Rigid disk with concentrated mass m_i and transverse I_i and polar I_{0i} moments of inertia. The equation of the two-dimensional motion of such disk rotating with the constant angular velocity ω in the fixed reference and matrix form:

$$\begin{bmatrix} [m_d] & [0] \\ [0] & [m_d] \end{bmatrix} \begin{Bmatrix} \{\ddot{x}\} \\ \{\ddot{y}\} \end{Bmatrix} + \omega \begin{bmatrix} [0] & [g_d] \\ -[g_d] & [0] \end{bmatrix} \begin{Bmatrix} \{\dot{x}\} \\ \{y\} \end{Bmatrix} = \begin{Bmatrix} \{f_x\} \\ \{f_y\} \end{Bmatrix},$$

where

$$\begin{Bmatrix} \{x\} \\ \{y\} \end{Bmatrix} = \{x_i \quad \varphi_{yi} \quad y_i \quad \varphi_{xi}\}^{\mathrm{T}}, \begin{Bmatrix} \{f_x\} \\ \{f_y\} \end{Bmatrix} = \{P_{xi} \quad P_{\varphi yi} \quad P_{yi} \quad P_{\varphi xi}\}^{\mathrm{T}}, [m_d] = \begin{bmatrix} m_i & 0 \\ 0 & I_i \end{bmatrix}, [g_d] = \begin{bmatrix} 0 & 0 \\ 0 & I_{0i} \end{bmatrix}.$$

The transverse oscillations of rotors occur in a form of circular or elliptical precession, therefore it is necessary to take into account the generalized forces and displacements of the rotor elements in two mutually perpendicular directions.

2. Rotating beam element with length l and radius r and distributed inertial-elastic parameters and internal damping of the hysteresis type. The motion equation of such a beam element in the adopted form [10]:

$$\begin{bmatrix}[m] & [0]\\ [0] & [m]\end{bmatrix}\begin{Bmatrix}\{\ddot{x}\}\\ \{\ddot{y}\}\end{Bmatrix}+\omega\begin{bmatrix}[0] & [g]\\ -[g] & [0]\end{bmatrix}\begin{Bmatrix}\{\dot{x}\}\\ \{\dot{y}\}\end{Bmatrix}+\left(\begin{bmatrix}[k] & [0]\\ [0] & [k]\end{bmatrix}+\frac{\eta_{\mathrm{h}}}{\sqrt{1+\eta_{\mathrm{h}}^2}}\begin{bmatrix}[0] & [k]\\ -[k] & [0]\end{bmatrix}\right)\begin{Bmatrix}\{x\}\\ \{y\}\end{Bmatrix}=\begin{Bmatrix}\{f_x\}\\ \{f_y\}\end{Bmatrix},$$

where $\begin{Bmatrix}\{x\}\\ \{y\}\end{Bmatrix}=\{x_i \quad \varphi_{yi} \quad x_{i+1} \quad \varphi_{yi+1} \quad y_i \quad \varphi_{xi} \quad y_{i+1} \quad \varphi_{xi+1}\}^{\mathrm{T}}$, $[m]=[m_t]+[m_r]$,

$\begin{Bmatrix}\{f_x\}\\ \{f_y\}\end{Bmatrix}=\{F_{xi} \quad M_{yi} \quad F_{xi+1} \quad M_{yi+1} \quad F_{yi} \quad M_{xi} \quad F_{yi+1} \quad M_{xi+1}\}^{\mathrm{T}}$, $[g]=2[m_r]$,

$$[m_t]=\frac{m_0 l}{420(1+s)^2}\begin{bmatrix}c_1 & & & \text{sym.}\\ c_2 & c_5 & & \\ c_3 & -c_4 & c_1 & \\ c_4 & c_6 & -c_2 & c_5\end{bmatrix},[m_r]=\frac{m_0 r^2}{120l(1+s)^2}\begin{bmatrix}36 & & & \text{sym.}\\ c_7 & c_8 & & \\ -36 & -c_7 & 36 & \\ c_7 & c_9 & -c_7 & c_8\end{bmatrix}, \quad s=\frac{12EJ}{0.9\mathrm{G}\pi r^2 l^2}, \quad \mathrm{G}\approx 0.4E$$

$$[k]=\frac{EJ}{l^3(1+s)}\begin{bmatrix}12 & & & \text{sym.}\\ 6l & c_{10} & & \\ -12 & -6l & 12 & \\ 6l & c_{11} & -6l & c_{10}\end{bmatrix}, \quad \begin{aligned}c_1&=156+294s+140s^2\\ c_2&=(22+38.5s+17.5s^2)l\\ c_3&=54+126s+70s^2\\ c_4&=-(13+31.5s+17.5s^2)l\\ c_5&=(4+7s+3.5s^2)l^2\\ c_6&=-(3+7s+3.5s^2)l^2\end{aligned}, \quad \begin{aligned}c_7&=(3-15s)l\\ c_8&=(4+5s+10s^2)l^2\\ c_9&=-(1+5s-5s^2)l^2\\ c_{10}&=(4+s)l^2\\ c_{11}&=(2-s)l^2\end{aligned}.$$

Here any element of the matrices is a generalized force at i node arising from the unit acceleration or velocity or displacement in this or i+1 node.

In this general case the element is a Timoshenko beam i.e. besides the classical processes of elasticity (submatrices $[k]$) and precession inertia including translational displacements (submatrices $[m_t]$) and gyroscopic action (submatrices $[g]$) are taken into account rotation inertia (submatrices $[m_r]$ added to $[m_t]$) and shear deformation (s-members). If a ratio of the length to diameter is greater than one and less than three, then it is permissible to neglect the influence of shear deformation i.e. to use the Rayleigh beam. If $l/_{2r}>3$, then consideration without inertia of rotation i.e. of the Bernoulli-Euler beam is possible.

It is assumed that the beam element has a dissipation that does not depend on the frequency. The stress is ahead of the corresponding deformation on an angle ϕ determined by the material. The hysteresis loop in the stress-deformation plane repeats an ellipse which area is proportional $\sin\phi$ and characterizes the value of the dissipated energy so that $\sin\phi=\eta_{\mathrm{h}}\big/\sqrt{1+\eta_{\mathrm{h}}^2}$ where η_{h} is the coefficient of hysteresis losses in the beam material.

3. Linearized short bearing characterized by the equation of the form:

$$\begin{bmatrix}\eta_{\mathrm{v}}k_b & 0\\ 0 & \eta_{\mathrm{v}}k_b\end{bmatrix}\begin{Bmatrix}\dot{x}_j\\ \dot{y}_j\end{Bmatrix}+\begin{bmatrix}k_b & \eta_{\mathrm{h}b}k_b\\ -\eta_{\mathrm{h}b}k_b & k_b\end{bmatrix}\begin{Bmatrix}x_j\\ y_j\end{Bmatrix}=\begin{Bmatrix}Q_{xj}\\ Q_{yj}\end{Bmatrix},$$

i.e. with isotropic transverse stiffness and zero angular stiffness and viscous damping expressed by coefficient η_{v} and hysteresis losses in the bearing materials expressed by coefficient $\eta_{\mathrm{h}b}$.

In general, the bearing also prevents the angular displacements of the rotor and is determined by 16 stiffness coefficients and 16 inelastic resistance coefficients proportional to speeds of the transverse-angular deformation. Angular coefficients are essential for a rigid rotor or for the higher vibration forms of a flexible rotor. However in such cases, the so-called "long bearings" are often replaced by a pair of closely spaced short bearings.

As a rule, only the radial coefficients play a significant role:

$$\begin{bmatrix} c_{xx} & c_{xy} \\ c_{yx} & c_{yy} \end{bmatrix} \begin{Bmatrix} \dot{x}_j \\ \dot{y}_j \end{Bmatrix} + \begin{bmatrix} k_{xx} & k_{xy} \\ k_{yx} & k_{yy} \end{bmatrix} \begin{Bmatrix} x_j \\ y_j \end{Bmatrix} = \begin{Bmatrix} Q_{xj} \\ Q_{yj} \end{Bmatrix},$$

but even in this classical bearing model not all of them are used. The essential side coefficients of stiffness k_{xy}, k_{yx} and viscous damping c_{xy}, c_{yx} are characteristic for a fluid film bearing. At the same time, it should be borne in mind that usually such bearing is significantly non-conservative i.e. its rigidity matrix is significantly asymmetric. The difference of the side stiffness coefficients $\Delta k_{ij}=k_{xy}-k_{yx}$ determines an action direction of "circular" resultant on the rotor which in the case of a lot of Δk_{ij} in comparison with the main stiffness coefficients k_{xx}, k_{yy} leads to a rotor circulation (whirl), at $k_{xy}<k_{yx}$ to the reverse, at $k_{xy}>k_{yx}$ to the forward. The latter is directly related to the appearance of the phenomenon of dynamic instability of the rotor on bearings of liquid friction (in the form of its significant direct asynchronous whirl).

It should be noted here that a bearing with equal main stiffness coefficients ($k_{xx}=k_{yy}$) and small side stiffness coefficients ($k_{xy}\approx k_{yx}\approx 0$) in comparison with k_{xx}, k_{yy} is called isotropic. Bearing with equal main and also with essential and almost identical side stiffness coefficients ($k_{xy}\approx k_{yx}\neq 0$) is called orthotropic. A bearing with unequal main ($k_{xx}\neq k_{yy}$) and close side ($k_{xy}\approx k_{yx}$) stiffness coefficients is called anisotropic, and if $k_{xy}\neq k_{yx}$ then it is simply non-conservative.

It is also important to note that the contact place of the rotor with stator is formally analogous to the short support devoid of viscous friction, therefore:

$$\begin{bmatrix} k_k & \eta_{hk} k_k \\ -\eta_{hk} k_k & k_k \end{bmatrix} \begin{Bmatrix} x_k \\ y_k \end{Bmatrix} = \begin{Bmatrix} Q_{xk} \\ Q_{yk} \end{Bmatrix},$$

where η_{hk} is hysteresis losses coefficient in the materials of contact surfaces of the rotor and the stator.

Since the coupling of rotor elements is made in the axial direction, a transition from the local coordinates of elements to the common coordinates of the rotor system is trivial. It does not contain geometric transformations and reduces to simple sequential numbering of generalized displacements/forces. A composing rule of the rotor system is that if two elements are connected, then the displacements of adjacent ends must be equal, and the corresponding generalized forces must be summed. This technique is called the elements assembly and discussed in many manuals on matrix structural analysis.

The finished result is a system of linear differential equations with constant coefficients of the form:

$$[\mathrm{M}]_{2n\times 2n}\{\ddot{q}\}_{2n\times 1} + [\mathrm{C}]_{2n\times 2n}\{\dot{q}\}_{2n\times 1} + [\mathrm{K}]_{2n\times 2n}\{q\}_{2n\times 1} = \{\mathrm{F}\}_{2n\times 1}, \qquad (1)$$

where $[\mathrm{M}]_{2n\times 2n} = \begin{bmatrix} [M]+[M_d] & [0] \\ [0] & [M]+[M_d] \end{bmatrix}$, $[\mathrm{C}]_{2n\times 2n} = \begin{bmatrix} \eta_{\mathrm{v}}[K_b] & \omega[G]+\omega[G_d] \\ -\omega[G]-\omega[G_d] & \eta_{\mathrm{v}}[K_b] \end{bmatrix}$,

$$[\mathrm{K}]_{2n\times 2n} = \begin{bmatrix} [K]+[K_b] & \eta_{\mathrm{h}}[K]+\eta_{\mathrm{h}b}[K_b] \\ -\eta_{\mathrm{h}}[K]-\eta_{\mathrm{h}b}[K_b] & [K]+[K_b] \end{bmatrix},\ \{q\}_{2n\times 1} = \begin{Bmatrix} \{q_x\} \\ \{q_y\} \end{Bmatrix},\ \{\mathrm{F}\}_{2n\times 1} = \begin{Bmatrix} \{F_x\} \\ \{F_y\} \end{Bmatrix}.$$

A number of freedom degrees n in the plane is $2p$, where p is the nodes number and $(p-1)$ is the number of beam elements. Vectors of order $n\times 1$ are formed from the translational and angular displacements of nodes occurring in a plane or from components of unbalanced forces acting in the plane to certain nodes. Submatrices of size $n\times n$ mean an assembly of submatrices or finite element constants.

IOP Conf. Series: Materials Science and Engineering **489** (2019) 012041 doi:10.1088/1757-899X/489/1/012041

Inertia $[M]$ and gyroscopic moments $[G]$ and stiffness $[K]$ submatrices for beam elements are the band matrices:

$$[C]_{n\times n}=\begin{bmatrix} c_{11} & c_{12} & c_{13} & c_{14} & 0 & 0 & \\ c_{21} & c_{22} & c_{23} & c_{24} & 0 & 0 & \\ c_{31} & c_{32} & c_{33}+c_{11} & c_{34}+c_{12} & c_{13} & c_{14} & \cdots \\ c_{41} & c_{42} & c_{43}+c_{21} & c_{44}+c_{22} & c_{23} & c_{24} & \\ 0 & 0 & c_{31} & c_{32} & c_{33}+c_{11} & c_{34}+c_{12} & \\ 0 & 0 & c_{41} & c_{42} & c_{43}+c_{21} & c_{44}+c_{22} & \\ & & \cdots & & & & \cdots \end{bmatrix},$$

where each two-digit index means the row number and column number of the matrices $[m]$, $[g]$ и $[k]$.

Inertial $[M_d]$ and gyroscopic $[G_d]$ disk submatrices, as well as the bearing stiffness submatrix $[K_b]$, are diagonal $[\dot{\cdot}\cdot]_{n\times n}$ so their union with the submatrices of the beam elements can be expressed as:

$$[C]_{n\times n}+[\dot{\cdot}\cdot]_{n\times n}=\begin{bmatrix} \cdots & & & & & & \\ & c_{2i-1\,2i-1}+C_i & & & & & \\ & & c_{2i\,2i}+C_{i+1} & & & & \\ & & & \cdots & & & \\ & & & & c_{2j-1\,2j-1}+C_j & & \\ & & & & & \cdots & \\ & & & & & & c_{2k-1\,2k-1}+C_k \\ & & & & & & \cdots \end{bmatrix}.$$

Here under each two-digit index is meant the row number and column number of the combined matrices which are determined by one or another node number. The constant C_i equals m_i, the constant C_{i+1} equals to I_i or I_{0i}, the constant C_j equals to k_{bj}.

The constant C_k equaled to k_k is added to the stiffness matrix of the rotor system $[\mathrm{K}]_{n\times n}=[K]+[K_b]$ when the k-th or other cross-section of the flexible rotor contacts with the elastic surface of the stationary stator (Figure 3). The resulting stiffness matrix of the contact rotor system is $[\mathrm{K}]_{n\times n}=[K]+[K_b]+[K_k]$, where $[K_k]$ is the κ-filled diagonal matrix, κ is the number of contact points.

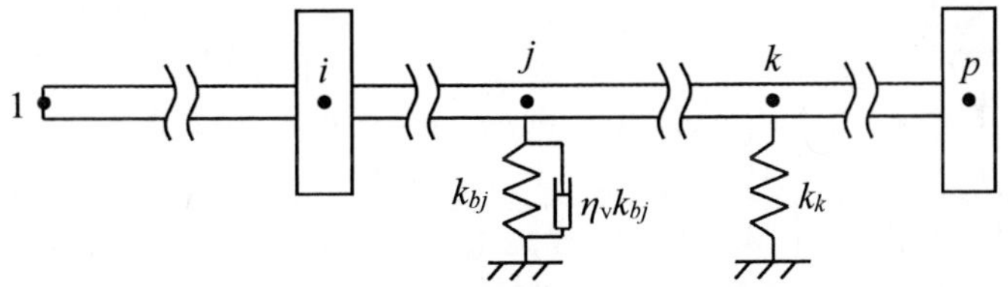

Figure 3. The arbitrary rotor system and its p nodes.

Since the system is axisymmetric a more compact form (1) is admissible by combining the matrix equations

$$\begin{gathered}[\mathrm{M}]_{n\times n}\{\ddot{q}_x\}+\eta_{\mathrm{v}}[K_b]\{\dot{q}_x\}+\omega[\mathrm{G}]\{\dot{q}_y\}+[\mathrm{K}]_{n\times n}\{q_x\}+(\eta_{\mathrm{h}}[K]+\eta_{\mathrm{h}b}[K_b])\{q_y\}=\{F_x\}\\ \mathrm{i}[\mathrm{M}]_{n\times n}\{\ddot{q}_y\}+\mathrm{i}\eta_{\mathrm{v}}[K_b]\{\dot{q}_y\}-\mathrm{i}\omega[\mathrm{G}]\{\dot{q}_x\}+\mathrm{i}[\mathrm{K}]_{n\times n}\{q_y\}-\mathrm{i}(\eta_{\mathrm{h}}[K]+\eta_{\mathrm{h}b}[K_b])\{q_x\}=\mathrm{i}\{F_y\}\end{gathered},$$

where $[\mathrm{M}]_{n\times n}=[M]+[M_d]$, $[\mathrm{G}]=[G]+[G_d]$.

As a result, determining displacements in the xz plane are real and displacements in the yz plane are imaginary, i.e. considering the vectors of complex generalized coordinates $\{q\}_{n\times 1}=\{q_x\}+\mathrm{i}\{q_y\}$ and forces $\{\mathrm{F}\}_{n\times 1}=\{F_x\}+\mathrm{i}\{F_y\}$, it will be:

$$[\mathrm{M}]_{n\times n}\{\ddot{q}\}_{n\times 1}+\left(\eta_{\mathrm{v}}[K_b]-\mathrm{i}\omega[\mathrm{G}]\right)\{\dot{q}\}_{n\times 1}+\left\langle[\mathrm{K}]_{n\times n}-\mathrm{i}\left(\eta_{\mathrm{h}}[K]+\eta_{\mathrm{h}b}[K_b]\right)\right\rangle\{q\}_{n\times 1}=\{\mathrm{F}\}_{n\times 1}. \quad (2)$$

In the above equations, the elements displacements are assumed to be small (gyroscopic moments and positional forces are linearized). Although the theory is limited to small amplitudes (of elements vibrations), it gives surprisingly good results for large amplitudes (of rotor oscillations in general). In this case, the basic rule is simple: the number of used beam elements must be one higher than the number of the interesting highest frequency and mode of the natural rotor oscillations. The sufficiency of a small number of freedom degrees for accurate simulation is explained by the weak dependence of the dynamic rotor reaction on higher forms of rotor oscillations.

3.2. Rotor-stator system or the case of the elastically fixed stator

In modeling of practically used axisymmetric stator subsystem, i.e. including a relatively rigid frame and uniformly flexible elements of fastening to the foundation, it is appropriate to choose its discrete submission as the concentrated mass m_s and transverse moment of inertia I_s at the p+1 node with linear constraints by total for the plane stiffness k_s=k_1+k_2 and viscous damping $\eta_{vs}k_s$=$\eta_{vs}k_1$+$\eta_{vs}k_2$ (Figure 4). Usually, in rotor machines, the stator stiffness on deflections of its cross-sections from planes perpendicular to the rotation axis of undeformed rotor exceeds in two or more times the bending stiffness of the rotor.

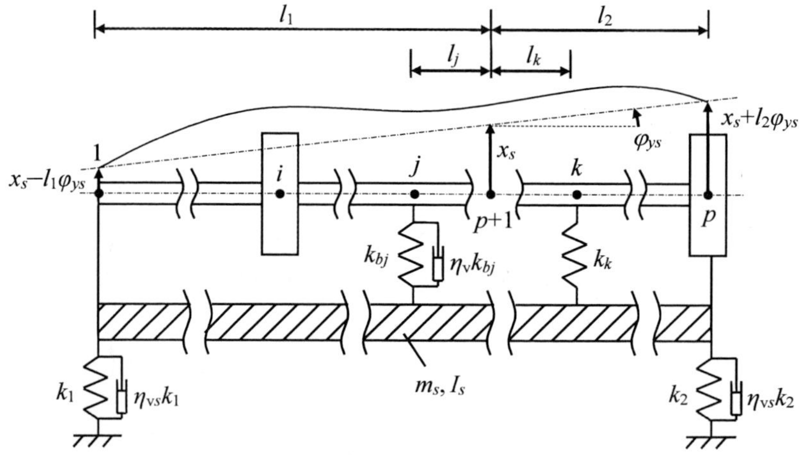

Figure 4. The arbitrary rotor-stator system and its p+1 nodes.

Let coordinates of the stator inertia center are x_s, y_s, φ_{ys}, φ_{xs}, and its translational/angular oscillations as a rigid body on the compliant supports are small, then the equations of its motion in transverse directions and around the mass center (the p+1 point) will be:

$$\begin{aligned}
&m_s\ddot{x}_s+\eta_{vs}k_1\left(\dot{x}_s-l_1\dot{\varphi}_{ys}\right)+\eta_{vs}k_2\left(\dot{x}_s+l_2\dot{\varphi}_{ys}\right)+k_1\left(x_s-l_1\varphi_{ys}\right)+k_2\left(x_s+l_2\varphi_{ys}\right)=0\\
&m_s\ddot{y}_s+\eta_{vs}k_1\left(\dot{y}_s-l_1\dot{\varphi}_{xs}\right)+\eta_{vs}k_2\left(\dot{y}_s+l_2\dot{\varphi}_{xs}\right)+k_1\left(y_s-l_1\varphi_{xs}\right)+k_2\left(y_s+l_2\varphi_{xs}\right)=0\\
&I_s\ddot{\varphi}_{ys}-\eta_{vs}k_1\left(\dot{x}_s-l_1\dot{\varphi}_{ys}\right)l_1+\eta_{vs}k_2\left(\dot{x}_s+l_2\dot{\varphi}_{ys}\right)l_2-k_1\left(x_s-l_1\varphi_{ys}\right)l_1+k_2\left(x_s+l_2\varphi_{ys}\right)l_2=0\\
&I_s\ddot{\varphi}_{xs}-\eta_{vs}k_1\left(\dot{y}_s-l_1\dot{\varphi}_{xs}\right)l_1+\eta_{vs}k_2\left(\dot{y}_s+l_2\dot{\varphi}_{xs}\right)l_2-k_1\left(y_s-l_1\varphi_{xs}\right)l_1+k_2\left(y_s+l_2\varphi_{xs}\right)l_2=0
\end{aligned}.$$

Obviously, this approach ensures the smallest number of necessary freedom degrees for a rotor-stator system. System vectors and matrices are:

$$\{q_x\}_{n+2\times1}=\begin{Bmatrix}\{q_x\}\\ x_s\\ \varphi_{ys}\end{Bmatrix},\ \{q_y\}_{n+2\times1}=\begin{Bmatrix}\{q_y\}\\ y_s\\ \varphi_{xs}\end{Bmatrix},\ \{F_x\}_{n+2\times1}=\begin{Bmatrix}\{F_x\}\\ 0\\ 0\end{Bmatrix},\ \{F_y\}_{n+1\times1}=\begin{Bmatrix}\{F_y\}\\ 0\\ 0\end{Bmatrix},\ [\mathrm{M}]_{n+2\times n+2}=\begin{bmatrix}[\mathrm{M}]_{n\times n}&&\\&m_s&\\&&I_s\end{bmatrix}$$

$$\omega[\mathrm{G}]_{n+2\times n+2}=\omega\begin{bmatrix}[\mathrm{G}]&&\\&0&\\&&0\end{bmatrix},\ \eta_{\mathrm{h}}[K]_{n+2\times n+2}=\eta_{\mathrm{h}}\begin{bmatrix}[K]&&\\&0&\\&&0\end{bmatrix},\ \eta_{\mathrm{h}b}[K_b]_{n+2\times n+2}=\eta_{\mathrm{h}b}\begin{bmatrix}[K_b]&&\\&0&\\&&0\end{bmatrix},$$

$$\eta_{\mathrm{v}}[K_b]_{n+2\times n+2}=\eta_{\mathrm{v}}\begin{bmatrix}[K_b]&\dots&\\ \dots&\mathrm{k}_{2(p+1)-1\,2(p+1)-1}&\mathrm{k}_{2(p+1)-1\,2(p+1)}\\ &\mathrm{k}_{2(p+1)\,2(p+1)-1}&\mathrm{k}_{2(p+1)\,2(p+1)}\end{bmatrix}=$$

$$=\eta_{\mathrm{v}}\begin{bmatrix}\ddots&&&&\\ &k_{bj}&&-k_{bj}&\\ &&\ddots&&\\ &-k_{bj}&&\mathrm{k}_{2(p+1)-1\,2(p+1)-1}&\mathrm{k}_{2(p+1)-1\,2(p+1)}\\ &&&\mathrm{k}_{2(p+1)\,2(p+1)-1}&\mathrm{k}_{2(p+1)\,2(p+1)}\end{bmatrix},\quad \begin{aligned}&\mathrm{k}_{2(p+1)-1\,2(p+1)-1}=\frac{\eta_{\mathrm{v}s}}{\eta_{\mathrm{v}}}k_1+\frac{\eta_{\mathrm{v}s}}{\eta_{\mathrm{v}}}k_2+k_{bj}+..\\ &\mathrm{k}_{2(p+1)-1\,2(p+1)}=\mathrm{k}_{2(p+1)\,2(p+1)-1}=\\ &=\frac{\eta_{\mathrm{v}s}}{\eta_{\mathrm{v}}}k_2l_2-\frac{\eta_{\mathrm{v}s}}{\eta_{\mathrm{v}}}k_1l_1-k_{bj}l_j+..\\ &\mathrm{k}_{2(p+1)\,2(p+1)}=\frac{\eta_{\mathrm{v}s}}{\eta_{\mathrm{v}}}k_1l_1^2+\frac{\eta_{\mathrm{v}s}}{\eta_{\mathrm{v}}}k_2l_2^2+k_{bj}l_j^2+..\end{aligned},$$

$$[\mathrm{K}]_{n+2\times n+2}=\begin{bmatrix}[\mathrm{K}]&\dots&\\ \dots&k_{2(p+1)-1\,2(p+1)-1}&k_{2(p+1)-1\,2(p+1)}\\ &k_{2(p+1)\,2(p+1)-1}&k_{2(p+1)\,2(p+1)}\end{bmatrix}=$$

$$=\begin{bmatrix}\ddots&&&&&\\ &k_{2j-1\,2j-1}+k_{bj}&&&-k_{bj}&\\ &&\ddots&&&\\ &&k_{2k-1\,2k-1}+k_k&&-k_k&\\ &&&\ddots&&\\ &-k_{bj}&-k_k&&k_{2(p+1)-1\,2(p+1)-1}&k_{2(p+1)-1\,2(p+1)}\\ &&&&k_{2(p+1)\,2(p+1)-1}&k_{2(p+1)\,2(p+1)}\end{bmatrix},\quad \begin{aligned}&k_{2(p+1)-1\,2(p+1)-1}=k_1+k_2+k_{bj}+k_k+..\\ &k_{2(p+1)-1\,2(p+1)}=k_{2(p+1)\,2(p+1)-1}=\\ &=k_2l_2-k_1l_1+k_kl_k-k_{bj}l_j+..\\ &k_{2(p+1)\,2(p+1)}=k_1l_1^2+k_2l_2^2+k_{bj}l_j^2+k_kl_k^2+..\end{aligned}.$$

Conclusion

Equations (1) or (2) can be directly used to determine the response of synchronously whirling rotor on unbalanced forces. So, in the case of (2):

$$[\mathrm{M}]\{\ddot{q}\}+\left(\eta_{\mathrm{v}}[K_b]-\mathrm{i}\omega[\mathrm{G}]\right)\{\dot{q}\}+\left\langle[\mathrm{K}]-\mathrm{i}\left(\eta_{\mathrm{h}}[K]+\eta_{\mathrm{h}b}[K_b]\right)\right\rangle\{q\}=\omega^2\mathrm{e}^{i\omega t}\left(\{F_n\}+\mathrm{i}\{F_\tau\}\right),$$

where $\{F_n\}=\left\{..\quad m_ie_i\quad \left(I_{0i}-I_i\right)e_{1i}\cos e_{2i}\quad ..\right\}^{\mathrm{T}}$, $\{F_\tau\}=\left\{..\quad 0\quad \left(I_{0i}-I_i\right)e_{1i}\sin e_{2i}\quad ..\right\}^{\mathrm{T}}$.

Acknowledgment

The research work is was financially supported by the Ministry of Education and Sciences of the Russian Federation within the federal target program "Studies and Developments of Promising Trends in Russia's Science and Technology Sector for 2014–2020". Subsidy agreement No. 14.607.21.0191 of September 26, 2017, project RFMEFI60714X0191.

References

[1] Khulief Y A and Mohiuddin M A 1999 *J. of Sound & Vibration* **223(2)** pp 297-316

IOP Conf. Series: Materials Science and Engineering **489** (2019) 012041 doi:10.1088/1757-899X/489/1/012041

[2] Torkhani M et al 2012 *Mech Syst Signal Process* **29** pp 45-66
[3] Jacquet-Richardet G et al 2013 *Mech Syst Signal Process* **40** pp 401-20
[4] Liu X et al 2013 *Nonlinear Dyn* **72** pp 61-77
[5] Vlajic N et al 2014 *Int J Mech Sci* **83** pp 65-75
[6] Miranda W M and Faria M T C 2014 Finite Element Method Applied to the Eigenvalue Analysis of Flexible Rotors Supported by Journal Bearings *Engineering* vol. 6 pp 127-37
[7] Zilli A et al 2015 Nonlinear dynamics of a simplified model of an overhung rotor subjected to intermittent annular rubs *Journal of Engineering for Gas Turbines and Power* vol. 137 065001
[8] Shaw A D et al 2016 Asynchronous partial contact motion due to internal resonance in multiple degree-of-freedom rotordynamics *Proc R Soc A* vol. 472: 20160303 http://dx.doi.org/10.1098/rspa.2016.0303
[9] Krestnikovskiy K V et al 2018 Frequency of backward whirl or whip for the rotor in stator depending on the value of clearances between them *Handbook An Engineering Journal* vol. 8 pp 24-38 (in Russian) DOI:10.14489/hb.2018.08.pp.024-038
[10] Nelson H D 1980 *J. of Mechanical Design* **102 (4)** pp 793-803

IOP Conf. Series: Materials Science and Engineering **489** (2019) 012042 doi:10.1088/1757-899X/489/1/012042

Thermographic bench tests of rubber-cord pneumatic spring

A V Pozdeev, D A Chumakov, V V Novikov, I A Golyatkin, K V Chernyshov, A E Gavrilov and A V Leonard

Volgograd State Technical University, Volgograd, Russian Federation

pozdeev.vstu@gmail.com

Abstract. The article is dedicated to thermographic bench inspection of the rubber-cord pneumatic spring with or without air damper at different operating volumes. It also describes an experimental model, a procedure and results of thermographic bench tests of the rubber-cord pneumatic spring.

1. Introduction

One of the ways to increase the damping properties of pneumatic springs (PS) with rubber-cord casings (RCC) is air damping application [1…14]. However, it is known that constant casing bends and air throttling cause additional heating of the spring. In order to assess heating rates the Chair of "Automated units" of the Volgograd State Technical University performed bench tests of the pneumatic spring with and without air damper at a different operational volume. The purpose of these tests was to define the zones of maximum heating for different PS surfaces, as well as the peak temperature and the time for reaching a stabilized temperature mode.

2. Description of experimental model

Tests have been performed on a dynamic stand with servo-hydraulic actuator co-developed with *Biss ITW* (India) (figure 1) [15]. The test PS *1* is installed on a bench between the support plate *3* of hydropulsator *2* and adjusting channels *6* of the sprung mass *5*. The test PS has a sleeve type rubber-cord casing (RCC) *1* VL 260-340, a bored piston *4* and an upper cover *7* with a filler connector *8*. The bore of RCC *1* is connected with a bore *10* of piston *4* through the air damper with body *11* axially installed on the top end of piston *4* and fixed using a clamp *12* and bolts *13* with a sealing *14*. Inside the body *11* there is a throttle *15* interconnecting bores *9* and *10*, two rows of radial ports *16* and elastic non-return valve *17* shutting down these ports during the rebound and connecting bore *9* with bore *10* of the piston *4* only during the compression. The top end of piston *4* has an installed buffer of maximum compression *18* fixed with a clamp *12*.

The PS volume above the piston in a static position V_{st} = 11 L, as well as the volume of an extended bored piston V_b = 7.6 L. The PS height in a static position (from the top RCC cover to the bottom RCC cover) H_{st} = 450 mm. Static pressure in PS P_{st} = 0.2 MPa. Throttle diameter d = 6 mm. Hose length l_h = 1 m, and its cross section d_h = 20 mm. Receiver volume V_r = 20 L.

3. Methodology for bench testing

The methodology for bench testing is as follows.

Before starting the test, the traverse and sprung mass are fixed against bench guides *33*. To ensure the required operational volume in the PS piston its axial port either hermetically sealed or remained open. In order to increase an operational PS volume we connect receiver *30* to the bore *9* of RCC *1* through the inlet port *29* and hose *32*. This receiver is installed on SM *5*.

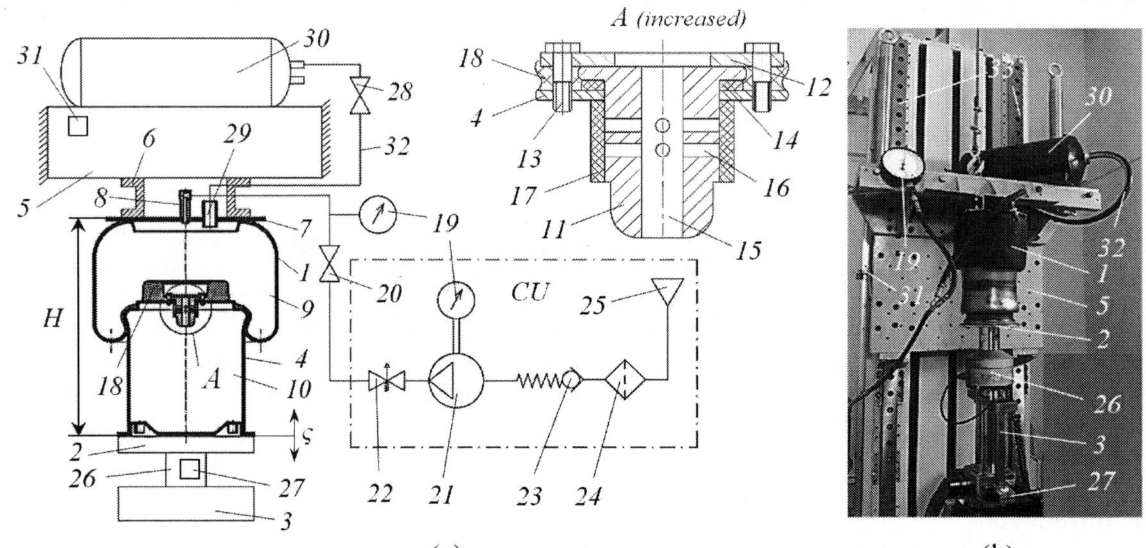

(a) (b)

Figure 1. Tested pneumatic spring: (a) – scheme of experimental pneumatic spring; (b) – general view of experimental unit on the stand; *1* – sleeve type RCC; *2* – hydropulsator with support plate *3*; *4* – bored piston; *5* – sprung mass (SM); *6* – adjusting channels; *7* – upper cover; *8* – filler connector; *9* – bore of RCC *1*; *10* – bore of piston *4*; *11* – body of air damper; *12* – clamp; *13* – bolts; *14* – sealing; *15* – throttle; *16* – radial ports; *17* – elastic non-return valve; *18* – buffer of maximum compression; *19* – manometers; *20, 28* – valves; *21* – compressor; *22* – bypass valve; *23* – check valve; *24* – filter; *25* – intake; *26* – force sensor; *27* – position sensor; *29* – connecting pipe with receiver *30*; *31* – displacement and speed sensor; *32* – hose; *33* – guides; *H* – height of PS in static position; *ς* – displacement of the hydropulsator plate; *CU* – compressor unit

While studying the process of PS heating with an air damping, damping node is installed in the piston. The pneumatic spring is put in a static position corresponding to its height of 450 mm by air supply from the external compressor to the filler connector and hydropulsator rod extension to the required height. The air pressure inside the PS is brought to the required value, while the operating pressure is controlled by pressure gauge *19*.

Further, using control panel we set up an amplitude, rate and time period of hydropulsator rod extension. During tests we selected harmonic loading modes with a frequency of 1 Hz and amplitude of 50 mm which are close to resonant modes of sprung mass vibrations of the car.

Using *Testo 890 V1* thermal imaging camera we measure the temperature in three zones of PS outer surface every two minutes: in the middle zone of RCC wall; on the RCC bend zone; in the zone of the bored piston. The tests have been performed in the laboratory at ambient temperature + 22 ... + 23 °C without forced air cooling of the PS.

In the process of testing the PS had the following operational volumes: 11, 18.6, 31 and 38.6 liters. With a volume of 11 and 31 L, the piston bore was covered by a plug. With a volume of 18.6 and 38.6 L, the bored piston, with a volume of 7.6 L, was either open, or was connected by a throttle or a throttle with a non-return valve opened during the compression. Also, with a volume of 31 and 38.6 liters, a 20 L receiver was connected to the top cover of the PS through a hose.

IOP Conf. Series: Materials Science and Engineering **489** (2019) 012042 doi:10.1088/1757-899X/489/1/012042

4. Results of bench tests

Figure 2 shows examples of PS thermograms at different operational volumes with and without damping node. Brighter colors correspond to greater heating of PS surfaces.

Figures 3 ... 5 show the process graphs of heating the PS components (at different operational volumes, with and without air damper) depending on time.

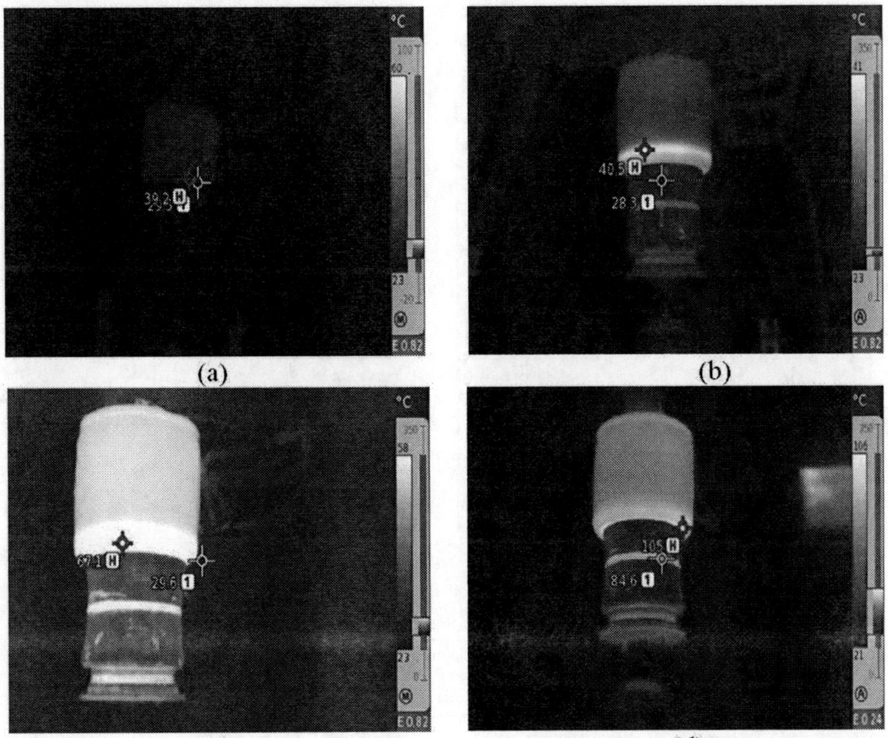

(a) (b) (c) (d)

Figure 2. Thermograms of pneumatic spring: (a) and (b) – with operational volume of 11 and 18.6 liters (without damping node); (c) – with operational volume of 18.6 liters and throttle of 6 mm; (d) – with operational volume of 18.6 liters, throttle of 6 mm and non-return valve

Figure 3(a) shows that maximum heating temperature +40 °C of RCC bend zone at operational volume of 11 liters is reached during 0...20 min with an increasing average rate of 1 °C·min^{-1}, and between 20…40 min PS temperature remains unchanged (curve *1*).

An operational volume increase to 18.6 liters does not affect the maximum heating temperature of the bend zone, however, the time taken to reach the mode increases from 20 to 40 minutes. Whereas, between 0 ... 15 min, the rate of temperature increase averages 1.2 °C·min^{-1}, and then the rate slows down by a factor of 10 (curve *2*).

The throttle of 6 mm diameter installed in PS of 18.6 liters causes the increase of maximum heating temperature for RCC bend zone from +40 up to +67 °C, while the heating time to stabilize the temperature grows from 40 to 60 minutes. The maximum temperature increase from +22 to +58 °C at a rate of 2.2 °C·min^{-1} is observed in the interval of 0 ... 16 minutes, a the interval of 16 ... 30 minutes the temperature remains practically unchanged, and in the interval of 30 ... 60 minutes the temperature rises to +67 °C at a rate of about 0.3 °C·min^{-1} (curve *3*).

The throttle with a non-return valve installed in PS of 18.6 liters increases the maximum heating temperature of RCC bend zone from +67 up to +110 °C within 60 minutes. Whereas, the increase of maximum temperature from +22 to +100 °C at a rate of about 2.6 °C·min^{-1} is observed in the interval of 0 ... 30 minutes, in the interval of 30 ... 50 min the temperature increase slows down from 2.6 to 0.5 °C·min^{-1}, and then the temperature slightly increases (curve *4*).

The increase of PS operational volume from 11 and 18.6 liters to 31 and 38.6 liters has a slight effect on the decrease in the heating temperature in the RCC bend zone (curves *5* and *6*).

The increase in operational volume of PS with the throttle, as well as with the throttle including non-return valve, from 18.6 to 38.6 liters leads to a decrease in the heating temperature of the RCC bend zone from +67 to +39.3 °C (curve *7*) and from +110 to +38.6 °C (curve *8*).

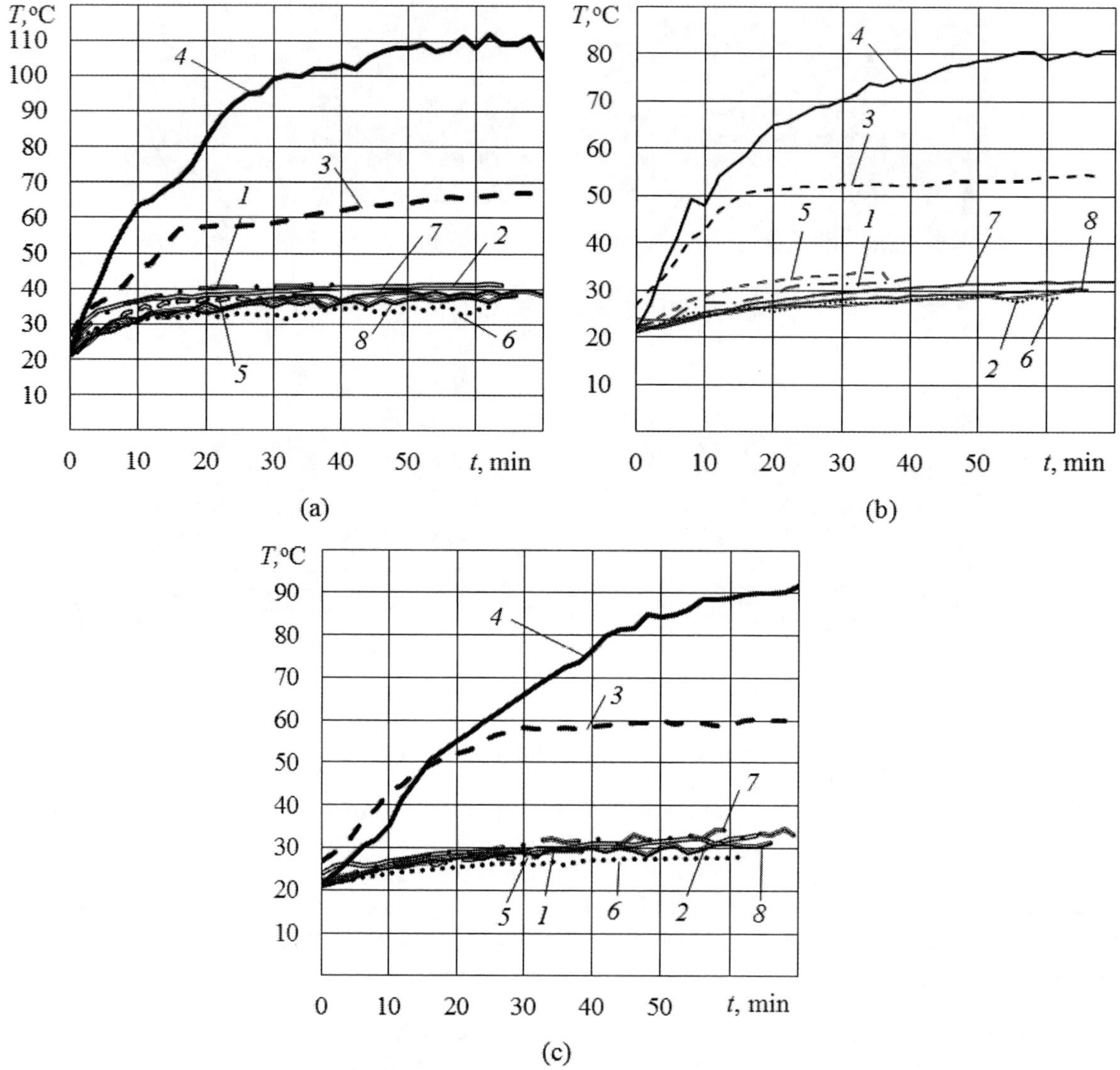

Figure 3. Process graphs of heating the PS components at different operational volumes, with and without damping node: (a) – RCC bend zone heating; (b) – RCC middle zone heating; (c) – PS piston wall heating; *1*, *2*, *5*, *6* – PS with operational volume of 11, 18.6, 31 and 38.6 liters; *3*, *4*, *7*, *8* – PS with operational volume 18.6 и 38.6 liters, throttle and throttle with non-return valve

Figure 3(b) indicates that the maximum temperature of +33 °C for heating the RCC middle zone with operating volume of 11 liters is reached within an interval of 0 ... 30 minutes with an average increase rate of 0.5 °C·min^{-1}, then the PS goes to the balanced state and the temperature of this zone remains almost unchanged (curve *1*).

Operating volume increase to 18.6 liters reduces the heating temperature of RCC middle zone from +32 to +30.5 °C, however, the ramp-up time increases from 30 to 50 minutes, while the temperature increases extremely slowly at the average rate of 0.1°C·min^{-1} (curve *2*).

Throttle installation in the PS with an operating volume of 18.6 liters causes an increase in the maximum heating temperature of the middle zone from +33 to +54 °C. Whereas, the heating time to reach the balanced temperature is 30 minutes. The maximum temperature increase from +22 to +52 °C at a rate of about 1.5 °C·min^{-1} is observed in the interval of 0 ... 20 minutes, in the interval of 16 ... 30 minutes the temperature practically does not increase following by the decrease of temperature growth rate almost by 30 times (curve *3*).

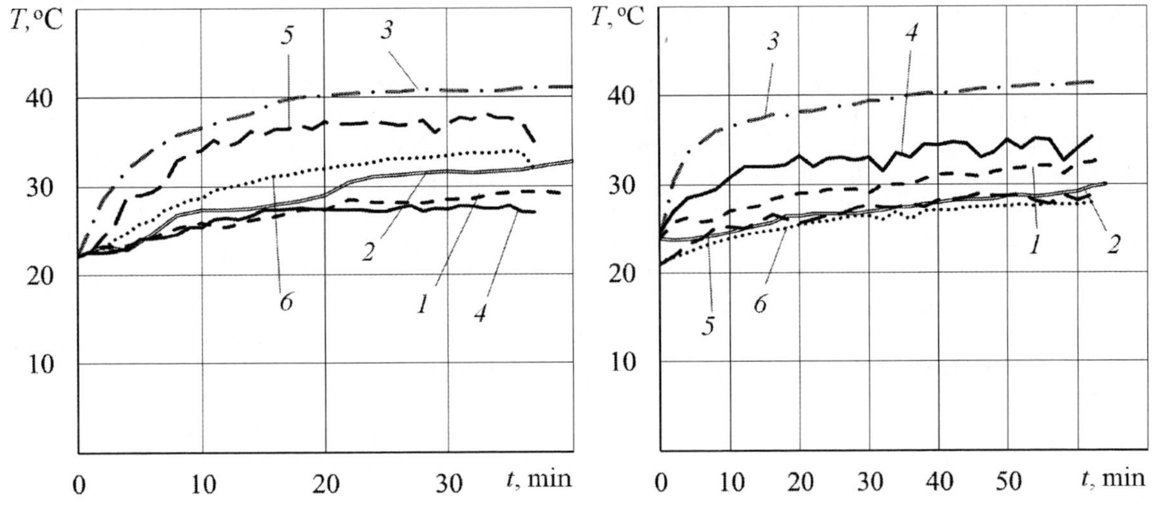

(a) (b)

Figure 4. Process graphs of heating the PS components at different operational volumes, without damping node: (a) – PS with operational volume of 11 liters – curves *1…3* and 31 liters – curves *4…6*; (b) – PS with operational volume of 18.6 liters – curves *1…3* and 38.6 liters – curves *4…6*; *1, 4* – PS piston wall heating; *2, 5* – RCC middle zone heating; *3, 6* – RCC bend zone heating

The throttle with a non-return valve installed in PS of 18.6 liters increases the maximum heating temperature of RCC middle zone from 54 up to +80°C within 60 minutes. Whereas, the increase of maximum temperature from +22 to +65 °C at a rate of about 2 °C·min^{-1} is observed in the interval of 0 ... 20 minutes, in the interval of 20 ... 60 min the temperature increase slows down from 2 to 0.4° C·min^{-1}, and then the temperature slightly increases (curve *4*).

The increase of PS operational volume from 11 and 18.6 liters to 31 and 38.6 liters has a slight effect on the heating temperature in the RCC middle zone (curves *5* and *6*).

The increase in operational volume of PS with the throttle, as well as with the throttle including non-return valve, from 18.6 to 38.6 liters leads to a decrease in the heating temperature of the RCC middle zone from +54 up to +32 °C (curve *7*) and from +80 up to +30.5 °C (curve *8*).

Figure 3(c) indicates that the maximum temperature of +30°C for heating the PS piston with operating volume of 11 liters is reached within 25 minutes with an average increase rate of 0.25°C·min^{-1}, then the PS remains almost unchanged (curve *1*).

Operating volume increase to 18.6 liters practically does not affect the maximum heating temperature of PS piston, however, the ramp-up time increases from 25 to 50 minutes, while the temperature increases extremely slowly at the average rate of 0.15 °C·min^{-1} (curve *2*).

The throttle installed in PS of 18.6 liters causes the increase of maximum heating temperature of the bored piston from +30 up to +60 °C, while the heating time to stabilize the temperature is 50 minutes. The maximum temperature increase from +22 to +58 °C at a rate of 1.3 °C·min^{-1} is observed in the interval of 0 ... 28 minutes, in the interval of 28 ... 50 minutes the temperature increase remains rate slows down up to 0.1 °C·min^{-1}, and then the temperature remains practically unchanged (curve *3*).

The throttle with a non-return valve installed in PS of 18.6 liters increases the maximum heating temperature of the bored piston from +60 up to +90 °C within 60 minutes. Whereas, the increase of maximum temperature from +22 to +90 °C at a rate of about 1.1 °C·min^{-1} before reaching the balanced temperature is observed in a whole time interval of tests performed (curve *4*).

The increase of PS operational volume from 11 and 18.6 liters to 31 and 38.6 liters has a slight effect on the heating temperature of the bored piston (curves *5* and *6*).

The increase in operational volume of PS with the throttle, as well as with the throttle including non-return valve, from 18.6 to 38.6 liters leads to a decrease in the heating temperature of the bored piston from +60 up to +33.2 °C (curve *7*) and from +90 up to +30.6 °C (curve *8*).

The interval for reaching the balanced temperature of RCC bend zone, RCC middle zone and bored piston during PS tests with a receiver (figure 3, (a)....(c); curves *5-8*) is 30 minutes.

Figure 4 shows that with an operating volume of 11 and 18.6 liters and the lack of air damper, the maximum temperature of +41 °C for heating the RCC bend zone is reached within 30 minutes (curve *3*). However, the maximum temperature of the piston wall (curve *1*) and RCC (curve *2*) is significantly lower, which is caused by the low rate of air heating in the operating volume and the high thermal metal conductivity of the piston. It should be noted that with an increase in operating volume from 11 liters (figure 4(a)) to 18.6 liters (figure 4(b)), the walls of the piston heat up slightly more than RCC walls. With an increase in the operating volume from 11 and 18.6 liters to 31 and 38.6 liters, the heating temperature for the piston wall, RCC middle zone and the bend zone decreases by an average of 10 °C.

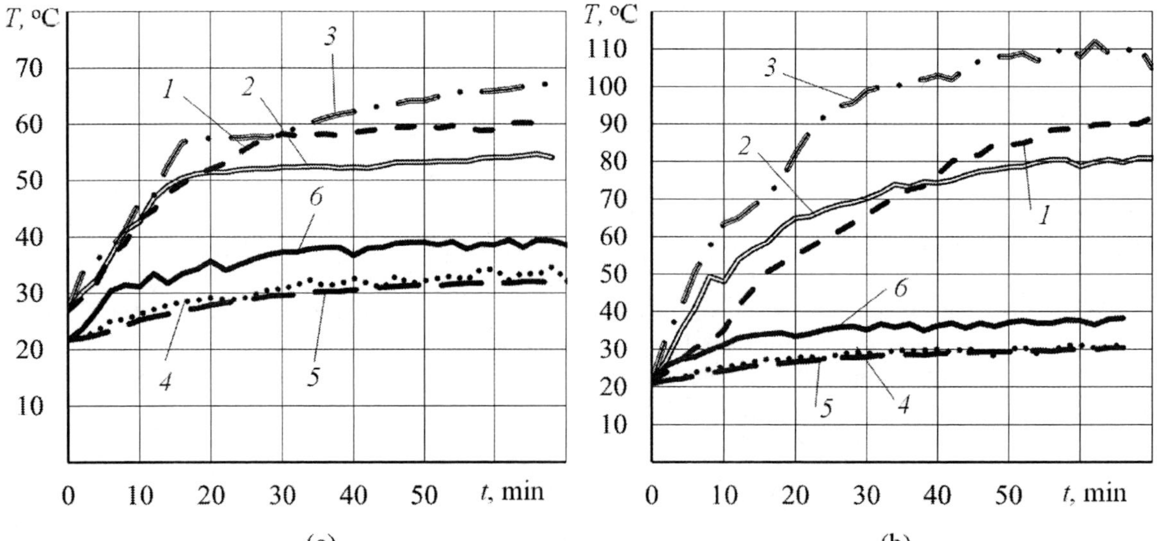

Figure 5. Process graphs of heating the PS components at different operational volumes, with damping node: (a) – PS with throttle; (b) – PS with throttle and non-return valve; *1…3* – PS with operational volume of 18.6 liters; *4…6* – PS with operational volume of 38.6 liters; *1, 4* – PS piston wall heating; *2, 5* – RCC middle zone heating; *3, 6* – RCC bend zone heating

If we compare the graphs in figure 5, (a) and (b), we will see that the highest maximum heating temperature of the PS is achieved when installing a throttle including non-return valve. The maximum heating temperature of the piston wall is higher (curves *1*) than RCC (curves *2*), due to the higher thermal conductivity of the metal. With an operating volume increase for PS with a throttle from 18.6 to 38.6 liters, the heating temperature of +60 °C for the piston wall, +54 °C for RCC middle zone and +67.2 °C for RCC bend zone decreases to +34.7, +33.2 and +39.3 °C, respectively (figure 5(a)), that is, almost twice. With operating volume increase for PS with a throttle including non-return valve from 18.6 to 38.6 liters, the heating temperature of +90 °C for the piston wall, +80 °C for RCC middle zone and +110 °C for RCC bend zone decreases to +30 , +30 and +37.2 ºC, respectively, that is, almost three-fold (figure 5(b)).

The experimental data provided above were obtained for the most severe resonant test mode without a forced air cooling of the PS. Therefore, when the receiver is connected to the PS in the suspension/sprung design and when it is blown with an opposing air flow and its random vibrations, the maximum heating temperatures of the PS with an air damper installed in its piston will be much lower. This allows recommending the use of air damping instead of hydraulic shock absorbers or combination of both to improve the smooth ride and speed of vehicles with a pneumatic spring system.

5. Conclusions

- In the absence of a throttle in the piston of the pneumatic spring with an operating volume of 11 L, the maximum temperature of +40 °C of the bend zone of the rubber-cord casing is reached within 20 minutes. PS operating volume increase from 11 to 18.6 L does not affect the maximum temperature, but it causes a double increase of the heating time from 20 to 40 minutes. In this case, the maximum heating temperature of the piston and RCC remains lower approximately by 25%, i.e. +30 ºC.
- Installation of the throttle of 6mm in diameter in the PS piston of 18.6 L, leads to a maximum heating temperature increase in the piston up to +60°C within 50 minutes, and the installation of a throttle including non-return valve raises its maximum temperature up to +90 °C within 60 min.
- Maximum heating up to +110 °C is achieved in the RCC bend zone when testing the PS with a throttle including non-return valve. This temperature increases within 60 minutes due to the air flow through the throttle and non-return valve, internal intermolecular friction of RCC and its friction against the piston wall.
- An increase in the operating volume of the PS without air damper leads to a slight decrease in the heating temperature: the piston walls – by 1 ... 2 °C, the bend zone and the middle zone of the RCC wall – by 5°C.
- An increase in the operating volume of the PS with a throttle leads to a decrease in the heating temperature of the piston wall from +60 to +35 °C, the bend zone and the middle zone of the RCC wall from +67 to +33 °C and from +54 to +39 °C, respectively.
- An increase in the operating volume of the PS with a throttle including non-return leads to a decrease in the heating temperature of the piston wall from +90 to +30 °C, the bend zone and the middle zone of the RCC wall from +110 to +37 °C and from +80 to +30 °C, respectively.
- In order to increase the vibration-protective properties of the vehicle's spring when using pneumatic springs with high operating volumes, it is possible to recommend installation of an air damper in the pneumatic spring with no risk of strong heating of its elements.

References

[1] Pevzner Ya M, Gorelik A M 1963 *Pneumatic and hydropneumatic suspension* (Moscow, Mashgiz)
[2] Novikov V V 2007 Bench testing of the air suspension of the bus "VZTM-32731" with hydraulic shock absorbers of different capacity *Truck: transport complex, special equipment* **6** 41-4

IOP Conf. Series: Materials Science and Engineering **489** (2019) 012042 doi:10.1088/1757-899X/489/1/012042

[3] Novikov V V 2007 *Truck: transport complex, special equipment* **7** 43-6
[4] Novikov V V, Bukaev S O, Dyakov A S *Automotive industry* **1** 20-2
[5] Kalashnikov B A 2008 *Shock-absorbing systems of the installations with discontinuous switching of the elastic elements* (Omsk, Omsk State Technical University)
[6] Novikov V V, Ryabov I M, Chernyshov K V 2009 *The vibration isolation properties of the motor vehicles suspensions* (Volgograd, Volgograd State Technical University)
[7] Novikov V V, Lapynin Yu G, Ryabov I M, Gorobtsov A S, Chernyshov K V, Diakov A S, Bukaev S O, Pozdeev A V, Nikolaev D A 2009 *Rear suspension of vehicle wheels* (RU Utility model 85403)
[8] Novikov V V, Pozdeev A V, Diakov A S, Karlov V I, Cherkashina E A 2011 *Air suspension* (RU Utility model 109698)
[9] Riabov I M, Pozdeev A V, Novikov V V, Diakov A S, Chernyshov K V 2011 *Vehicle wheel suspension* (RU Utility model 109697)
[10] Pozdeev A V, Novikov V V, Diakov A S, Pokhlebin A V, Ryabov I M, Chernyshov K V 2013 *Adjustable pneumatic and hydro-pneumatic springs of motor vehicles suspensions* (Volgograd, Volgograd State Technical University)
[11] Pozdeev A V, Diakov A S, Novikov V V, Ryabov I M 2013 Studies of dual-chamber pneumatic spring with switching of chambers *Truck: transport complex, special equipment* **1** 35-37
[12] Pozdeev A V, Diakov A S, Novikov V V, Riabov I M 2013 *Truck: transport complex, special equipment* **9** pp 2-5
[13] Pozdeev A V 2015 Vibro-protecting two-cavity pneumatic springs properties determined on the basis of cavity commutation optimal algorithms synthesis *Tekhnologiya kolesnykh i gusenichnykh mashin – Technology of Wheeled and Tracked Machines* **1** 27-31
[14] Korneev S A, Korneev V S, Zubarev A V, Klimentiev E V 2016 *Basics of the technical theory of pneumatic shock absorbers* (Omsk, Omsk State Technical University)
[15] Novikov V V, Pozdeev A V, Diyakov A S 2015 Research and testing complex for analysis of vehicle suspension units *Procedia Engineering* (*International Conference on Industrial Engineering (ICIE-2015)* vol 129) ed A A Radionov (Elsevier publishing) 465-70

IOP Conf. Series: Materials Science and Engineering **489** (2019) 012043 doi:10.1088/1757-899X/489/1/012043

Force driven nonlinear vibrations of a thin plate with 1:1 internal resonance in a fractional viscoelastic medium

M V Shitikova, V V Kandu

Voronezh Technical State University, Research Center on Dynamics of Solids and Structures, 20-letija Oktjabrja Street 84, Voronezh 394006, Russian Federation.

MVS@vgasu.vrn.ru

Abstract. In the present paper, force driven vibrations of a physically nonlinear plate are studied by the method of multiple time scales when the plate is subjected to the conditions of the one-to-one internal resonance. The damping features of the surrounding medium are described by the fractional derivative Kelvin-Voight model. The influence of viscosity on the energy exchange mechanism between interacting nonlinear modes has been analyzed. For the one-to-one internal resonance, the nonlinear set of resolving equations in terms of amplitudes and phase differences has been obtained, and a comparative analysis of numerical calculations of free and forced vibrations are presented.

1. Introduction

It is well known that the analysis of forced nonlinear vibrations of plates is the important area of applied mechanics, since plates, which could be made of different materials, are used as structural elements in many fields of industry and technology [1]. Moreover, nonlinear vibrations could be accompanied by such a phenomenon as the internal resonance, resulting in multimode response with a strong interaction of the modes involved [2]. The internal resonance could be found within a certain combination of natural frequencies of one and the same type of vibrations, when vibratory motions are described by one equation [3], or between natural frequencies belonging to different types of vibrations, when two or more equations are used for the description of dynamic behaviour of a structure [4].

Nonlinear free damped vibrations of a rectangular plate described by three nonlinear differential equations have been studied in [5], wherein the procedure resulting in decoupling linear parts of equations has been proposed with the further utilization of the generalized method of multiple scales [6] for solving nonlinear governing equations of motion, in so doing the amplitude functions are expanded into power series in terms of the small parameter and depend on different time scales. All possible types of the internal resonance within the considered model of nonlinear plate's behaviour have been revealed in [5].

In the present paper, the approaches proposed in [5] for free vibrations of plates and in [7] for forced vibrations of a nonlinear oscillator with weak fractional damping have been extended to the force driven vibrations of a thin plate under the one-to-one internal resonance with the force frequency approximately equal to a certain natural frequency of vertical vibrations.

2. Problem formulation and the method of solution

Let us consider the dynamic behavior of a free supported nonlinear thin rectangular plate (figure 1), forced vibrations of which in a viscoelastic fractional derivative medium are described by the following three differential equations in the dimensionless form (free damped equations presented in [5] are supplemented herein by the vertical harmonic force applied at the point with the coordinates x_0, y_0):

$$u_{xx} + \frac{1-\nu}{2}\beta_1^2 u_{yy} + \frac{1+\nu}{2}\beta_1 v_{xy} + w_x\left(w_{xx} + \frac{1-\nu}{2}\beta_1^2 w_{yy}\right) + \frac{1+\nu}{2}\beta_1^2 w_y w_{xy} = \ddot{u} + æ_1 D^\gamma u, \quad (1)$$

$$\beta_1^2 v_{yy} + \frac{1-\nu}{2} v_{xx} + \frac{1+\nu}{2}\beta_1 u_{xy} + \beta_1 w_y\left(\beta_1^2 w_{yy} + \frac{1-\nu}{2} w_{xx}\right) + \frac{1+\nu}{2}\beta_1 w_x w_{xy} = \ddot{v} + æ_2 D^\gamma v, \quad (2)$$

$$\begin{aligned}
\frac{\beta_2^2}{12}\left(w_{xxxx}\right. &+ \left. 2\beta_1^2 w_{xxyy} + \beta_1^4 w_{yyyy}\right) - w_{xx}\left(u_x + \nu\beta_1 v_y\right) - w_x\left(u_{xx} + \nu\beta_1 v_{xy}\right) \\
&- \frac{1-\nu}{2}\beta_1\left[w_{xy}\left(\beta_1 u_y + v_x\right) + w_y\left(\beta_1 u_{xy} + v_{xx}\right)\right] \\
&- \beta_1^2\left[w_{yy}\left(\nu u_x + \beta_1 v_y\right) + w_y\left(\nu u_{xy} + \beta_1 v_{yy}\right)\right] \\
&- \frac{1-\nu}{2}\beta_1\left[w_{xy}\left(\beta_1 u_y + v_x\right) + w_x\left(\beta_1 u_{yy} + v_{xy}\right)\right] \\
&- F\cos\left(\Omega_F t\right)\delta(x - x_0)\delta(y - y_0) = -\ddot{w} - æ_3 D^\gamma w, \quad (3)
\end{aligned}$$

subjected to the initial

$$u|_{t=0} = v|_{t=0} = w|_{t=0} = 0, \quad (4)$$

$$\dot{u}\Big|_{t=0} = \varepsilon U^0(x,y), \quad \dot{v}\Big|_{t=0} = \varepsilon V^0(x,y), \quad \dot{w}\Big|_{t=0} = \varepsilon W^0(x,y), \quad (5)$$

as well as the boundary conditions

$$w|_{x=0} = w\Big|_{x=1} = 0, \quad v|_{x=0} = v\Big|_{x=1} = 0, \quad u_x\Big|_{x=0} = u_x\Big|_{x=1} = 0, \; w_{xx}\Big|_{x=0} = w_{xx}\Big|_{x=1} = 0, \quad (6)$$

$$w|_{y=0} = w\Big|_{y=1} = 0, \quad u|_{y=0} = u\Big|_{y=1} = 0, \quad v_y\Big|_{y=0} = v_y\Big|_{y=1} = 0, \; w_{yy}\Big|_{y=0} = w_{yy}\Big|_{y=1} = 0, \quad (7)$$

where $u = u(x,y,t)$, $v = v(x,y,t)$, and $w = w(x,y,t)$ are the displacements of points located in the plate's middle surface in the $x-$, $y-$, and $z-$directions, respectively, ν is the Poisson's ratio, $\beta_1 = a/b$ and $\beta_2 = h/a$ are the parameters defining the dimensions of the plate, a and b are the plate's dimensions along the $x-$ and $y-$axes, respectively, h is the thickness, t is the time, an overdot denotes the time-derivative, lower indices label the derivatives with respect to the corresponding coordinates, $U^0(x,y)$, $V^0(x,y)$ and $W^0(x,y)$ are functions describing the distribution of the initial velocities of the points lying in the middle surface of the plate, ε is a small dimensionless parameter of the same order of magnitude as the amplitudes, F is the amplitude of the harmonic force with the frequency Ω_F, δ is the Dirac delta function, $æ_i = \varepsilon\mu_i\tau_i^\gamma$ $(i = 1,2,3)$ are damping coefficients, μ_i are finite values, τ_i are the relaxation times, and D_{0+}^γ is the Riemann-Liouville fractional derivative of the γ-order [8].

The set of equations (1)-(3) admits the solution of the Navier type for the simply-supported plate (figure 1), which could be represented in terms of the time-dependent generalized

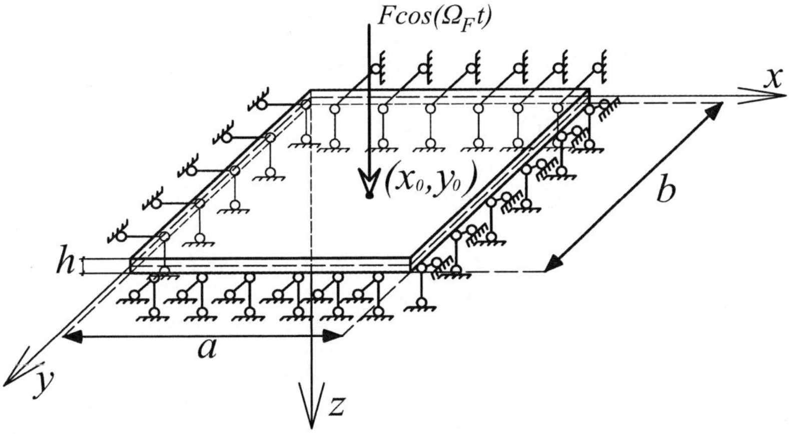

Figure 1. Scheme of a freely supported rectangular plate

displacements $x_{imn}(t)$ $(i = 1, 2, 3)$ and eigen-functions of the mn-th natural modes of linear vibrations with the natural frequencies ω_{imn} of the plate. The procedure resulting in decoupling linear parts of nonlinear equations of motion has been proposed in [5] with the further utilization of the generalized method of multiple scales [6] for solving nonlinear governing equations of motion, in so doing the amplitude functions are expanded into power series in terms of the small parameter and depend on different time scales with $T_0 = t$ as a fast time characterizing oscillatory motions with eigenfrequencies, and $T_1 = \varepsilon t$, $T_2 = \varepsilon^2 t$ as slow scales characterizing the modulation of amplitudes and phases of nonlinear vibrations.

It has been revealed [5] that nonlinear vibrations of the plate could be accompanied by different types of the internal resonance when two or more modes could be coupled. Moreover, its type depends on the order of smallness of the viscosity involved into consideration. Thus, at the ε^2-order, damped vibrations could be accompanied by the one-to-one, one-to-one-to-one, and combinational resonances of the additive and difference types .

In [5] it has been shown that the one-to-one internal resonance could be of two types: (1) when the natural frequencies of two modes of in-plane vibrations are close to each other, i.e., when $\omega_1 = \omega_2 = \omega$, and (2) when the natural frequency of the out-of-plane mode is close to the natural frequency of one of the in-plane modes, i.e., $\omega_3 = \omega_1 = \omega$ or $\omega_3 = \omega_2 = \omega$. From hereafter the indices mn are omitted for the ease of presentation.

2.1. One-to-one internal resonance, when $\omega_1 = \omega_2 = \omega$, but $\omega_3 \neq \omega$

Now let us consider the case of the one-to-one internal resonance $\omega_1 = \omega_2 = \omega$ when coupled with the external resonance $\omega_3 \approx \Omega_F \neq \omega$, in so doing the external force, following [7], is assumed to be $F \sin \pi x_0 \sin \pi y_0 = \varepsilon^3 f$.

Using the system of solvability equations to eliminate secular terms in the case of free vibrations [5], and adding the external resonance, we obtain the following solvability equations for the case of force driven vibrations:

$$2i\omega_1 D_2 A_1 + \mu_1 (i\omega_1 \tau_1)^\gamma A_1 + 2\zeta_1 \left[(k_5 + k_7) A_1 A_3 \bar{A}_3 + (k_6 + k_8) A_2 A_3 \bar{A}_3 \right] = 0, \tag{8}$$

$$2i\omega_2 D_2 A_2 + \mu_2 (i\omega_2 \tau_2)^\gamma A_2 + 2\zeta_2 \left[(k_6 + k_8) A_2 A_3 \bar{A}_3 + (k_5 + k_7) A_1 A_3 \bar{A}_3 \right] = 0, \tag{9}$$

$$\begin{aligned} 2i\omega_3 D_2 A_3 &+ \mu_3 (i\omega_3 \tau_3)^\gamma A_3 + \zeta_{13} \left[(k_5 + k_7) A_1 \bar{A}_1 A_3 + (k_1 + 2k_2) A_3^2 \bar{A}_3 \right] \\ &+ \zeta_{23} \left[(k_6 + k_8) A_2 \bar{A}_2 A_3 + (k_3 + 2k_4) A_3^2 \bar{A}_3 \right] + (\zeta_{13} k_6 + \zeta_{23} k_7) \bar{A}_1 A_2 A_3 \end{aligned}$$

IOP Conf. Series: Materials Science and Engineering **489** (2019) 012043 doi:10.1088/1757-899X/489/1/012043

$$+ \quad (\zeta_{13}k_8 + \zeta_{23}k_5)\, A_1\bar{A}_2A_3 - 2f = 0, \tag{10}$$

where $A_{j\,mn}(T_2)$ and $\bar{A}_{j\,mn}(T_2)(j = 1,2,3)$ are complex conjugate functions to be found, ζ_2, ζ_{13}, and ζ_{23} are coefficients depending on the plate dimensions and numbers of excited modes [5], while coefficients k_i $(i = 1, 2, ..., 8)$ have the form

$$k_1 = \frac{\zeta_1}{4\omega_3^2 - \omega_1^2}, \quad k_2 = -\frac{\zeta_1}{\omega_1^2}, \quad k_3 = \frac{\zeta_2}{4\omega_3^2 - \omega_2^2}, \quad k_4 = -\frac{\zeta_2}{\omega_2^2}, \quad k_5 = \frac{\zeta_{13}}{\omega_1(\omega_1 + 2\omega_3)},$$

$$k_6 = \frac{\zeta_{23}}{\omega_2(\omega_2 + 2\omega_3)}, \quad k_7 = \frac{\zeta_{13}}{\omega_1(\omega_1 - 2\omega_3)}, \quad k_8 = \frac{\zeta_{23}}{\omega_2(\omega_2 - 2\omega_3)}.$$

Let us multiply equations (8)–(10), respectively, by $\bar{A}_1$, $\bar{A}_2$, and $\bar{A}_3$ and find their complex conjugates. Adding every pair of the mutually adjoint equations with each other and subtracting one from another, and considering that the functions $A_i(T_2)$ could be represented in the polar form $a_i(T_2)e^{i\,\varphi_i(T_2)}$ $(i = 1, 2, 3)$, where a_i and φ_i are amplitudes and phases, as a result we arrive at a set of six coupled nonlinear differential equations

$$\left(a_1^2\right)^{\cdot} \quad + \quad s_1a_1^2 + 2\omega_1^{-1}\zeta_1\,(k_6 + k_8)\,a_1a_2a_3^2\sin\delta = 0, \tag{11}$$
$$\left(a_2^2\right)^{\cdot} \quad + \quad s_2a_2^2 - 2\omega_2^{-1}\zeta_2\,(k_5 + k_7)\,a_1a_2a_3^2\sin\delta = 0, \tag{12}$$
$$\left(a_3^2\right)^{\cdot} \quad + \quad s_3a_3^2 = -2f\omega_3^{-1}a_3\sin\varphi_3, \tag{13}$$
$$\dot{\varphi}_1 \quad - \quad \frac{1}{2}\,\sigma_1 - \omega_1^{-1}\zeta_1\,(k_5 + k_7)\,a_3^2 \;\; - \omega_1^{-1}\zeta_1\,(k_6 + k_8)\,a_1^{-1}a_2a_3^2\cos\delta = 0, \tag{14}$$
$$\dot{\varphi}_2 \quad - \quad \frac{1}{2}\,\sigma_2 - \omega_2^{-1}\zeta_2\,(k_6 + k_8)\,a_3^2 - \omega_2^{-1}\zeta_2\,(k_5 + k_7)\,a_2^{-1}a_1a_3^2\cos\delta \;= 0, \tag{15}$$
$$\dot{\varphi}_3 \quad - \quad \frac{1}{2}\,\sigma_3 - \frac{1}{2}\,\omega_3^{-1}\{\zeta_{13}\,(k_1 + 2k_2) + \zeta_{23}\,(k_3 + 2k_4)\}a_3^2 + f(\omega_3\,a_3)^{-1}\cos\varphi_3$$
$$- \quad \frac{1}{2}\,\omega_3^{-1}\zeta_{23}\,(k_6 + k_8)\,a_2^2 - \frac{1}{2}\,\omega_3^{-1}\zeta_{13}\,(k_5 + k_7)\,a_1^2 = 0, \tag{16}$$

where $\delta = \varphi_2 - \varphi_1$ is the phase difference, an overdot denotes the differentiation with respect to T_2, $s_i = \mu_i\tau_i^\gamma\omega_i^{\gamma-1}\sin\psi$, $\sigma_i = \mu_i\tau_i^\gamma\omega_i^{\gamma-1}\cos\psi$, and $\psi = \pi\gamma/2$ $(i = 1, 2, 3)$.

2.2. The case of free damped vibrations

In the absence of the external force equations (13) and (16) are reduced to

$$\left(a_3^2\right)^{\cdot} + s_3a_3^2 = 0, \tag{17}$$

$$\dot{\varphi}_3 \quad = \quad \frac{1}{2}\,\sigma_3 + \frac{1}{2}\,\omega_3^{-1}\left[\zeta_{13}\,(k_1 + 2k_2) + \zeta_{23}\,(k_3 + 2k_4)\right]a_3^2$$
$$+ \quad \frac{1}{2}\,\omega_3^{-1}\zeta_{23}\,(k_6 + k_8)\,a_2^2 - \frac{1}{2}\,\omega_3^{-1}\zeta_{13}\,(k_5 + k_7)\,a_1^2. \tag{18}$$

From equation (17) it is seen that it is independent of all other five equations, and its solution has the form

$$a_3^2 = c_3e^{-s_3T_2}, \tag{19}$$

where c_3 is a constant of integration to be determined from the initial conditions, while reference to equation (18) shows that the phase φ_3 depends only on squared amplitudes of all three interacting modes of vibrations.

Introducing new functions $\xi_1(T_2)$ and $\xi_2(T_2)$ such that

$$a_1^2 = \frac{\zeta_1 (k_6 + k_8)}{\omega} \xi_1 \exp(-s_1 T_2), \quad a_2^2 = \frac{\zeta_2 (k_5 + k_7)}{\omega} \xi_2 \exp(-s_2 T_2), \tag{20}$$

and adding equations (11) and (12) with due account for (20) yield

$$\dot{\xi}_2 + \dot{\xi}_1 e^{(s_2 - s_1)T_2} = 0, \tag{21}$$

while subtracting (14) from (15) we obtain

$$\dot{\delta} = \dot{\varphi}_2 - \dot{\varphi}_1 = \Sigma + \omega^{-1} \left[\zeta_2 (k_6 + k_8) - \zeta_1 (k_5 + k_7)\right] a_3^2$$
$$+ \omega^{-1} \left[\zeta_2 (k_5 + k_7) a_1^2 - \zeta_1 (k_6 + k_8) a_2^2\right] \frac{a_3^2}{a_1 a_2} \cos\delta, \tag{22}$$

where $\Sigma = \frac{1}{2}(\sigma_2 - \sigma_1)$.

Considering (20), equation (22) could be reduced to

$$\dot{\delta} - \Sigma = K e^{-s_3 T_2} + \frac{1}{2}\left(\frac{\dot{\xi}_1}{\xi_1} + \frac{\dot{\xi}_2}{\xi_2}\right) \cot\delta, \tag{23}$$

where $K = \omega^{-1} \left[\zeta_2 (k_6 + k_8) - \zeta_1 (k_5 + k_7)\right] c_3$.

2.3. The case $K = 0$

For this case at $\Sigma = 0$, the first integral of the set of equations (11)–(16) has been found in [5] in a form of the stream-function $G(\xi, \delta)$

$$G_1(\xi, \delta) = \xi^{1/2} (1 - \xi)^{1/2} \cos\delta = G_1^0(\xi_0, \delta_0). \tag{24}$$

where ξ_0 and δ_0 are the initial magnitudes of the relative amplitude and phase difference, respectively.

The stream-lines of the phase fluid in the phase plane $\xi - \delta$ are presented in figure (2). Magnitudes of G are indicated by digits near the curves which correspond to the stream-lines; the flow direction of the phase fluid elements are shown by arrows on the stream-lines.

Reference to figure (2) shows that the phase fluid flows within the circulation zones, which tend to be located around the perimeter of the rectangles bounded by the lines $\xi = 0$, $\xi = 1$, and $\delta = \pm(\pi/2) \pm 2\pi n$ $(n = 0, 1, 2, ...)$. As this takes place, the flow in each such rectangle becomes isolated. On all four rectangle sides $G = 0$ and inside it the value G preserves its sign. The function G attains its extreme magnitudes ± 0.5 at the points with the coordinates $\xi = 0.5$, $\delta = \pm\pi n$ $(n = 0, 1, 2, ...)$.

The transition of fluid elements from the points $\xi = 0$, $\delta = \pi/2 \pm 2\pi n$ to the points $\xi = 0$, $\delta = -\pi/2 \pm 2\pi n$ $(n = 0, 1, 2...)$ and from the points $\xi = 1$, $\delta = -\pi/2 \pm 2\pi n$ to the points $\xi = 1$, $\delta = \pi/2 \pm 2\pi n$ $(n = 0, 1, 2...)$ proceeds instantly, because according to the distribution of the phase velocity $\mathbf{v}$ along the section $\delta = 0$ its magnitude tends to infinity as $\xi \to 0$ and $\xi \to 1$, respectively, [5].

The transition of fluid elements from the points $\xi = 1$, $\delta = -\pi/2 \pm 2\pi n$ to the points $\xi = 1$, $\delta = \pi/2 \pm 2\pi n$ $(n = 0, 1, 2...)$ proceeds instantly as well, because according to the distribution of the phase velocity $\mathbf{v}$ along the section $\delta = 0$ its magnitude [5] tends to infinity as $\xi \to 1$.

Along the lines $\delta = \pm(\pi/2) \pm 2\pi n$ $(n = 0, 1, 2, ...)$ pure amplitude modulated aperiodic motions are realized, since with an increase in time t the value ξ increases from ξ_0 to 1 (along

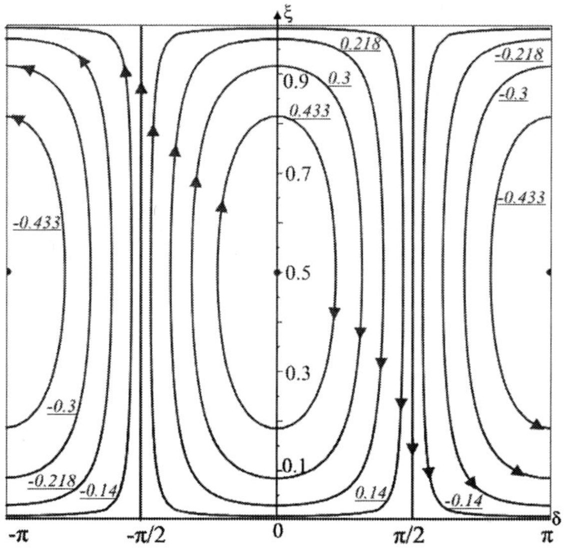

Figure 2. Phase portrait for the case of 1:1 internal resonance $\omega_1 = \omega_2$ at $K = 0$

the line $\delta = -\pi/2$) or decreases from ξ_0 to 0 (along the line $\delta = \pi/2$), and the solution could be written in the form of a soliton-like [5].

Stream-lines give a pictorial estimate of the connection of G with all types of the energy-exchange mechanism. Thus, in the case of undamped vibrations, i.e., when the damping coefficient is equal to zero and $s = 0$, the points with the coordinates $\xi_0 = 1/2$, $\delta_0 = \pm\pi n$ ($n = 0, 1, 2, \ldots$) correspond to the stationary regime, since $\dot{\xi} = 0$ and $\dot{\delta} = 0$. The stationary points $\xi_0 = 1/2$, $\delta_0 = \pm\pi n$ are centers, as with a small deviation from a center, a phase element begins to move around the stationary point along a closed trajectory. Closed stream-lines correspond to the periodic change of both amplitudes and phases.

3. Numerical investigations

Equations (11)–(16) have been solved numerically using the Runge-Kutta fourth-order algorithm. The results calculated for the cases of free ($f = 0$) and force driven ($f \neq 0$) vibrations are presented in figure 3 for different magnitudes of the fractional parameter γ: 0 (the case of undamped vibrations), 0.25 and 0.5 (fractionally damped vibrations).

From figure 3, wherein the T_2-dependence of the amplitudes a_1 (dashed lines) and a_2 (solid lines) is shown for the initial amplitudes $a_{i0} = 0.5$, it is seen that the presence of small viscosity results in damping of the energy exchange between two coupled modes of in-plane vibrations due to the one-to-one internal resonance $\omega_{1m_1n_1} \approx \omega_{2m_2n_2}$ and in the increase of the energy cycle with time. In so doing the damping character of the energy interchange is amplified with the increase in the fractional parameter. The points of tangency of the envelopes in figure 3a allows one to calculate the value characterizing the energy decay as $\ln(a_{i+1}/a_i) / (T_{i+1} - T_i)$.

Reference to figures 3b-d shows the influence of the external force on the amplitudes of vibrations. It is evident that even in the case under consideration when the frequency of the external vertical harmonic force is close to one of the frequencies of plate's vertical vibrations $\omega_F \approx \omega_3$ which is not involved in the internal resonance $\omega_1 \approx \omega_2$, the external force increases the amplitudes of plate's in-plane vibrations which are coupled due to the internal resonance between the frequencies of horizontal modes. With the increase of the magnitude of force amplitude, the

IOP Conf. Series: Materials Science and Engineering **489** (2019) 012043 doi:10.1088/1757-899X/489/1/012043

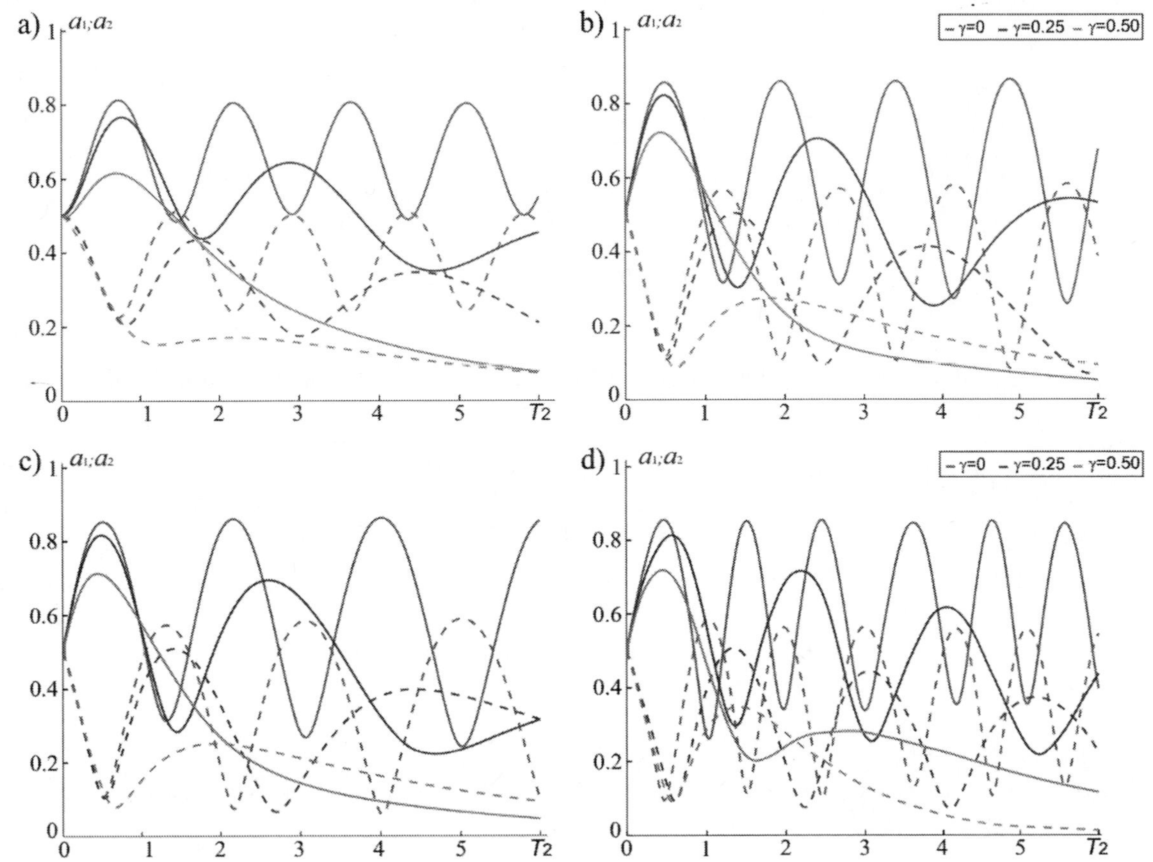

Figure 3. The T_2-dependence of the amplitudes a_1 (dashed lines) and a_2 (solid lines) at $\nu = 0.27$, $\beta_1 = 4.75$, $\beta_2 = 0.067$, $m_1 = m_2 = m_3 = 1$, $n_1 = n_3 = 3$, $n_2 = 5$, $\omega_{1m_1n_1} = 44.88$, $\omega_{2m_2n_2} = 45.11$, and $\omega_{3m_3n_3} \approx \Omega_F = 22.45$: (a) free vibrations, and forced vibrations (b) $f = 3$, (c) $f = 10^3$, (d) $f = 5 \cdot 10^3$

energy exchange process has been amplified, and the periods of energy interchange cycles have been decreased.

The case of stationary vibrations corresponding to the stationary center-like point with the coordinates $\xi_0 = 0.5$, $\delta_0 = 0$ in the phase portrait (figure 2) is shown in figure 4 for the initial amplitudes $a_{10} = 0.05$ and $a_{20} = 0.03$. In the case of free vibrations (figure 4a), there is no energy exchange between the modes. However, the presence of the small external vertical harmonic force initiates the process of energy exchange between the modes of in-plane vibrations what is seen in figure 4b.

4. Conclusions

In the present paper, nonlinear force driven vibrations of thin plates in a viscoelastic medium have been studied, when the motion of the plate is described by a set of three coupled nonlinear differential equations subjected to the condition of the one-to-one internal resonance accompanied by the external resonance, resulting in the interaction of three modes corresponding to the mutually orthogonal displacements. Nonlinear sets of resolving equations in terms of amplitudes and phase differences have been obtained. The influence of viscosity and external

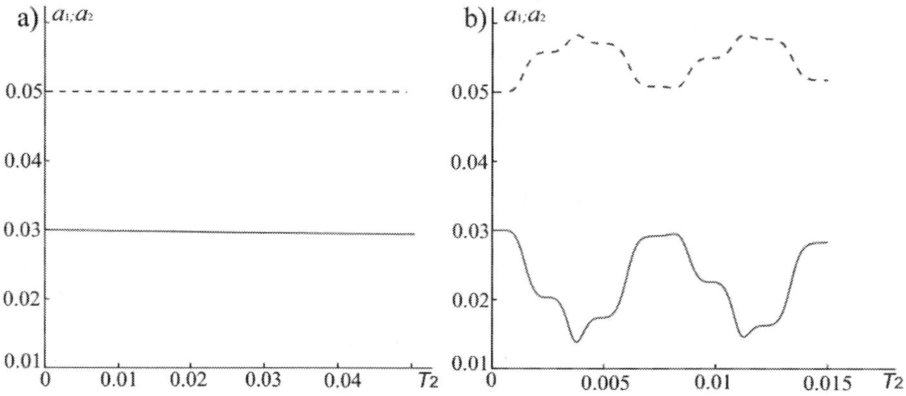

Figure 4. Steady-case vibrations (a) free vibrations, and (b) forced vibrations, at $\nu = 0.27$, $f = 2 \cdot 10^4$, $\beta_1 = 2.75$, and $\beta_2 = 0.0386$, $m_1 = m_2 = m_3 = 1$, $n_1 = n_3 = 3$, $n_2 = 5$, $\gamma = 0$, $\omega_{1m_1n_1} = 26.108$, $\omega_{2m_2n_2} = 26,166$, and $\omega_3 \approx \Omega_F = 7.60$: $a_2(T_2)$ - solid line, $a_1(T_2)$ - dashed line

vertical harmonic force on the energy exchange mechanism between the coupled modes has been analyzed.

Acknowledgments
This research has been supported by the Ministry of Education and Science of the Russian Federation (Project 9.5138.2017/8.9).

Reference
[1] Breslavsky I D, Amabili M and Legrand M. 2014 Physically and geometrically non-linear vibrations of thin rectangular plates *Int. J. Non-Linear Mech.* **58** 30–40
[2] Nayfeh A H 2000 Nonlinear Interaction: Analytical, Computational, and Experimental Methods (New York: Wiley)
[3] Ribeiro P and Petyt M 2000 Non-linear free vibration of isotropic plates with internal resonance *Int. J. Non-Linear Mech.* **35** (2) 263–78
[4] Rossikhin Yu A and Shitikova M V 2003 Free damped nonlinear vibrations of a viscoelastic plate under two-to-one internal resonance *Materials Science Forum* **440–441** 29–36
[5] Rossikhin Yu A, Shitikova M V and Ngenzi J Cl 2015 A new approach for studying nonlinear dynamic response of a thin plate with internal resonance in a fractional viscoelastic medium *Shock Vibr.* **2015** 795606
[6] Rossikhin Yu A and Shitikova M V 2012 On fallacies in the decision between the Caputo and Riemann–Liouville fractional derivatives for the analysis of the dynamic response of a nonlinear viscoelastic oscillator *Mech. Res. Com.* **45** 22–7
[7] Rossikhin Yu A, Shitikova M V and Shcheglova T A 2009 Forced vibrations of a nonlinear oscillator with weak fractional damping *J. Mech. Mat. Struct.* **4** (9) 1619–36
[8] Rossikhin Yu A and Shitikova M V 2010 Application of fractional calculus for dynamic problems of solid mechanics: Novel trends and recent results *Appl. Mech. Rev.* **63** (1) 010801

IOP Conf. Series: Materials Science and Engineering **489** (2019) 012044 doi:10.1088/1757-899X/489/1/012044

On self-synchronization of inertial vibration exciters in a chain-type oscillatory system

A E Shokhin, A N Nikiforov and K V Krestnikovskii

Mechanical Engineering Research Institute of the Russian Academy of Sciences, Moscow, Russia

shohinsn@mail.ru

Abstract. A model of a vibratory jaw crusher in the form of four-mass chain-type oscillating system, which oscillations are excited by self-synchronizing inertial vibration exciters installed on the working bodies is considered. Modes of the crusher motion without separating cheeks from processed medium are analyzed. Possible types of synchronization of the vibration exciters rotation and corresponding modes of the system's oscillations depending on the inertia and elasticity characteristics of the processed medium are identified.

1. Introduction

For operation of vibratory jaw crushers with two movable cheeks, which oscillations are excited by self-synchronizing inertial exciters, it is necessary to ensure their synchronous oscillations in opposite direction. Type of synchronization of cheeks oscillations and of vibration exciters rotation in such a system is ensured by the effect of self-synchronization, and is determined by the ratio between mechanical parameters of the system, characteristics of the processed medium, parameters of electric drive of the vibration exciters, and frequency of their rotation [1, 2]. It is especially difficult to account for the interaction of the cheeks with the material being processed when studying the dynamics of such crushers. When developing mathematical models, force interaction is usually taken into account in the form of equivalent viscous friction forces [2, 3]. It usually does not take into account the mass of the processed medium, its elastic characteristics, as well as the possible asymmetry of the forces of interaction of the cheeks with the processed medium, due to the heterogeneity of its properties distribution in the occupied volume.

In the work in order to identify possible influence of the processed medium properties on the synchronization of the debalances rotation and oscillations of the crusher's cheeks, the dynamics of a jaw crusher with two translationally movable cheeks excited by unbalance exciters installed on the cheeks are analyzed based on the model of a four-mass chain type oscillating system. It is assumed that the crusher operates in the non-impact mode, i.e. without breaking contact between the cheeks and processed medium. Such modes are characteristic of a number of technological processes which requires creation of a shock-free high-frequency force effect on the material being processed.

2. Mathematical model and simulation methods

The design scheme of the crusher model with translational oscillations of the cheeks is shown in figure 1. The model consists of a rigid frame modeled by a rigid body of mass m_1, and two movable cheeks modeled by rigid bodies of mass m_2 and m_4, respectively. The frame is attached to the fixed base by a linear spring with stiffness c_{01}. The left-hand-side and right-hand-side cheeks are attached to the frame by means of linear springs with stiffness c_{12} and c_{14}, respectively. The processed medium is modeled by a rigid body of mass m_3, which interacts with the cheeks by means of linear viscoelastic elements with stiffness and viscosity coefficients, respectively, c_{23} and b_{23} for the left-hand-side element, and c_{34} and b_{34} for the right-hand-side element. Each of these solids has one degree of freedom corresponding to the translational motion along the horizontal axis Ox.

Oscillation of the mechanical system are excited by two unbalance exciters with imbalanced mass m_{ej} and eccentricity r_j (j = 1,2 – the exciter number), rigidly fixed on the cheeks. Rotation of each of the debalances is provided by an asynchronous electric motor, with the moment of inertia J_j reduced to the debalance shaft. The torque L_j of each motor is described by its static characteristic [4]. Friction in the supports of the debalance shafts is taken into account in the form of the moments R_j of the dry frictional forces (not shown in figure 1).

Displacements of the bodies m_i (i = 1,2..4) are described by coordinates x_i of their mass centers, respectively, measured from their equilibrium position. Angular positions of the imbalances are described by the angles of rotation φ_j (j = 1,2), measured from the negative direction of the axis Ox.

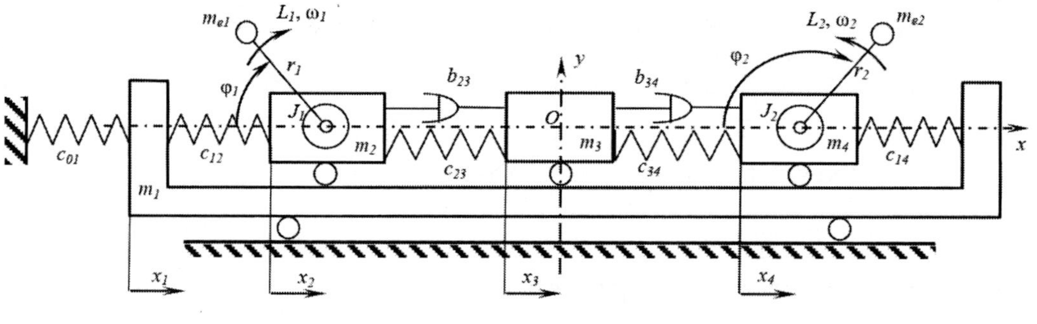

Figure 1. Design scheme.

The equations of system's motion in dimensionless form are:

$$\begin{cases} \mu_1\ddot{y}_1 + (1+\zeta_{14}+\zeta_{01})y_1 - y_2 - \zeta_{14}y_4 = 0 \\ \ddot{y}_2 + 2\eta\dot{y}_2 - 2\eta\dot{y}_3 - y_1 + (1+\zeta_{23})y_2 - \zeta_{23}y_3 = \ddot{\varphi}_1 sin\varphi_1 + \dot{\varphi}_1^2 cos\varphi_1 \\ \mu_3\ddot{y}_3 - 2\eta\dot{y}_2 + 2\eta(1+\beta_{34})\dot{y}_3 - 2\eta\beta_{34}\dot{y}_4 - \zeta_{23}y_2 + (\zeta_{23}+\zeta_{34})y_3 - \zeta_{34}y_4 = 0 \\ \mu_4\ddot{y}_4 - 2\eta\beta_{34}\dot{y}_3 + 2\eta\beta_{34}\dot{y}_4 - \zeta_{14}y_1 - \zeta_{34}y_3 + (\zeta_{14}+\zeta_{34})y_4 = \ddot{\varphi}_2 sin\varphi_2 + \dot{\varphi}_2^2 cos\varphi_2 \\ \ddot{\varphi}_1 = d_1((\tilde{L}_1 - \tilde{R}_1)L_1^* + \ddot{y}_2 sin\varphi_1) \\ \ddot{\varphi}_2 = d_2((\tilde{L}_2 - \tilde{R}_2)L_2^* + \ddot{y}_4 sin\varphi_2) \end{cases} \quad (1)$$

where $\mu_s = m_s/(m_2 + m_{e1})$ $(s = 1,3)$, $\mu_4 = (m_4 + m_{e2})/(m_2 + m_{e1})$, $\zeta_{01} = c_{01}/c_{12}$, $\zeta_{23} = c_{23}/c_{12}$, $\zeta_{34} = c_{34}/c_{12}$, $\zeta_{14} = c_{14}/c_{12}$, $2\eta = \frac{b_{23}}{m_2+m_{e1}}T_*$, $\beta_{34} = b_{34}/b_{23}$, $y_i = x_i/X_*$ - dimensionless coordinates, $X_* = \frac{m_{e1}r_1}{c_{12}T_*^2}$, $T_* = \sqrt{\frac{m_2+m_{e1}}{c_{12}}}$ – time scale factor, $d_j = \frac{m_{ej}r_jX_*}{(J_j+m_{ej}r_j^2)}$, $L_j^* = \frac{T_*^2}{m_{ej}r_jX_*}$, $\dot{}$ - denoted differentiation by dimensionless time $\tau = t/T_*$. The system of equations allows to analyze the modes of the crusher motion taking into account possible asymmetry of the forces of interaction of the processed medium with its cheeks.

Consider the problem of simple self-synchronization of the exciters in case of non-resonant oscillations of the crusher. The problem solution will be obtained using the method of separation of motions, which is described in detail in [5]. Following this method, we will look for solutions corresponding to almost uniform rotation of the exciter's debalances with the same angular velocity $\omega_1 = \omega_2 = \omega$ in the form $\varphi_j = \sigma_j(\omega\tau + \alpha_j + \psi_j(\tau, \omega\tau))$ and $x_k = x_k(\tau, \omega\tau)$ (k = 1.2..4), where α_j are slow, and ψ_j and

x_k are fast 2π period components by $\omega\tau$ periodic, the average of which is zero over a period, σ_j=1 when the debalance rotates in the positive direction of the angle φ measurement and σ_j=-1 when rotating in the opposite direction. In this case, the stationary values of the slow variables α_j correspond to the synchronous motions. The rotation frequency and the phase shift $\alpha = \alpha_1 - \alpha_2$ are determined from the system of transcendental equations:

$$\begin{cases} P_1 = d_1((\tilde{L}_1(\omega) - \tilde{R}_1(\omega))L_1^* - W_1) = 0 \\ P_2 = d_2((\tilde{L}_2(\omega) - \tilde{R}_2(\omega))L_2^* - W_2) = 0\text{'} \end{cases} \quad (2)$$

where W_1 and W_2 – vibrational moments, determined as follows:

$$W_1 = -\tfrac{1}{2}\omega^2(A_{22}sin\alpha + B_{22}cos\alpha + B_{21}),$$
$$W_2 = -\tfrac{1}{2}\omega^2(-A_{41}sin\alpha + B_{41}cos\alpha + B_{42}),$$

which coefficients A_{ij} and B_{ij} are determined from the solution of the problem of forced oscillations of system (1) under conditions of the vibration exciters uniform rotation with constant angular velocity ω, which is sought in the form $x_i = \sum_{j=1}^{2}(A_{ij}cos(\omega\tau + \alpha_j) + B_{ij}sin(\omega\tau + \alpha_j))$, $(i = 1,2..4;\ j = 1,2)$, where index j corresponds to the number of the vibration exciter.

System (2) allows the solution $\alpha = \pi/2$, at which the debalances rotate in opposite directions synchronously in antiphase, thus exciting synchronous antiphase oscillations of the cheeks, that corresponds to the normal operation of the crusher. The stability of the solutions found is determined from the analysis of the roots of the equation:

$$\left|\frac{d(P_1-P_2)}{d\alpha} - \chi\right|_{\alpha=\frac{\pi}{2}} = 0,$$

wherein the stable solutions correspond to negative values of χ.

Consider the effect of the processed medium parameters of elasticity ζ_{23} and ζ_{34} as well as damping β_{34} on the stability of the solution $\alpha = \pi / 2$. In this case, we will assume that the parameters of the motors, the debalances, inertia of the cheeks and elasticity of their suspension are identical, i.e. $(\tilde{L}_1 - \tilde{R}_1)L_1^* = (\tilde{L}_2 - \tilde{R}_2)L_2^*, J_1 = J_2, m_{e1} = m_{e2} = m_e, r_1 = r_2 = r,\ m_2 = m_4,\ \zeta_{14} = 1$. Then:

$$\chi = \frac{X_* m_e r\omega^2}{2}(A_{22} + A_{41}),$$

and the stability of the solution will be determined by the sign of the expression in parentheses $\chi_* = A_{22} + A_{41}$. The difference value of χ_* from zero characterizes the magnitude of the stability margin of the synchronous antiphase mode of the debalances rotation.

3. Simulation results

Fig. 2 shows the change of χ_* depending on the debalances rotational speed for different ratios of the processed medium stiffness $\lambda = \zeta_{34}/\zeta_{23}$. The following values of the system parameters were taken for the calculations: $\mu_1 = 5, \mu_3 = 0.1,\ \mu_4 = 1,\ \eta = 0.05,\ \zeta_{01} = 0.1,\ \zeta_{23} = 10,\ \beta_{34} = 1$. One can see that the stability condition of synchronous antiphase rotation is satisfied between the first and second eigenfrequencies of the system in the frequency range (p_1, p_*), as well as in the frequency range bounded by the second and third eigenfrequencies of the system. Note that the stability condition is also satisfied in the frequency range above the fourth eigenfrequency, which, however, is not usually used due to low energy efficiency, and therefore is not shown in the figure. Decrease in λ leads to a decrease in the third eigenfrequency and, as a consequence, to a narrowing the frequency range of stability of the debalances synchronous antiphase rotation.

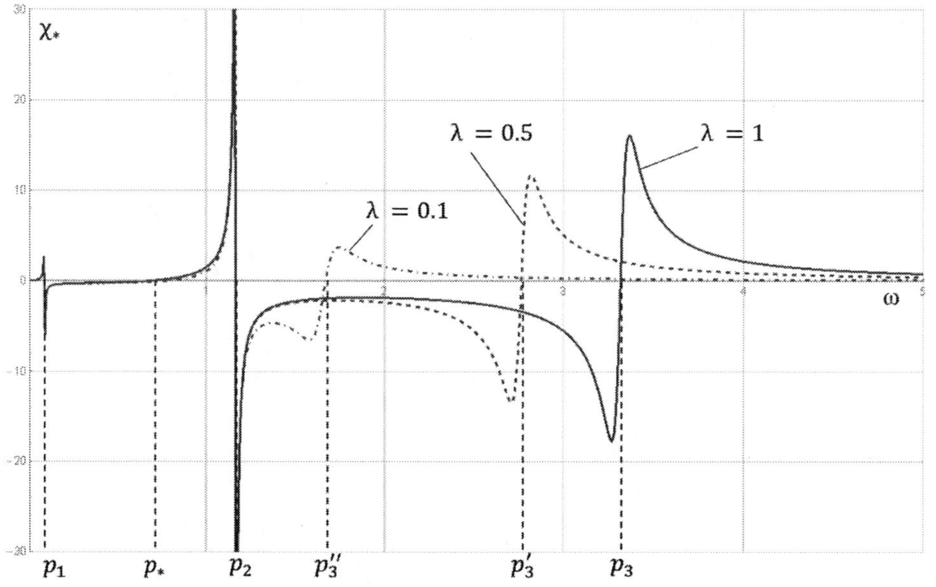

Figure 2. Change of χ_* depending on the frequency at different ratios of the processed medium stiffness.

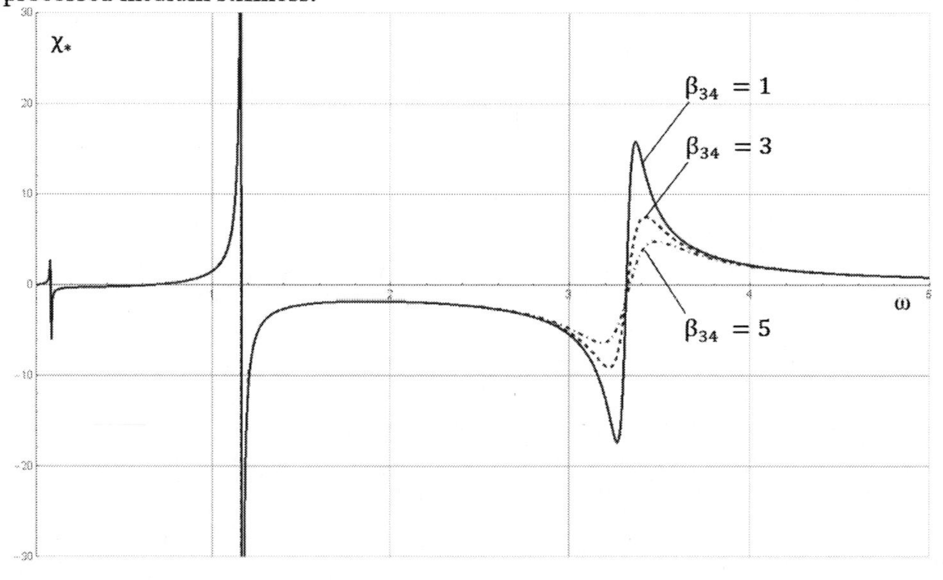

Figure 3. Change of χ_* depending on the frequency at different dissipation of the processed medium.

Figure 3 shows the change of χ_* depending on the debalances rotational speed for different values of the parameter β_{34}. The calculations were carried out at $\lambda_1 = 1$ and the same values of the other parameters as indicated above. One can see that a change in the dissipative properties of the processed medium leads to a change in the stability margin near the third natural frequency of the oscillations of the system.

Analysis of the effect of the processed medium mass on the stability regions of synchronous antiphase rotation of the debalances showed that an increase in mass by 10 times leads to a slight expansion of the stability region between the second and third natural frequencies due to a decrease in the second natural frequency.

IOP Conf. Series: Materials Science and Engineering **489** (2019) 012044 doi:10.1088/1757-899X/489/1/012044

4. Conclusion

The proposed model of a jaw crusher with unilateral movements of the cheeks allows to take into account possible asymmetry in the properties of the forces of interaction between the processed medium and the cheeks at impactless operational mode. It is found that a decrease in the contact stiffness of the treated medium with one of the crusher's cheeks leads to a narrowing the frequency range of the stability of the debalances synchronous antiphase rotations. Herewith a change in dissipative characteristics of the material leads to a change only in the stability margin near the third natural frequency of the system.

Acknowledgments

The reported study was funded by RFBR according to the research project № 18-08-01491_a.

References

[1] Lavendelis E E 1981 *Vibration in the Technique, Handbook, Vibration Processes and Machines* vol. 4 (Moscow: Mashinostroenie)
[2] Vaisberg L A, Zarogatsky L P and Turkin V Y 2004 *Vibrating crushers. Basics of calculation, design and technological application* (St.Peterburg: Izdatelstvo VSEGEI)
[3] Shishkin E V and Safronov A N 2016 *Obogashchenie Rud* **6** 39-43.
[4] Kolovskiy M Z 1989 *Dynamics of machines* (St.Peterburg: Mashinostroenie)
[5] Blekhman I I 2000 *Vibrational Mechanics: Nonlinear Dynamic Effects, General Approach, Applications (*Singapore: World Scientific)

IOP Conf. Series: Materials Science and Engineering **489** (2019) 012045 doi:10.1088/1757-899X/489/1/012045

Normal vibrations and the stability of plates with moving borders

T I Zhdan

Lomonosov Moscow State University, Moscow, Russia

tzhdan1992@mail.ru

Abstract. In this paper vibrations of a thin elastic moving plate are investigated. The plate is fixed between vibrating mill rolls. Cases of cylindrical and general bending are considered. Natural frequencies and natural modes of the plate are obtained numerically. It is shown that a resonance is possible in this system under certain boundary conditions.

1. Introduction

Plates are widely applied in modern construction and engineering. Investigating the problem of movable plates between mill rolls is related to work of the rolling mill. Sometimes the rolling strip may be deflected during its movement. This occurs when the system frequencies are equal to the natural frequencies of the elastic plate. Detailed study of this problem is needed in order to avoid structural failures. The problem of plate and membrane vibrations is investigated in a lot of papers, e.g. [1], [2]. Paper [1] is devoted to the study of vibrations and the resonance of plates under constant tension. Paper [2] is devoted to free vibrations of the rectangular membrane.

In the current paper, the problem of the vibrations of the plate with moving borders under constant tension is considered. The expression of plate deflection is obtained and a case of the resonance is shown.

2. Setting of the problem

Consider a rectangular plate moving along the axis x at a constant velocity V as shown in figure 1. The plate is fixed between rotating mill rolls. It is assumed to be under constant tension T_0 along the axis x.

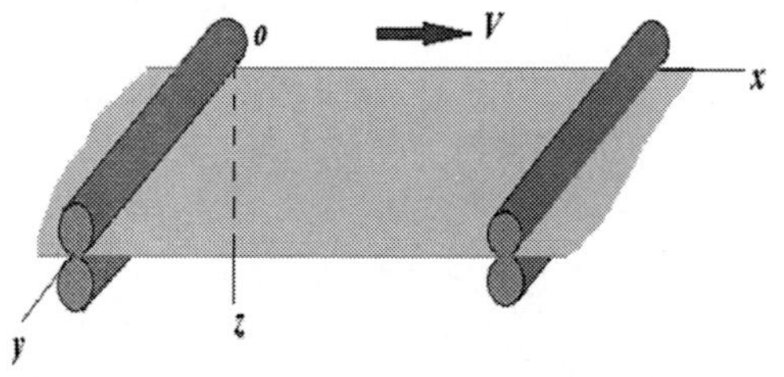

Figure 1. Moving plate fixed between rolls.

In the case of cylindrical bending the motion equation in the moving coordinate system ξ related to the plate can be written as follows:

$$\frac{\partial^4 w}{\partial \xi^4}-\frac{T_0}{D}\frac{\partial^2 w}{\partial \xi^2}=-\frac{m}{D}\frac{\partial^2 w}{\partial t'^2}, \tag{1}$$

here w is the transverse displacement, ξ is the coordinate that is co-directional with the x axis, D is the bending stiffness of the plate, m is the mass per unit length of plate, t' is the time.
Transform to a fixed coordinate system $(x;t)$ by using formulas:

$$x=\xi+Vt';\ t=t';\quad \frac{\partial}{\partial \xi}=\frac{\partial}{\partial x};\ \frac{\partial}{\partial t'}=V\frac{\partial}{\partial x}+\frac{\partial}{\partial t}.$$

The equation of motion in dimensionless variables takes the following form:

$$\frac{\partial^4 w^*}{\partial x^{*4}}-r^2\frac{\partial^2 w^*}{\partial x^{*2}}+p^2\left(\frac{\partial^2 w^*}{\partial x^{*2}}+2\frac{\partial^2 w^*}{\partial x^*\partial t^*}+\frac{\partial^2 w^*}{\partial t^{*2}}\right)=0, \tag{2}$$

where $p^2=\frac{12(1-\nu^2)mV^2L^2}{Eh^3}$, $r^2=\frac{12(1-\nu^2)T_0}{LEh^3}$, $x^*=\frac{x}{L}$, $t^*=\frac{Vt}{L}$, $w^*=\frac{w}{h}$, L is the distance between mill rolls, h is the plate thickness. For convenience we shall drop the sign *. We assume that the plate is fixed along two opposite edges parallel to the x axis:

$$w(x,t)\big|_{x=0}=0,\ \frac{\partial w}{\partial x}\bigg|_{x=0}=0,\ w(x,t)\big|_{x=1}=0,\ \frac{\partial w}{\partial x}\bigg|_{x=1}=0.$$

We shall look for a solution in the form of harmonic vibrations $w(x,t)=Y(x)e^{-i\omega t}$. The boundary conditions for $Y(x)$ take the following form: $Y(x)$: $Y(0)=Y(1)=0, Y'(0)=Y'(1)=0$. Substituting this expression into equation (2), we obtain a forth-order differential equation for $Y(x)$

$$\frac{d^4Y}{dx^4}-r^2\frac{d^2Y}{dx^2}+p^2\left(\frac{d^2Y}{dx^2}+2(-i\omega)\frac{dY}{dx}+(-i\omega)^2Y\right)=0.$$

Its characteristic equation is of the following form:

$$\lambda^4+\lambda^2(p^2-r^2)+2(-i\omega)p^2\lambda+(-i\omega)^2p^2=0.$$

Denote the characteristic roots by $\lambda_k(\omega)$, $k=\overline{1,\dots,4}$. Consider the case when these roots are distinct. Then general solution $Y(x)$ takes the form:

$$Y(x)=\sum_{k=1}^{4}C_k e^{\lambda_k(\omega)x}. \tag{3}$$

Boundary conditions along the edges perpendicular to x axis provide a system of homogeneous equations in C_k:

$$\sum_{k=1}^{4}C_k=0, \sum_{k=1}^{4}e^{\lambda_k}C_k=0,\ \sum_{k=1}^{4}\lambda_k C_k=0,\ \sum_{k=1}^{4}\lambda_k e^{\lambda_k}C_k=0. \tag{4}$$

For the non-trivial solution, the determinant of the coefficient matrix of this set of equation must vanish. On equating this determinant to zero, the frequency equation for the plate can be obtained.

$$F(\omega)=\begin{vmatrix}1 & 1 & 1 & 1\\ e^{\lambda_1} & e^{\lambda_2} & e^{\lambda_3} & e^{\lambda_4}\\ \lambda_1 & \lambda_2 & \lambda_3 & \lambda_4\\ \lambda_1 e^{\lambda_1} & \lambda_2 e^{\lambda_2} & \lambda_3 e^{\lambda_3} & \lambda_4 e^{\lambda_4}\end{vmatrix}=0. \tag{5}$$

For given values of the parameters p and r, the real roots of (5) can be found numerically. For example, if $E=2\cdot10^{11}\ Pa,\ \nu=0.33,\ h=0.05\ m,\ m=770\ \frac{kg}{m},\ V=2.35\ \frac{m}{s},\ L=10\ m,\ T_0=1000\ N$, then the corresponding dimensionless frequencies
ω_k =123.259, 339.926, 666.44805, 1101.70935, 1645.79038, 2298.6876730... From ω_k values, the values C_k can be obtained from (4). Then the $Y_k(x)$ can be find by relation (3).

2.1. Setting of the inhomogeneous problem
We shall assume plate is fixed along two opposite edges parallel to the x axis and the deflection is a time-dependent function. Physically this arises because of roll instability and certain other reasons. In this case boundary conditions can be expressed as follows:

$$w(x,t)\big|_{x=0}=w_0(t),\ \left.\frac{\partial w}{\partial x}\right|_{x=0}=0,\ w(x,t)\big|_{x=1}=w_1(t),\ \left.\frac{\partial w}{\partial x}\right|_{x=1}=0.$$

Perform a change of variable in order to obtain homogeneous boundary conditions:

$$w(x,t)=u(x,t)+\left(-2w_1(t)+2w_0(t)\right)x^3+\left(3w_1(t)-3w_0(t)\right)x^2+w_0(t).$$

This enables us to reduce (2) to the following form:

$$\frac{\partial^4 u}{\partial x^4}-r^2\frac{\partial^2 u}{\partial x^2}+p^2\left(\frac{\partial^2 u}{\partial x^2}+2\frac{\partial^2 u}{\partial x\partial t}+\frac{\partial^2 u}{\partial t^2}\right)=f(x,t). \quad (6)$$

Assuming that the plane is in a state of equilibrium and immovable at the initial moment, initial conditions can be expressed as follows:

$$u(x,t)\big|_{t=0}=0,\ \left.\frac{\partial u(x,t)}{\partial t}\right|_{t=0}=0.$$

The solution to equation (6) can be taken in the form:

$$u(x,t)=\sum_{k=1}^{\infty}Y_k(x)T_k(t). \quad (7)$$

Recall that $Y_k(x)$ is a natural mode.

Substituting (7) into equation (6), multiply this one by $Y_k(x)$ and integrate from zero to one. After some reduction, we obtain a differential equation for $T(t)$ subject to the initial conditions $T_k(0)=0,\ T'_k(0)=0$.

$$\left.\frac{Y_k^{\prime 2}(x)}{2}\right|_0^1\left(2i\omega p^2T_k(t)+2p^2T'_k(t)\right)+\left(\omega^2p^2T_k(t)+p^2\ T''_k(t)\right)\int_0^1 Y_k^2(x)dx=\int_0^1 Y_k(x)f(x,t)dx \quad (8)$$

Solving this equation, obtain functions $T_k(t)$ and form of solution $u(x,t)=\sum_{k=1}^{\infty}Y_k(x)T_k(t)$.

3. The resonance case
The resonance case may be of particular interest. It occurs when the characteristic root of (8) is equal to frequency in the right part of the equation under consideration. It may be observed that $-i\omega_k$ is the characteristic root of (8), consequently if one of the boundary functions $w_1(t)$ and $w_0(t)$ has the same frequency there is a resonance in the system. For definiteness we shall consider a case when $w_1(t)=\cos(\omega_1 t)$ and $w_0(t)=0$. Set the parameters of the problem are the same as done above. It

gives the expression of $T_k(t)$ and $W(x,t)=Y_k(x)T_k(t)$. The shape of the real part $W(x,t)$ for $x=0.5$ is shown in figure 2.

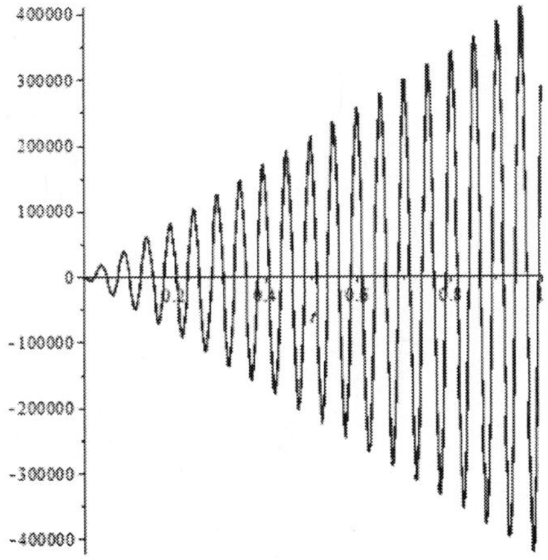

Figure 2. Shape of the real part of $W(x,t)$.

From the graphic, it is gathered that the deflection increases linearly with the time, hence the resonance occurs in the system.

4. General case of plate bending

The governing partial differential equation for the free vibration of a plate in a moving system can be written as

$$\frac{\partial^4 w}{\partial \xi^4}+2\frac{\partial^4 w}{\partial \xi^2 \partial \eta^2}+\frac{\partial^4 w}{\partial \eta^{24}}=-\frac{-\mu}{D}\frac{\partial^2 w}{\partial t'^2}. \tag{9}$$

As shown above, equation (9) in the fixed coordinate system takes the form

$$\frac{\partial^4 w}{\partial x^4}-r^2\frac{\partial^2 w}{\partial x^2}+2q^2\frac{\partial^4 w}{\partial x^2 \partial y^2}+q^4\frac{\partial^4 w}{\partial y^4}+p^2\left(\frac{\partial^2 w}{\partial x^2}+2\frac{\partial^2 w}{\partial x \partial t}+\frac{\partial^2 w}{\partial t^2}\right)=0.$$

Where $p^2=\frac{12\left(1-\nu^2\right)mV^2L^2}{Eh^3}, r^2=\frac{12\left(1-\nu^2\right)T_0}{LEh^3}, q^2=\frac{L^2}{a^2}$, a is the band width. Assume that the plate is fixed along two opposite edges parallel to the x axis and that it is simply supported along other two edges. In this case, the boundary conditions along x and y are

$$x=0,\ x=1: w=0,\ \frac{\partial w}{\partial x}=0, y=0,\ y=1: w=0,\ \frac{\partial^2 w}{\partial y^2}=0.$$

Then the solution of the motion equation can be taken in the form

$$w(x,y,t)=\sum_n \sin(\pi n y)Y(x)e^{-i\omega t}.$$

Solving this problem for certain parameters in the same way, the expression of the deflection can be find in terms of double series:

$$w(x,y,t)=\sum_n \sin(\pi n y)\left(\sum_m Y_{nm}(x)e^{-i\omega_m t}\right).$$

The first natural frequencies are of great importance in practice. Thus, this is crucial to find the nature of their dependence on parameters of the problem. For certain values of parameters the numerical results are obtained and presented here in the graphical form in Fig 3 to 4. In Fig. 3 the dimensional ω_1 is plotted against the dimensional L and in Fig. 4 the dimensional ω_1 is plotted against the dimensional V.

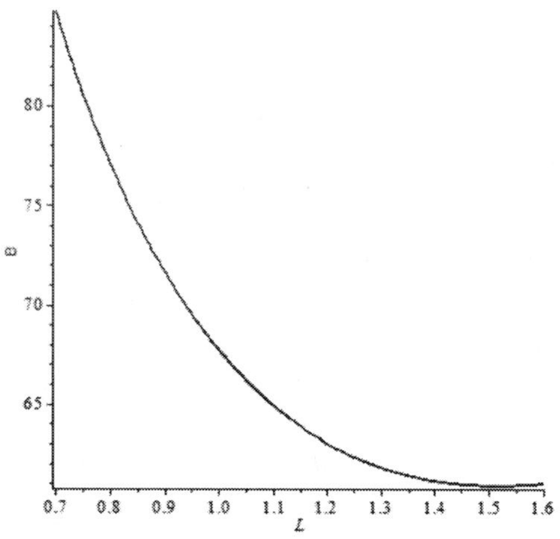

Figure 3. Shape of $\omega_1(L)$

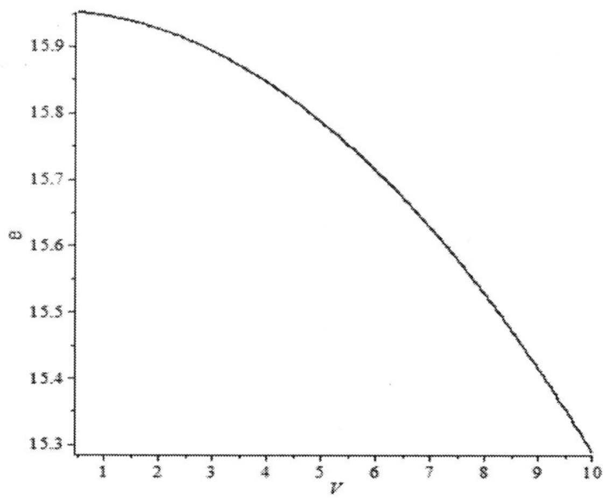

Figure 4. Shape of $\omega_1(V)$

5. Conclusions

The natural frequencies are also resonance frequencies. Thus their investigating can help to avoid resonance occurrence and structural failures. It was found in this paper that the first natural frequency remains static for growing V and T_0. The first value of natural frequency ω_1 decreases with growth of the distance between rolls and band width and increases with growth of the plate thickness.

References

[1] Chernyshev H A, Chernyshev A D 1998 Vibrations of the triangle elastic plate under a transverse load and uniform tension *J. News of the Eng. and Tech. Chuvash Rep. Ac* **11** 87-95

[2] Atamuratov A ZH 2013 Investigating methods of solving mathematical physics problems using an example of vibrations of a rectangular membrane *J. Young scientist* **10** 1-5

IOP Conf. Series: Materials Science and Engineering **489** (2019) 012046 doi:10.1088/1757-899X/489/1/012046

Influence of mechanical impact on bone tissue regeneration in porous scaffold

E E Iliyna[1], L B Maslov[12], N A Sabaneev[1]

[1] Department of Theoretical and Applied Mechanics, Ivanovo State Power Engineering University, 34 Rabfakovskaya, Ivanovo, 153003, Russia
[2] Institute of Advanced Manufacturing Technologies, Peter the Great St. Petersburg Polytechnic University, 29 Politekhnicheskaya, St. Petersburg, 195251, Russia

maslov@tipm.ispu.ru

Abstract. The three-dimensional finite element model of biomechanical system was built. The model includes the fixer, tibia bone's simulator, callus and porous scaffold. By using author's program code the bone tissue regeneration was calculated in the volume of porous scaffold with increasing external loading. The obtained results allow us to estimate the influence of mechanical loading on the process of regeneration. The presented data can be used in a clinical condition for treatment of the injured bone.

1. Introduction

Bone fractures and fracture healing are so common in our lives that it is easy to lose sight awesomeness of this biomechanical phenomenon. Unlike other adult biological tissues that regenerates with the formation of scar tissue, the bones regenerates with the formation of bone. A new bone is formed and continuously remodeled until its properties are restored and the initial part of the fracture is not distinguishable from the healthy part of the bone. It is known that bone substance is a highly structured biological composite material formed by a matrix of collagen and hydroxyapatite crystals and a complex hierarchical system of pores filled with vascular and interstitial fluid. Since the bulk of peculiar bone cells are located on the walls of the tubules and the surfaces of the lagoons forming the micropore system, it is assumed that perturbations introduced by an external mechanical load into the steady fluid motion in the bone transport system can ensure the transfer of control signals between the bone cells during its structural adjustment.

Structural adjustment of undeveloped soft substance into dense bone tissue occurs as a result of differentiation of bone cells, for example, during restoration of bone integrity after fracture and implantation of skeletal implants into bone substance, which leads to the start of reparative bone regeneration in the area of contact with the surface of a foreign object or between bone fragments.

There is a hypothesis that has clinical evidence that, after the initial differentiation of the soft matter of bone callus to successfully complete the process of osteosynthesis of fractures of long tubular bones, dosed loading on the limb is necessary [1]. The lack of mechanical stress leads to inhibition of the process of ossification of jelly-like callus and the formation of predominantly fibrous or cartilage tissue from it. It was experimentally shown that decisive for the initiation of growth processes is not the peak value of tissue deformation, but a certain value of the lower limit of deformation, above which the tissue neoplasm processes prevail over resorption [2]. It has also been established that other characteristics of mechanical stress, such as the distribution of stresses and fluxes of interstitial fluid in the pores of the bone substance, the duration and frequency of force are of certain importance [3, 4]. It has been suggested that dynamic, in particular periodic, loads are more effective for stimulating reparative bone regeneration than static [5, 6].

Mechanical action may cause destruction of fracture, or change its biological path. However, the mechanisms that are caused by mechanical excitation, passing into a biological respond, remain partially unknown. A deeper understanding of these processes will enable the development of more accurate and rational fracture treatment strategies and will open up an unlimited area of research in other disciplines of regenerative medicine.

2. Materials and methods

The article describes the three-dimensional model of the tibia, the formulation of the problem of tissue regeneration and the tests conditions. The results of the study of bone tissue regeneration processes in the volume of scaffold with different porosity under the dynamic influence of increasing frequency are also presented.

The model includes a tibial bone, a fixation frame, a porous scaffold and a callus (Figure 1). For construction, a spatial finite element with parabolic interpolation of geometry and functions was used [5]. By studying the convergence of the test numerical solution, an adequate grid was selected, consisting of 3,600 finite elements and 16,276 nodes.

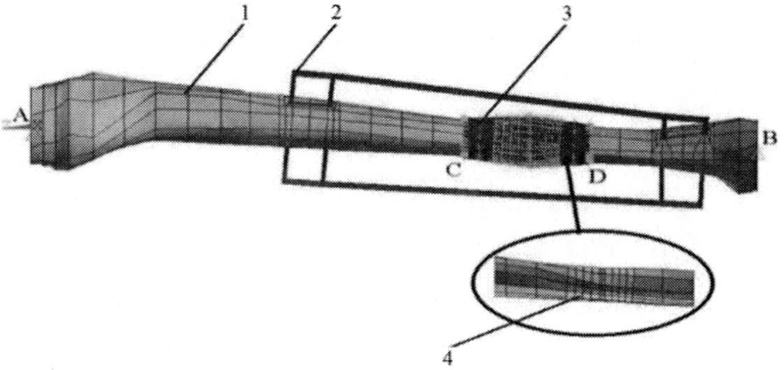

Figure 1. Model: 1) a tibial bone; 2) a fixation frame; 3) a callus; 4) a porous scaffold

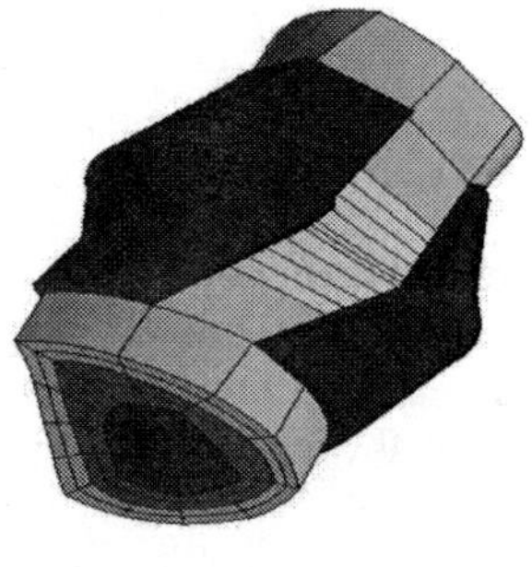

Figure 2. The area is considered: "scaffold - callus"

The loading conditions of the model can also be seen in Figure 1. In section A the bone is rigidly fixed, at the location CD the pore fluid pressure of the poroelastic medium acts on the callus. Static and harmonic loads act perpendicular to section B. The developed mathematical model and program allow us to investigate the influence of the type and parameters of the mechanical factor on the structural adjustment of bone tissue in the volume of a scaffold filled with biological tissue (Figure 2) [7, 8]. As the load applied in section B of the bone, consider the sum of a slowly varying compressive force, constant at each time step, and a fast component varying harmonically with a relatively high frequency:

$$F(t,\tau,\omega) = F_{sta}(t) + F_{dyn}(t)e^{i\omega\tau}, \quad F_{dyn} = kF_{sta} \tag{1}$$

F – the total force depending on the slow t and fast τ times and frequency ω; F_{sta} – a static force that depends only on the slow time t (a graph of the static force versus time is shown in Figure 3); F_{dyn} – the amplitude of the harmonic force, also dependent on the slow time t.

IOP Conf. Series: Materials Science and Engineering **489** (2019) 012046 doi:10.1088/1757-899X/489/1/012046

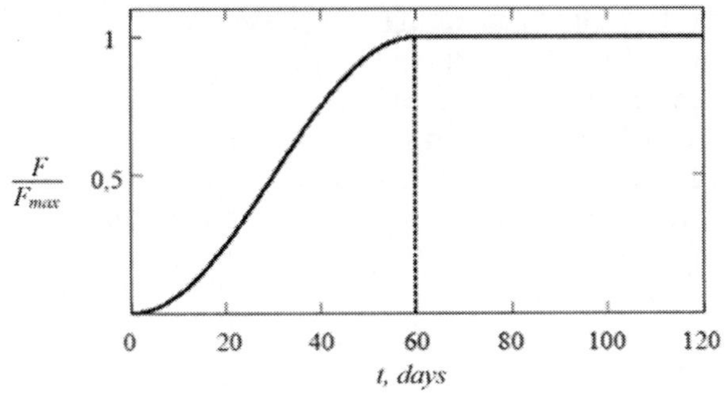

Figure 3. Graph of the normalized harmonic force change. T_1 = 60 days - time for which the force F_{sta} reaches a maximum

The task of penetration of mesenchymal stem cells in the zone of a scaffold with a porosity of 50 or 90% was solved for the period of 120 days. The diffusion coefficient is J = 0.65 mm^2/day. The initial conditions are the concentration of active cells in the volume of callus C_{vc} = 80% and in the volume of scaffold C_{vs} = 30%. The boundary condition is the concentration of cells at the border of the callus C_b = 100%. To study the effect of the harmonic load a solution is given for values of the frequency f equal to 10, 100, 200 Hz and the harmonic load factor $k = 0.01$.

The results of the tissue structural adjustment calculation with a time step equal to one day are displayed in the form of graphs that show the dependence of the physical-mechanical properties of the bone on time at specified points in the area under study (Figure 4) with variation of key parameters.

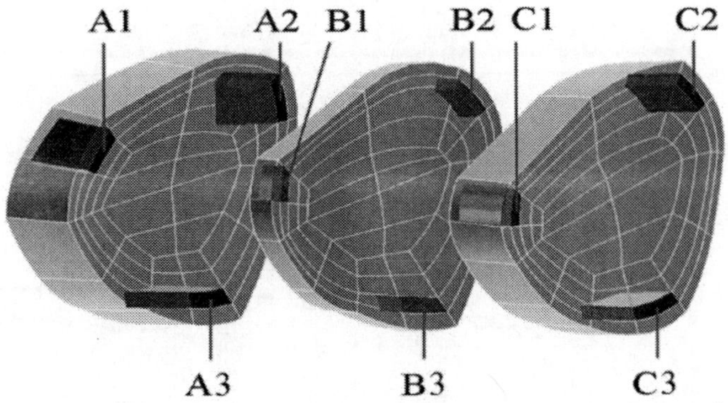

Figure 4. Points in the scaffold volume to output the results

The density of the tissue is considered as the main variable characterizing the process of regeneration. Successful is the process of regeneration with the yield of density curves at values close to the maximum - 1416 kg·m^{-3} [6].

3. Results and discussion

Graphs of changes in tissue density (for scaffold with a porosity of 50% - figure 5-7, with a porosity of 90% - Figure 8-10) show that the frequency of the applied harmonic load influences the convergence of the results and, consequently, the formation of mature bone tissue.

Analyzing the results, it can be seen that the recovery process of tissue properties depends on scaffold porosity: for a more rigid scaffold (with less porosity), a successful recovery requires a higher frequency load than for a less hard scaffold. It is not shown on the graphs, but it follows from the calculation, that a purely static load, which is not capable of generating forced flows of interstitial fluid in the system of pore channels, does not lead to the formation of bone tissue.

Analyzing the results for a scaffold with a 90% porosity it can be seen that at low frequencies the recovery process is unstable and leads to the formation of tissues of the fibro-cartilage type with low mechanical characteristics. An increase in the harmonic load plays a positive role in the formation of bone tissue in the scaffold volume.

After all experiments we can conclude that the task of regenerating active stem cells brings the model of the "bone – scaffold" system considered in computational experiments to the actual processes of bone restoration during structural adjustment by porous scaffolds.

According to the obtained graphs, it is clear that the density of the formed bone tissue in the scaffold increases throughout the entire volume of the area under consideration, which confirms the beneficial effect of the harmonic load on the tissue regeneration processes in the volume of the porous scaffold.

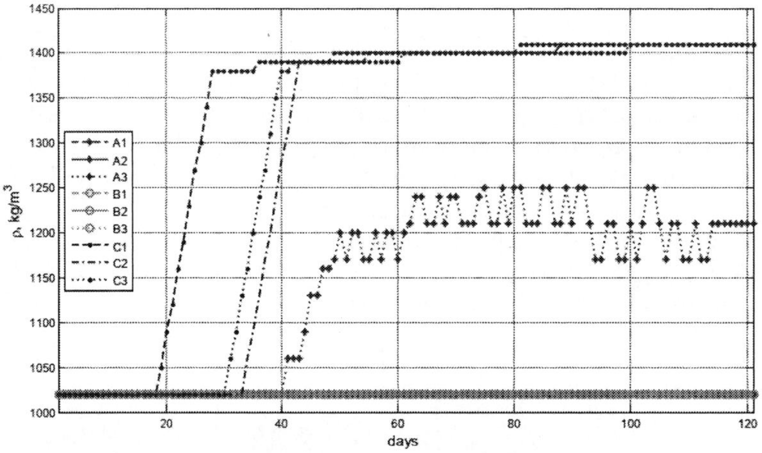

Figure 5. Bone density at selected points of a scaffold with a 50% porosity at a loading frequency 10 Hz

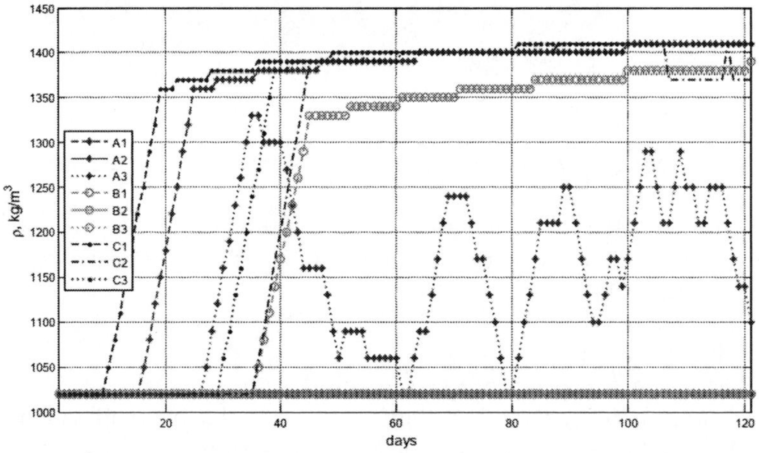

Figure 6. Bone density at selected points of a scaffold with a 50% porosity at a loading frequency 100 Hz

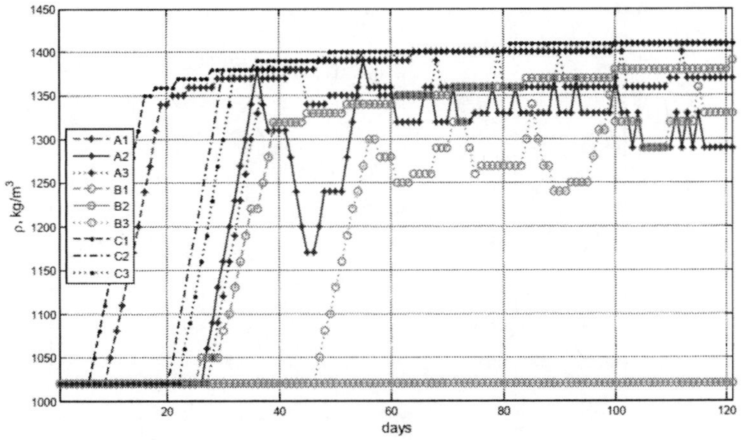

Figure 7. Bone density at selected points of a scaffold with a 50% porosity at a loading frequency 200 Hz

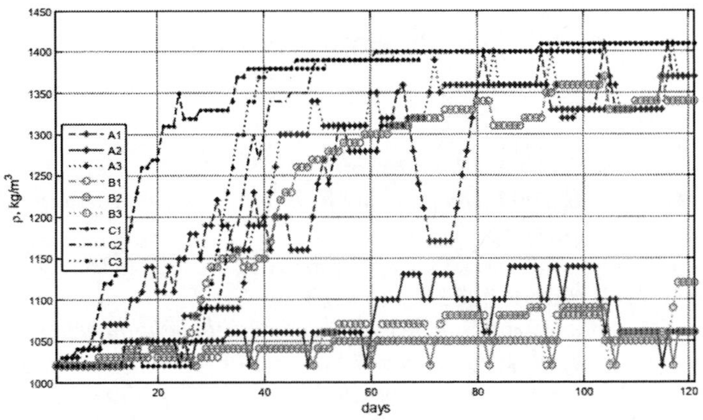

Figure 8. Bone density at selected points of a scaffold with a 90% porosity at a loading frequency 10 Hz

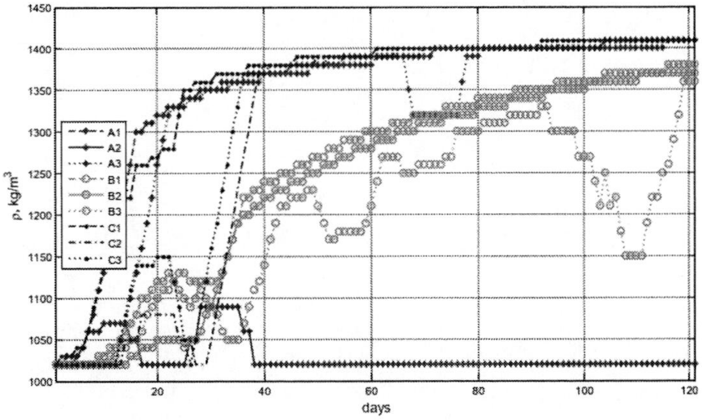

Figure 9. Bone density at selected points of a scaffold with a 90% porosity at a loading frequency 100 Hz

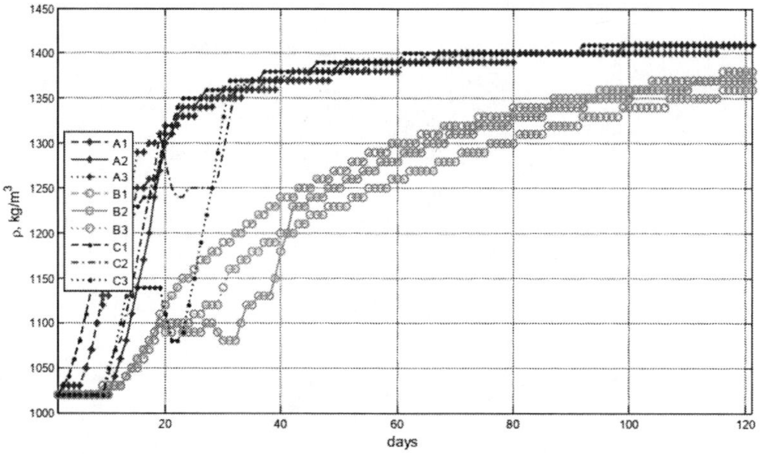

Figure 10. Bone density at selected points of a scaffold with a 90% porosity at a loading frequency 200 Hz

Acknowledgments
The work is supported by the Russian Ministry of Education and Science grant № 2.7557.2017/Basic Part

References

[1] Kenwright J Gardner T 1998 *Clin. Orthop. Rel. Res.* **355** S 179
[2] Burr D B Martin R B 1992 *Mechanisms of bone adaptation to the mechanical environment* // Triangle. vol 31 no 2/3 pp 59–76
[3] Volozhin A I Oganov V S 1978 *Osteoporosis* M.: Science p 336
[4] Van der Meulen, Huiskes R 2002 *J. Biomech.* **35** 401
[5] Maslov L B Etheve J – B and Sabaneev N A 2016 *Vibroengineering PROCEDIA* **8** pp 334-9
[6] Maslov L B 2015 *J. Appl. Math. Mech.* **79** pp 195-206
[7] Maslov L B 2017 *Mechanics of Composite Materials* **53** pp 399–414
[8] Program of the finite-element calculation of the bone tissue physical-mechanical characteristics during reconstruction by porous implants: Certificate № 2017612467 Russian Federation / Maslov L B; reg. in the Register of computer programs 22.02.17.

IOP Conf. Series: Materials Science and Engineering **489** (2019) 012047 doi:10.1088/1757-899X/489/1/012047

Distributed control system for mobile medical complexes.

Daniil V Khmyachin[1], Ekaterina A Khmyachina[1], Maxim V Archipov[1], Gleb A Shumeiko[1], Uriu P Panov[2], Sergei A Ippolitov[1]

[1]Moscow Polytechnic University. Bolshaya Semyonovskaya str., 38, Moscow, Russia
[2]Moscow State Geological Prospecting University named after Sergo Ordzhonikidze Mikluho-Maklaya str., 23, Moscow, Russia

m.bagrow@gmail.com

Abstract. Currently, high-tech service robotics is increasingly developing. Service robots perform various tasks: patient care, condition monitoring and alert medical personnel and patient registration. In connection with the urgency of the task, this article will consider a distributed control system for mobile complexes, which task is to control, manage, and distribute tasks for complexes for the operation of re-equipping hospital wards and transporting mobile equipment in a medical institution. For more accurate route building, a local positioning system can be used. Now there are no ready-made algorithms for solving such problems. There are algorithms for managing single complexes, complexes in a closed space, and the problems of transition from the initial state to the final state have not been considered either. In this connection, a number of problems arise, which are considered in this article: transfer of medical complexes from the initial state to the final state, determination of initial parameters and conditions, determination of final parameters and conditions. During the task creation with a distributed control system, the following factors should be taken into account: the availability of the resource required for delivery on the mobile complex, the priority of the task being performed. The practical significance of the work will be determined by the result of the development of a mathematical model that will take into account all current problems and factors. As a result, a theory was determined by which the theoretical aspects were analysed and the closest toolkit of interactions within the control system in a distributed environment was selected.

1. Disclosure of the concept of a mobile service robot

Now the role of medical service robots in the world is increasing, but this is mainly manifested in the development of service information systems. This article describes an automated medical system that is responsible for transporting service medical equipment to wards, based on RFID positioning system-using triangulation. The robotic system described in the article is intended to facilitate the work of medical personnel, since there is often a shortage of skilled workers in medical institutions. Using the local positioning system, the robotic complex, coupled with mobile equipment and mobile cabinets for medicines, will be able to move them around the medical institution in an automatic mode, as well as to re-equip the operating rooms and other chambers.

Accuracy of positioning at the optimum (from 4x to 1 room) number of access points is 30 cm, so the sensor system is present in the complex for accurate positioning of the object in the room. Also, this system is responsible for the exclusion of cases of the human factor.

The algorithm for moving the robotic complex work as follows: RFID tag is installed on the complex that receives data from the application server, requesting data about its own location, a signal about movement to the required area is sent from the control panel or in the automatic mode. In the Arduino IDE programming environment (Arduino C++ code), an algorithm was implemented that allowed the exchange of data between the server and the robotic complex.

Piece Arduino IDE C++ library for control RTLS System

```
void GetCoord()
{
 HttpClient client;
 char c;
 //Get coordinates
 client.get("http://192.168.99.5:9002/location/B0B448D4E282/get");
 String coordString = "";
 while(coordString == ""){
  coordString = "";
  while (client.available()) {
   c = client.read();
   coordString += c;  } }
```

Testing of this program was carried out on the model of the robot. When assembling of which were used boards ShieldBot and Seeduino CLOUD, Optical distance sensors, Gyroscope. As a result of testing, the following practical results were obtained:

- the possibility of determining the robot coordinates in automatic mode and using them for movement;
- test routes were compiled, on which the robot performed the movement in real time. Real-time route change by the operator's command.

2. The principle of operation of the local positioning system RTLS.

The positioning system consists of several components:

Access points that contain RFID and WIFI modules, as well as one or several points, are "hosts" connecting to other points in an independent mode, exchanging the received information, and transmitting it to the server of the local positioning control system as well as the API application server [6].

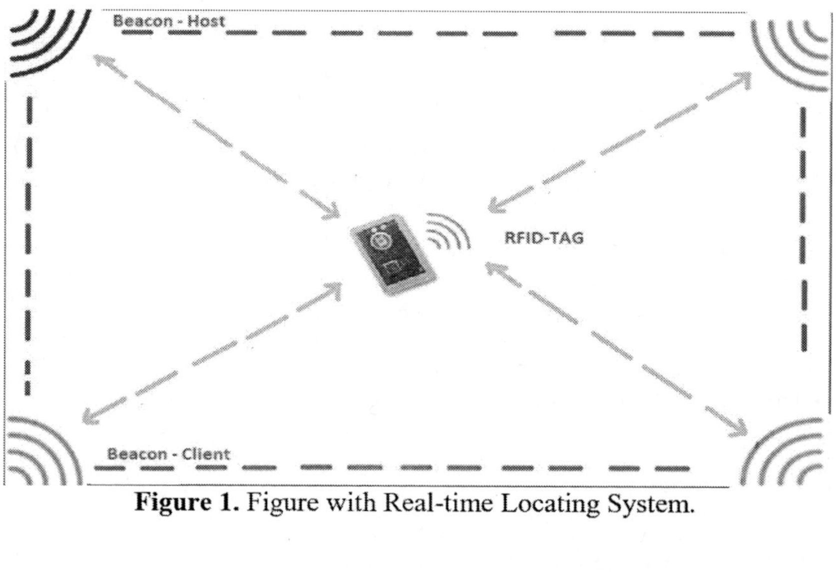

Figure 1. Figure with Real-time Locating System.

The location server described above performs the reception and transmission of signals from access points, is responsible for visualizing the information necessary for the operation of security systems and other needs, the server is based on the RTLS technology (Real-time Locating Systems - Figure 1) contact robot and local positioning system. The basis for this article is the RealTrac local positioning system, but any other similar system based on the principles described above can be used. Servers use the open operating system Linux Debian 8.

3. Mathematical algorithm for mobile service robots movement.

Control block diagram of a multiple system of *A*1,. . . , *Ak* as a whole. OU - control object (equipment position). (see figure 2)

CU is the set of control devices of all equipment, *U* is the control generated by equipment based on parameters *g* and measurement results; *E* reflects an error in the formation of a system corresponding to the target geometric structure of the system.

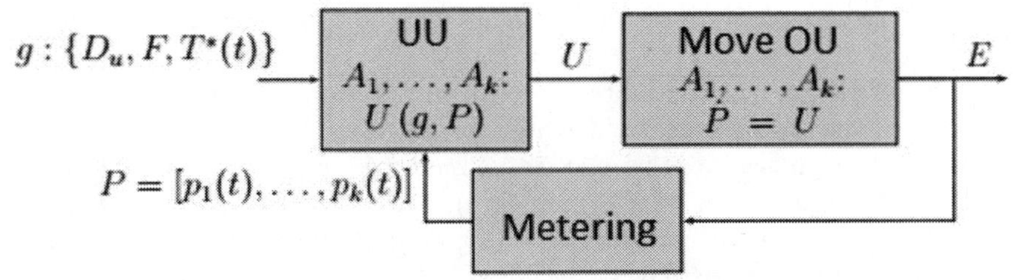

Figure 2. Block diagram multi-system control.

The practical significance of the work is determined by the result of the development of a mathematical model that will take into account all current problems and factors.

4. Distributed control system for mobile medical complexes.

To improve the accuracy of work and obtain the location data of medical complexes, a local positioning system (LPS) can be used for confined spaces.

The RTLS consists of several components: Access points that contain RFID and WIFI modules, as well as one or several points are “leading” connecting to other points in an independent mode and exchanging the received information and transmitting it to the server of the local positioning control system, as well as to the API application server (see figure 3), as well as radio tags that are part of a single information and management positioning system and are designed to determine the exact location of the object in the medical area facilities.

The application server provides the possibility of contacting the management system, the medical service complex and the RTLS. This article is based on the RealTrac RTLS, but any other similar system based on the same principles can be used.

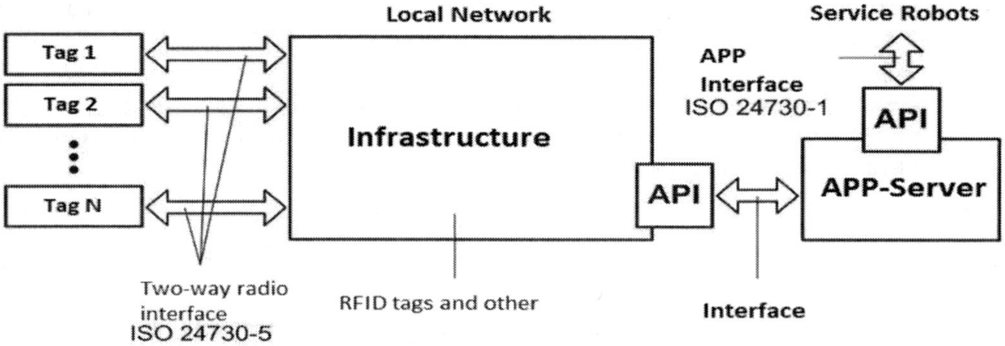

Figure 3. Schematic diagram of the operation of the RTLS and the Distributed Control System through the API.

After authorization, it is possible to send service complexes in a medical institution by points [5] But only if the map is loaded and route graphs are added. (see figure 4) As for controlling the control points, the web application is used with the subsequent query of the coordinates of the points when pressing the buttons in the program and then using the coordinates data for sending to the mobile equipment.

The route of the graph can be edited directly in the application window, just pull the vertices of the graph. All settings of the map, routes and walls are performed in a web form. Interface elements appear if the mouse cursor placed on the map of the medical institution. To send equipment you need to select it in the list or click on it on the map (pin). Press the "Dot 1" button to confirm. Select a destination and click "Dot 2". To confirm, click the route button. The route graph selected by the system (the shortest path) will be displayed.

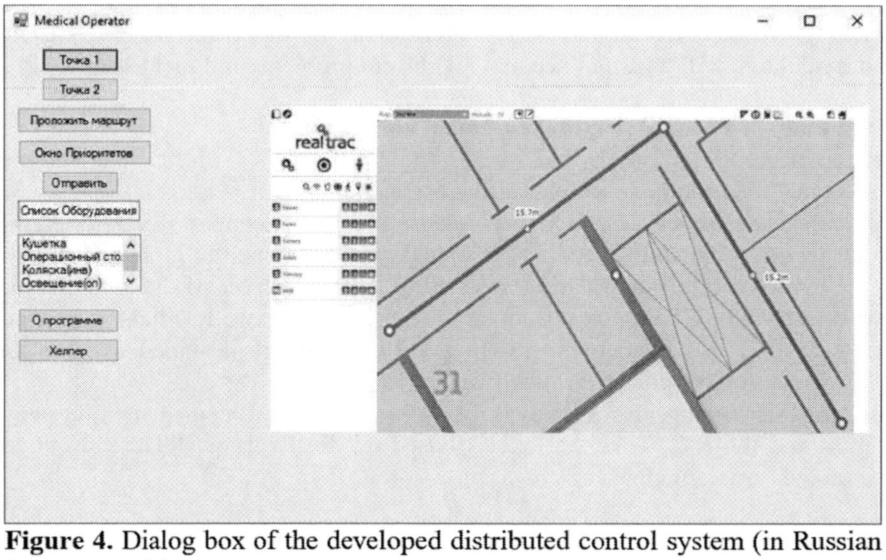

Figure 4. Dialog box of the developed distributed control system (in Russian language only)

After pressing the "Send" button, the array of coordinates of the selected route will be transferred to the equipment via a wireless connection, and the service complex will start moving to the specified point. Also the equipment priority will be changed in the priority window. In the graph editing window it is possible to edit the track route of the mobile complex. All possible options for interaction with the program are indicated in the dialog box (see figure 5).

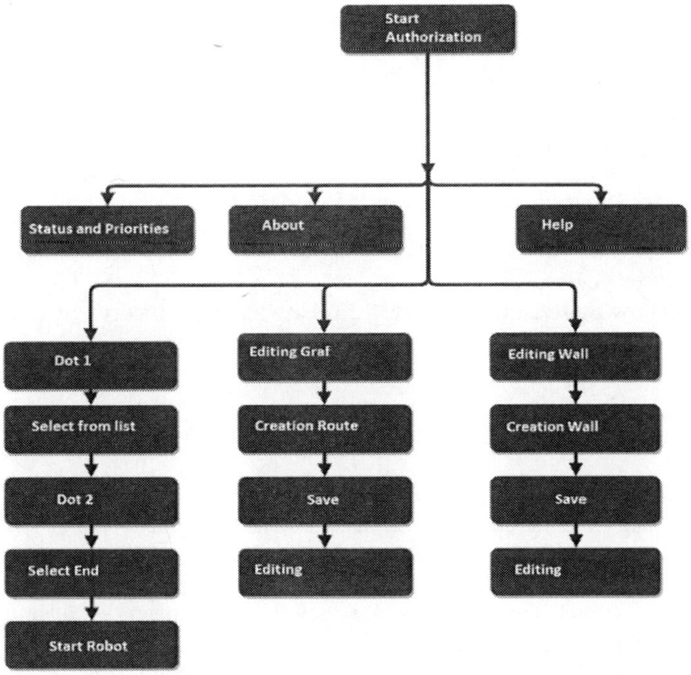

Figure 5. Graph of the distributed control system dialogue

As a result, a theory was determined by which the theoretical aspects were analyzed and the closest toolkit of interactions within the control system in a distributed environment was selected.

References

[1] Lin C-Y, Hung M-T, Huang W-H 2012 A Location-Based Personal Task Management Application for Indoor and Outdoor Environments. Network-Based Information Systems (NBiS), 2012 15th International Conference on DOI: 10.1109/NBiS.2012.108

[2] Adler S, Schmitt S, Wolter K, Kyas M 2015 A survey of experimental evaluation in indoor localization research.Indoor Positioning and Indoor Navigation (IPIN), 2015 International Conference on, DOI: 10.1109/IPIN.2015.7346749

[3] Deak G, Curran K, Condell J 2012 A survey of active and passive indoor localisation systems. Computer Communications **35** 1939–54

[4] Rüßmann M, Lorenz M, Gerbert P, Waldner M, Justus J, Engel P and Harnisch M 2015 Industry 4.0: The Future of Productivity and Growth in Manufacturing Industries https://www.bcgperspectives.com/content/articles/engineered_products_project_business_industry_40_future_productivity_growth_manufacturing_industries (acc. 10 November 2015)

[5] Liu C, Xiong H, Papadimitriou S, Ge Y and Xiao K 2017 A Proactive Workflow Model for Healthcare Operation and Management. IEEE Transactions on Knowledge and Data Engineering **29**

[6] Li H, Chan G, KwokWaiWong J and Skitmore M 2016 *Automation in Construction* **63** 37–47

[7] Ma X, Liu T 2011 *Proceeding of the IEEE International Conference on Automation and Logistics (ICAL)* Chongqing China, pp 64-9

[8] Posdorfer W, Maalej W 2016 *Procedia Computer Science* **83** pp 42-9, doi.org/10.1016/j.procs.2016.04.097

IOP Conf. Series: Materials Science and Engineering **489** (2019) 012048 doi:10.1088/1757-899X/489/1/012048

Genetic alphabets, unitary matrixes and quantum-algorithmic genetics

S Petoukhov [1,2], E Petukhova[1], V Svirin [1,2]

[1]Mechanical Engineering Research Institute of RAS, 4, M. Kharitonyevskiy Lane, Moscow, 101990, Russia
[2]Moscow State Conservatory, 13/6 Bolshaya Nikitskaya Street, Moscow, 125009, Russia

spetoukhov@gmail.com

Abstract. Unitary operators are the basis of all calculations in quantum computers and play an important role in quantum mechanics. The presented results show that matrix forms of representation of structured DNA and RNA alphabets can be considered as a set of sparse unitary matrices. These results are useful for the development of models of quantum-algorithmic genetics, from the point of view of which living organisms are quantum-information entities.

1. Introduction
Information molecules of DNA and RNA follow the principles of quantum mechanics. It can be assumed that genetic informatics uses quantum computing algorithms based on unitary (or orthogonal, in the case of real numbers) matrices. Any unitary matrix can be used as a logic gate in quantum computers [1]. The article presents some authors' results for the development of quantum-algorithmic genetics models [2-8].

2. The genetic code and binary oppositions
Why DNA alphabet is built by nature on the basis of only four letters: adenine A, guanine G, cytosine C and thymine T? Why is this alphabet based on these very simple molecules? Modern science does not know the answer to these questions. But it knows that this set of 4 "letters" has symmetric properties because it is endowed with three pairs of binary-opposition signs or indicators:

1) two letters are purines (A and G) and the other two are pyrimidines (C and T). In terms of these features the following can be taken: C = T = 0, A = G = 1;
2) two complementary letters (A and T) are connected by 2 hydrogen bonds, and other complementary letters (C and G) are connected by 3 hydrogen bonds. From the point of view of this feature, the following can be taken: A = T = 0, C = G = 1;
3) two letters are keto-molecules (G and T) and the other two are amino-molecules (A and C).

From the point of view of this feature, the following can be taken A = C = 0, G = T = 1. With regard to these three types of binary-oppositional traits in DNA letters any DNA text can be presented as a combination of three parallel messages in three binary languages. At the same time, these messages are mutually connected on the basis of the logical operation of modulo-2 addition used in quantum and conventional computers: the sum of modulo-2 addition of any two of these three binary representations of the DNA text always gives its third binary representation [4].

It is easy to check that each of the letters A, C, G, T is uniquely determined by any two of these three indicators. This allows presenting DNA alphabets of 4 nucleotides, 16 doublets, 64 triplets, etc. in a convenient form of square tables (Fig. 1-3), in which lines and columns are numbered with binary numbers according to the following principle. The components of each column are binary numbered in accordance with the binary opposition signs "purine or pyrimidine" (for example, the CAG triplet and all other triplets in the same column are a bunch of "pyrimidine-purine-purine", and therefore this column is numbered 011). The components of each line are numbered with binary numbers in accordance with the signs of "keto- or amino-" (for example, the same CAG triplet and all other triplets in its line are a bunch of "amino-amino-keto", and therefore this line is numbered 001). In such tables (Fig. 1-3) each of the members of the alphabets of 4 nucleotides, 16 doublets, 64 triplets, ... takes its individual place in a strict order

	000	001	010	011	100	101	110	111
000	CCC	CCA	CAC	CAA	ACC	ACA	AAC	AAA
001	CCT	CCG	CAT	CAG	ACT	ACG	AAT	AAG
010	CTC	CTA	CGC	CGA	ATC	ATA	AGC	AGA
011	CTT	CTG	CGT	CGG	ATT	ATG	AGT	AGG
100	TCC	TCA	TAC	TAA	GCC	GCA	GAC	GAA
101	TCT	TCG	TAT	TAG	GCT	GCG	GAT	GAG
110	TTC	TTA	TGC	TGA	GTC	GTA	GGC	GGA
111	TTT	TTG	TGT	TGG	GTT	GTG	GGT	GGG

Figure 1. Table of DNA alphabet of triplets with the arrangement of all members in accordance with the binary-oppositional signs of the bases A, C, G, T..

	0	1
0	C	A
1	T	G

Figure 2. Table of DNA alphabet of nucleotides with the arrangement of all members in accordance with the binary-oppositional signs of the bases A, C, G, T

	00	01	10	11
00	CC	CA	AC	AA
01	CT	CG	AT	AG
10	TC	TA	GC	GA
11	TT	TG	GT	GG

Figure 3. . Table of DNA alphabet of doublets with the arrangement of all members in accordance with the binary-oppositional signs of the bases A, C, G, T

It is essential that these 3 separate genetic tables form a single tensor family of matrices based on the tensor (or Kronecker) product [2, 4]. Indeed, the second tensor degree of (2x2)-nucleotide matrix [C, A; T, G] automatically gives the (4x4)-matrix of 16 doublets, and its third tensor degree gives the (8x8)-matrix of 64 triplets with the same arrangement of components as in the tables in Fig. 1-3, which appear now as matrices. Thus, the structural organization of the system of DNA alphabets is connected with matrices and algebraic operation of the tensor product of matrices, which plays an important role in quantum mechanics and quantum computing in connection with the following: the state space of a multicomponent quantum system is the tensor product of the state spaces of its components [1, p. 71].

It is well known that the degeneracy of the genetic code is associated with the division of the whole set of 64 triplets by the nature into two equal subsets:

- 32 triplets have "strong roots" (indicated in black in the matrixes in Fig. 1), i.e. start with 8 "strong" doublets AC, CC, CG, CT, GC, GG, GT, TC;
- 32 triplets have "weak roots" (white in Fig. 1), i.e. start with 8 “weak” doublets CA, AA, AG, AT, GA, TA, TG, TT.

Code values of triplets with strong roots do not depend on the letters in the third position, but triplets with weak roots depend on the letters in the third position.

What is the location of these" black "and" white " members of the DNA alphabets in the genetic matrixes of 16 doublets and 64 triplets in Fig. 1? An unexpected phenomenological fact is their symmetrical arrangement in these matrices, built quite formally without mentioning the degeneracy of the genetic code. The symmetric properties of the resulting mosaics in these matrices are as follows:

- the left and right halves of the matrices are mirror-antisymmetric to each other;
- both quadrants along each diagonal are the same on the mosaic;
- the mosaic of each line has a meander character of the known Rademacher functions, which are a special case of the Walsh functions and contain only the numbers +1 and -1.

Through this analogy with the Rademacher-Walsh functions from the theory of noise-immunity coding of information, the symbolic genomatrix from Fig. 1-3 are converted to numerical matrixes V and M with components +1 and -1, where components +1 (-1) represent black (white) doublets and triplets respectively (Fig. 4).

V=

1	-1	1	-1
1	1	-1	-1
1	-1	1	-1
-1	-1	1	1

W=

1	1	-1	-1	1	1	-1	-1
1	1	-1	-1	1	1	-1	-1
1	1	1	1	-1	-1	-1	-1
1	1	1	1	-1	-1	-1	-1
1	1	-1	-1	1	1	-1	-1
1	1	-1	-1	1	1	-1	-1
-1	-1	-1	-1	1	1	1	1
-1	-1	-1	-1	1	1	1	1

Figure 4. Numerical representations V and W of the genetic matrices of 16 doublets (Fig. 2) and 64 triplets (Fig. 1). Cells with numbers +1 (-1) correspond to cells with black (white) doublets and triplets in Fig. 1 and Fig.3. Each line of these numerical matrices coincides with one of the Rademacher-Walsh functions.

3. On unitary matrices

These matrices V and W (Fig. 4) represent the phenomenological features of the degeneracy of the genetic code. Below the properties of genetic matrices V and W will be analyzed more deeply. The analysis begins with (4x4)-matrix V, which turns out to be the sum of four sparse unitary (orthogonal) matrixes $V = V_0 + V_1 + V_2 + V_3$ (Fig. 5), for which the unitarity condition is fulfilled: $V_0{*}V_0^T = I$, $V_1{*}V_1^T = I$, $V_2{*}V_2^T = I$, $V_3{*}V_3^T = I$, where I is the unit matrix (this decomposition is not arbitrary, but is based on the principle of dyadic shifts from the theory of digital signal processing [2]).

$$\begin{bmatrix} 1 & -1 & 1 & -1 \\ 1 & 1 & -1 & -1 \\ 1 & -1 & 1 & -1 \\ -1 & -1 & 1 & 1 \end{bmatrix} = \begin{vmatrix} 1 & 0 & 0 & 0 \\ 0 & 1 & 0 & 0 \\ 0 & 0 & 1 & 0 \\ 0 & 0 & 0 & 1 \end{vmatrix} + \begin{vmatrix} 0 & -1 & 0 & 0 \\ 1 & 0 & 0 & 0 \\ 0 & 0 & 0 & -1 \\ 0 & 0 & 1 & 0 \end{vmatrix} + \begin{vmatrix} 0 & 0 & 0 & -1 \\ 0 & 0 & -1 & 0 \\ 0 & -1 & 0 & 0 \\ -1 & 0 & 0 & 0 \end{vmatrix} + \begin{vmatrix} 0 & 0 & 1 & 0 \\ 0 & 0 & 0 & -1 \\ 1 & 0 & 0 & 0 \\ 0 & -1 & 0 & 0 \end{vmatrix}$$

Figure .5. The dyadic-shift decomposition of the matrix V (figure 4) into the sum of four unitary matrices.

Unitary transformations preserve the norms of vectors. The tensor product of two unitary matrixes always generates a new unitary matrix. According to the described result, the degeneracy of the genetic code is associated with sparse unitary matrices, which can serve as quantum gates in biological quantum computers.

A set of these unitary matrices V_0, V_1, V_2 and V_3 (Fig. 5) is closed with respect to multiplication. Their multiplication table (Fig. 4) coincides with the multiplication table of Cockle's split-quaternions (http://en.wikipedia.org/wiki/Split-quaternion), known in particular in connection with the Poincare disk model of Lobachevski geometry.

*	V_0	V_1	V_2	V_3
V_0	V_0	V_1	V_2	V_3
V_1	V_1	$-V_0$	V_3	$-V_2$
V_2	V_2	$-V_3$	V_0	$-V_1$
V_3	V_3	V_2	V_1	V_0

Figure 6. Table of multiplication of unitary genetic matrixes V_0, V_1, V_2, V_3 from Fig. 5

It is to be added that the dyadic-shift decomposition of (8x8)-matrix W (Fig. 4) gives a set of 8 sparse unitary matrices, which is also closed with respect to multiplication and has a Cockle's bi-split quaternion multiplication table [4].

The reasons for the connection of structured DNA and RNA alphabets in their different matrix representations with Cockle's split-quaternions should be explained. It seems to be related to the connection of split-quaternions and Poincare disc model of Lobachevski geometry with known experimental data on Lobachevski geometry of the space of visual perception, which has resulted in some authors' model approaches to the brain on the basis of Clifford algebra of quantum computers (see details in [4]). Our results on the relationship between the genetic system and split-quaternions support the importance of developing algebra-geometric approaches to understanding the inherited properties of sensory systems and biological informatics in general.

The lines and columns of each of the unitary matrices V_0, V_1, V_2 and V_3 form an orthogonal system of functions that can be used for spectral analysis of physiological and other processes in their vector representations. The construction of each of these matrices to tensor powers gives new unitary matrixes of increased order with orthogonal systems of functions in their lines and columns.

4. Conclusions

The results presented in the article together with our other works [2-9] allow us to develop the elements of quantum-algorithmic genetics as one of the promising new directions of science. This scientific direction concerns also works of some other authors about quantum biology, which suggests that biological phenomena such as photosynthesis, enzyme catalysis, avian navigation or olfaction may use of a number of the non-trivial features of quantum mechanics, such as coherence, tunnelling and entanglement.

References

[1] Nielsen M A and Chuang I L 2010 *Quantum Computation and Quantum Information.* (New York: Cambridge University Press)
[2] Petoukhov S V 2008 *Matrix Genetics, Algebras of the Genetic Code, Noise Immunity* (Moscow: RHD)
[3] Petoukhov S V 2017 The rules of long DNA-sequences and tetra-groups of oligonucleotides *Preprint* 1709.04943
[4] Petoukhov S V 2018 The genetic coding system and unitary matrices *Preprint* 201804.0131
[5] Petoukhov S V and Svirin V I *Biologia Serbica*, **40**, 51

IOP Conf. Series: Materials Science and Engineering **489** (2019) 012048 doi:10.1088/1757-899X/489/1/012048

[6] Petoukhov S, Petukhova E and Svirin V New Symmetries and Fractal-Like Structures in the Genetic Coding System. In: Z. Hu, S. Petoukhov, Dychka I., He, M. (eds) *Advances* in Computer Science for Engineering and Education. ICCSEEA 2018. Advances in Intelligent Systems and Computing, vol. 754. Springer, Cham. DOI: https://doi.org/10.1007/978-3-319-91008-6_59
[7] Petoukhov S V January 2016 The system-resonance approach in modeling genetic structures *Biosystems* **139** 1-11
[8] Petoukhov S. V. and Petukhova E. S. 2017 Symmetries in genetic systems and the concept of geno-logical coding. - *Information*, **8,** https://doi.org/10.3390/info8010002
[9] . Darvas G. 2018 Petoukhov's rules on symmetries in long DNA-texts *Symmetry: Culture and Science* **29** 318-20 https://doi.org/10.26830/

IOP Conf. Series: Materials Science and Engineering **489** (2019) 012049 doi:10.1088/1757-899X/489/1/012049

The inherited information geometric analysis and the genetic sub-alphabets invariant system

I Stepanyan[1,2,3] and V Svirin[1,3]

[1]Mechanical Engineering Research Institute of RAS, 4, M. Kharitonyevskiy Lane, Moscow, 101990, Russia
[2]Izmerov Research Institute of Occupational Health of the RAS, 31, Prospect Budennogo, Moscow, 105275, Russia
[3]Moscow State Conservatory, 13/6 Bolshaya Nikitskaya Street, Moscow, 125009, Russia

neurocomp.pro@mgail.com

Abstract. Examples of invariants in the system of molecular genetic coding are given. Models and methods of algebraic biology based on matrix genetics (Petukhov, 2008) with the transition from matrix algebra to discrete geometry and visualization are presented. Algorithms based on genetic sub-alphabets allow displaying geometrically the biochemical composition of polynucleotide chains in spaces of different dimensions using Walsh orthogonal functions of physicochemical parameters. The visualization method makes it possible to substantiate the relationship between the parameters of DNA and RNA molecules with fractal clusters and geometric mosaics, reveals the orderliness and symmetry of polynucleotides and the noise immunity of their visual representations in the orthogonal coordinate system. A scale-parametric model of nucleic acid visualization is proposed. The developed methods can serve to simplify the perception by researchers of polynucleotide chains by visualizing them in spaces of various dimensions. Examples of visualization of the nucleotide composition of the various species of living organisms genomes are given

1. Introduction

Many papers are devoted to symmetries in genetic coding. In recent years, interest in symmetries in genetic coding has increased due to the intensive study of CRISPR-C as systems. Symmetries are closely related to invariance.

In [1] it is applied to study molecular evolution in vivo and in vitro. A representation of the genetic code as a six–dimensional Boolean hypercube is described. This structure is the result of the interaction energies hierarchical order of the bases in codon– anticodon recognition. Based on dynamical systems theory convergence properties of genetic algorithms are investigated at [2]. A classification procedure is proposed for genetic algorithms based on a conjecture: the entropy and the fractal dimension of trajectories produced by them are quantities that characterize the classes of the algorithms. The role of these quantities as invariants of the algorithm classes is presented.

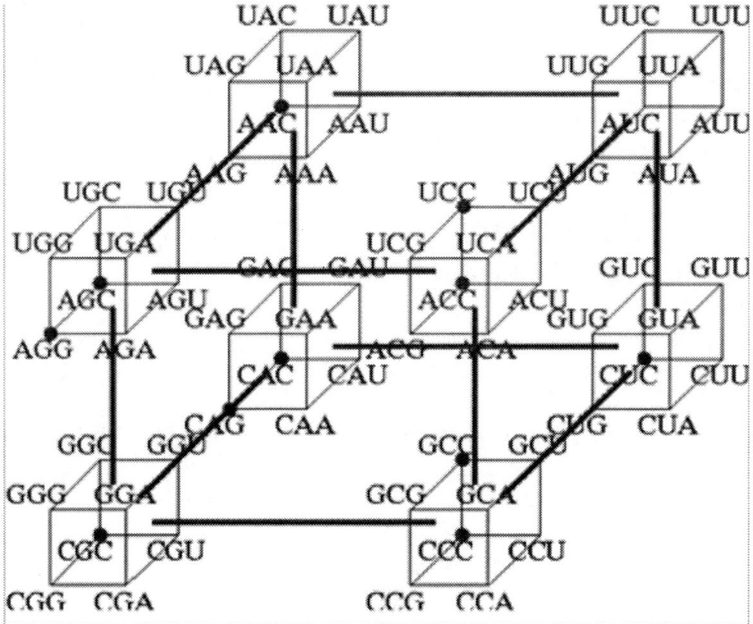

Figure 1. The hypercube representation of the genetic code from [3]

The aim of [3] was to present and analyze the probabilistic models of mathematical phylogenetics. In other cases new objects appear, such as the remarkable quintic `squangle' invariants for quartet tree discrimination and DNA data, with their own unique interpretation in the phylogenetic modelling context. The hypercube representation of the genetic code from [3] is shown at figure 1.

This paper considers geometrization as the methodological principle for the development of bioinformatics. The emergence of sound methods for comparing geometric models of genotypes with certain phenotypic traits contributes to the expansion of the circle of researchers in the field of molecular genetics.

2. Materials and methods. Invariant coding of physicochemical parameters of nucleotides

In [5], it was demonstrated that each nitrogenous base of the genetic code has three variants of its binary representation. These variants of representations, called by S.V. Petuhov the binary sub-alphabets are distinguished according to the types of binary oppositional properties in the set of nitrogenous bases:

- G = C "3 hydrogen bonds" / A = T "2 hydrogen bonds";
- C = T "pyrimidines" / A = G "purines";
- A = C "amino" / G = T "keto" [24];
- A = T = G = C (presence of phosphate residue).

Taking into consideration the additional fourth feature, which is not oppositional, the system of genetic sub-alphabets can be represented in the form of the Hadamard matrix, demonstrated in figure 2.

IOP Conf. Series: Materials Science and Engineering **489** (2019) 012049 doi:10.1088/1757-899X/489/1/012049

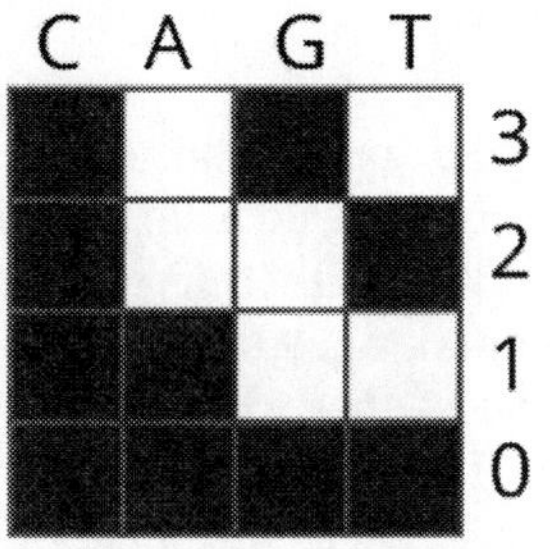

Figure 2. The variant of the Hadamard matrix displaying the encoding of nucleotide sub-alphabets. Shaded cells +1, white cells -1 (or vice versa, depending on the encoding method). Numbers of sub-alphabets are denoted as 1, 2, 3 and 0.

This matrix is symmetric, since nucleotides can be replaced by corresponding sub-alphabets without changing the structure of the matrix [6].

Each row and column of the resulting symmetric Hadamard matrix is a Walsh function [7]. The Walsh functions are a complete set of orthogonal functions that can be used to represent any discrete function by analogy with the use of trigonometric functions in Fourier analysis [10]. They are widely used in digital computing, in coding noise-resistant communications, as well as in quantum mechanics and quantum computer science. Information on Hadamard symmetries and matrices in genetic coding is described in detail in [5,13,18].

Physical holography, which possesses the highest properties of noise-immunity, is based on a record of standing waves from two coherent physical waves of the object beam and of the reference beam. But physical waves can be modelled digitally.

Correspondingly noise-immunity and other properties of optical and acoustical holography can be digitally modelled, in particular, with using Walsh functions and logic operations concerning dyadic groups of binary numbers because Walsh transforms are Fourier transforms on the dyadic groups. This can be made on the base of discrete electrical or other signals without any application of physical waves. In this digital Walsh-holography, objects, whose digital holograms should be made, are represented in forms of 2 n -dimensional vectors.

3. Algorithm of scale-parametric modelling of nucleic acids

1) a sequence of single-letter symbols encoding nitrogen bases in a nucleic acid is divided into fragments of equal length N, where N is a free parameter of the algorithm. The resulting fragments of equal length will be called N-measures or N-lashes [8];

2) taking into account the system of genetic sub-alphabets, the sequence of nitrogenous bases can be represented in the form of three binary sequences consisting of zeros and ones. The choice of coding method (which is considered to be zero or one) affects the turns and other transformations of the final visualization;

3) the resulting binary record of these fragments is a representation in the form of three sequences of decimal or other uniquely identifying values. Converting binary N-plets to decimal numbers allows their use in one or another coordinate system. These values define the coordinates of points in the parameter space, taking into account the chosen coding method (hereinafter, parametric space or visualization space).

Steps 1 and 2 can be rearranged (first parameterization, then cutting), which can significantly affect the computational load when calculating long sequences.

The given parametric space is finite, discrete, and three-dimensional in terms of the number of binary opposition signs. Its combinatorial properties allow you to display any polynucleotides for an arbitrary finite N. Ordered numerical values on the coordinate axes uniquely reflect the physicochemical characteristics of N-mers, since they are precisely determined by the properties of binary opposition sub-alphabets.

4. Results and discussion.

The set of three binary opposition signs can be compared with the {X, Y, Z} axes of the Cartesian coordinate system. The N coefficient performs the functions of the resolution of geometric visualization: large N gives a small number of points, small N gives a small coordinate grid. This circumstance allows us to speak of multi-scale analysis in multidimensional parametric spaces.

In accordance with the methods of Harmut's theory of sequential analysis [9], additional visualizations were constructed by the number of elements (zeros or ones) that were encountered in binary representations of N-plets in sequences of nitrogenous bases. Due to the fact that such a visualization method is based on the total number of certain parameters in an N-plet, the corresponding visualization spaces will be called integrals.

Visualization algorithms can be applied to any nucleic acids, including RNA and chromosomes of living organisms. In this work, genomes from the NCBI bioinformation database [17] were used for visualization, as well as materials kindly provided by the laboratory of prof. N.S. Zenkina Center of Bacterial Cell Biology of the Newcastle University (UK).

On the basis of the developed computer program, it was found that the chromosomes of various types of organisms have individual structural features in various spaces. Genomic imaging of various organisms may have a pronounced fractal character, which is identical for all chromosomes of the organism under study, as well as for arbitrary fragments of these chromosomes.

To visualize nucleic acids in the three-dimensional parameter space, we use the orthogonal basis {X, Y, Z}, chosen as the three-dimensional Cartesian coordinate system. It gives the visualization, the example of which is shown in Fig. 3. Each point represents the characteristic of the binary opposition of the corresponding fragment of the sequence, summarized in accordance with the algorithm, which allows one to display the nucleotide composition of the molecule.

Figure 3. Illustration of a three-dimensional representation of the nucleotide composition on the example of the chromosome of a living organism in various projections. The X, Y, and Z axes correspond to the ordered decimal representations of the binary coding of N-plets based on three binary opposition sub-alphabets. Each point of the figure corresponds to N-measures, the coordinate of which is given by its proton-numerical features.

The resulting geometric shape, which is the three-dimensional analogue of the Sierpinski triangle, is typical for the three-dimensional visualization of any long nucleotide sequences. The shape of the resulting figure is determined by the properties of binary sub-alphabets. The coordinates of each point in the three-dimensional visualization space are completely defined by any two coordinates, since the third coordinate is calculated by modulo two remaining coordinates. This algebraic feature is associated with the redundancy of binary sub-alphabets used by nature for storing and transmitting genetic information in the chains of generations. Fig.4. demonstrates an integral three-dimensional representation of the nucleotide composition.

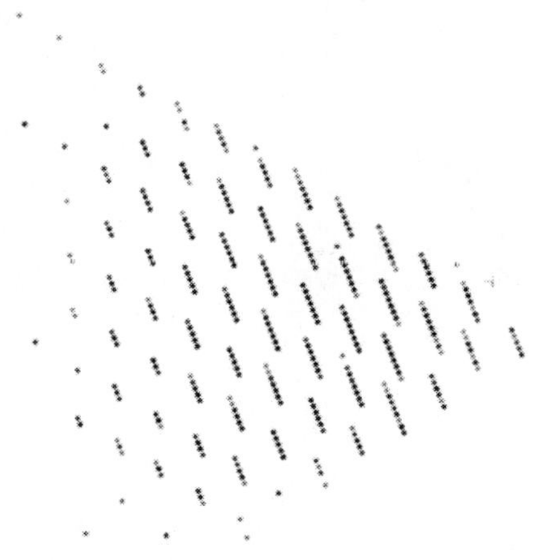

Figure 4. Integral-three-dimensional representation of the nucleotide composition. The X, Y, and Z axes correspond to the number of units in the decimal binary encoding of each N-plet using three binary opposition sub-alphabets.

An example of the nucleotide composition integral three-dimensional representation of the living organism chromosome is presented in Figure 3. This is a discrete pyramid, each point of which corresponds to a set of N-measures of nucleic acid, combined by the number of units in binary coding for each of the signs.

As can be seen, in connection with the geometric properties of a given space, three-dimensional visualizations are not very convenient for perceiving and analyzing the characteristics of a particular nucleic acid. However, the available three two-dimensional projections of these visualizations are well suited to display the specificity of the structure of long polynucleotide chains. In the bases {X, Y}, (X, Z} and {Y, Z}, chosen as Cartesian coordinate systems, three-dimensional visualization gives three different two-dimensional projections based on the corresponding sub-alphabets of physicochemical signs of nucleotides. Based on the property of the genetic coding, according to which a triple of binary opposition sub-alphabets is connected with each other by addition modulo two, to determine an arbitrary nucleic acid, any pair of its binary representations is sufficient. Therefore, for two-dimensional visualization of a nucleotide composition it is sufficient to have any axes pair.

The question of determining the most informative pair of coordinate axes and, accordingly, the parameters taken into account may depend on the type of organism being analyzed and requires additional research. As a result of the analysis of some genomes, it was found that the three variants of two-dimensional visualization, mosaics based on information about the external structure of the molecule, i.e. built on the elements of the structures encoding the signs of amino / keto and purine / pyrimidine. Such mosaics had a detailed fractal pattern consisting of squares or rectangles. In some genomes, mosaics based on types of hydrogen bonds appeared to be the most symmetrical. They are usually characterized by triangular elements or pattern diagonals and are found in plant DNA. At the same time, diagonal or other elements of the pattern can be directed in various directions in different organisms while maintaining the structure of the pattern (this feature can also be modelled by reading a complementary strand of DNA).

IOP Conf. Series: Materials Science and Engineering **489** (2019) 012049 doi:10.1088/1757-899X/489/1/012049

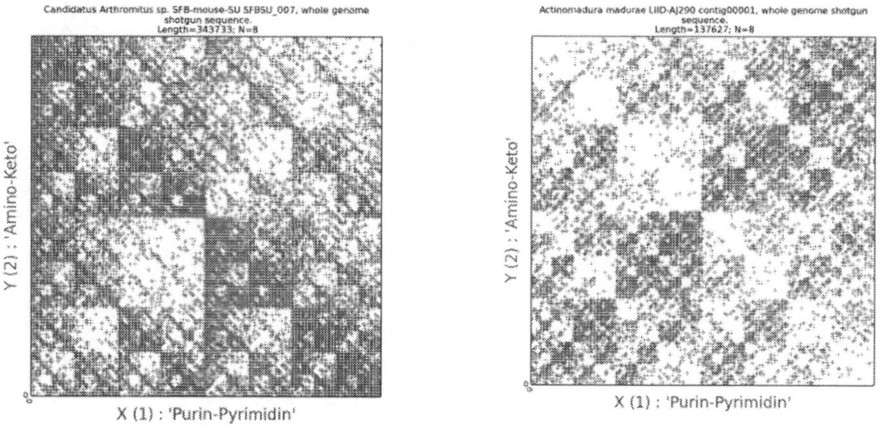

a) b)

Figure 5. a) Illustration of a two-dimensional representation of the nucleotide composition of genomes: Candidatus Arthromitus bacterium; b) the genus Actinomadura (one of the four genes). The abscissa and ordinate axis correspond to the decimal representations of the binary encoding of each 8-plet.

General scientific studying methods of nucleic acids usually focus their attention on those fragments that are present in them. The proposed algorithms allow visualization of the phenomenology and features of the deficiency and presence of various types of N-mers.

Two-dimensional models, which are obtained using the method described, sometimes resemble the fractal patterns of nucleic acids, which in [11] were obtained using the CGR method. However, these methods are quite different in their algorithmic nature. In particular, the CGR method is missing information about binary sub-alphabets.

In parametric visualization and comparison of the human and monkey genomes, as well as the genomes of other species of living organisms, two-dimensional mosaics of arbitrary fragments of the genomes with the whole genome were observed. This dismantles the fractal properties of the structure of nucleic acids and indicates the existence of both interspecific and intraspecific variants of nucleotide composition.

The preliminary results of applying two-dimensional visualization algorithms allow us to conclude that the final mosaics are highly stable when the original sequence is noisy, including when shifting the reading frame of the sequences, when arbitrary fragments of the sequence are removed (distortion), when reversing the entire analyzed circuit or its fragments permutations of N-mers and nucleotides (in some cases up to a complete permutation of all nucleotides in the sequence).

In particular, we observed the stability of mosaic patterns in cases of every second nucleotide deletion, every third nucleotide, etc. At the same time, the visualization of nucleic acids in two-dimensional spaces in some cases is characterized by pronounced symmetries and stability not only to different noise in the original data, but also to different values of the scale parameter N within a certain range.

IOP Conf. Series: Materials Science and Engineering **489** (2019) 012049 doi:10.1088/1757-899X/489/1/012049

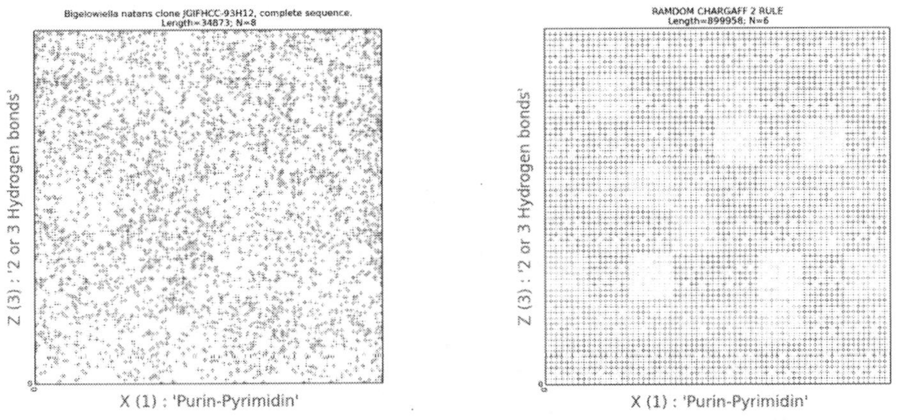

a) b)

Figure 6. .a) Illustration of a two-dimensional representation of the nucleotide composition of the organism bigelowiella natans (Eukaryota). Natans are a key resource for the supergroup of mostly unicellular eukaryotes. The abscissa and ordinate axis correspond to the decimal representations of the binary encoding of each 8-plet; b) illustration of a two-dimensional representation of the nucleotide composition of an algorithmically generated chromosome in compliance with the second Chargaff rule. The abscissa and ordinate axis correspond to the decimal representations of the binary encoding of each 6-plet.

For further studies, sequences of nitrogenous bases 100,000 nucleotides in length were randomly generated using a computer program and divided into N-laps on 8, 16, and 28. Randomly generated sequences in visualization give a pattern, all points of which are scattered randomly. Such visual representations are irregular, chaotic in the complete absence of any mosaics and patterns in all sub-alphabets, which significantly distinguishes them from real genomes.

Visual representations of the genomes of various penicillin kinds were constructed. The results suggest that the genomes of this group of antibiotics usually generate high-density mosaics that resemble mosaics of random sequences, which indicates a high diversity of their nucleotide composition. Perhaps the medical value of penicillin kinds is associated with this their feature.

Pseudorandom nucleic acids are modelled following the rules of Chargaff, valid for each of the two DNA strands [4,14]. A special type of patterns for these generated sequences was visualized at N = 6 in Fig. 6b. Some fractals of real genomes were also modelled by generating pseudo-random nucleic acids with selection of probability coefficients.

Two-dimensional visualization algorithms appear to be useful for studying hidden patterns in chromosomes, as well as for the classification and comparative analysis of various genomes with possible applications in biotechnology and medicine. Examples of genetic mosaics built in a nonpositional number system are given in [12].

As a result of the analysis, we can conclude that the two-dimensional visualization of the nucleotide composition of both algorithmically generated and real genomes makes it possible to display the implementation options for the quantitative rules of Chargaff [4] using elements of combinatorial final geometry [25]. This fact can help in the study of internal symmetries and other characteristics of nucleic acids, transferring their analysis from the field of statistics to the field of geometry to study the complex relationships between biological objects.

As noted, binary sub-alphabets are interconnected by modulo-two addition and de-fine a parametric space with properties at which the coordinates of all points are connected or “glued” together. In this

IOP Conf. Series: Materials Science and Engineering **489** (2019) 012049 doi:10.1088/1757-899X/489/1/012049

regard, for a one-dimensional analysis, it makes sense to consider each dimension separately and independently

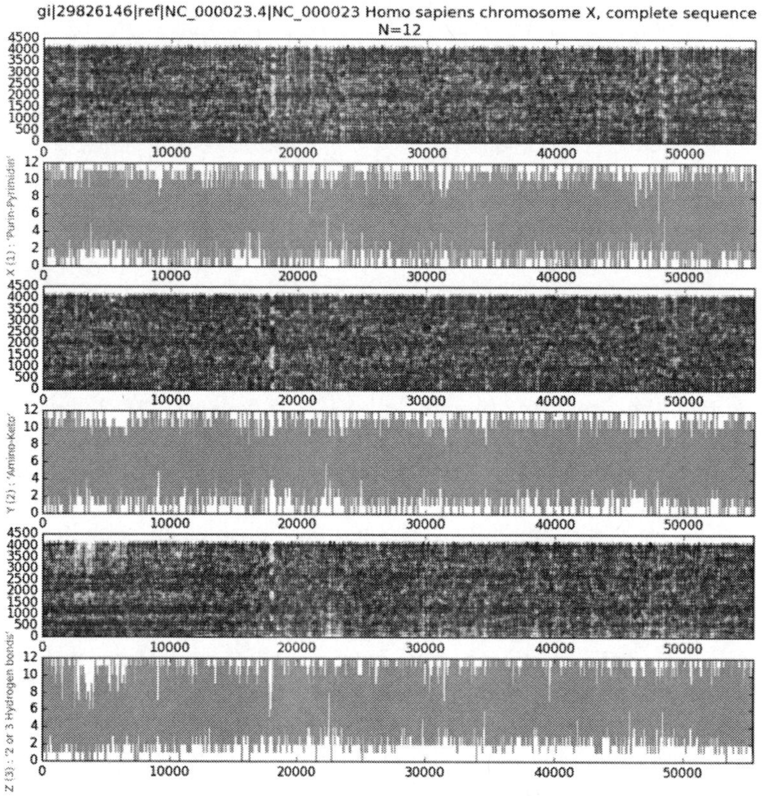

Figure 7. One-dimensional visualization of the three-channel representation of the nucleotide composition of a fragment of the 22nd chromosome of Homo Sapiens. Each of the three rows corresponds to a binary opposition sub-alphabet. In each channel, the abscissa axis encodes the ordinal number of the N-plet, the ordinate axis encodes the ordered ascending decimal values of the binary representation of the N-plets.

Taking into consideration the presence of three binary opposition signs, there are three one-dimensional visualization spaces. At the same time, the mathematical relationship between their coordinate axes is preserved. Thus, the use of one-dimensional coordinate axes {X}, {Y} and {Z} gives a triplet of various mappings using the corresponding sub-alphabets.

In Fig. 7 in the 22nd chromosome of Homo Sapiens, sub-alphabets clearly show areas with different nucleotide composition. These specific regions can be visualized in two-dimensional parametric spaces for further comparison and analysis.

The obtained graphs allow us to estimate changes in the nucleotide composition when reading a fragment of a molecule from beginning to the end. The depth of the recorded changes is determined by the scaling parameter N of the algorithm.

5. Conclusions

Parametric visualization of both fragments and whole DNA and RNA molecules made it possible to substantiate their connection with fractal and other mosaics. The correlation of molecular genetic systems with binary code and Hadamard matrices are demonstrated. The stated principles of visualization allow visual assessment of the correlation types between the N-measures present and absent in the genomes of various organisms and viruses, which in some cases are characterized by fractal-cluster organization in spaces of binary-orthogonal functions of their physical and chemical parameters. The obtained geometric patterns are associated with the quantitative phenomenological rules of Chargaff [4,14,22].

The scaling parameter N allows exploring the genome at a variety of different levels of detail to search for hidden symmetries and patterns when visualizing in spaces of different dimensions.

The proposed methods can serve to simplify the perception by researchers of long polynucleotide chains, and also serve as an additional criterion for the classification and identification of interspecific interactions. Modern ontologies and thesauri for organizing and storing biological and molecular genetic data can be supplemented with visualization options for educational purposes, as well as for the presentation and retrieval of biological information.

The results obtained are related to the theory of number systems, self-organizing systems and correlate with the theory of dynamic systems. Parametric modelling and visualization of nucleotide composition can contribute to an in-depth understanding of genetic phenomena, not only by simplifying the mechanisms of perception, but also by the possibility of using adaptive neural network technologies, since the structure of chromosomes of living organisms, represented in binary form, fits well with the format of binary artificial neural networks [16] . In addition, DNA geometrization is associated with logical holography.

As known, holographic methods in engineering allow quick detecting of individual elements in a huge image. The theory of genetic logical holography allows assuming that one of secrets of noise-immunity of genetic informatics is based on the similar possibilities of genetic logical holography. In an appropriate modelling approach, if one of DNA molecules mutates, the genetic logical holography – by analogy with classic physical holography - allows quick detecting of the mutated DNA in the whole logical hologram of a set of DNA molecules. In the result, the genetic information of this individual DNA molecule could be found to be incorrect for further using in organism automatically.

Conducted studies on the comparative analysis of nucleotide sequences visualization of various types of living organisms (protozoa, plants, fungi, animals, viruses) confirm that the nucleotide composition may be identical in organisms that are not related in the phylogenetic tree and different in related organisms [15].

It opens up the possibility of modelling pseudorandom nucleotide sequences with the observance of the phenomenological rules of Chargaff for their visualization and further research in parametric spaces of various dimensions. The proposed tool for visualization of nucleic acids can help advance the understanding of the immune system functioning principles in recognizing the nucleotide composition of viruses, parasite DNA, as well as in food chains and ecosystems.

Part of the calculations was performed on the MVS-10P supercomputer (MSC RAS).

References

[1] Jiménez-Montaño M A, Mora-Basánez C R and Pöschel T 1996 *Biosystems* **39(2)** 117-25.

[2] Kotowski S., Kosiński W., Michalewicz Z., Nowicki J., Przepiórkiewicz B. 2008 Fractal Dimension of Trajectory as Invariant of Genetic Algorithms. In: Rutkowski L., Tadeusiewicz R., Zadeh L.A., Zurada J.M. (eds) Artificial Intelligence and Soft Computing – ICAISC 2008. ICAISC 2008. Lecture Notes in Computer Science, vol 5097. Springer, Berlin, Heidelberg

[3] Jarvis P D , Sumner J G 2018 Systematics and symmetry in molecular phylogenetic modelling: perspectives from physics *Preprint* 1809.03078

[4] Chargaff E, Lipshitz R, Green C 1952 *J Biol Chem.* **195 (1)** 155–60

IOP Conf. Series: Materials Science and Engineering **489** (2019) 012049 doi:10.1088/1757-899X/489/1/012049

[5] Petoukhov S.V. and He M 2010 *Symmetrical Analysis Techniques for Genetic Systems and Bioinformatics: Advanced Patterns and Applications* (Hershey, USA: IGI Global)

[6] Balonin N A, Balonin Y N, Djokovic D Z, Karbovskiy D A, Sergeev M B 2017 Construc-tion of symmetric Hadamard matrices *Preprint* 1708.05098

[7] Georgiou S, Koukouvinos C, Seberry J 2003 *Further computational and con-structive design theory* (Boston: Kluwer.) pp 133–205. ISBN 1-4020-7599-5

[8] Stepanian I V, Petoukhov SV 2013 The matrix method of representation, analysis and classification of long genetic sequences *Preprint* 1310.8469

[9] Harmuth H 2016 *Applying of Methods of Theory of Information in phisics* (Moscow.: Mir)

[10] 10.Ferleger, Sergei V. (March 1998). RUC-Systems In Non-Commutative Symmetric Spaces (Technical report). MP-ARC-98-188.

[11] Jeffrey H J 1990. *Nucleic Acids Re-search* **18(8)** pp 2163-70

[12] Feldman D P 2012 *Chaos and Fractals: An Elementary Introduction* (Oxford: OUP) pp 178–80. ISBN 9780199566440

[13] 13.Darvas G, Koblyakov A A, Petoukhov S V, Stepanyan I V. 2012 *Symmetry: Culture and Science* **23(3-4)** 343-75

[14] Rudner R, Karkas J D and Chargaff E 1968 Separation of B. subtilis DNA into complementary strands. 3. Direct analysis *PNAS* **60(3)** 921-22; https://doi.org/10.1073/pnas.60.3.921

[15] Townsend JP, Su Z, Tekle Y 2012 *Systematic Biology* **61(5)** 835–49. https://doi.org/10.1093/sysbio/sys036

[16] Stepanyan I V, Ziep N N 2018 *Neurocomputers: development, application* **5** pp 4-11

[17] ftp://ftp.ncbi.nlm.nih.gov/

[18] Petukhov S 2008 *Matrix Genetics, Algebra of Genetic Code, Noise Immunity* (Moscow: RHD)

[19] Takuyuki I, Wakana T, Masafumi M, Masaki E, Hiro-Yuki H 2015 *Genes & Genetic Systems* **90(4)** 231-35. https://doi.org/10.1266/ggs.15-00030

[20] Crick F H C, Wang J C and Bauer W R 1979 *jmb* **129(3)** 449-61. https://doi.org/10.1016/0022-2836(79)90506-0

[21] Wilkins M H F, Stokes A R and Wilson H R 1953 *Nature* **171** 738–40. https://doi.org/10.1038/171738a0

[22] Chargaff E, Lipshitz R and Green C 1952 *J Biol Chem.* **195 (1)** 155–60

[23] Sharov A. 1992 *Biosemiotics: The Semiotic Web 1991.* Ed T.A.Sebeok and J.Umiker-Sebeok. (New York :Mouton de Gruyter) pp 345-373

[24] 24.Waterman M S 1995 *Introduction to Computational Biology. Map, Sequences and Genomes* (London: Chapman & Hall)

[25] Batten L M 1997 *Combinatorics of Finite Geometries* (Cambridge: CUP) ISBN 0521590140

IOP Conf. Series: Materials Science and Engineering **489** (2019) 012050 doi:10.1088/1757-899X/489/1/012050

On the stability of trajectories in the task of pursuit

M V Andreeva[1,2] and V E Pavlovsky[1,2]

[1] Moscow Automobile and Road Construction State Technical University (MADI), 64, Leningradsky prospect, Moscow, Russia
[2] Keldysh Institute of Applied Mathematics of the Russian Academy of Sciences (KIAM) 4, Miusskaya sq., Moscow, Russia

point15@ro.ru

Abstract. In two-dimensional and three-dimensional statements, a generalization of the pursuit problem is considered, when the object being pursued moves in a circle (plane case) or in a helical line (three-dimensional case), and various cases of dependence of object velocities in the problem are considered. Conditions are analytically obtained for the stability of the trajectories of motion of objects in the problem. Possible realization of this problem - the pursuit of one drone by another one as flying objects.

1. Introduction

Let a simple movement be a movement with a limited speed, and the direction of movement can be changed arbitrarily. Let two material points *A* (catching up) and *B* (running away) move in the same plane with speeds u and v of constant value. The velocity of point *B* does not change direction, and the velocity vector of point *A* rotates and is always directed to *B*. The task of describing the movement of these points is called the pursuit problem, and the trajectory of the movement of the catcher is called the pursuit curve.

2. Description of a problem

The trajectory of the object-chaser in a coordinate system rotating around the origin with angular velocity $\omega = \frac{V_0}{R_0}$ of the target object is the solution of an autonomous system of differential equations:

$$\begin{cases} \dot{x} = \frac{V_0}{R_0} y + V \frac{R_0 - x}{\sqrt{(R_0 - x)^2 + y^2}} \\ \dot{y} = -\frac{V_0}{R_0} x - V \frac{y}{\sqrt{(R_0 - x)^2 + y^2}} \end{cases} \quad (1)$$

where V is the velocity of the object-chaser $V = f(x, y) < V_0$ and the target object in this system has coordinates$(R_0; 0)$.

There are two natural problem statements:

The first statement: the velocity of the object-chaser depends on the distance to the origin: $V = f\left(\sqrt{x^2 + y^2}\right) = f(r)$

The second statement: the velocity of the object of the pursuer depends on the distance to the object – goal $V = f\left(\sqrt{(R_0 - x)^2 + y^2}\right)$

For both statements, we have:

IOP Conf. Series: Materials Science and Engineering **489** (2019) 012050 doi:10.1088/1757-899X/489/1/012050

The matrix of the linear part of system (1) at a singular point
$(x = \frac{R_0}{V_0^2} V^2;\ y = -\frac{R_0}{V_0^2} V\sqrt{V_0^2 - V^2})$:

$$\frac{V_0}{R_0}\begin{pmatrix} -\frac{\frac{V}{V_0}\left(\left(\frac{V}{V_0}\right)^2 - \left(1-\left(\frac{V}{V_0}\right)^2\right)\frac{R_0}{V_0}f'(r)\right)}{\sqrt{1-\left(\frac{V}{V_0}\right)^2}} & 1+\left(\frac{V}{V_0}\right)^2 - (1-\left(\frac{V}{V_0}\right)^2)\frac{R_0}{V_0}f'(r) \\ -1+\left(\frac{V}{V_0}\right)^2 + \left(\frac{V}{V_0}\right)^2\frac{R_0}{V_0}f'(r) & -\left(\frac{V}{V_0}\right)^2\sqrt{1-\left(\frac{V}{V_0}\right)^2}\,(1+\frac{R_0}{V_0}f'(r)) \end{pmatrix}$$

3. For the first statement in the plane problem
The eigenvalues of the matrix of the linear part of the system in the equilibrium position:

$$\lambda_{1,2} = \frac{-VV_0 \pm \sqrt{(VV_0)^2 + 4\left(1-\left(\frac{V}{V_0}\right)^2\right)(\frac{R_0}{V_0}f'(r) - 1)}}{2R_0\sqrt{V_0^2 - V^2}}$$

If $f'(r) - \frac{V_0}{R_0} > 0$, then $\lambda_1 > 0$ and the equilibrium position is unstable. ($V_0^2 - V^2 > 0$)

If a $f'(r) - \frac{V_0}{R_0} < 0$, then $\lambda_{1,2} < 0$ or $\operatorname{Re}\lambda_{1,2} < 0$ and the equilibrium position is stable.

4. For the second statement in the plane problem
The matrix of the linear part of system (1) at a singular point:

$$\begin{pmatrix} \dot{x}'_x & \dot{x}'_y \\ \dot{y}'_x & \dot{y}'_y \end{pmatrix} \rightarrow \begin{vmatrix} \dot{x}'_x - \lambda & \dot{x}'_y \\ \dot{y}'_x & \dot{y}'_y - \lambda \end{vmatrix}$$

The trajectory will be stable:

- $\frac{-\dot{x}'_x - \dot{y}'_y}{2} > 0$
- $\dot{x}'_x \cdot \dot{y}'_y - \dot{y}'_x \cdot \dot{x}'_y > 0$
- $\frac{-\dot{x}'_x - \dot{y}'_y}{2} < 0$

The trajectory will be unstable:

- $\begin{cases} \frac{-\dot{x}'_x - \dot{y}'_y}{2} > 0 \\ \dot{x}'_x \cdot \dot{y}'_y - \dot{y}'_x \cdot \dot{x}'_y < 0 \end{cases}$
- $\begin{cases} \frac{-\dot{x}'_x - \dot{y}'_y}{2} = 0 \\ \dot{x}'_x \cdot \dot{y}'_y - \dot{y}'_x \cdot \dot{x}'_y < 0 \end{cases}$

It is not clear what will happen with sustainability:

IOP Conf. Series: Materials Science and Engineering **489** (2019) 012050 doi:10.1088/1757-899X/489/1/012050

- $$\begin{cases} \frac{-\dot{x}'_x-\dot{y}'_y}{2} > 0 \\ \dot{x}'_x \cdot \dot{y}'_y - \dot{y}'_x \cdot \dot{x}'_y = 0 \end{cases}$$

Due to the difficulties of expression of stability conditions in an explicit form through function of dependence of speed on distance to the target it is necessary to provide explicit examples showing in what cases the trajectory of the movement will be stable and in what will be not (from what parameters and as far as it depends) [1-3].

$$x'(t) = \frac{(-x+\cos(t))\left(\frac{1}{2}+\sqrt{(-x+\cos(t))^2+(-y+\sin(t))^2}\right)}{\sqrt{(-x+\cos(t))^2+(-y+\sin(t))^2}}$$

$$y'(t) = \frac{(-y+\sin(t))\left(\frac{1}{2}+\sqrt{(-x+\cos(t))^2+(-y+\sin(t))^2}\right)}{\sqrt{(-x+\cos(t))^2+(-y+\sin(t))^2}}$$

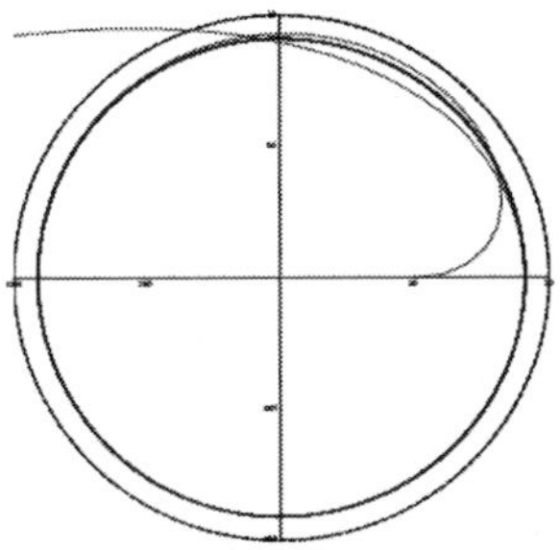

Figure 1. Stable trajectory.

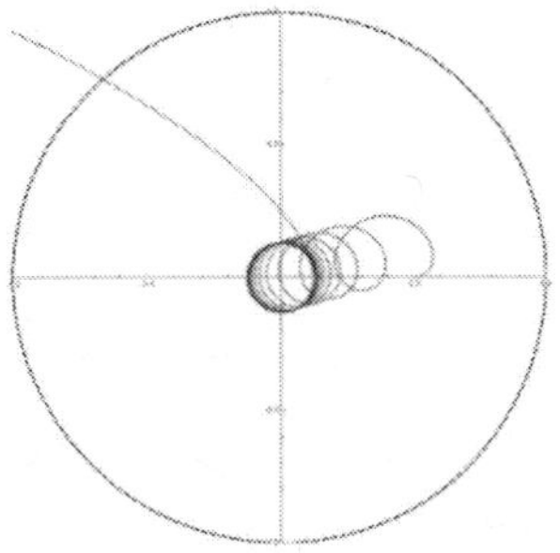

Figure 2. Unstable trajectory.

The equation to find the speed limit:

$$\mathrm{NSolve}[x == \mathrm{VMin} + \sqrt{(1-x^2)\hat{}2 + (1-x^2)x^2}, x]$$

Solution: $(\mathrm{VMin} + \sqrt{2-\mathrm{VMin}^2})/2$

With VMin = 0.5 we get speed limit equal to 0.9114378277661477

5. Study of the trajectory of the pursuer when moving along a helical path in 3D-space

Statement. If the trajectory of the target does not contain straight segments, then the distance between the points *P(t)* of different paths of the pursuer decreases with increasing time *t*.

The target moves along a helix.

In figure 3 trajectories of the target and pursuer depending on time are shown. From the visual representation, it can be concluded that pursuers coordinates *x (t)*, *y (t)*, z (t) are the same as target ones (the distance between them decreases) with increasing *t*.

The target moves along a closed line with a periodic ascent and descent.

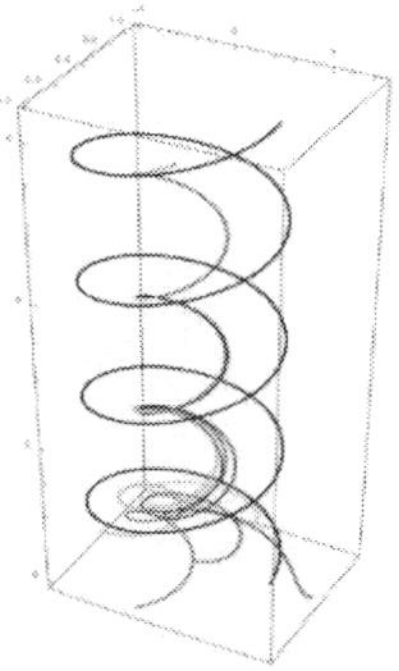

Figure 3. Movement of the target (black trajectory) and the pursuer.

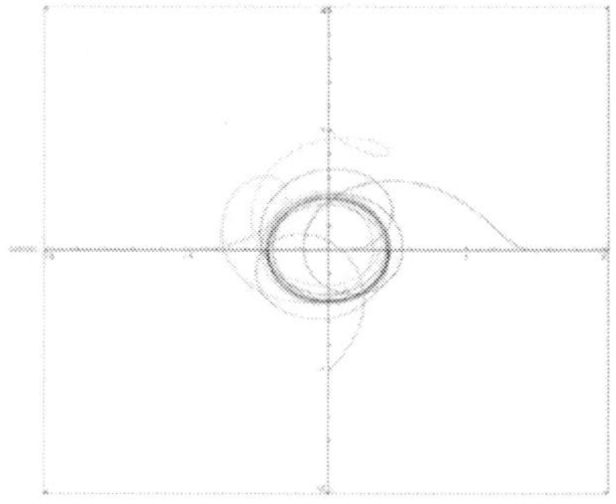

Figure 4. Analog to figure 3 top view.

Figures 5 and 6 show trajectories of the target and the pursuer depending on the time. From the visual representation we conclude that the coordinates of the pursuers x *(*t), *y (*t), *z (*t) visually coincide with increasing *t*.

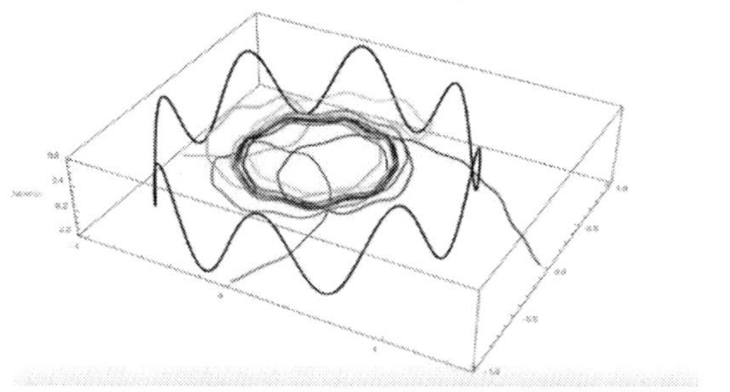

Figure 5. Spatial picture of the goal and the pursuer.

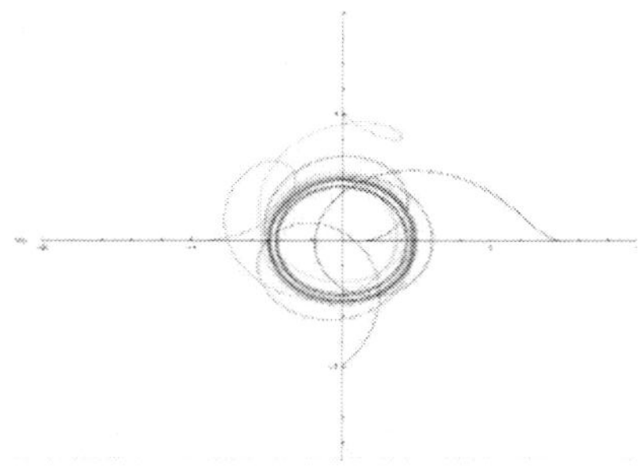

Figure 6. Analogue figure 5 top view.

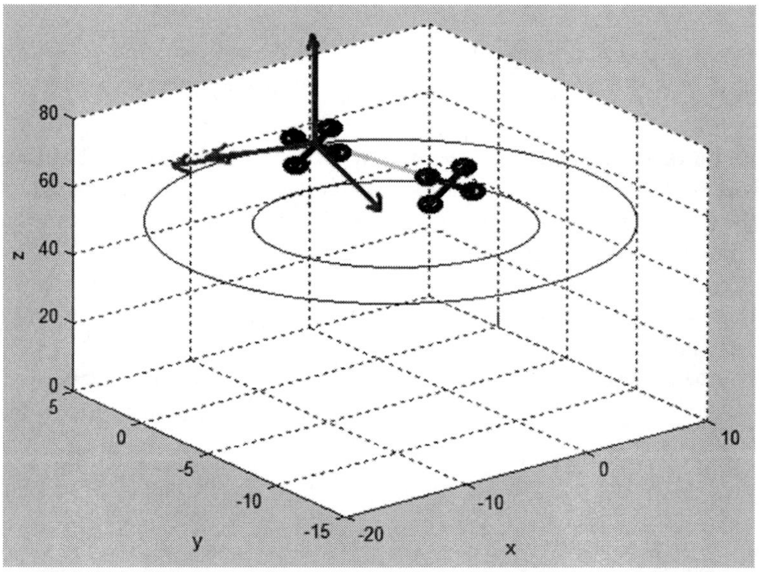

Figure 7. The scheme of the pursuit problem for two drones in space.

IOP Conf. Series: Materials Science and Engineering **489** (2019) 012050 doi:10.1088/1757-899X/489/1/012050

6. Conclusion

Under certain conditions imposed both on the trajectory of the pursuer object and the target object, we can obtain both a stable trajectory and an unstable one. Using Wolfram Mathematica Package shows that the behavior of numerical solutions of systems of differential equations of motion corresponds to the following property of solutions given earlier: the trajectories of the pursuer with different initial data at $t = 0$ approach each other as t increases. Moreover, from figures 3-6, it can be concluded that the distance between points of different trajectories taken at the same t tends to zero when $t \to \infty$. From the behavior of the trajectories, we can conclude that when the target moves along a helix, the trajectory of the pursuer also approaches the helical line with a reduced radius; when the target moves periodically, the trajectory of the pursuer also approaches the periodic one with reduced amplitude of oscillations.

References

[1] Savelov A A 1960 *Flat curves. Systematics, properties, applications (reference manual)* (Moscow: Nauka)
[2] Amelkin V V 1987 *Differential equations in applications* (Moscow: Nauka)
[3] Troshin P I 2015 *Computer geometry and geometric modeling (tasks). Teaching guide* (Kazan: Kazan Federal University)

IOP Conf. Series: Materials Science and Engineering **489** (2019) 012051 doi:10.1088/1757-899X/489/1/012051

Tensometry of soft biological tissues with manipulation robot

M V Arkhipov[1], E A Vzhesnevsky[1], M A Eremushkin[2], E A Samorukov[2] and L B Kocherevskaya[3]

[1] Moscow Polytechnic University, Bolshaya Semyonovskaya str., 38, Moscow, 107023, Russian Federation
[2] National Medical Research Center for Rehabilitation and Balneology, Novy Arbat, 32, Moscow, 121099, Russian Federation
[3] Moscow Aviation Institute National Research University, Volokolamskoje shosse 4, Moscow, 125993, Russian Federation

evgesnev@gmail.com

Abstract. A method of automating the process of controlling strain gauge measurements of parameters of elastic objects using a manipulation robot is considered. The subjects of the study are elastic objects whose structure is multilayered, and individual layers have the properties of heterogeneity and nonlinearity. An example of such objects is soft biological tissue. Currently available diagnostic tools have insufficient accuracy because of the fact that the measurement results significantly affected by the errors arising from the explorer's errors during automated diagnostics and accidental interference on the part of the research object. Strain gauge measurements relate to the subject area of assessment of the state of elastic structures. The paper discusses the implementation of the algorithm for controlling the procedure of tensometric measurement of soft biological tissues, proposes engineering solutions to eliminate these drawbacks. Elimination of inaccuracies in the development of force and speed tasks causing the greatest errors achieved by using a manipulation robotics equipped with a strain gauge measurement device. A prototype of a robotic system equipped with a power sensor, a compensation mechanism and a software control algorithm is given. The results of the study are presented in the form of characteristic dependences of forces on displacements.

1. Introduction

In problems, where position-force control (PFC) is applied, there are pneumatic, potentiometric, binary force sensors of known design [1]. These sensors are used in feedback in case of force control. While we establish feedback on the force, such a sensor can be used not only for the PFC, but also for strain gauge measurements of soft biological tissues [1, 2, 3]. From the point of view of mechanics, the task of controlling a manipulation robot is to deform certain sections of soft biological tissues (SBT). They are skin, muscle and tendons. The force-controlled robotic system is equipped with a manipulator drive system which guides the tool flange along the tool axis until the force measured by the sensor along the tool axis is equal to the set value [4]. However, unlike the indirect effect on the SBT through the levers of bones and joints, when the deformation directed along the muscles, when manipulating the SBT, the tool directly affects the muscle, mainly in the transverse direction, which discussed in detail in the article.

2. Control object model at tensometry

To measure the parameters of soft biological tissue, there is currently a variety of sensors called myotonometers. There are electromechanical myotonometer Uflyanda, Myoton-2 and mechanical myotonometer Sirmai, Dubrovsky [5]. In the case of measuring the force with a compliance of the analyzed material in a few millimeters, the deformation of the elastic element of the force sensor can be measured with a strain gauge sensor. Figures 1 and 2 show the design of a single-component force sensor with spring elastic elements and a strain gauge of the spring deformation [6].

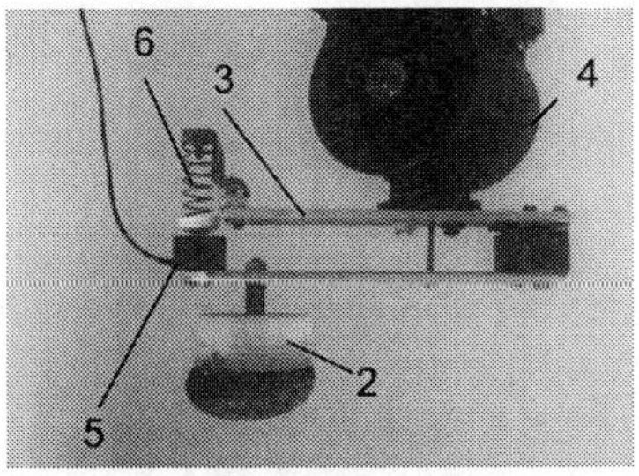

Figure 1. Muscle tone measurer (external view of the device installed on the robot flange)

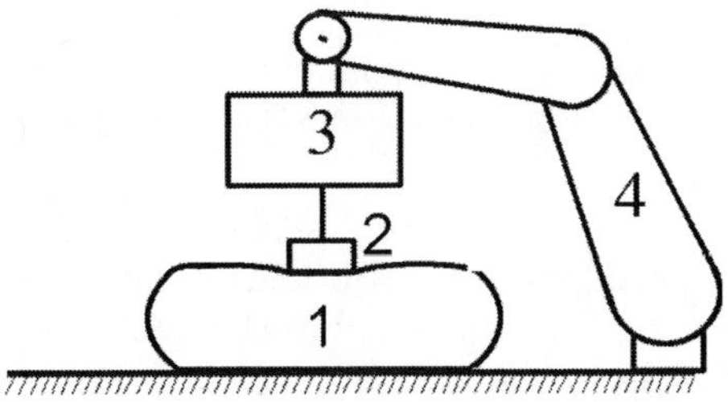

Figure 2. Instrument of the measurer in contact with soft tissues, where: 1-biological tissue, 2-tool, 3-force sensor, 4-robot, 5 – strain gauge sensor, 6 – spring

Soft biological tissue has been chosen as an object of analysis, and muscle tone (MT) chosen as a registered parameter. Muscle tone is one of the fundamental characteristics of a muscle, reflecting its properties and the state of the neuromuscular apparatus. Muscle tone is a long-lasting muscle tension, which is not accompanied by fatigue and provides a certain position of the body in space. There are muscle tone in a calm state (plastic tone) and tense state (contractile tone) [7].

Muscle tone characterizes the elastic properties of soft tissue and therefore defined as its "stiffness", from the dependence on the displacement *Δ* at deformation on increasing effort F on the SBT, which causes this deformation (figure 3). For the SBT and the muscles, it is a non-linear dependence with a variable coefficient of elasticity (stiffness). The dimension of this coefficient is N/m.

If the SBT on the linear sector has elasticity k_2 (figure 3 (a), and the force sensor in the linear sector has a slope k_1, then their serial connection in the design of the force sensor is represented by the characteristic as in figure 3 (b), and in the linear sector the equivalent slope k is equal to:

$$k = \frac{k_1 \cdot k_2}{k_1 + k_2} \text{ [8].} \qquad (1)$$

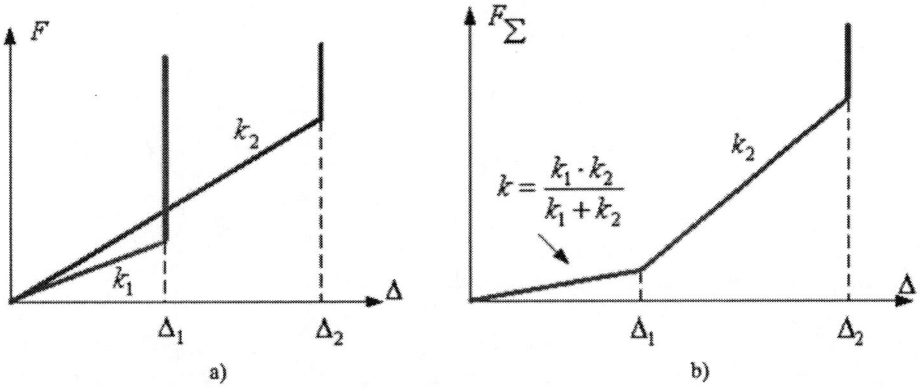

Figure 3. Models of contact interaction between the tool of the force sensor and the SBT and

characteristics: a) k_1 - SBT, k_2 - springs, b) SBT and springs.

Skeletal muscles in statics are a visco-elastic material by mechanical behavior. The calculated value k can be fixed through an elastic measuring element (spring) attached to the movable element of the strain gauge (figure 1). The signal from the strain gauge U, proportional to the deformation of the spring, converted into a code one, inputting to the robot control system. To describe the processes occurring in ST as in the elastic element, Hooke's law is valid [1]:

$$k = \frac{F}{\Delta}, \tag{2}$$

where F is force equal to external load applied perpendicular to cross section of SBT; k is coefficient of cross section; Δ is deformation of SBT. The growth rate of curve $F(\Delta)$ is another characteristic feature.

$$tg\alpha = \frac{dF}{d\Delta}, \tag{3}$$

where dF is the increment of effort; $d\Delta$ is the deformation increment; α is the angle of inclination of the tangent to $F(\Delta)$. The steepness of the characteristic $F(\Delta)$ is variable and depends on the deformation of the SBT, its thickness, on the dimensions of the contact surface of the tool. The elasticity of the tissue in terms of thickness is non-uniform, the upper part - the fat layer is the softest. In addition, the steepness of the characteristic depends on the degree of muscle tension. The range of values of the steepness for different MBT is wide. For a wide soft layer of tissue, the values of the coefficient of elasticity on the linear section are of the order of 0.01 $N \cdot mm^{-1}$, for the stressed thin-layer section are 100 $N \cdot mm^{-1}$.

3. The model of the control system at tensometry of SBT

The registration of muscle tone is performed by installing a sensor instrument on an SBT sector, with continuous (or delayed) tool immersion in the SBT in a static mode, when steady-state values of forces and strains are being measured (figures 4 and 5).

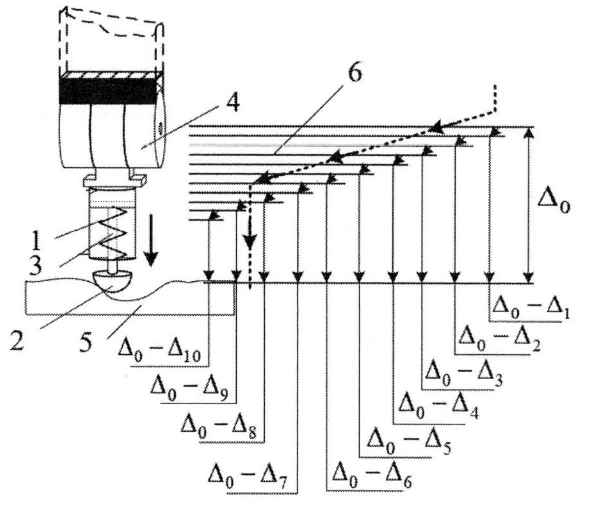

Figure 4. Measurement scheme of MT registration, 1-spring, 2-tool, 3-inductive transducer, 4-link of robot, 5-biological tissue, 6- trajectory of the robot;

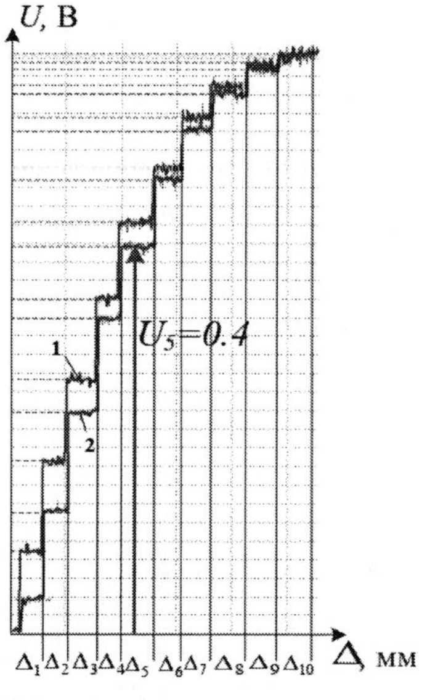

Figure 5. Registration graph of SBT, 1-solid body, 2-soft body.

Immersion levels are set by software offset by Δin mm. With each immersion in the memory of the measuring system the voltage level U fixed on the strain gauge. While studying the SBT and the muscles, a miotonogram with the coordinates is obtained: on the abscissa axis - the depth of immersion (muscle

IOP Conf. Series: Materials Science and Engineering **489** (2019) 012051 doi:10.1088/1757-899X/489/1/012051

hardness), on the ordinate axis the increasing force, for each projection on the miotonogram you can get the coefficient of elasticity. The dimension of this coefficient is $N \cdot mm^{-1}$.

The depth of immersion of the tool in the muscle is not more than 18 mm. The characteristics of the registration process were determined on the basis of the characteristics of a single-component force sensor FC2231. Depth measurement error would be up to 0.5 mm. The error in determining the force is up to 3 grams with a force of up to 750 grams and no more than 10 grams with a force of up to 2000 grams. Power supply is 5 volts. The measurement results are inputting into a computer through the COM port and processed with a program that represents them in graphical and tabular forms.

On the graph in figure 5 it is possible to obtain information about different values of the elasticity of the tissues. Track the dynamics of changes in elasticity can be on the values of MT, as well as the deviation of the amplitudes of the voltages at each step.

The sensitive element of the force sensor is a spring. For a linear spring, its deformation Δ in statics is proportional to the force of the $F_{ELASTIC}$ with the coefficient of elasticity of the K_{SPRING} [9]:

$$F_{ELASTIC} = -\Delta \times K_{SPRING}, \tag{4}$$

After spring calibration, for example, using a dynamometer, you can recalculate its deformations into force. To diagnose the elasticity of the soft tissue, the arm of the robot must move by the value of x, the spring is compressed by the value of Δ, and the soft tissue by the value of δ. For the calculated force $F_{ELASTIC}$, soft tissue deformation δ is a measure of the elasticity of soft tissue K_{ST}:

$$K_{ST} = \frac{-F_{ELASTIC}}{\delta}, \tag{5}$$

Let us establish the relationship between the values of x, δ and Δ, since the program shift is determined by the formula:

$$x = \Delta + \delta, \tag{6}$$

Therefore,

$$x = \Delta + \frac{F}{K_{ST}} = \Delta + (\frac{\Delta \times K_{SPRING}}{K_{ST}}) = \Delta \times (1 + \frac{K_{SPRING}}{K_{ST}}), \tag{7}$$

where x is the programmed movement of the robot;
Δ is the deformation of the spring measured by the induction sensor and is a measure of the force;
K_{SPRING} - spring stiffness, calculated by calibration.
Let us calculate the coefficient of elasticity of soft tissue:

$$K_{ST} = \Delta \cdot (\frac{K_{SPRING}}{x - \Delta}). \tag{8}$$

Due to the straight-line nature of the robot's movement at tensometry, one can consider a drive, whose movement in the horizontal direction causes a tool pressure on the SBT along the Z axis of the tool coordinate system [10]. Then the force of gravity will not affect the pressure. Excluding the viscous component, due to the low approach speeds, putting $q = z$:

$$U = iR + C_v \cdot \frac{dz}{dt}, \tag{9}$$

$$M \cdot \frac{d^2 z}{dt^2} = F_d - F, \tag{10}$$

$$F_d = C_f \cdot i, \tag{11}$$

IOP Conf. Series: Materials Science and Engineering **489** (2019) 012051 doi:10.1088/1757-899X/489/1/012051

$$U = K_z \cdot (z_0 - z), \tag{12}$$

$$z_0 = K_f \cdot (F - F_0), \tag{13}$$

$$F = f(z - z_e). \tag{14}$$

where U is the voltage applied to the DC motor, i is the armature winding current of the DC motor, R is the resistance of the armature winding, C_v is the coupling coefficient between speed and electrical driving force, z is the real displacement of the manipulator along the tool axis, z_e is the surface relief ST, M is the mass of moving parts of the manipulator, F_d is the driving force, F is the real force of interaction with the external environment, C_f is the coupling coefficient between the current and the driving force, K_z is the coefficient of the proportional regulator for the position, K_f is the coefficient of the proportional controller for force. Accordingly, the block scheme of a similar model is shown in figure 6.

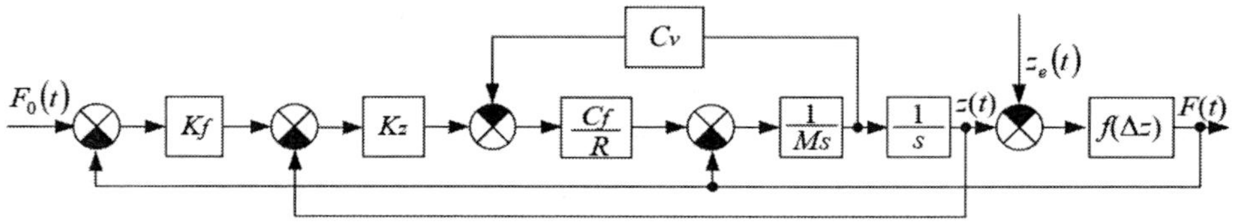

Figure 6. The structure of the model of control of the manipulator at tensometry.

At the initial moment of time, the tool touches the surface of the MBT and then immerses into it. The presence of an elastic element in the structure of the control system provides compensation in the distribution of sharp forces.

4. Measurement of the parameters of the SBT by manipulation robot

Investigating the elasticity of the soft tissue with the use of the modernized robot RM-01, an experiment on tensometry has been carried out on the quadriceps muscle of the thigh. Immersing of a tool was carried out in two stages with repetition. The first stage was in the study of the relaxed muscle, the second stage was in the study by the same method of strained muscle.

An example of this can be the experimental curves $F=f(\Delta)$ obtained on soft tissue while immersing and leaving the tool in areas with different elasticity, reflected by the distance of the graphs from each other (figure 7).

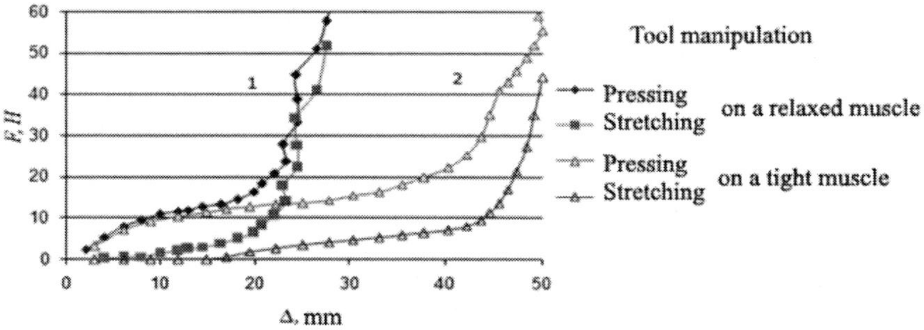

Figure 7. Graph of dependencies $F=f(\Delta)$, relaxed (1) and strained (2) quadriceps thigh muscle.

IOP Conf. Series: Materials Science and Engineering **489** (2019) 012051 doi:10.1088/1757-899X/489/1/012051

The graphs reflect the main difference between the studied strained and relaxed muscles, different levels of amplitude values in elasticity.

The strained muscle curve (highlighted in red) reaches its maximum stress value in the same number of steps Δ. Tensometry was held for a short time period, close to the actual operation of the tool on the surface of the soft tissue at the manual method of measurement.

In order to estimate changes in the parameters of the SBT, there are not necessary absolute measurements, but differential, considering set of points at different periods of time associated with rehabilitation processes (for example, rehabilitation therapy).

With the improvement of the functional state the amplitude and muscle tone index increase (the difference between strain and relaxation). With fatigue the amplitude decreases, the tone of rest increases.

There are many empirical indicators [11] applied to integral estimates for a given curve. Such a methodological approach allowed us to formulate a semantic function that adequately reflects the degree of MT at rest and during the mobilization of the SBT.

One of the stages of the continuation of work is the development of a tensometry methods using a robot that would allow an accurate estimation of the parameters of the SBT [12, 13].

5. Conclusion

The objectivity of tensometry of soft biological tissues is due to given forces and speeds, which will allow you to work out a controlled multi-link manipulation robot using the last drive as a force sensor. The presence of a compensation mechanism can improve the safety of tensometry. The obtained research results allow to analyze static and dynamic indicators, including various integral evaluations.

References

[1] Gorinevsky D M, Formalsky A M and Shneider A Yu 1994 *Control of manipulation systems based on information on forces* (Moscow: Phys.-math. Lit.) p 365

[2] Skovoroda A R 2006 *Problems of the theory of elasticity in the diagnosis of pathologies of soft biological tissues* (Moscow: Physical and mathematical literature) p 232

[3] Sadovnichy V A, Goryacheva I G, Akaev A A, Martynenko Yu G, Okunev Yu M, Vlakhova A V and Bogdanovich I Yu 2009 *Application of methods of conthediagnosis of pathological conditions of soft biological tissues* (Moscow: Publishing house of MSU) p 309

[4] Arkhipov M, Golovin V and Vzhesnevsky E 2016 Research of operating modes of service manipulation robot interacting with a deformable elastic medium *Proc. Int. Scientific and Technological Conf. «EXTREME ROBOTICS 2016»* (Saint-Petersburg) pp 134-8

[5] Vzhesnevsky E A, Arhipov M V and Orlov I A 2017 *Problems and tasks of touching in manipulating robotics when working with compliant objects* (Sevastopol: Intellectual systems, management and mechatronics) pp 184-9

[6] Arhipov M V, Rachkov M Yu, Samorukov A E, Golovin V F, Zhuravlev V V, Polonskiy M E, and Vzhesnevsky E A 2013 Mechatronic Myotonometer *Patent RU 130826*

[7] Golovin V, Arhipov M and Zhuravlev V 2012 *Robotics in restorative medicine. Robots for mechanotherapy* (Saarbrücken: LAP LAMBERT Academic Publishing, GmbH & Co.) p 280

[8] Golovin V F, Zhuravlev V V, Razumov A N and Rachkov M Yu 2005 Adaptive force module for medical robots *Proc. Workshop on Adaptive and Intelligent robots: Present and Future* (Moscow: IARP) pp 70-82

[9] Arkhipov M V, Rachkov M Yu, Golovin V F and Kocherevskaya L B 2018 *Mechatronics, Automation, Control* **4** pp 243-50

[10] Vukobratovic M 1997 *Journal of Intelligent and Robotic System* **19** pp 119-152

[11] Zhbankov O V and Tolstoy E V 1997 *Theory and Practice of Physical Culture* **8** pp 38-42

[12] Arkhipov M, Leskov A, Golovin V, Gercik Ya and Kocherevskaya L 2017 Prospects of Robotics Development for Restorative Medicine *Proc, 25th Conf. on Robotics in Alpe-Adria-Danube Region (RAAD16)* (Belgrade: Springer) pp 499-506

[13] Gerasimenko M Yu, Eremushkin M A, Arkhipov M V, Kolyagin Yu I and Antonovich I V 2017 *Physiotherapy, balneology and rehabilitation* **16** pp 65-9

IOP Conf. Series: Materials Science and Engineering **489** (2019) 012052 doi:10.1088/1757-899X/489/1/012052

3DoF Parallel robot analysis

D V Artemov, V M Masyuk, S Y Orekhov and I V Pchelkina
Kaluga Branch of Bauman Moscow State Technical University, Bazhenova st. 4, Kaluga, 248004, Russia

colt50rus@mail.ru

Abstract. This paper addresses a 3DoF parallel robot. A kinematic scheme and joint types are proposed. Kinematic analysis of the mechanism is done. Aspects of robot control are considered: a power supply, a controller.

1. Introduction

Parallel robots are known by stiffness of their mechanical structure, high position accuracy and high load carrying capacity. At the same time they have some drawbacks such as limited working space and difficult forward position kinematics because of the closed structure. In contrast, the inverse position kinematics is quite simple. Parallel mechanisms have been studying since the 60s of the last century but in this field there is a large area that needs to be enhanced, namely, dynamic analysis.

When designing a parallel mechanism, a designer has many ways for assembly organization. Most common structures have been described in detail in [1], [2]. The main requirements that the equipment must meet are the possibility of easy installation and controlling and using of universal components for assembly. For practical reasons, in this work a parallel robot was developed with the ability to work in three and six degrees of freedom (DoF) depending on the modification of the structure. The 3D CAD models in SolidWorks are shown in figure 1. Below we consider a mechanism with 3DoF.

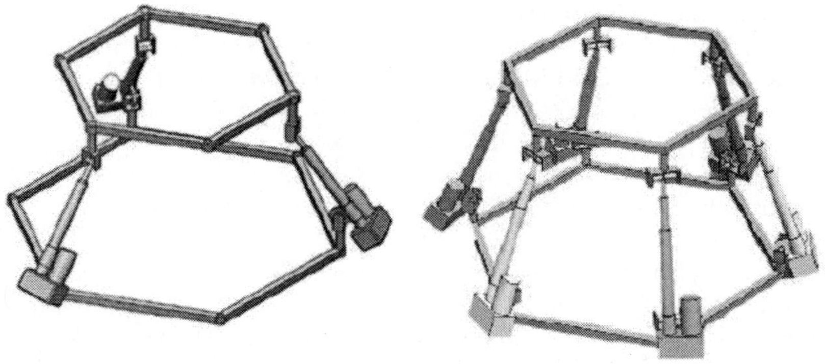

a) b)

Figure 1. Robot modelling in SolidWorks: tripod (a) and hexapod (b).

IOP Conf. Series: Materials Science and Engineering **489** (2019) 012052 doi:10.1088/1757-899X/489/1/012052

Although 3DoF robots are considered well known [3], the proposed mechanism is comprehensible and suitable for educational and research purposes because it allows to illustrate the robot modelling and control tasks.

2. Assumptions

In this research, the objective is a 3DoF robot based on three legs. It consists of a fixed base and a moving platform connected by three serial chains. We consider two cases:
1) each leg has a revolute-revolute-translational-revolute-revolute (RRTRR) joint structure (figure 2, a);
2) each leg has a revolute-translational-spherical (RTS) joint structure (figure 2, b).
The revolute joints are passive, only the three translational joints are actuated. According to the kinematic schemes, in the first case the mechanism has 15 kinematic pairs of the fifth class (12 rotational and 3 translational). In the second case the robot has 6 kinematic pairs of the fifth class (3 rotational and 3 translational) and 3 kinematic pairs of the third class (3 rotational). The DoF for a closed-loop mechanism is examined by using the Somov-Malyshev formula:

$$W_1 = 6 \cdot n - 5 \cdot p_5 = 6 \cdot 13 - 5 \cdot 15 = 3, \tag{1}$$

$$W_2 = 6 \cdot n - 5 \cdot p_5 - 3 \cdot p_3 = 6 \cdot 7 - 5 \cdot 6 - 3 \cdot 3 = 3,$$

where n is a number of all links, p_5 and p_3 denote the class of a kinematic pairs, W_1 is DoF for the RRTRR case, W_2 is DoF for RTS case. Interestingly, the first proposed kinematic scheme is very similar to the universal-prismatic-universal (UPU) structure [4].

If we fix the all three translational kinematic joints in RRTRR and RTS joint structures, the derived mechanisms will have the following DoF:

$$W_{1,\,FIX} = 6 \cdot n - 5 \cdot p_5 = 6 \cdot 11 - 5 \cdot 12 = 0, \tag{2}$$

$$W_{2,\,FIX} = 6 \cdot n - 5 \cdot p_5 - 3 \cdot p_3 = 6 \cdot 4 - 5 \cdot 3 - 3 \cdot 3 = 0.$$

Two prototypes are built (see figure 4), and it turned out that the RRTRR prototype collapses while RTS one behaves properly. It is clear that a RRTRR mechanism has unaccounted DoF. Therefore, further a RTS mechanism is considered.

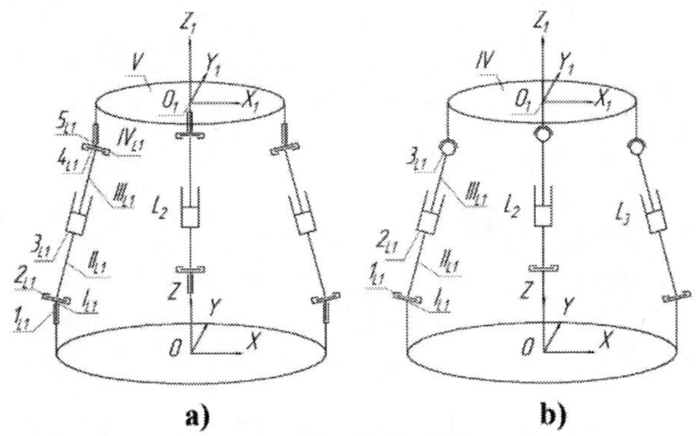

a) b)

Figure 2. Kinematic schemes of the robot with RRTRR (a) and RTS (b) joint structure: specification of kinematic pairs (Greek numerals), links (Roman numerals) and axes. OXYZ – fixed coordinate system (CS), $O_1X_1Y_1Z_1$ – moving CS.

IOP Conf. Series: Materials Science and Engineering **489** (2019) 012052 doi:10.1088/1757-899X/489/1/012052

For the RTS robot the position of the i-th support chain is given by the values of L_i and angles of the upper and lower hinges $\alpha_i^{up}=\alpha_i^{low}=\alpha_i$, $\beta_i^{up}=\beta_i^{low}=0$ (figure 3). The angles α_i,β_i determine the orientation of the i-th support chain relative to the plane of the base, L_i – the length of the support chain.

3. Kinematics analysis

Let's introduce a fixed coordinate system (CS) $Oxyz$ and a moving CS $O_1x_1y_1z_1$. The latter is rigidly connected to the moving platform (figure 3). Three rods of variable length A_iB_i are modelling the translational kinematic pairs. They are attached to the base platform at points A_i and to the mobile platform at points B_i. Points A_i,B_i lie on circles of radii R_A and R_B, respectively. Wherein points A_i,B_i lie in planes parallel to the platforms planes, but separated from them by some distance d_s, so the size of the hinge mountings has been taking into account.
Position of A_i,B_i are fixed in the corresponding CS:

$$A^0=\begin{bmatrix}A_1\\A_2\\A_3\end{bmatrix}=\begin{bmatrix}-\frac{R_A}{2}\sqrt{3} & -\frac{R_A}{2} & d_s\\ 0 & R_A & d_s\\ \frac{R_A}{2}\sqrt{3} & -\frac{R_A}{2} & d_s\end{bmatrix};B^1=\begin{bmatrix}B_1\\B_2\\B_3\end{bmatrix}=\begin{bmatrix}-\frac{R_B}{2}\sqrt{3} & -\frac{R_B}{2} & d_s\\ 0 & R_B & d_s\\ \frac{R_B}{2}\sqrt{3} & -\frac{R_B}{2} & d_s\end{bmatrix}, \quad (3)$$

where the upper index shows the coordinate system relative to which the vectors are considered: 0 for the base CS $OXYZ$ and 1 for CS of the moving platform $O_1X_1Y_1Z_1$. In the nominal position, the robot legs are symmetrical with a symmetry angle of $2\pi/3$, and the planes in which the points A_i,B_i are located are parallel, the distance between them is z_{nom}.

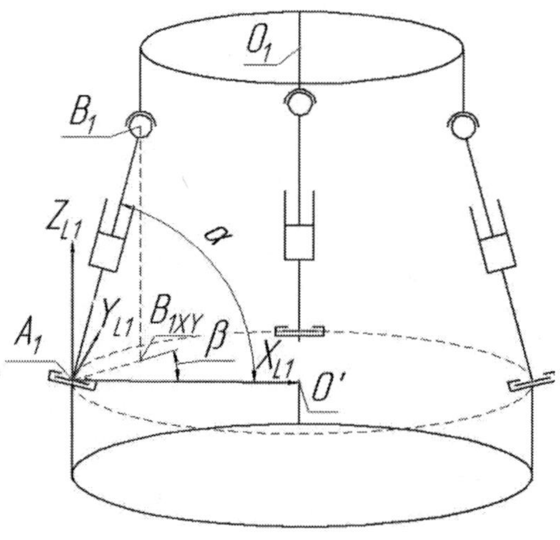

Figure 3. Kinematic scheme of the robot with RTS joint structure.

3.1. Inverse Kinematics

Let's take for output values a vector:

IOP Conf. Series: Materials Science and Engineering **489** (2019) 012052 doi:10.1088/1757-899X/489/1/012052

$$\xi = \left[x, y, z, \phi, \theta, \psi\right]^T, \tag{4}$$

where $\{x, y, z\}$ is the position of the point O_1 (the origin of CS $O_1X_1Y_1Z_1$); $\{\phi, \theta, \psi\}$ are the yaw, roll and pitch angles defining the orientation of CS $O_1X_1Y_1Z_1$ relative to CS $OXYZ$. The inputs are the rod lengths A_iB_i:

$$L = \left[L_1; L_2; L_3\right], \ L_i = A_iB_i, \ i = 1,2,3. \tag{5}$$

Usually the inverse kinematics of parallel mechanism can be accomplished by the transformation matrix [5]. Introduce the transformation matrix T_0^1:

$$T_0^1 = \begin{bmatrix} R_{\phi\theta\psi} & p \\ 000 & 1 \end{bmatrix}, \tag{6}$$

Where $R_{\phi\theta\psi}$ is the 3×3 submatrix describing rotation by angles $\{\phi, \theta, \psi\}$, and $p = \left[x_{O_1}; y_{O_1}; z_{O_1}\right]$ is the 3×1 submatrix describing translation.

As far as we know the position of the points B_i in CS $O_1x_1y_1z_1$, and A_i in CS $Oxyz$, we can find the position of the i-th support leg:

$$L_i = \left|\overline{A_iB_i}\right|, \ \alpha_i = \cos^{-1}\left(\frac{A_iB_{i,XY}^2 + A_iB_i^2 - B_iB_{i,XY}^2}{2\cdot A_iB_{i,XY}\cdot A_iB_i}\right), \ \beta_i = \cos^{-1}\left(\frac{A_iB_{i,XY}^2 + A_iO'^2 - O'B_{i,XY}^2}{2\cdot A_iB_{i,XY}\cdot A_iO'}\right), \tag{7}$$

Where $B_{i,XY}$ is the projection of B_i on the plane $O'A_1A_2A_3$ that is parallel to the base plane with distance d_s between them; position of the point O' in CS $O_1x_1y_1z_1$ is $O' = \left[0;0;d_s\right]$; $A_iB_i, B_iB_{i,XY}, A_iO', O'B_{i,XY}$ and $A_iB_{i,XY}$ are the lengths of corresponding vectors.

3.2. Forward Kinematics

The direct kinematic problem is solved by establishing the kinematics constraint equations for the mechanism. Thus, the output vector $\xi = \left[x, y, z, \phi, \theta, \psi\right]^T$ is obtained from equations

$$\left|\overline{A_iB_i}\right|^2 = L_i^2, \ \left|\overline{B_iO'_1}\right|^2 = R_B^2, \ i = 1,2,3. \tag{8}$$

where $L = \left[L_1; L_2; L_3\right]$ is the input vector; the point O_1' lies in the plane $B_1B_2B_3$ with its position $O_1' = \left[0;0;-d_s\right]$ in CS $O_1X_1Y_1Z_1$.

3.3. Modelling

Mathematical modelling of (3)-(8) was developed by using the MATLAB software and Optimization Toolbox package. The geometric parameters of the parallel robot are given as $R_A = 0.7$, $R_B = 0.990$, $d_s = 0.15$, $z_{nom} = 0.610$. Results are shown in figure 4.

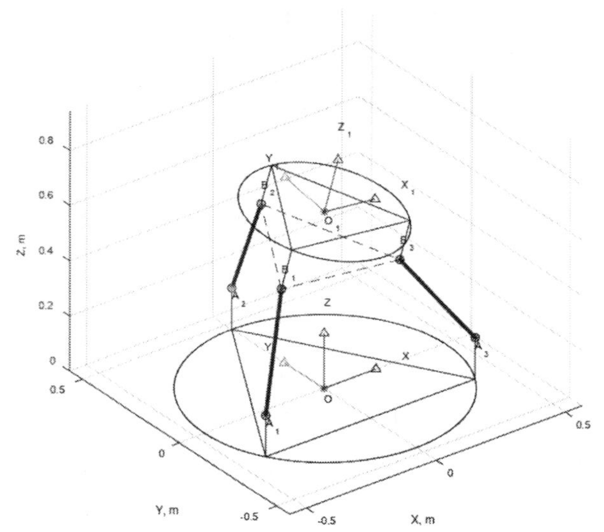

Figure 4. Simulation results of robot kinematics in Matlab.

4. Prototype development

4.1. Robot

The prototype of 3DoF parallel manipulator has been developed and assembled at the department "Mechatronic and Robotics" at Kaluga Branch of BMSTU (see figure 5). As a linear drive joint electric drives SuperJack HARL3612+ were used. These linear actuators have high load characteristics. Factory mountings of the actuators are presented by two spherical hinges that do not correspond to the proposed kinematic scheme. Therefore two sets of metal bushings for each actuator were designed and manufactured. The bushings are installed on spherical hinges by special fasteners and block two unwanted degrees of rotation. The fasteners are installed in specially designed brackets which act as universal joints. Thus, the final mechanism meets the proposed kinematic scheme.

As the framework of the robot universal 25 mm metal pipes and Joker/Uno fastening systems were used which include various connectors, clamps and flanges; this ensures uncomplicated and universal installation of the robot.

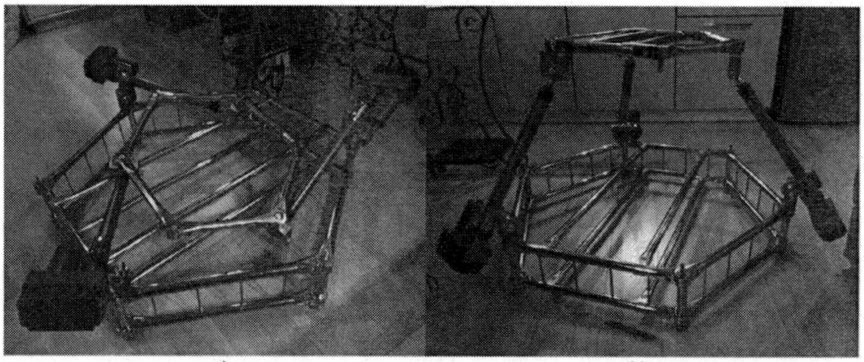

a) b)

Figure 5. Prototypes of the robot with RRTRR (a) and RTS (b) joint structure.

4.2. Control System

In this study, the task was to develop and create an H-bridge debug board based on MOSFET transistors with an allowable power of 110 W for the implementation of a DC motor control stand. As part of the project, we considered existing methods for constructing mechatronic control systems for the H-bridge [6], and analysed the advantages and disadvantages of various approaches. Also a patent search has been conducted, and relevant literature has been studied.

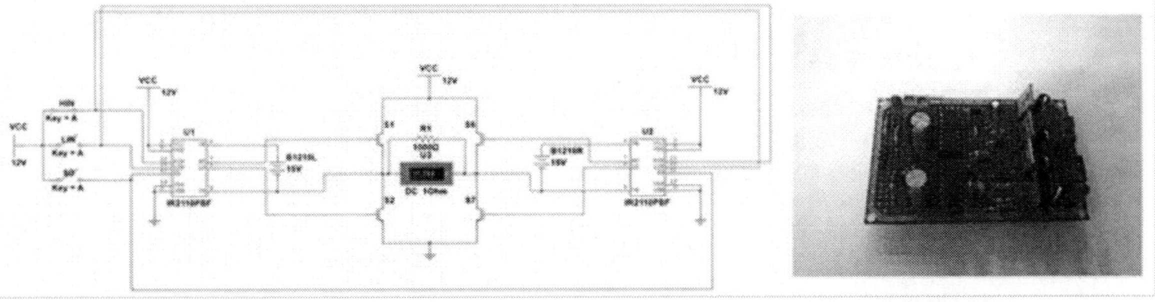

a) b)

Figure 6. Model diagram of the control system (a) and its implementation (b).

At the current stage of the work, the H-bridge control circuit was designed using the Bootstrap circuit as power supply for the upper parts of the bridge. The system is designed on two IR2110 controllers and high-power field-effect transistors IRF 3205, a model diagram of the system is shown in figure 6.

5. Conclusions

In this paper the design of 3DoF parallel robot based on three kinematic chains is proposed. It is supposed to have the ability of changing the robot to 6DoF kinematic scheme by mounting three additional kinematic chains.

For a mechanism with 3DoF: kinematic analysis of two schemes (RRTRR and RTS) was done; an experimental setup was developed; control system for the linear actuator was designed. It was found that a RRTRR robot collapses while a RTS robot behaves properly. The next step of the study is to implement dynamic analysis of 3DoF model with RTS joint structure and refine the control system. In the future, we plan to develop a 6-UPS parallel robot.

References

[1] Merlet J P 2006 *Parallel Robots* (Sophia-Antipolis: Springer)

[2] Glazunov V A, Koliskor A Sh and Krainev A F 1991 *Spatial mechanisms of the parallel structure* (Moscow: Nauka)

[3] Tsai L-W 1999 *Robot Analysis: The Mechanics of Serial and Parallel Manipulators* (New York: John Wiley & Sons)

[4] Tsai L-W and Joshi S A 2000 Kinematics and optimization of a spatial 3-UPU parallel manipulator *ASME Journal of Mechanical Design* **122** 439

[5] Jadran L and Parenti-Castelli V 1996 *Recent Advances in Robot Kinematics* (Dordrecht: Springer)

[6] Bogatenkov V A and Masyuk V M 2017 Research and development of high efficiency control system for DC motor with a multi-level protection system *Regional Scientific Conference Kaluga, Bauman MSTU*

IOP Conf. Series: Materials Science and Engineering **489** (2019) 012053 doi:10.1088/1757-899X/489/1/012053

Clearance control of vehicle with hydropneumatic suspension and the wheel formula 8x8

D N Demyanov[1], V S Karabtsev[1,2] andA I Malinov[2]

[1]Kazan Federal University, Naberezhnye Chelny Institute, Naberezhnye Chelny, Prospekt Syuyumbike 10A 423812, Russia
[2]Scientific Technical Center of KAMAZ PTC, Naberezhnye Chelny, Transportnyi Proezd 70423815, Russia

alexander.malinov@mail.ru

Abstract. In the present work the problem of controlling the elements of a hydropneumatic suspension of a multi-axle vehicle is considered. Control algorithms are proposed that provide the required law of clearance change in various conditions. A computer simulation was carried out, which showed the correctness of the work of the presented algorithms.

Introduction

One of the main requirements for high-speed vehicles is a high average speed on the ground. This speed depends on many factors, including the efficiency of the suspension system. In case of insufficiently effective operation of the suspension system, the driver is forced to reduce the speed of the vehicle due to overload or fatigue. Thus, the improvement of vehicle suspension systems can be considered a very important task [1].

Depending on the operating conditions and requirements for the vehicle, various suspension systems can be used. One of the possible variants of such systems is a hydropneumatic suspension (HPS), the main advantages of which are high smoothness, the ability to adjust the position of the vehicle body relative to the road surface, effective reduction of vibrations, adaptation to the driving style of a particular person [2, 3].

In modern vehicles the main units are controlled by an onboard information control system, one of the main elements of which is an electronic suspension control unit. Software development for electronic suspension control unit is one of the most important stages of design, since it determines the main characteristics of the vehicle and its compliance with technical requirements.

The main results in the design and implementation of algorithms and control systems for HPS are currently obtained for cases of two-axis and three-axis vehicles (these types of vehicles are most often used in practice). However in some cases there is a need for the development of the control system for HPS for multi-axle vehicles intended for the transportation of non-standard cargoes in non-standard operating conditions. Hydraulic circuits of such vehicles are usually characterized by the absence of diagonal connections to control any wheel in case of main line breakdown, and the number of HPS's functions is minimized to ensure the greatest simplicity, reliability and maintainability.

The purpose of this work is to create algorithms for controlling HPS of a vehicle with a wheel formula 8x8, ensuring the process of uniform raising and lowering of the platform when changing the clearance and leveling, as well as the correct process of hanging the wheel ("jack" mode). In the process, computer simulation methods were used, and the results were verified by testing the "Model-in-the-Loop" (MiL) and "Hardware-in-the-Loop" (HiL).

1. Clearance control algorithms

Changing the clearance is effected by moving the rod up or down for one or more hydropneumatic springs. Based on the technical requirements, the following functions should be implemented in the algorithms:

- • setting the clearance value;
- • leveling of the vehicle body within the working area of the suspension travel;
- • hanging of individual wheels.

The input signals for the control algorithms embedded in the electronic suspension control unit will be wheel movements, roll and trim values of the vehicle body. The wheel position sensors are designed to determine the current value of the clearance and are included in the feedback circuit of the control loop for changing the clearance of each hydropneumatic spring. The readings of the roll and trim angle sensors are used for the leveling.

The executive elements used to change the clearance are:

- • one-way hydraulic control valves for injection and discharge from the cavities of the hydropneumatic springs;
- • one-way hydraulic control valves on suspension control units for injection and discharge from piston and rod cavities;
- • proportional valves on suspension control units for injection and discharge from piston and rod cavities.

Structurally, vehicle is divided into 2 trolleys; each one is controlled by a single suspension control unit. On the front trolley wheels are located from 1 to 4, and on the back – from 5 to 8, respectively.

To implement the control algorithms, a deterministic finite state automaton was developed in the Stateflow environment of the MATLAB system. A generalized scheme of its states and transitions is presented in the figure 1.

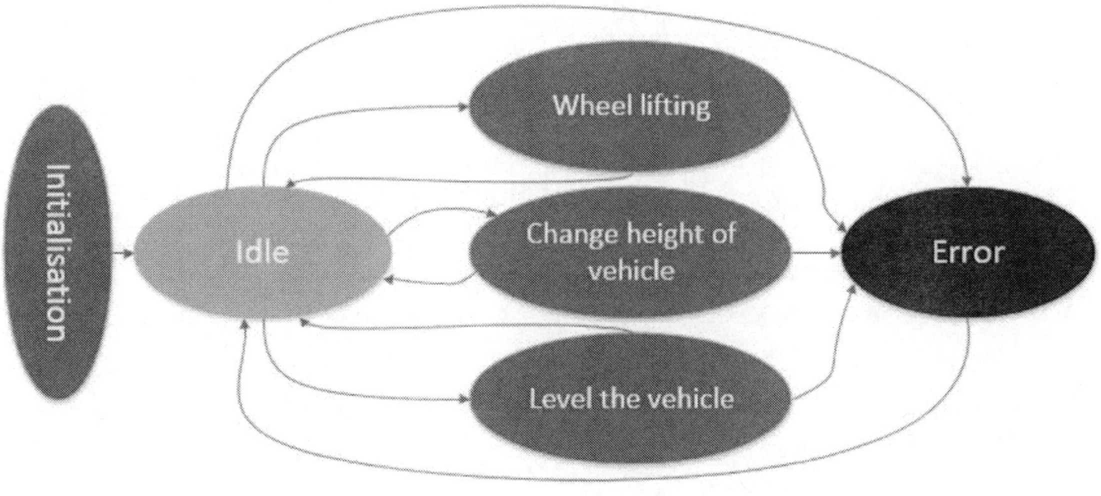

Figure 1. Generalized state machine scheme.

The control process is carried out simultaneously for two trolleys. In each of them there is a choice between lifting and lowering the vehicle, with descent being of higher priority. The HPS control algorithms were implemented in the MATLAB / Simulink system using a model-based approach. An example of the implementation of the lifting function using the Stateflow tool is shown in figure 2, an example of the controller structure is shown in figure 3.

The increase or decrease in clearance is achieved by injecting or discharging the working fluid from the piston / rod cavity of hydropneumatic springs due to the coordinated operation of one-way control valves and proportional valves on the suspension control unit. Opening / closing of the hydraulic locks occurs when the specified height of the rack is reached. Similarly the vehicle body levelingand wheel hanging are realized.

For uniform descent and lifting of the vehicle PI-regulators are used which determine the duty cycle of the PWM signal for opening proportional valves.

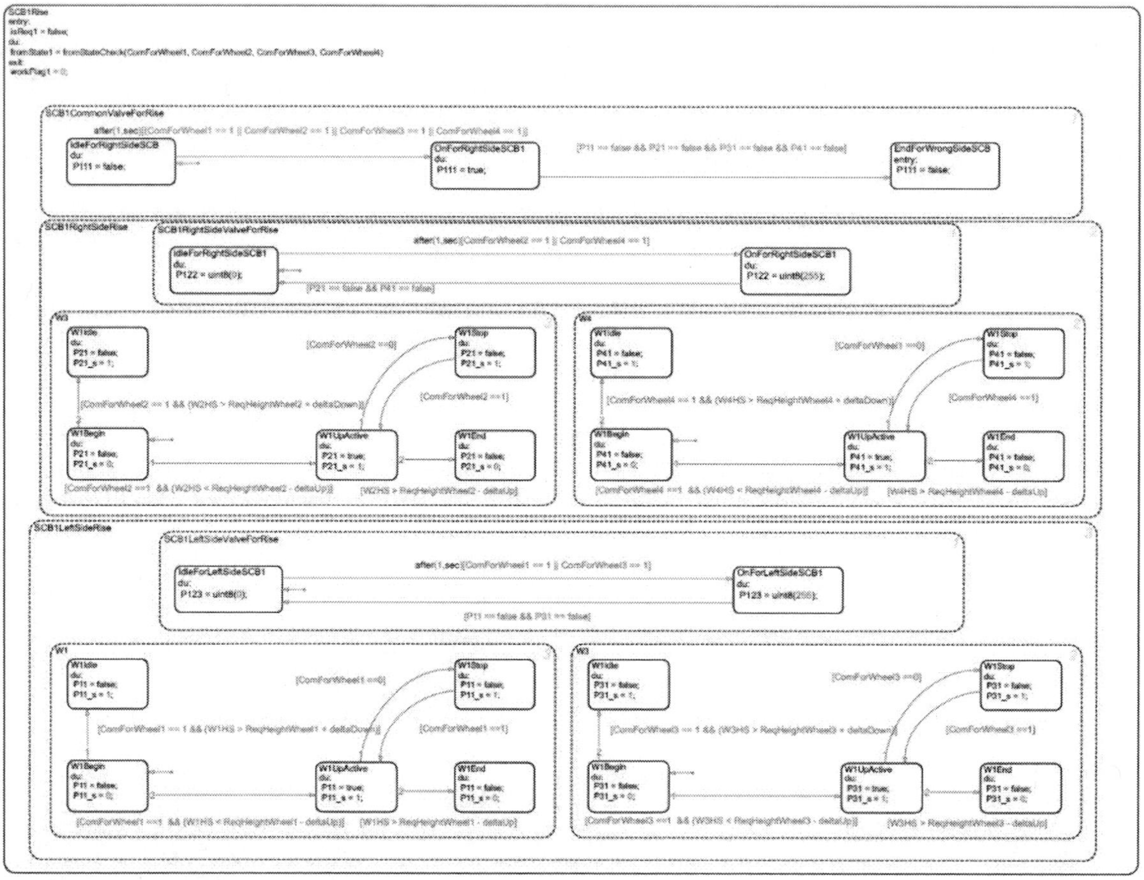

Figure 2. Implementation of finite state automaton for lifting function.

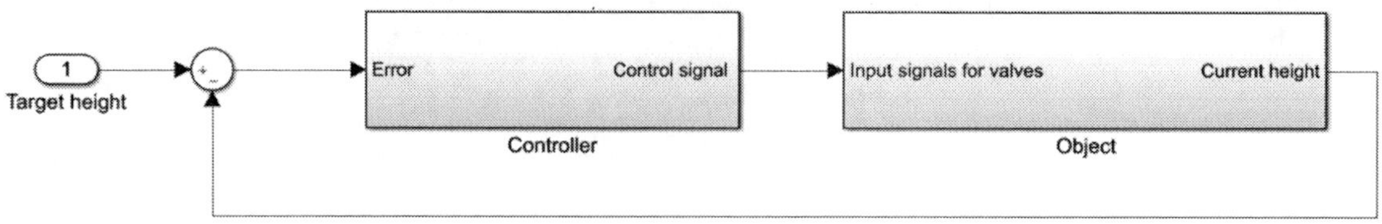

Figure 3. Controller structure for lowering and lifting.

2. Algorithm testing

Verification of the developed clearance control algorithms took place in two stages. At the first stage, MiL-testing was carried out, in which the control system model was integrated with the vehicle model in Simulink. In this case, the output signals of the first model and the input signals of the second model were the control signals coming to the valves. In this model, it is possible to track the correctness of the algorithms of lifting and lowering the vehicle without feedback (i.e., without position sensors), as well as to estimate the change in pressure in the rod and piston.

At the second stage, the control algorithms were tested by HiL-testing. From the developed model in Simulink was generated code in C. Then the code was loaded into the rapid prototyping unit, replacing the real electronic control unit in the process of the control system debugging. The input signals for this model were the position sensor readings coming from the CAN bus, and the output signals were the control signals coming to the valves. For example, figure 4 shows graphs of changes in output values with an increase in clearance from 300 mm to 700 mm.

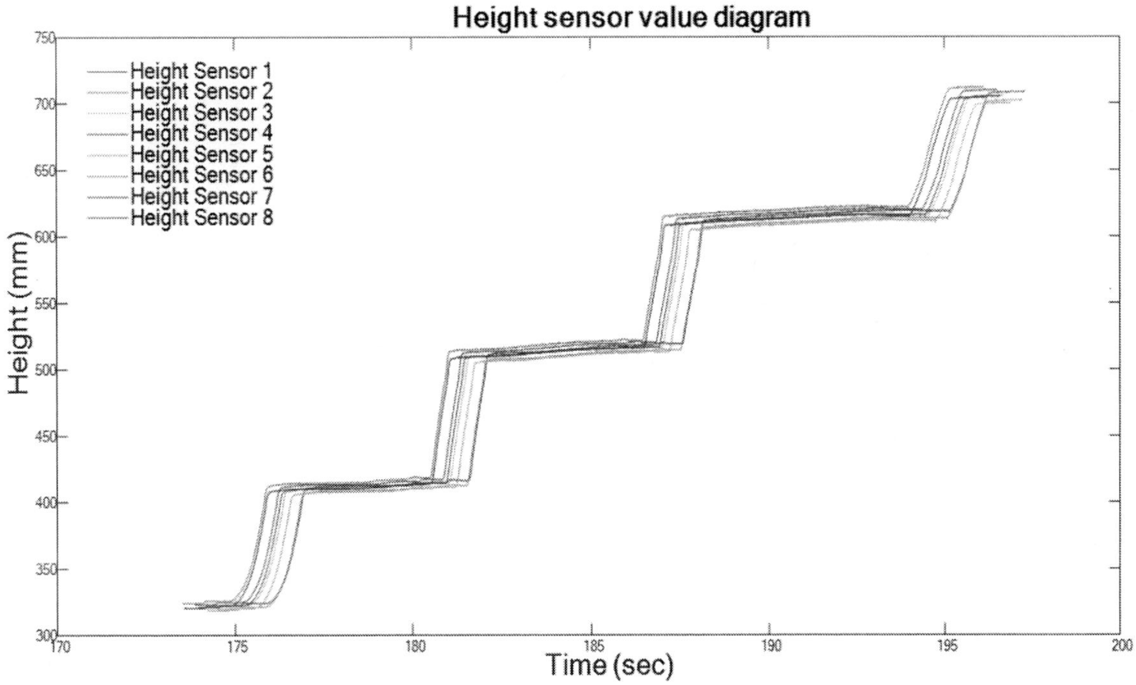

Figure 4. Readings of the position sensors with a step-like increase in clearance.

3. Conclusions

In the course of the work, the control algorithms for the hydropneumatic suspension were formulated and their software implementation in the Simulink environment was proposed.The conducted two-stage verification allowed to draw a conclusion about the correctness of their work: all algorithms ensure the fulfillment of the stated requirements; the vehicle platform rises and descends evenly (the readings of the position sensors are close to each other, which means no distortions).

Acknowledgments

The work is executed at financial support of Russian Foundation for Basic Research (grant № 17-08-00516).

References

[1] Kotiev G O and Padalkin B V 2013 Prediction of mobility of special wheeled chassis at the design stage *Inzhenernyzhurnal: Naukai Innovaciy* **3** 29

[2] HodesI V, DerbentsevA N and Egorov P G 2009 Influence of suspension on vehicle handling *Gruzovik* **9** pp 2–6

[3] Gorobtsov A S, Polyakov U A and Lebedev A I 2014 Influence of damping coefficients of adaptive hydropneumatic suspensions on the vehicle vibration load *VestnikMGTU im. N.E. Baumana. Seriya: Mashinostroenie* **2** pp 122–32

IOP Conf. Series: Materials Science and Engineering **489** (2019) 012054 doi:10.1088/1757-899X/489/1/012054

Features of the mathematical model for the control forces calculation of the four-wheeled robot

A V Leonard, S A Tverdohlebov, A E Gavrilov, S A Ustinov and A A Burhanov,

Automatic Units Department, Volgograd State Technical University, Volgograd, Lenin avenue, 28, 400005, Russia

alex-leonard@yandex.ru

Abstract. The article discusses the issue of determining program forces in drives of a four-wheel robotic platform - the issue of opening the uncertainty of the equations system. The motion laws are synthesized on the basis of application of the calculus of variations. The method of determining the program forces is based on the compilation of equations of the dynamics of a rigid body motion and additional equations - "zero-functions" (displacements) of the points of the application of redundant reactions.

1. Introduction

Formation of the laws of motion, both for walking machines and for traditional vehicles, can be based on methods of the calculus of variations – minimization of criteria in the form of linear convolutions of integral indicators with weighting coefficients [1, 2]. In general, the formula of the criteria is a functional as follows:

$$I = \sum_{i=1}^{n} k_i \int_0^{\tau} f_i\left(x,\dot{x},\ddot{x},..,y,\dot{y},\ddot{y},..,\varphi,\dot{\varphi},\ddot{\varphi},..,x_1,..,x_j,..,y_1,..,y_j,..,p_j,..\right)dt \to \min, \quad (1)$$

where k_i – the weights that are set by experts or determined by the robot as a result of the learning algorithm (other conditions); x, y – the coordinates of the robot's characteristic point when it moves along a plane; x_j, y_j – the coordinates characteristic points of the obstacles (moving or static); φ – the angular coordinate defining the orientation of the robot body on the plane; p_j – the parameters (in particular, the radii of the circles, describing the obstacles); τ – the time of performing the task by robot. In this case, the integrands are written in the form, which contains variables or their linear combinations in powers of two. It's necessary that criteria have the ordinary minimum and method is worked too. And then, the conditions of criterion's minimum are:

IOP Conf. Series: Materials Science and Engineering **489** (2019) 012054 doi:10.1088/1757-899X/489/1/012054

$$\begin{cases} \dfrac{\partial \Phi}{\partial \varphi} - \dfrac{d}{dt}\left(\dfrac{\partial \Phi}{\partial \dot{\varphi}}\right) + \dfrac{d^2}{dt^2}\left(\dfrac{\partial \Phi}{\partial \ddot{\varphi}}\right) - \dfrac{d^3}{dt^3}\left(\dfrac{\partial \Phi}{\partial \dddot{\varphi}}\right) + \ldots = 0, \\ \dfrac{\partial \Phi}{\partial x} - \dfrac{d}{dt}\left(\dfrac{\partial \Phi}{\partial \dot{x}}\right) + \dfrac{d^3}{dt^3}\left(\dfrac{\partial \Phi}{\partial \ddot{x}}\right) - \dfrac{d^3}{dt^3}\left(\dfrac{\partial \Phi}{\partial \dddot{x}}\right) + \ldots = 0, \\ \dfrac{\partial \Phi}{\partial y} - \dfrac{d}{dt}\left(\dfrac{\partial \Phi}{\partial \dot{y}}\right) + \dfrac{d^2}{dt^2}\left(\dfrac{\partial \Phi}{\partial \ddot{y}}\right) - \dfrac{d^3}{dt^3}\left(\dfrac{\partial \Phi}{\partial \dddot{y}}\right) + \ldots = 0, \end{cases} \tag{2}$$

$$\Phi = \sum_{i=1}^{n} k_i f_i\left(x, \dot{x}, \ddot{x}, \ldots, y, \dot{y}, \ddot{y}, \ldots, \varphi, \dot{\varphi}, \ddot{\varphi}, \ldots, x_1, \ldots, x_j, \ldots, y_1, \ldots, y_j, \ldots, p_j, \ldots\right), \tag{3}$$

Here are the program laws of robot's motion in differential form. After that, the problem of integrating the obtained system of equations is solved for given boundary conditions. To obtain the values of the motion laws, the numerical method of integration is mainly used. The method of integration is included in the method of multi-criteria optimization to achieve exact values of the parameters set at the boundaries. The question of numerical integration at boundaries is a separate task and will be discussed in details in another article. As a result of solving the boundary task, tables of program laws values are obtained. In order to adapt the proposed approach to the control of walking machines [3-5], control refinement based on the variational calculus is implemented on a laboratory model of a wheeled platform (see figure 1).

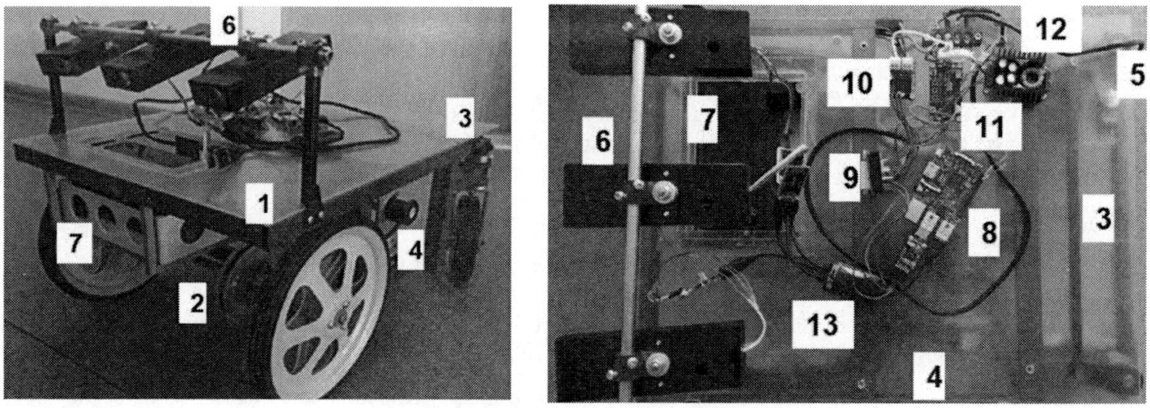

Figure 1. Robotic wheeled platform.

In Figure 1: 1 – wheeled platform; 2 – drive course movement; 3 – rotational drive; 4 – angular encoder; 5 – servo; 6 – a group of laser range finders; 7 – 12 V battery; 8 – on-board computer Raspberry Pi 3; 9 – Arduino Nano controller; 10 – Driver amplifier; 11 – voltage converter 5 V; 12 – servo current limiter; 13 – USB hub.

2. Determination of additional force equations

To implement the motion laws obtained from the extremum conditions (2) of integral target functions (1), it is necessary to take into account the features of robot design, the organization features of its drive and steering. Taking this into account, the features of the robot are reduced to the definition of a force factors system (see figure 2) describing the effect of the environment and actuators on the device. Ultimately, a system of algebraic equations is compiled, from which the programmed drive and steering forces necessary to implement the synthesized motion laws are determined [6, 7]. Using

IOP Conf. Series: Materials Science and Engineering **489** (2019) 012054 doi:10.1088/1757-899X/489/1/012054

the example of a robotic platform with four wheels, the number of required force factors determining the dynamics of the robot's movement is reduced to ten in advance: F_1^τ, F_2^τ, N_1, N_2, N_3, N_4, F_1^n, F_2^n, F_3^n, F_4^n, a $L = L^* = L_{if}$. For the closure of the dynamic equations system of robot's motion four equations are missing. These equations can be obtained from another hypotheses that relate to the object of research.

Three possible ways to obtain additional forces ratios include:

- The rejection of the hypothesis of hardness of wheels and the support base - taking into account their elastic (ductile) properties.
- Applying control to the supports reactions.
- Taking into account the elastic properties of the platform design.

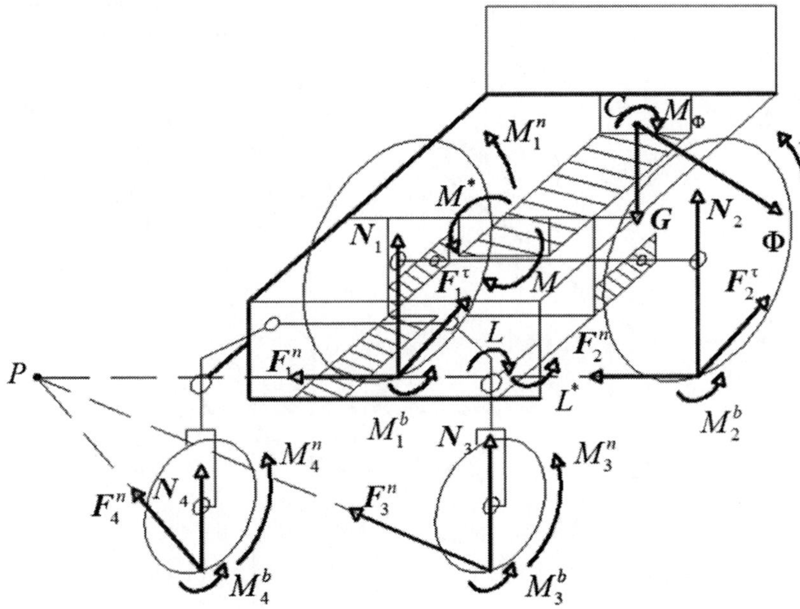

Figure 2. A complete system of force factors determining the movement of the platform on the plane.

For example, the first case (soft wheels and base) requires preliminary identification of the wheel stiffness and installation of four sensors into the control system (excluding the effect of vibrations) that can measure the vertical deformations of four wheels rubber with high precision (in fractions of millimeters). Since the robot control system is already configured specifically: servo, angle encoder, inertial navigation device, terrain laser scanners, the first approach of determining unknown reactions is not considered.

The second approach also requires changes in the robot's design - installation of pneumatic wheels with sensors and devices for distribution and continuous maintenance of pressure integrated into the control system. In the case of recording four additional relations between forces and pressures, the system of equations of motion dynamics will close and it will be possible to determine all unknowns. The obtained pressure values together with the control moments must be reproduced by the robot in the process of implementing the program motion.

The third way to obtain additional relations between forces does not require changes in the physical design of the robot and modernization (complication of the control system). This approach is based on the application of the method of forces - the method of opening the static uncertainty of structures through the deformations recording (functions equal to zero / canonical equations). A robot moving with small accelerations (with real deformations close to zero) can be considered to be in equilibrium under the action of the considered force factors system (see figure 2). To accurately compile canonical equations by the forces method, it is necessary to take into account the gravity of each core element of the structure. Release from internal relations in the core elements of the frame will lead to the compilation of about 50 canonical equations. To simplify the mathematical model for calculating the internal forces of the platform, it is assumed that the platform consists of rods with the same cross section and material, and the gravity force of the entire platform is concentrated in one point belonging to the frame. As a result, after analyzing the conditions of the kinematic immutability of an equivalent construction (discarding redundant bonds), a calculation scheme is obtained for compiling additional 5 equations to 6 known and independent of them (see figure 3).

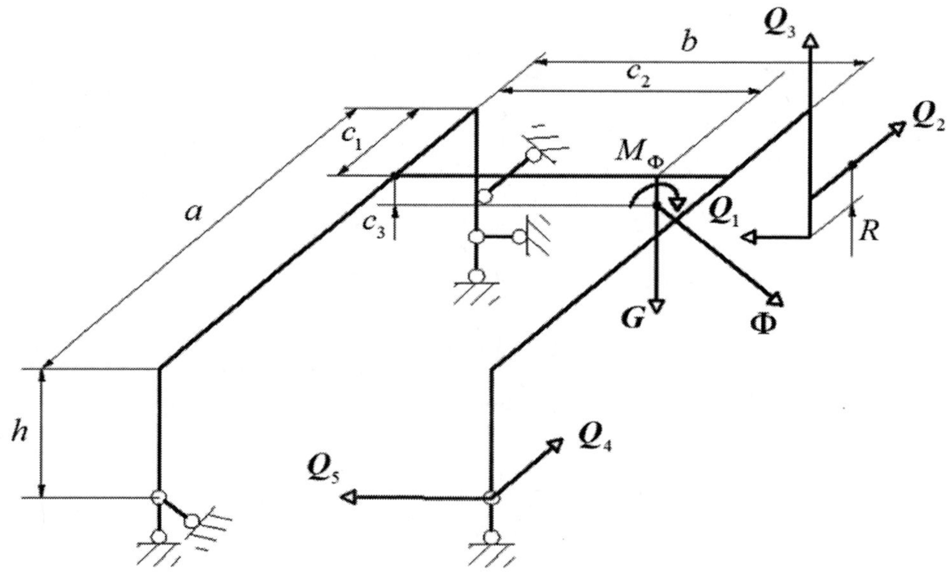

Figure 3. Calculation scheme for obtaining additional forces equations.

For the convenience of compiling canonical equations, the designation of unknown forces varies: $Q_1 = F_2^n$, $Q_2 = F_2^\tau$, $Q_3 = N_2$, $Q_4 = T_3$, $Q_5 = T_4$, and: $F_3^n = \sqrt{T_3^2 + T_4^2}$. Since the equivalent construction is at rest, the displacement of the points of application forces along the directions of these forces from the all forces action must be zero:

$$\begin{cases} \Delta_1 (Q_1, Q_2, Q_3, Q_4, Q_5, G, \Phi, M_\Phi) = 0, \\ \Delta_2 (Q_1, Q_2, Q_3, Q_4, Q_5, G, \Phi, M_\Phi) = 0, \\ \Delta_3 (Q_1, Q_2, Q_3, Q_4, Q_5, G, \Phi, M_\Phi) = 0, \\ \Delta_4 (Q_1, Q_2, Q_3, Q_4, Q_5, G, \Phi, M_\Phi) = 0, \\ \Delta_5 (Q_1, Q_2, Q_3, Q_4, Q_5, G, \Phi, M_\Phi) = 0. \end{cases} \quad (4)$$

Then, using the principle of independence of the forces action, the sums of displacements are recorded:

$$\begin{cases} \Delta_{1Q_1} + \Delta_{1Q_2} + \Delta_{1Q_3} + \Delta_{1Q_4} + \Delta_{1Q_5} + \Delta_{1G} + \Delta_{1\Phi} + \Delta_{1M} = 0, \\ \Delta_{2Q_1} + \Delta_{2Q_2} + \Delta_{2Q_3} + \Delta_{2Q_4} + \Delta_{2Q_5} + \Delta_{2G} + \Delta_{2\Phi} + \Delta_{2M} = 0, \\ \Delta_{3Q_1} + \Delta_{3Q_2} + \Delta_{3Q_3} + \Delta_{3Q_4} + \Delta_{3Q_5} + \Delta_{3G} + \Delta_{3\Phi} + \Delta_{3M} = 0, \\ \Delta_{4Q_1} + \Delta_{4Q_2} + \Delta_{4Q_3} + \Delta_{4Q_4} + \Delta_{4Q_5} + \Delta_{4G} + \Delta_{4\Phi} + \Delta_{4M} = 0, \\ \Delta_{5Q_1} + \Delta_{5Q_2} + \Delta_{5Q_3} + \Delta_{5Q_4} + \Delta_{5Q_5} + \Delta_{5G} + \Delta_{5\Phi} + \Delta_{5M} = 0. \end{cases} \quad (5)$$

Since each partial displacement is proportional to the force of its generating, then:

$$\begin{cases} \delta_{11}Q_1 + \delta_{12}Q_2 + \delta_{13}Q_3 + \delta_{14}Q_4 + \delta_{15}Q_5 = -\Delta_{1G} - \Delta_{1\Phi} - \Delta_{1M}, \\ \delta_{21}Q_1 + \delta_{22}Q_2 + \delta_{23}Q_3 + \delta_{24}Q_4 + \delta_{25}Q_5 = -\Delta_{2G} - \Delta_{2\Phi} - \Delta_{2M}, \\ \delta_{31}Q_1 + \delta_{32}Q_2 + \delta_{33}Q_3 + \delta_{34}Q_4 + \delta_{35}Q_5 = -\Delta_{3G} - \Delta_{3\Phi} - \Delta_{3M}, \\ \delta_{41}Q_1 + \delta_{42}Q_2 + \delta_{43}Q_3 + \delta_{44}Q_4 + \delta_{45}Q_5 = -\Delta_{4G} - \Delta_{4\Phi} - \Delta_{4M}, \\ \delta_{51}Q_1 + \delta_{52}Q_2 + \delta_{53}Q_3 + \delta_{54}Q_4 + \delta_{55}Q_5 = -\Delta_{5G} - \Delta_{5\Phi} - \Delta_{5M}, \end{cases} \quad (6)$$

where δ_{ik} – partial displacement of the i-th point of force's application from the action of a single value force in the direction of the k-th force's factor. The proportionality coefficients are determined through Mohr integrals:

$$\delta_{ik} = \int_l \frac{M_{Kk}M_{Ki}}{GJ_K} dl + \int_l \frac{M_{xk}M_{xi}}{EJ_x} dl + \int_l \frac{M_{yk}M_{yi}}{EJ_y} dl + \int_l \frac{N_k N_i}{EF} dl + \int_l \frac{k_x L_{xk} L_{xi}}{GF} dl + \int_l \frac{k_y L_{yk} L_{yi}}{GF} dl, \quad (7)$$

where M_{Ki} – internal torsional moment; M_{xi} – internal bending moment relative to the main axis of inertia of the cross section of the core element of the structure parallel to the support base; M_{yi} – internal bending moment relative to the main axis of inertia of the cross section of the core element, perpendicular to the support base; N_i – internal tensile force; L_{xi} – internal shear force parallel to the section axis x; L_{yi} – internal shear force parallel to the section axis y.

All the above, internal force factors arise from the action of the i-th unit force. Given the orthogonality of forces, the system of canonical equations will take the following form:

$$\begin{cases} \delta_{11}Q_1 + \delta_{15}Q_5 = -\Delta_{1\Phi} - \Delta_{1M}, \\ \delta_{22}Q_2 + \delta_{24}Q_4 = -\Delta_{2\Phi}, \\ \delta_{33}Q_3 = -\Delta_{3G}, \\ \delta_{42}Q_2 + \delta_{44}Q_4 = -\Delta_{4\Phi}, \\ \delta_{51}Q_1 + \delta_{55}Q_5 = -\Delta_{5\Phi} - \Delta_{5M}. \end{cases} \quad (8)$$

Mohr integrals are calculated using the Vereshchagin method. For this purpose, all the diagrams of internal force factors are constructed from the action of each external force factor (reactions and active forces). Analysis of internal forces begins with the definition of a complete system of external forces (reactions). For example: the equilibrium equations determine the reactions from the action of a single force in the direction of the force Q_1 (see Figure 4):

IOP Conf. Series: Materials Science and Engineering **489** (2019) 012054 doi:10.1088/1757-899X/489/1/012054

$$
\begin{aligned}
&\sum M_{DD_1} = 0: F_1^n a - 1a = 0 \Rightarrow F_1^n = 1;\\
&\sum Y_i = 0:\ -1 + F_1^n - F_4^n f_y = 0 \Rightarrow F_4^n = 0;\\
&\sum X_i = 0:\ F_4^n f_x - F_1^\tau = 0 \Rightarrow F_1^\tau = 0;\\
&\sum M_{D_1A_1} = 0: \left(F_1^n - 1 - F_4^n f_y\right)h + N_3 b = 0 \Rightarrow N_3 = 0;\\
&\sum M_{A_1B_1} = 0: \left(N_3 + N_4\right)a - F_1^\tau\left(h - R\right) = 0 \Rightarrow N_4 = 0;\\
&\sum Z_i = 0: N_1 + N_3 + N_4 = 0 \Rightarrow N_1 = 0.
\end{aligned} \tag{9}
$$

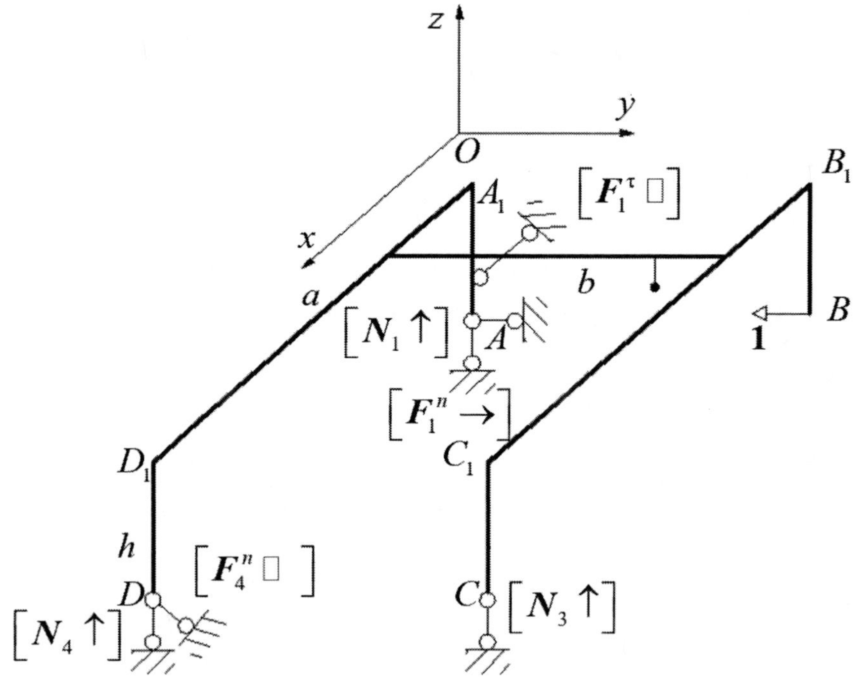

Figure 4. The scheme for calculating reactions from the action of a single force in the direction of Q_1.

From the analysis of equations (9) it follows that the structure is in equilibrium under the action of only two: external and internal unit forces. Since the structure is symmetrical and symmetrical with the load, to determine the δ_{11} internal forces diagrams are constructed for a half frame (see figure 5). For the second half of the diagram will be symmetric.

IOP Conf. Series: Materials Science and Engineering **489** (2019) 012054 doi:10.1088/1757-899X/489/1/012054

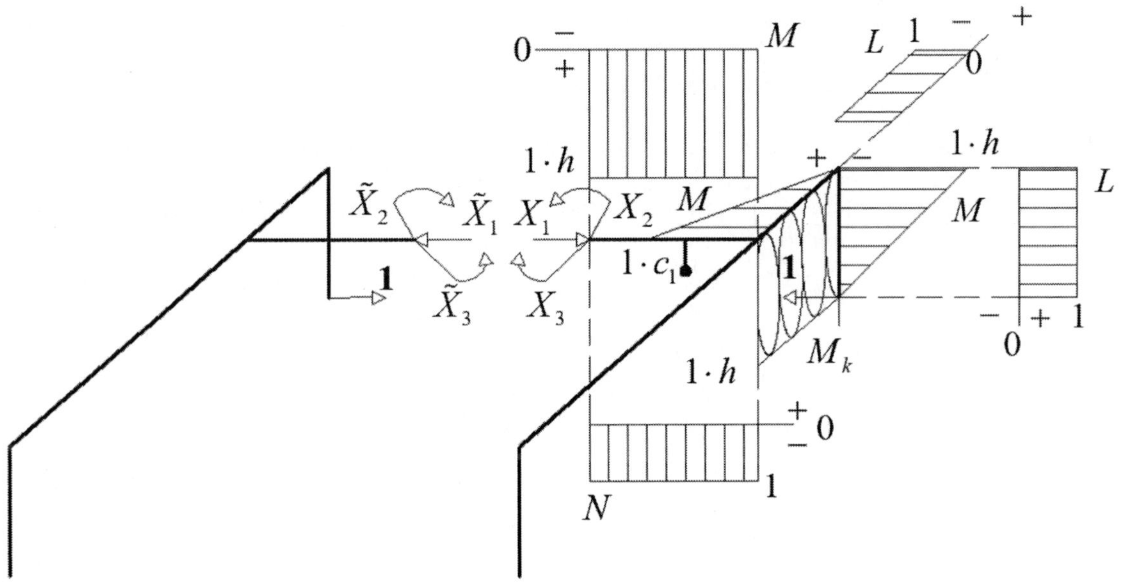

Figure 5. – Diagrams of internal efforts from the action of a single force in the direction Q_1

Considering symmetry diagrams of internal forces, δ_{11} is:

$$\delta_{11} = 2\frac{1}{GJ_{л}}\left[1hc_1 1h\right] + 2\frac{1}{EJ}\left[\frac{1}{2}hh\frac{2}{3}h + \frac{1}{2}c_1c_1\frac{2}{3}c_1 + \frac{b}{2}hh\right] + 2\frac{1}{EF}\left[1\frac{b}{2}1\right] + 2\frac{k}{GF}\left[1h1 + 1c_1 1\right]. \quad (10)$$

3. Conclusion

The proposed approach of compiling additional equations (by the method of forces) allows one to determine software control forces for multiwheel transport robotized systems that move along a fairly rigid foundation without adjusting the pressure in the wheels (with a small deformation of the rubber-casing shell). The mathematical model does not describe fully the design features of the robot, so the inaccuracy of the forces calculation can be compensated (partially or completely) with the control system for the values of the parameters coming from the sensors.

Acknowledgments

The research was supported by Russian Foundation for Basic Research (Project No. 16-38-60086 mol_a_dk).

References

[1] Briskin E S, Maloletov A V, Sharonov N G, Kalinin Ya V, Leonard A V and Serov V A, Shurygin V A 2016 Walking robot «character» as element of intelligent system *Advances in Cooperative Robotics : Proc. of the Int. Conf. on CLAWAR 2016* ed. by Mohammad O. Tokhi, Gurvinder S. Virk (London, UK, 12-14 September) pp 386-94

[2] Statnikov R B, Matusov J B 1995 *Multicriteria Optimization and Engineering* Springer p 236

[3] Briskin E S, Leonard A V 2013 *J. of Computer and Systems Sciences Int.* **52** pp 972-9

[4] Briskin E S et al 2014 Problems of Increasing Efficiency and Experience of Walking Machines Elaborating *Advances on theory and practice of robots and manipulators : Proc. of ROMANSY 2014 XX CISM-IFToMM Symp. on Theory and Practice of Robots and Manipulators* ed. by Marco Ceccarelli, V.A. Glazunov (Moscow, 23-26 June) 22 pp 383-90

[5] Briskin E S, Maloletov A V, Sharonov N G, Fomenko S S, Kalinin Ya V, Leonard A V 2016

IOP Conf. Series: Materials Science and Engineering **489** (2019) 012054 doi:10.1088/1757-899X/489/1/012054

Development of Rotary Type Movers Discretely Interacting with Supporting Surface and Problems of Control Their Movement *ROMANSY 21* ed. by V. Parenti-Castelli, W. Schiehlen (Udine, Italy, June 20-23) 569 pp 351-9

[6] Briskin E S, Leonard A V 2016 Energy profile and the open-loop control of the translational motion of the walking machine Cyclone *J. of Computer and Systems Sciences Int.* **55** pp 1001-9

[7] Okhotsimskii D E and Golubev Yu F 1984 *Mechanics and Motion Control of Walking Machine* (Nauka, Moscow) p 310

IOP Conf. Series: Materials Science and Engineering **489** (2019) 012055 doi:10.1088/1757-899X/489/1/012055

Energetically efficient modes of the walking robots movement at its displacement across an uneven surface

M V Miroshkina, E S Briskin and Y V Kalinin

Department of Theoretical Mechanics, Volgograd State Technical University, 28 Lenin Avenue, Volgograd 400005, Russia

dtm@vstu.ru

Abstract. There is the solution of one of the problems arising in the design of mobile walking robots: the problem of achievement of minimum of energy costs at displacement of a walking robot, increasing the efficiency of a multi-legged mechanism and a reduction of the developed power due to a change in the mode of motion of the multi-legged walking apparatus.

Currently, the need for mobile walking robots is quite large. This is due to the fact that the use of walking robots is the most effective compared to tracked and wheeled vehicles and is widely used at moving in difficult terrain, on a ground surface, the rescue operations in extreme situations, in space exploration in the military sphere.

One of the current challenges facing the developers of multi-legged walking machines is to increase their energy efficiency, as well as reduce the dynamic loads acting on the actuators and the machine body. Low energy efficiency and high dynamic loads are caused by the fundamental imbalance of walking movers of various types and their discrete periodic interaction with the supporting surface [7]. Methods for solving the problem can be based and are based on balancing walking movers as mechanisms, the introduction of energy recuperators [1], which convert the kinetic energy of the transported propulsion into the potential energy of some elastic element and vice versa. With the same purpose, one can purposefully control the gait [3] and the character of the carried movement of the walking mover [4, 5 and 6].

The representative of walking robots is the experimental apparatus "Ortonog" (figure 1). The walking apparatus with the original orthogonal walking movers is intended for testing control systems and demonstrating its capabilities.

Using this mobile robot as the example, we can examine in detail the nature of the motion of the walking machine. The undercarriage of the robot is represented by four blocks of orthogonal walking movers. Each of the four blocks of the undercarriage is conventionally divided into three elements: the mechanisms of vertical movement, the mechanisms of horizontal movement and rotation mechanisms. The first are composed of electric cylinders and are responsible for raising and lowering circular supports, placed on the sliding rods. Ortonog has eight such supports, two for each block of the chassis. Mechanisms of horizontal movements, consisting of guides and actuators, are designed to move the cylinder and rod with the support. Each block of the undercarriage has two sets of guides and drives, one for each cylinder with a support. Finally, the turning gears deploy the entire block of the undercarriage around a vertical axis. The entire undercarriage of the experimental machine is based on electric drives.

IOP Conf. Series: Materials Science and Engineering **489** (2019) 012055 doi:10.1088/1757-899X/489/1/012055

Figure 1. The walking robot "Ortonog".

The movement of the robot "Ortonog" on a flat surface can be performed as follows. During the rise and transfer to a new position of one group of four movers, the other group works in the traction mode and moves the robot body with a transferable group of movers forward. Thus, while driving, the Ortonog alternately uses vertical displacement drives, and the horizontal mechanisms operate continuously in traction or transferable mode. If necessary, the mechanisms of rotation are activated. Their synchronous or separate operation ensures rotation to the required angle. As can be seen from the design, the experimental walker can take turns with a minimum radius.

It is known that by choosing the driving mode, instead of the cubic dependence of heat losses in motor drives on the speed of movement [1], we obtain the dependence inversely proportional to speed [8]. This mode was obtained by studying the translational motion of the robot body with a rectilinear motion of the center of mass and propulsion as material points moving along the horizontal axis. However, when moving the walking mover to a new position, the latter, depending on the type and nature of the obstacles, rises to a certain height, in addition to the course movement, and then sinks to the ground. There is known view of the trajectory of the movement of support of walking movers, which was proposed by Professor N. Umnov (figure 2). This type of trajectory is due to the insufficient development of information-measuring systems in the 70s of the last century and the need to raise and lower the foot in any situation without “stumbling” and horizontal sliding on the ground, regardless of the unidentified profile of a ground.

However, if you have information about the profile of the support surface, you can purposefully change the laws of propulsion of the foot in the course and orthogonal directions to it. In particular, it is possible to reduce the distance traveled by the foot (figure 2, curve 2).

Figure 3 shows the translational, with a constant velocity of the body, movement of the robot with orthogonal walking movers and the movement of the foot of mass m, as a material point, in the vertical plane XOY.

The movement is supported by three drives: the drive of horizontal movement of the body, developing a force Q, that is equaled to the force R of interaction of the foot with the ground; the drive of horizontal movement of the transferable foot, which develops the force P; the drive of the vertical movement of the transferable foot, which develops the force T. The step length along the track is equaled to L, and the distance to a single obstacle of height H is equaled to S.

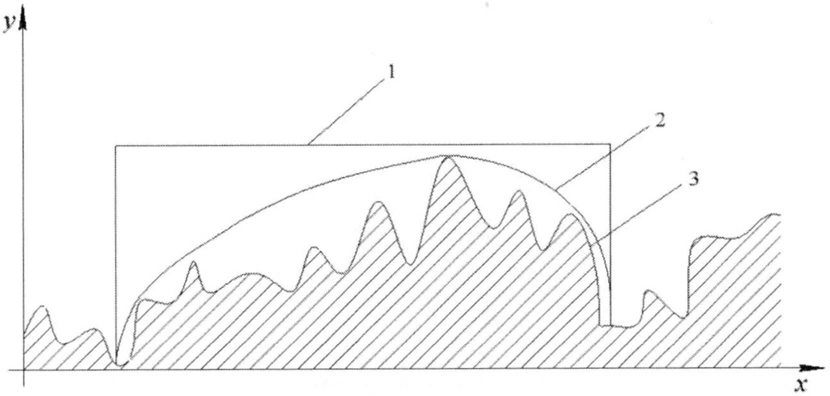

Figure 2. The trajectory of the movement of the foot of the walking mover
1 – absolute trajectory according to N. Umnov; 2 – possible trajectory;
3 – profile of the support surface.

The interaction of the transferable foot with the supporting surface is carried out at a height h. Among the indicators characterizing the quality of movement are taken into account: the level of heat loss W and the root-mean-square acceleration a_σ of the body in the translational course motion.

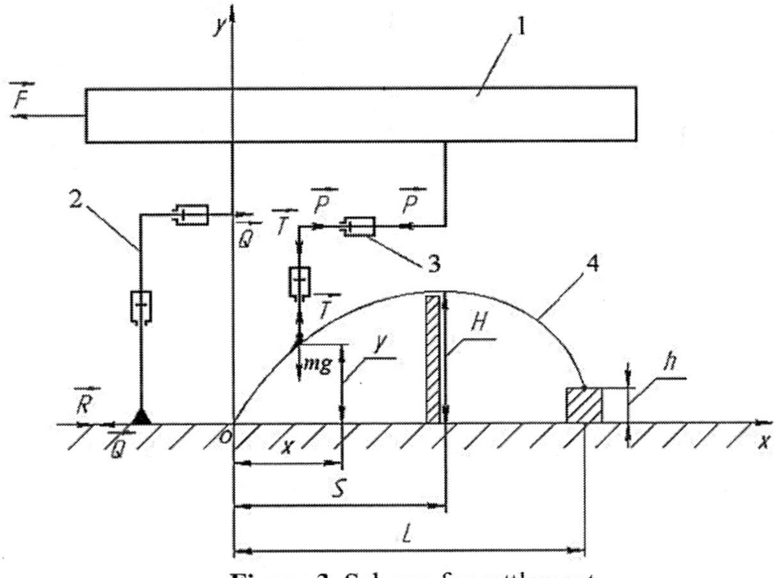

Figure 3. Scheme for settlement
1 – the body of the robot; 2 – the mover, which is interacting with the support surface; 3 – the transferable walking mover; 4 – the trajectory of the movement the foot of the walking mover.

The first indicator characterizes the energy efficiency of the mover, and the second - the comfort of movement. When studying the stationary modes of motion, these indicators are calculated on one period of motion τ (the period of the full cycle of the movement of the mover) [2].

The problem of determining such laws of motion x (t), y (t), which provide both overcoming obstacles and minimality either of level of heat losses W, or of root-mean-square acceleration of a body a_σ, or of complex criterion I, is posed and solved.

Thus, it follows that the movement, in which the path traveled by the foot can be reduced, is carried out by combining the operation of the horizontal and vertical movement mechanisms, therefore, with the simultaneous operation of the drives, the energy consumption should have the minimum value. To achieve the solution of the formulated task, it is necessary to determine the value of the speed with which the leg will be lifted depending on the distance to the obstacle and its height, to develop a control system for organizing this type of movement and to prove experimentally that during such movement of the walking robot, the energy consumption for moving the walking machine will be minimal.

Acknowledgments
The research was supported by Russian Science Foundation (project No. 18-71-10069).

References

[1] Lapshin V V 1985 Model estimates of the energy consumption of the walking apparatus *Mechanics of Solids* **5** pp 28
[2] Okhotsimskiy D E 1989 *Walking machines* (Moscow)
[3] Maloletov A V 2015 *Optimization of the structure, parameters and modes of motion of walking machines with twin movers : monograph* (Volgograd: VolgGTU)
[4] Briskin E S, Kalinin Y V et al 2017 Estimates of efficiency of cycle mechanisms *Mechanics of Solids* no 2 pp 13-9
[5] Briskin E S, Kalinin Y V et al 2017 Assessment of the performance of walking robots by multicriteria optimization of their parameters and algorithms of motion *J. Computer and Systems Sciences Int.* **2** pp 168-76
[6] Briskin E S, Kalinin Y V et al 2017 On controlling the adaptation of orthogonal walking movers to the supporting surface *J. Computer and Systems Sciences Int.* **3** pp 184-90
[7] Briskin E S, Kalinin Y V et al 2014 On the energy efficiency of cyclic mechanisms *Mechanics of Solids* **1** pp 18-25
[8] Okhotsimskiy D E 1984 *Mechanics and motion control of an automatic walking apparatus* (Moscow: Nauka)

IOP Conf. Series: Materials Science and Engineering **489** (2019) 012056 doi:10.1088/1757-899X/489/1/012056

Kinematic study of the weeding robot

A S Ovchinnikov, V S Bocharnikov, N S Vorob'yeva and A G Ivanov

Volgograd state agrarian university, Volgograd, University avenue, 26, 400002, Russia

vgsxa@mail.ru

Abstract. The paper presents the rationale for the structure, the calculation of the geometric and kinematic parameters of the robot for point removal of weeds. The working area of the output link of the weeding robot was obtained, and limitations were determined to prevent its falling into the dead position.

1. Introduction

With the mechanical destruction of weeds, the cultivator's working parts (paws) process only the areas of row spacing, without affecting the weeds located in rows between cultivated plants. Besides, the larger size of the cultivated plant, the greater the likelihood of its damage during continuous cultivation of row spacing.

2. Geometric synthesis

To solve the problem of complete weeding both in row spacings and in rows, the design of a robot has been developed for point weed removal (figure 1).

Figure 1. Mobile weeding robot.

IOP Conf. Series: Materials Science and Engineering **489** (2019) 012056 doi:10.1088/1757-899X/489/1/012056

The mechanism of the robot is a flat mechanism of a parallel-sequential structure and has five degrees of mobility of the working arm - milling cutters, which are also the generalized coordinates W = q = 5 [1, 2].

In figure 2 shows the structural scheme of the robot.

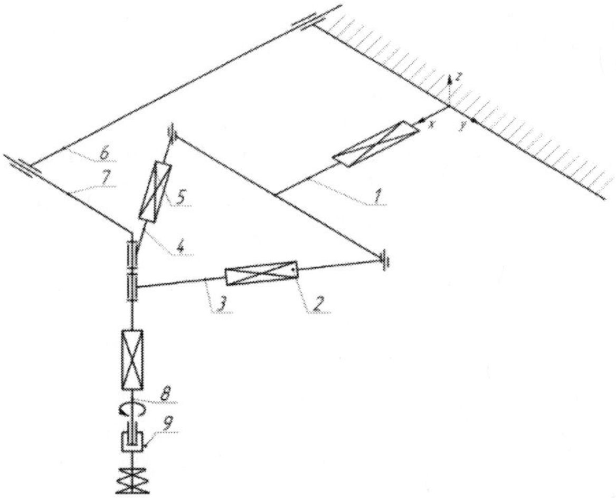

Figure 2. Structural scheme of the robot.

The degree of mobility of the output link 9 of the robot [2]

$$W=3n-2p5-p4 \quad = 3 \cdot 9 - 2 \cdot 11 = 5 \qquad (1)$$

The optimal geometrical dimensions of the executive links and their attachment points to the base of the robot are determined from the conditions for the implementation of the required movement, the required service area and the limitations imposed by the parameters [3, 4]: distance between plants (d) and protective zone (e) as shown in figure 2. To obtain the required service area, which would cover a certain number of rows - when designing a construction worker, it is necessary to take the distance (a) between the attachment points of the actuator links 2, 3.

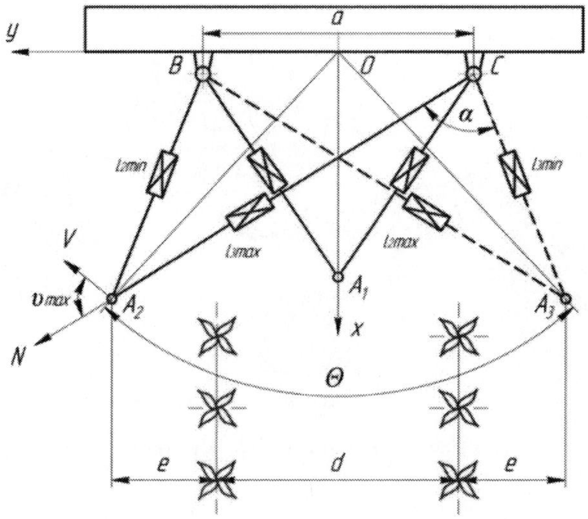

Figure 3. The design scheme of executive units 2, 3 (top view).

For the main geometric constraints in the synthesis of the robot on the basis of a flat mechanism we take the following indicators: the angle of the sector of coverage in the horizontalplane $\Theta \geq 90^0$; maximum departure of executive units l_{max}> [l].

The largest sector angle Θ is it turns out when the actuator length l_3 is minimal, and l_2 is maximum in the extreme right position, and vice versa l_3 is maximum, and l_2 is minimal in the extreme left position. Since the mechanism is symmetric with respect to the axis *Oy*, it is sufficient to consider its motion with a constant minimum length of the actuator l_2 and a change only l_3.

Taking the above assumptions (l_2=*const)*, we obtain a flat replacement circuit with a swinging actuator l_3 and a rocker l_2. Then, when going from one extreme position A_2 to the other, A_3, the length of the actuator l_2 changes from l_{2min} to l_{2Max}, turning the actuator l_3 through the angle α. The elongation of the actuator is denoted by $\Delta l = l_{2Max} - l_{2min}$, the ratio of the length of the stationary actuator to the rod stroke is the coefficient of the actuator elongation $k = l_2 / h$. For electric cylinders (actuators), the range of variation k =1,56 is subsequently taken.

It is important to consider that the performance of the mechanism and its efficiency factor largely depend on the pressure angle υ, the angle between the axis of the actuator and the velocity vector of the point of application of force. The allowable pressure angle in lever mechanisms, in general, should not exceed $[\upsilon] = 60^0$. A rational in terms of dimensions scheme of the mechanism is realized under the condition $\upsilon_{max} = [\upsilon]$, but on the other hand, the smaller [υ], the smaller the friction loss in hinges.

From the equation of the vector contour [5] for the scheme shown in Figure 3, we get the expression

$$(k+1)^2 \cdot \Delta l^2 - 4l_{3_{min}} \cdot \sin(\frac{\alpha}{2}) \cdot (k+1) \cdot \cos([\upsilon] - \frac{\alpha}{2}) \cdot \Delta l + 4l_{3_{min}}^2 \cdot \sin^2(\frac{\alpha}{2}) - l_{2_{min}}^2 = 0, \quad (2)$$

whose solution is

$$\Delta l = -\frac{b}{2} + \sqrt{\frac{b^2}{4} - c}, \quad (3)$$

$$b = -4l_{3_{min}}(k+1) \cdot \sin(\frac{\alpha}{2}) \cdot \cos([\upsilon] - \frac{\alpha}{2}) / (2k+1), \quad (4)$$

$$c = 4l_{3_{min}}^2 \cdot \sin^2(\frac{\alpha}{2}) / (2k+1). \quad (5)$$

After determining the solution of equation (2) is the center distance a

$$a = \sqrt{l_{3_{min}}^2 + (l_{2_{min}} + \Delta l)^2 - 2l_{3_{min}}(l_{2_{min}} + \Delta l) \cdot \sin[\upsilon]}.$$

Substituting l_{2min} = 700 mm and l_{3min} = 700 mm - the size of the actuator, as well as $\Delta l = l_{2Max} - l_{2min}$ = 450 mm - the stroke of the actuator rod, we get a = 646 mm.

The angle of rotation of the actuator α depends on the angle of the sector of the capture zone Θ as

$$\alpha = \frac{\Theta + [\upsilon] - \pi/6}{2}.$$

Based on all previously found parameters, we can determine the maximum departure of point A from the edge of the base [6]

$$L_{\max} = \sqrt{l_{2\max}^2 - 0{,}25a^2} = \sqrt{1150^2 - 0.25 \cdot 646^2} = 1103{,}7\,\text{мм}.$$

3. Definition of service area

One of the tasks of the kinematic analysis of robot is to determine the configuration of the service area, which is the spatial aggregate movement of the point M (figure 4).

The theoretical zone of possible movements of the point M depends on the ratio of the lengths of the links $l_k, k = 1 \div 4$, which are further taken as generalized coordinates, and on the relative position of the attachment points of the actuators l_2 and l_3 on the frame.

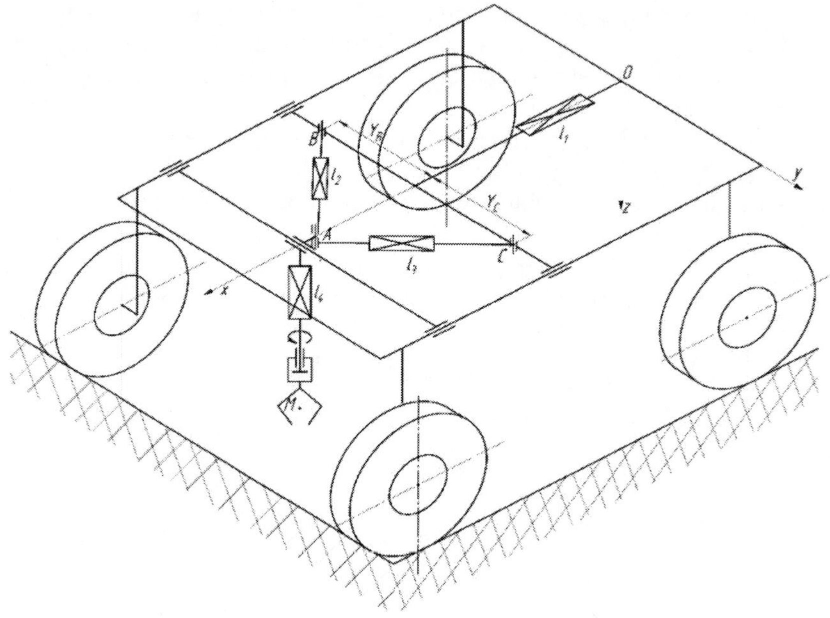

Figure 4. The kinematic scheme of the robot.

The coordinates of the point M of the working body in the absolute coordinate system O_{XYZ} are:

$$\begin{cases} X_M = l_1 + X_A \\ Y_M = Y_A \\ Z_M = l_4 \end{cases} \quad (9)$$

where X_A and Y_A - are the coordinates of point A, the connection of the actuator rods l_2, l_3.

The coupling equations (10) between the coordinates of point A, the lengths of the links l_2, l_3, and the relative positions of their attachment points are:

$$\begin{cases} l_2^2 = X_A^2 + (Y_A - Y_B)^2 \\ l_3^2 = X_A^2 + (Y_A + Y_B)^2 \end{cases} \quad (10)$$

IOP Conf. Series: Materials Science and Engineering **489** (2019) 012056 doi:10.1088/1757-899X/489/1/012056

where Y_B - the actuator mounting coordinate l_2.

Thus, the dependence of the coordinates of the point M (X_M, Y_M, Z_M) from the lengths of the links $l_k, k = 1 \div 4$, and the coordinates of the attachment points of the actuators l_2 и l_3 takes the form:

$$\begin{cases} X_M = l_1 + \sqrt{l_2^2 - \left(\dfrac{l_3^2 - l_2^2}{4Y_B}\right)^2 + \dfrac{l_3^2 - l_2^2}{2} - Y_B^2} \\ Y_M = \dfrac{l_3^2 - l_2^2}{4Y_B} \\ Z_M = l_4 \end{cases} \quad (11)$$

The system of equations (11) is a solution of the direct task of kinematics and completely determines the theoretical possible range of displacements of the point M in space, that is, the working area of service of the robot, and also allows to formulate the conditions excluding its falling into the dead position [7].

The restrictions are imposed on the coordinate Y_B of the point of attachment of the actuators l_2 and l_3.

$$l_2^2 - \left(\frac{l_3^2 - l_2^2}{4 \cdot Y_B}\right)^2 + \frac{l_3^2 - l_2^2}{2} - Y_B \geq 0 \quad (12)$$

$$Y_B \geq \pm \frac{\sqrt{l_2^4 + 2 \cdot l_2^2 \cdot l_3^2 + 4 \cdot l_2^2 + l_3^4 - 4 \cdot l_3^2}}{4} + \frac{l_2^2}{4} + \frac{l_3^2}{4} \quad (13)$$

Knowing the parameters of the lengths of actuators (l_{min} = 700 mm, l_{max} = 1150 mm) and a = 646 mm – the distance between attachment points of actuators l_2 and l_3 on the frame, coordinates Y_B = - Y_C = a/2, solving the system of equations (11) with the help of the Mathcad software package, we obtain the working area of the weeding robot, figure 5 [8].

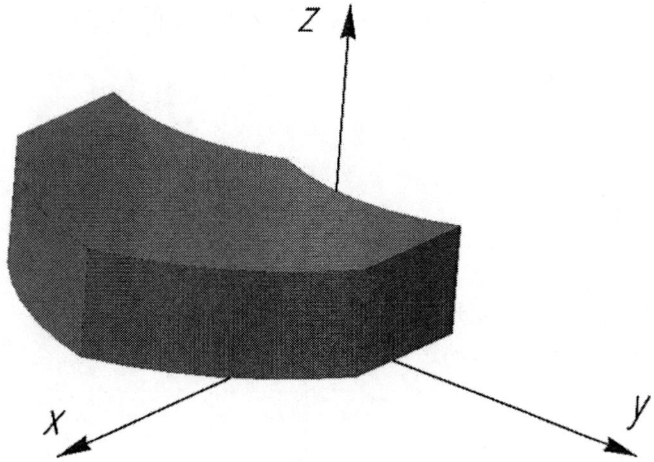

Figure 5. The working area of the robot.

Thus, from the shape of the working area of the robot in Figure 5, it can be seen that the point M moves along the X axis from min (X) = 1321 mm to max (X) = 2254 mm; Y-axis from min (Y) = -644 mm to max (Y) = 644 mm; on the Z axis from min (Z) = 700 mm to max (Z) = 1150 mm.

4. Conclusions

As a result, the degree of mobility of the weeding robot was determined, from the equations of geometric synthesis, a rational axial mounting distance of actuators 2, 3 and the maximum departure of the point of attachment of the working body were obtained. The working area of the robot was obtained, and limitations were determined to prevent its falling into the dead position. The application of the developed methods of calculation and design will allow to reasonably choose the design parameters of a flat manipulator of a parallel-sequential structure at the design stage.

Acknowledgments

The research was supported by Russian Foundation for Basic Research (Project No. 18-41-342002 r_mk).

References

[1] Reshetov L N 1972 Designing rational mechanisms *Ed. 2nd revised and add. Engineering* p 256.
[2] Semenov Yu A, Semenova N S 2003 *Theory of mechanisms and machines* **2** p 3-14.
[3] Gerasun V M et al 2012 The manipulators for mobile robots. Concepts and design of principles *Preprints IPM named by M V Keldysh* **44** p 24.
[4] Zhoga V V, Dyashkin-Titov V V, Dyashkin A V, Vorobyeva N S, Nesmiyanov I A, Ivanov A G 2017 Manipulator-tripod parallel-serial structure *Pat. 2616493 Russian Federation, IPC B66C 23/44* Bul. No. 11
[5] Vorobyeva N S, Dyashkin A V, Dyashkin-Titov V V, Nesmiyanov I A 2017 Structural-geometric synthesis of a manipulator of a parallel-sequential structure *Ecological and reclamation aspects of rational nature management. Materials of the Int. scientific-practical conf.* (Volgograd, 31 January -3 February) pp 71-7.
[6] Nesmiyanov I A, ZhogaV V, Vorobieva N S, Dyashkin-Titov V V 2016 *VP Vibroengineering PROCEDIA* **8** pp 512-6.
[7] Nesmiyanov I A, Zhoga V V, Skakunov V N, Vorob'eva N S, Dyashkin-Titov V V, Bocharnikov V S 2017 *J. of Machinery Manufacture and Reability* **46** pp 232-9 DOI: 10.3103/S1052618817030098.
[8] Gerasun, V M, Zhoga V V, Nesmiyanov I A, Vorobyeva N S, Dyashkin-Titov V V 2013 *Mechanical Engineering and Engineering Education* **3** pp 2-8.

IOP Conf. Series: Materials Science and Engineering **489** (2019) 012057 doi:10.1088/1757-899X/489/1/012057

Developing an algorithm to control the accuracy of the milling of aerospace parts with cellular structure by using copying machine-tools with CNC of "SVO" type

O Pas and N Serkov

Mechanical Engineering Research Institute of the Russian Academy of Sciences, Maliy Kharitonievskiy per. 4, 119334 Moscow, Russia

E-mail: o.pa88@yandex.ru

Abstract. Milling process of machining aerospace parts with cellular structure by using mirror milling system was examined. Different types of errors during machining were considered. Various techniques of reducing errors were defined.

1. Introduction

So called "waffle shells" are used in supporting structures of aerospace parts. Waffle shell (fig.1) is the part made of aluminium alloy with cylindrical, conical of ogival shape having "cellular" structure, which is the result of the milling of the rows of the cells.

Figure 1. Waffle shell.

There are several methods of producing waffle shells: 1) chemical milling; 2) mechanical milling of plate workpieces and their bending into shells; 3) mechanical milling of cylindrical workpieces.

The main technology to produce aerospace part with cellular structure is the milling of cylindrical workpieces by using specialized machine- tools of "SVO" type. By using this manufacturing method it is much easier to produce shells with cellular structure with non-uniform stiffness and it is not necessary to use a variety of bushes and fixtures for bending in this case.

2. Producing waffle shell with machine-tool of "SVO" type

There are two key features of the shells with cellular structure, that complicate the milling process by means of productivity and machining accuracy: the large size of the shell and their low thickness. One of the ways to solve this problem is to use the computer numerical control in combination with copying technique in the machine-tool of "SVO" type. On the fiq.2 shows the layout of machine-tool for producing the shells with cellular structure by using numerical-copy control of "SVO" type machine-tool. Here: CH1 – copying head, CH2 – cutting head, D1 - copying head drive, D2 – cutting head drive, E1 – position measuring encoder of the copying head drive, E2 – position measuring encoder of the cutting head drive, CNC – computer numerical control unit, W – workpiece. During the machining process of the waffle shells, the cutting head CH2 process the bottom of the cell, while the copying head CH1 traces the opposite surface of the workpiece W. In this case the pass strategy is determined by the part program, whereas the cell bottom thickness is determined by copying the opposite side of the bottom by copying head. The scatter of the cell bottom thicknesses, which is one of the main characteristics of the waffle shell, is significantly reduced by using this technique. Wherein using the computer numerical control together with copying technique can be considered as a new tendency of the development of the CNC systems for the machining aerospace parts with cellular structure.

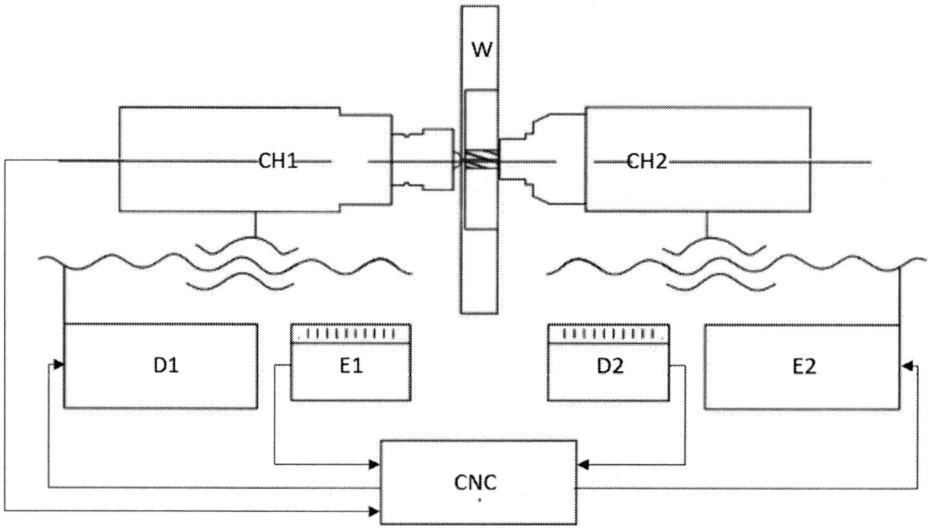

Figure 2. Layout of the machine-tool for producing waffle shells.

Several main factors, which can affect cell bottom thickness errors, can be distinguished (fig.3) for aforementioned machining process. After several tests it was concluded that cell bottom thickness and cell edge thickness values depend on the position within the cell, cell number within the row and the row number. As a result main factors which affect cell bottom and edge thicknesses although can be divided on two groups: 1st group – the factors that affect the thickness within the single cell, 2nd group – the factors that affect the thickness within the row and within the whole part.

3. Approaches of increasing accuracy

Several fundamental approaches of increasing accuracy of the machine-tools with CNC were considered in work[4]. All approaches were divided into two classes 1) improving the design of the supporting system and numerical control system of the machine-tool and improving the manufacturing technology of producing the machine-tool with CNC; 2) improving the machine-tool control process by developing the correction techniques.

Different correction techniques were examined in work [5], which can be used in improving the machine-tool control process:

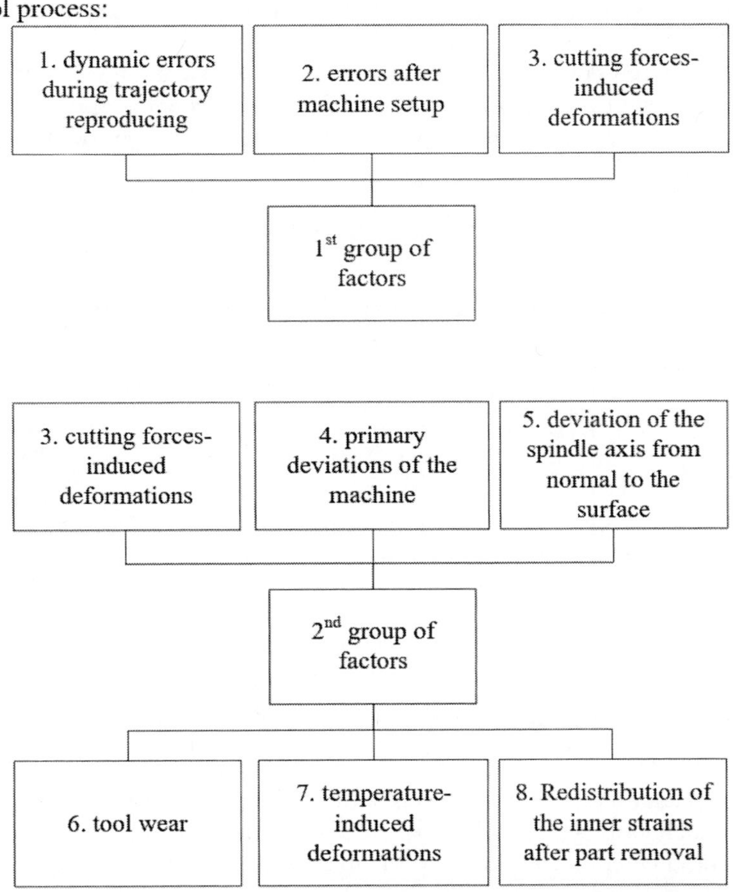

Figure 3. Main factors, which can affect cell bottom thickness errors.

- correction techniques that are based on a priory information (correction based on the results of the calibrating of the machine and part program predistortion)
- correction techniques that are based on the feedback loop (control is performed by measuring the output value error)
- correction techniques that are based on the control of the disturbances.

The cell bottom machining accuracy can be significantly increased by utilization of the aforementioned correction techniques with taking into account the main factors, which can affect cell bottom thickness errors, and numerical-copy control of “SVO” type machine-tools.

The latest models of “SVO” type machine-tools are equipped with cell bottom thickness automate ultrasonic measuring system. As a result the measured cell bottom thicknesses data can be used for performing the tool trajectory correction.

The best result could be achieved by usage of the ultrasonic measuring system as a feedback sensor in the real time mode during the machining process. In this case 6 of 8 factors (fig.3) can be compensated by means of their influence on the machining accuracy of the cell bottom thickness. Redistribution of the inner strains and temperature-induced variations of the machine-tool cannot be compensated by this approach because they are not covered by feedback loop.

It’s a difficult task to implement this kind of correction even by using up-to-date measuring equipment because of the following reasons:

- bad repeatability of measurements in “dynamic mode”(i.e. measurements during trajectory movement) due to lack of immunity to interference of the ultrasonic measurement system in the plant;
- too much time required for processing the measured data by computer of the ultrasonic measurement system, which does not allow to use this system in real time mode.

Therefore, the ultrasonic measurement system can be used as a part of the self-tuning system in combination with the disturbance control correction technique (fig.4).

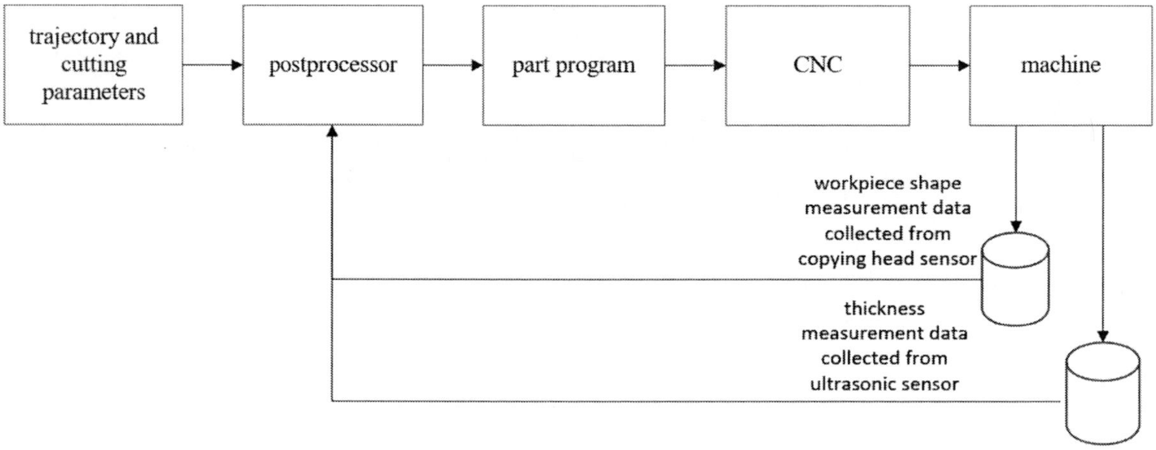

Figure 4. Self-tunning correction system.

Self-tunning correction system uses the cell bottom thicknesses data measured after machining the first cell (i=1) for machining the following cells with corrected part program. Thus the deviations from 1st group (fig.3) could be eliminated. During the machining of the following cells the errors, induced by the factors from the 2nd group will take effect. Disturbance control correction technique can compensate part of these errors (induced by the deviation of the spindle axis from normal to the surface and by the inaccuracies in the work of the copy head), the other part of errors can be compensated by self-tunning correction system based on the measurement data after processing the cell number i. Self-tunning correction system in this case calculates the corrections based on the thickness measurements after the machining of the ith cell and defines the number of the next cell, that will be used for taking the thicknesses measurements for the correction of the part program for machining the following cells.

It must be mentioned that the extension of the disturbance control correction techniques functions increases the amount of the cells, that could be machined by using the same corrected part program and thus increases the performance rate of the machining of the waffle shell. Search for the effective way of allocating the functions between the self-tunning correction technique and disturbance control correction technique is a subject of foregoing theoretical and experimental researches focused on developing the algorithm to control the accuracy of the milling of aerospace parts with cellular structure.

4. Conclusions

1. Computer numerical control of the machine-tools is one of the main ways to automate the machinery production, including the aerospace part production
2. Computer numerical control in combination with copying technique can be considered as a new tendency of the development of the CNC systems for the machining aerospace parts with cellular structure with high performance rate.

IOP Conf. Series: Materials Science and Engineering **489** (2019) 012057 doi:10.1088/1757-899X/489/1/012057

3. The researches focused on the creation of the effective design of the machines with numerical-copy control and algorithms to control the accuracy of the milling of aerospace parts with cellular structure need to be further developed.

References

[1] Computer-aided manufacturing system for creation of part programs for machining waffle shells. http://niigermes.ru/deyatelnost/reshenie-tekhnologicheskikh-problem-raketno-kosmicheskoj-tekhniki/item/55-avtomatizirovannaya-sistema-podgotovki-i-kontrolya-upravlyayushchikh-programm-dlya-frezerovaniya-vafelnykh-obechaek-na-stankakh-s-chpu-tipa-svo.html
[2] Bartrutdinov R and Sysoev S 2011 Manufacturing waffle shells for aircrafts *Actual problems of aviation and cosmonautics* vol 1 pp 7 - 8
[3] Kotov A, Chuikin S, Milekhin J, Makarov J, Korotkov A and Volodina S Metalworking machine http://bd.patent.su/2397000-2397999/pat/servl/servletd7c1.html
[4] Serkov N 2010 Fundamental approaches of increasing accuracy of the machine-tools with CNC *Engineering and Automation Problems* vol 2 pp 26 - 35
[5] Serkov N 2015 *Accuracy of the machine-tools with CNC: theoretical and practical fundamentals* (Moscow: Lenand)

On the features of the progressive movement of the solid body, controlled by two propulsion devices

V N Platonov, E S Briskin and N G Sharonov

Department of Theoretical Mechanics, Volgograd State Technical University, 28 Lenin Avenue, Volgograd 400005, Russia

E-mail: dtm@vstu.ru

Abstract. A calculation scheme and a mathematical model of the translational motion dynamics of a solid body, simulating the movement of the underwater platform with the help of anchor-cable thrusters, are proposed. Features of motor control, DC and stepper in the structure of the actuators of the anchor-rope propulsion have been set.

1. Introduction

The development of new hydrocarbon fields on the continental shelf requires the introduction of advanced technologies for their production and exploration. It is shown [1] that the use of heavy self-lifting bases of underwater drilling complexes with a significant number of drilling points (more than a hundred) is economically impractical, and light self-lifting bases are characterized by insufficient safety. The process of drilling from a self-propelled vessel is also fraught with difficulties, the main of which are pitching and drift.

One of the promising technologies for the production and exploration of new hydrocarbon fields on the continental shelf is the use of an autonomous underwater robotic complex with anchor-cable propulsion devices. The anchor-rope mover can be used to move an autonomous underwater or surface robotic complex (figure 1) due to the thrust arising from the tension of the cables going to the anchor [6, 7, 8].

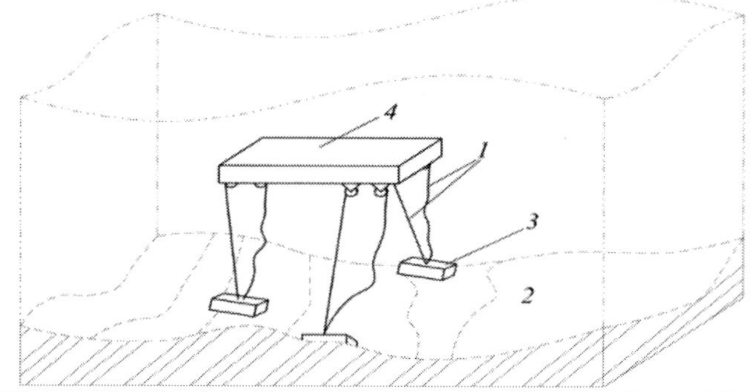

Figure 1. Moving the underwater platform with positive buoyancy.

The engine is an anchor. Together with two cables 1 it is associated with two flywheels, rotating in accordance with the relevant laws of motion. In the initial position, the cable 1 is weakened, the anchor in the absence of current occupies a position on the same vertical with the flywheel. When the flywheel is rotated, the cable 1 is stretched, and in the absence of slippage of the anchor along the bottom, the mobile robot is moving. The work of such an engine is similar to the work of a walking mover, and the difference is that the robot does not rely on mover, but clings to the support surface with its help.

A feature of such robotic systems is the use of excessive in terms of providing a given motion of the robot platform, the number of movers, which is explained by the need for reliable interaction of anchors with difficult to predict the bottom relief and its physical and mechanical properties. Indeed, if for an unambiguous description of the motion of an absolutely rigid body requires a minimum of 6 control actions (for the robot platform – 6 anchor-cable movers), then in the absence of reliable interaction of at least one anchor with the ground, the movement of the robot becomes unpredictable. Therefore, due to the excess of the number of control actions over the number of generalized coordinates, there is a dynamic uncertainty.

However, in solving the technical problem, this uncertainty is useful. It allows you to impose additional conditions on the control actions, optimal in one sense or another.

2. Statement of the model problem

The introduction of various additional conditions, the study and analysis of the results is convenient to carry out on a simple mechanical system with one degree of freedom, shown in figure 2.

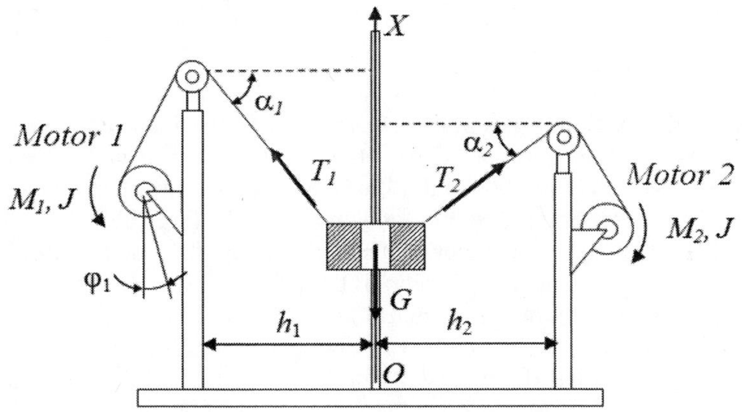

Figure 2. The settlement scheme of studies of the motion modes.

One drive is sufficient to move the load G along the vertical guide (figure 2). When using two drives, the task of the control system is to implement a given distribution of effort between these drives, as well as providing the required law of translational movement of load.

The control is reduced to the search for control actions on electric motors, in particular, to the determination of the voltage applied to the armature windings of a DC motor, to provide the required moments, or the determination of the switching frequency of windings for stepper motors.

To study the control system of anchor-cable thrusters in the translational motion of the underwater platform is considered a calculation scheme, which is shown in figure 2, where motor 1and motor 2 – electric motors; M_1, M_2 – the torque, developed by these electric motors; α_1, α_2 – the angles, formed by cables with a horizontal axis; φ_1, φ_2 – the angles of rotation of the flywheels of electric motors; T_1, T_2 – the tension force in the cables; J – the moment of inertia of the same flywheels drives; h_1, h_2 – the distance from the center of mass of the load up to flywheel axes; G/g – mass of the transported load; $x(t)$ - movement of the load.

The mathematical model, describing the movement of the load is based on geometric equations with simultaneous operation of electric motors 1 and 2 on the one hand, and on the equation of the dynamics of the forward movement of the load – on the other. The peculiarity of the solution of the control problem is the need for the subjective introduction of an additional ratio, both in the case of DC motors and in the case of stepper motors.

The control system can be implemented by software motion control, or using feedbacks to improve control accuracy. The block diagram of the control system is shown in the figure 3.

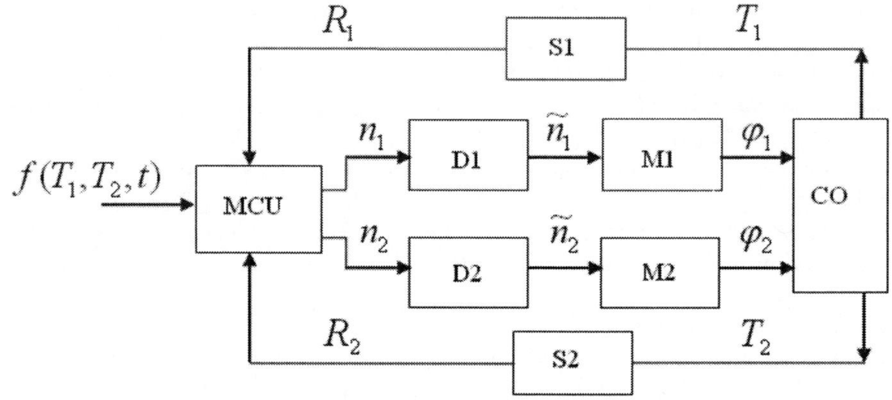

Figure 3. The block diagram of the control system: MCU – microcontroller; D1, D2 – DC-motor drivers; M1, M2 – DC-motors; S1, S2 – sensors; CO – control object.

The required ratio of the tension forces in the cables in the form of a set value is fed to the input of the control device, in this case the microcontroller (MCU). Control actions n_1, n_2 from the microcontroller are fed to the motor drivers *D1*and *D2*. Under the influence of control actions, electric motors *M1* and *M2* are installed at the appropriate angle φ_1 and φ_2. By adjusting the speed of electric motors, the necessary difference of angular accelerations calculated by the microcontroller according to the mathematical model and providing a given ratio of the tension forces of the cables are established. The determination of the engine for which the control action is required can be carried out by the value of the angle of rotation of its rotor, measured by the feedback circuit through the sensors *S1* and *S2*, or by direct measurement of the values of the tension forces.

3. Test stand development

The model of physical modeling allows us to study various modes of coordinated operation of the engines of the anchor-cable propulsion device, for example, the movement of the load with a given ratio of tension forces, the implementation of a cycle of one "step" with the movement of the load along the program trajectory. The interface of the program, controlling the stand, allows us to carry out control of electric motors, to set rotation angles of each electric motor separately, to measure and display sizes of the torques for each electric motor, and it allows us to monitor current values of tension of cables, calculated according to mathematical model. The developed stand allows you to simulate the movement of load with one degree of freedom and evaluate the effectiveness of the selected control actions.

The solution of the problem of synthesis of control signals for electric motors can be based on the conditions of simultaneous search for extreme values of several criteria - based on the solution of multi-criteria optimal control problem, for example, by the criterion of minimizing the effort in a particular cable, by the criterion of its minimum or maximum elongation, by the criterion of the load on a particular engine.

IOP Conf. Series: Materials Science and Engineering **489** (2019) 012058 doi:10.1088/1757-899X/489/1/012058

4. Results

During the process, a mathematical model of control of the translational motion of a solid body under the influence of two drives is developed, the criterion of optimal control, providing the required ratio of forces developed by the drives, is proposed, and the features of control of DC motors and stepper motors are considered.

References

[1] Serov V A, Kovshov I V and Ustinov S A 2017 *Izvestiya YUFU. Tekhnicheskie nauki.* **9** (**194**) pp 181-91

[2] Jin-Ho Kim, Tae-Kyeong Yeu, Suk-Min Yoon, Hyung-Woo Kim, Jong-Su Choi, Cheon-Hong Min and Sup Hong 2013 Electric-Electronic System of Pilot Mining Robot, MineRo-II *Proc. of the Tenth ISOPE Ocean Mining and Gas Hydrates Symp. Szczecin* pp 269-73

[3] Verichev S, Laurens de Jonge, Wiebe B and Rodney N 2014 Deep mining: from exploration to exploitation *Proc. Int. Conf. VNII Okeangeologia* pp 126-38

[4] Briskin E S , Chernyshev V V, Maloletov A V and Sharonov N G 2013 Comparative analysis of wheeled, tracked and walking machines *Robototekhnika i tekhnicheskaya kibernetika* **1** (**1**) pp 6-14

[5] Chernyshev V V and Arykancev V V 2015 MAK-1-underwater walking robot *Robototekhnika i tekhnicheskaya kibernetika* **2** pp 45-50

[6] Briskin E S, Penshin I S, Smirnaya L D and Sharonov N G 2017 Determination of forces in the engines of the anchor-cable type *Izvestiya VolgGTU.* Aktual'nye problemy upravleniya, vychislitel'noj tekhniki i informatiki v tekhnicheskih sistemah **14** (**209**) p. 87-90

[7] Briskin E S, Serov V A, Sharonov N G and Penshin I S 2017 About features of control of movement of mobile robots with movers of anchor and cable type *Ekstremalnaya robototekhnika* pp 337-43

[8] Sharonov N G, Penshin I S and Platonov V N 2017 Underwater platform motion control with anchor-rope propulsion devices *Proc. Int. Conf. MIKMUS* pp 353-5

[9] Kenio T. 1987 *Stepper motors and their microprocessor control sytems* (Moscow: Energoatmizdat) p 200.

IOP Conf. Series: Materials Science and Engineering **489** (2019) 012059 doi:10.1088/1757-899X/489/1/012059

Development of automated accuracy control tools for robotic equipment

A A Potapov, A A Fedorov, V M Masyuk and S Y Orekhov

Kaluga Branch of Bauman Moscow State Technical University, Russia, 248004, Kaluga, Bazhenova st., 4

andrewpotapov@yandex.ru

Abstract. This paper proposes a new method of visual automated control based on optical measurements. In this article, we investigate a number of questions concerning the organization of measurements, calibration of the data obtained, and interpretation of results. An experimental scheme is given, the coordinates of the gripper of a robotic manipulator are obtained by different methods, the accuracy is estimated. It is concluded that the use of the proposed method based on visual markers for obtaining coordinates based on visual markers has high accuracy, good repeatability of results and can be used to control the positioning of new robotic mechanisms at the testing and commissioning stage.

1. Introduction

Currently, the problem of the accuracy of robotic equipment is particularly relevant and the introduction of tools that use visual automated control can become an integral part of any robotic production, and with the necessary accuracy of the means used, can become the main accuracy control system. At this stage, in the way of the implementation of this technology, there is the inaccuracy of tools that use computer-aided visual control. The first step in solving this problem is to estimate the errors in determining the coordinates of the gripper of an industrial robot using the method of visual automated control. Using a more accurate method - a direct method, it is possible to determine the error of the results obtained by the method of visual automated control. By the direct method, we mean the coordinates of the gripper [1], obtained as a result of software testing of a fully intact calibrated industrial robot. To determine the accuracy of the results, it is necessary to compare the coordinates of the gripper of an industrial robot obtained by the direct method and by the method of visual inspection [2] when moving the gripper in preset positions.

For carrying out this experiment, evaluation of the methodology and calibration of further results, the industrial robot FANUC R-2000iB was used, at the end of the gripper of which the LED was fixed. A script was originally developed for the robot, as a result of which the gripper of the manipulator moves between 3 points in space along a predetermined path. During the execution of the program, a video fixation of this experiment was carried out, the data in the form of freeze frames were transmitted to a PC, then using the Matlab program, the coordinates of the LED were obtained from the received fragments. At this stage, three frames were considered, from which coordinates were obtained and compared with the data on the position of the LED relative to the program-specified ones.

IOP Conf. Series: Materials Science and Engineering **489** (2019) 012059 doi:10.1088/1757-899X/489/1/012059

Of all the projections employed in computer graphics, the perspective projection is one most widely used [3,6]. There are two stages to its computation: the first involves converting world coordinates to the camera's frame of reference, and the second transforms camera coordinates to the projection plane coordinates. Now we investigate how camera coordinates are transformed into a perspective projection. We begin by assuming that the camera is directed along the z-axis as shown in figure 1. Positioned d units along the z-axis is a projection screen, which is used to capture a perspective projection of an object. Figure 2 shows that any point (x_c, y_c, z_c) is transformed to (x_p, y_p, d). It also shows that the screen's x-axis is pointing in the opposite direction to the camera's x-axis, which can be compensated for by reversing the sign of x_p when it is computed. Figure 3 shows a plan view of the scenario depicted in figures. 1, and 2 a side view. Next, we reverse the sign of x_p and state:

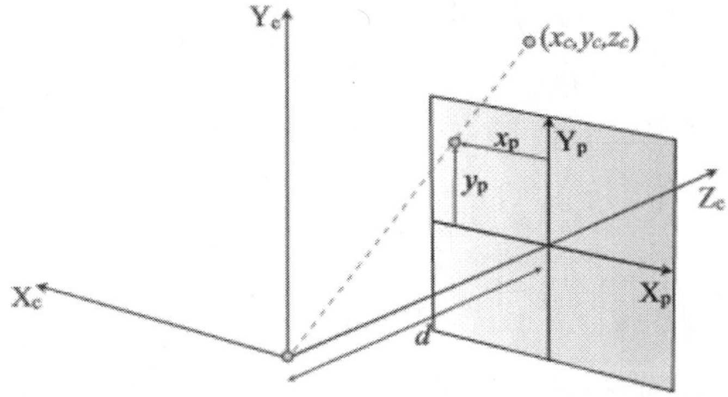

Figure 1. The axial system used to produce a perspective view.

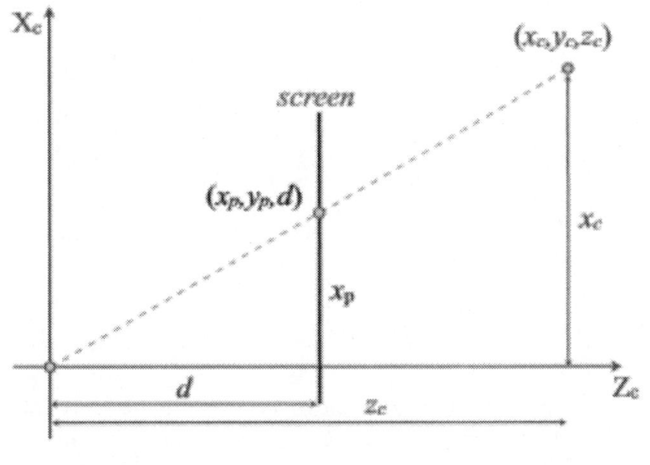

Figure 2. The plan view of the camera's axial system.

IOP Conf. Series: Materials Science and Engineering **489** (2019) 012059 doi:10.1088/1757-899X/489/1/012059

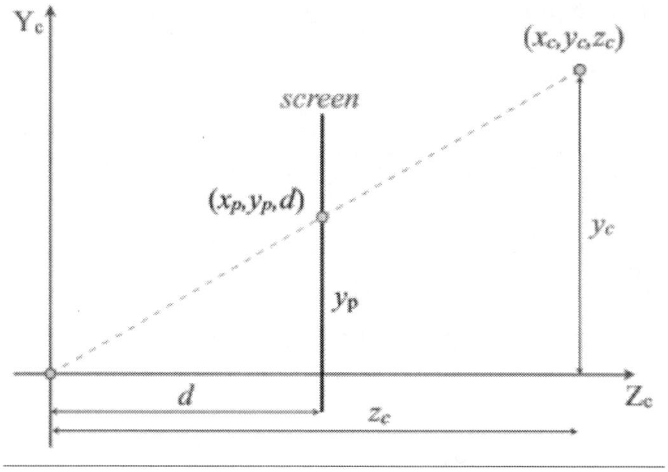

Figure 3 - The side view of the camera's axial system.

$$\frac{x_c}{z_c} = \frac{-x_p}{d} \rightarrow x_p = \frac{-x_c}{z_c / d}$$

$$\frac{y_c}{z_c} = \frac{y_p}{d} \rightarrow y_p = \frac{y_c}{z_c / d}$$

This is expressed in matrix form as $\begin{bmatrix} x_p \\ y_p \\ z_p \\ w \end{bmatrix} = \begin{bmatrix} -1 & 0 & 0 & 0 \\ 0 & 1 & 0 & 0 \\ 0 & 0 & 1 & 0 \\ 0 & 0 & 1/d & 0 \end{bmatrix} \begin{bmatrix} x_c \\ y_c \\ z_c \\ 1 \end{bmatrix}$ What can be represented as $\begin{bmatrix} x_p & y_p & z_p & w \end{bmatrix}^T = \begin{bmatrix} x_c & y_c & z_c & w/d \end{bmatrix}^T$ and if we remember the idea behind homogeneous coordinates, we must divide the terms (x_c, y_c, z_c) by w to get the scaled terms, which produces

$$x_p = \frac{-x_c}{z_c / d};$$

$$y_p = \frac{-y_c}{z_c / d};.$$

$$z_p = \frac{-z_c}{z_c / d}$$

2. Experiment

The first step in the experiment was the creation of coordinate systems for an industrial robot: Tool frame and User frame. The mapping of coordinate systems is presented in figure 4.

IOP Conf. Series: Materials Science and Engineering **489** (2019) 012059 doi:10.1088/1757-899X/489/1/012059

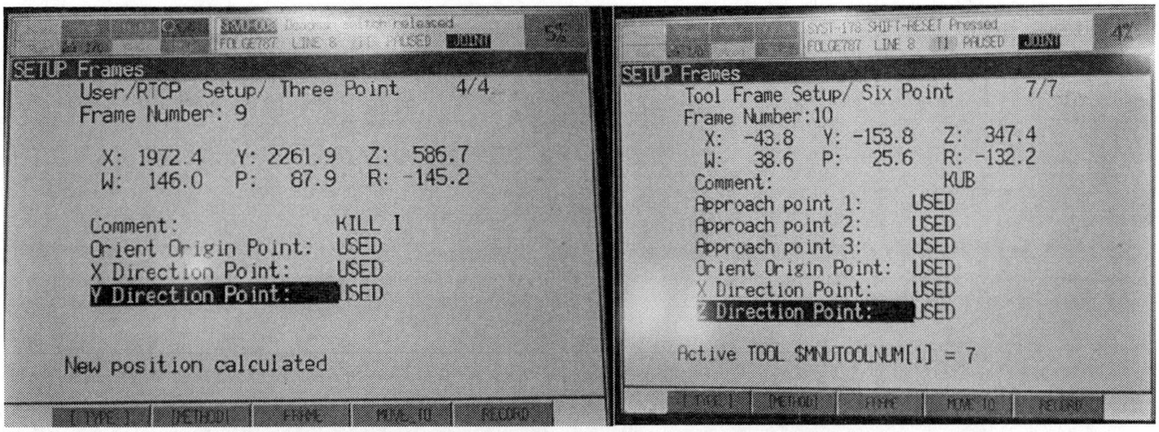

Figure 4. Coordinate systems of the robot.

The second step was the creation of a program for the FanucR-2000B 210F robot, which sets the movement of the LED along three points of the trajectory. The program code is presented in figure 5.

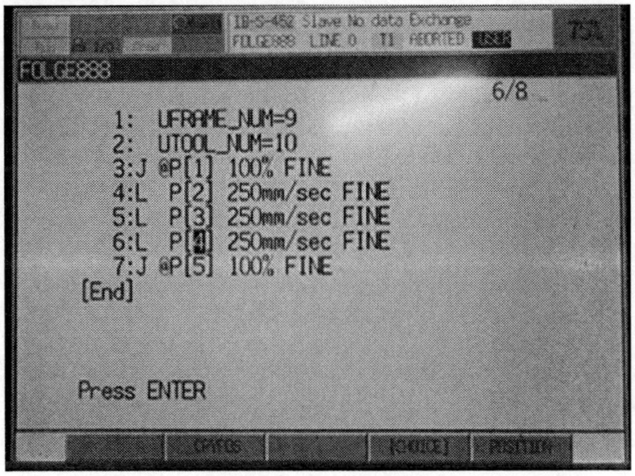

Figure 5. Program code.

According to the results of this program, the coordinates of three points were obtained, between which the LED moved linearly, table 1.

Table 1. Coordinates of points obtained by the direct method.

Point (position)	X, мм	Y, мм
1	313,494	344,995
2	158,235	613,859
3	331,657	833,832

The third step was to determine the coordinates of the three positions of the LED using the proposed method. The coordinates of the position of the gripper of an industrial robot in space are determined using the Matlab program, by using the Prewitt method.

To determine the position, we give the following algorithm of actions:

1.For the purpose of non-interference in the operation of equipment, we use a transparent panel.

2. We install the camera and adjust it relative to the panel.

3. We measure and set the working area of the system

IOP Conf. Series: Materials Science and Engineering **489** (2019) 012059 doi:10.1088/1757-899X/489/1/012059

4. For 3 given points, we obtain the necessary calculated relations.

The next step is the calculation of the positioning error.

At this stage of the research, one camera was used to obtain results in one plane to verify the method.

The result of this program was the recognition of the LED on the image presented in figure 6.

Figure 6. View of the work program window after the recognition of the light marker.

After conducting a mathematical calculation [4], the coordinates of three points were obtained using the method of visual automated control, table 2.

Table 2. Coordinates of points obtained by the method of visual automated control.

Point (position)	X, мм	Y, мм
1	312,268	346,754
2	156,770	615,542
3	330.148	835.49

Comparing the experimental results obtained by the direct method and the method of visual automated control, one can obtain the error of the developed method of visual automated control on the X-axis equal to $\Delta = 1.4$ mm and on the Y-axis equal to $\Delta = 1.7$ mm.

For the experiment, a camera with the following characteristics was used: Sensor model: Samsung S5K3P8 [5]; Sensor size: $4.74x3.56\,мм$; $4608x3456\ (15.93\ Mn)$

Photo accuracy with regard to the quality factor (optical resolution, diffraction, etc.); $k = 0,4 - 0,5$; $l = 1600мм / 4608 \approx 0,35мм; l / k = 0,875мм$ In the video mode, taking into account the quality factor (the resolution of the optics, diffraction, etc.); $k = 0,8$; $l = 1600мм / 4608 \approx 0,35мм; l / k = 0,4375мм$.

The positioning accuracy of the industrial robot FanucR-2000B 210F is 0.05 mm. This allows us to use it to calibrate the developed method.

3. Results

Ten experiments were conducted, in each of which the coordinates of the given points were determined. The mean values were chosen and the variance of the results obtained is given in the table below [7,8].

Table 3. Statistical indicators.

Indicator	Point X_1	Point X_2	Point X_3	Point Y_1	Point Y_2	Point Y_3
Average	311,9367	156,8085	330,1928	346,4388	615,6469	835,5226
Dispersion, $D[X]$	0,10485	0,1983	0,0841	0,2753	0,2176	0,3166
CKO, σ_X^2	0,3238	0,4453	0,2899	0,5247	0,4665	0,5627
$3\cdot\sigma_X^2$	0,9714	1,3359	0,8699	1,5741	1,3996	1,6882

Also of great interest is the removal of the spatial position of the gripper manipulator. A theoretical rationale is being worked out for 2 cameras and the use of stereo vision to estimate the spatial movement of a robotic gripper. To improve the accuracy of measurements, it is planned to retrofit the system with several light markers in mutually perpendicular planes.

As a result of the research, it can be concluded that the experimental data are in good agreement with the expected values. This method can be used to control various robotic systems that require positioning accuracy.

In the future, it is intended to generalize the method for different use cases. It is planned to increase the accuracy of the method by increasing the resolution of the camera, improving the camera.

References

[1] Zenkevich S L, Yushchenko A S 2004 *Fundamentals of manipulation robots management: A textbook for universities. - 2nd ed.,* (Moscow: BMSTU)

[2] An efficient orientation filter for inertial and inertial/magnetic sensor arrays/ Sebastian O H Madgwick. Date Views 07.10.2018: http://x-io.co.uk/res/doc/madgwick_internal_report.pdf

[3] John V 2017 *Mathematics for Computer Graphics, Fifth Edition* (London: Springer-Verlag)

[4] Rogers D, Adams J 2001 *Mathematical foundations of computer graphics* (Moscow: Mir)

[5] SAMSUNG, MOBILE IMAGE SENSOR S5K3P3. Data Views 07.10.2018: https://www.samsung.com/semiconductor/image-sensor/mobile-image-sensor/S5K3P8/

[6] Rogers D F, Adams J A 1990 *Mathematical Elements for Computer Graphics, 2nd edn.* (New York:McGraw-Hill).

[7] Geiger A, Lenz P, Stiller C, Urtasun R 2013 Vision meets robotics: The KITTI dataset. *The International Journal of Robotics Research* vol 32, issue 11, pp. 1231-7 DOI: https://doi.org/10.1177/0278364913491297

[8] Christian W, Stefan W, Stefan R, Konrad S, Bernt S 2013 Monocular Visual Scene Understanding: Understanding Multi-Object Traffic Scenes. *IEEE Transactions on Pattern Analysis and Machine Intelligence* vol 35, issue: 4 DOI: https://doi.org/10.1109/TPAMI.2012.174

IOP Conf. Series: Materials Science and Engineering **489** (2019) 012060 doi:10.1088/1757-899X/489/1/012060

The algorithm for planning the trajectory of the 3-RPR robot, taking into account the singularity zones based on the method of non-uniform covering

L A Rybak, E V Gaponenko, DI Malyshev and L G Virabyan

Belgorod State Technological University named after V. G. Shoukhov, 46, Kostyukov str., Belgorod, 308012, Russia

rlbgtu@gmail.com

Abstract. The article describes the method the trajectory planning of a 3-RPR robot, using the numerical solution of the forward kinematics. An algorithm for determining the position of a mobile robot platform based on the method of non-uniform coverings has been synthesized. The zones of singularity arising at some positions of the robot are investigated. The conditions for the emergence of singularity zones corresponding to its special provisions are revealed. It is shown that the solution of the forward kinematics is not unambiguous i.e. with the same change in the length of the rods there are different positions of the platform. The issues of planning the trajectory for a mobile platform are described while taking into account the zones of singularity. To implement the algorithm, a software package in the C++ language has been developed. The results of mathematical simulation are presented.

1. Introduction

Recently, parallel robots are increasingly used in the industry. Classical control system with feedback can be used to control such robots. However, to implement a mismatch control, a dynamic definition of the control result is required i.e., coordinates and orientation in the space of the output link of the robot. In most of the cases, the implementation of the position sensor of the mobile platform is difficult. So, it is necessary to organize the calculation of the feedback signal depending on the movement of the drive links i.e., solving the forward kinematics. The complexity of the forward kinematics is explained by the need to solve a system of nonlinear equations. Research in this area has led to the development of various ways to solve this problem [1-3]. It should be noted that there is an ambiguity of solving the forward kinematics for some of the parallel robots. This means that different positions of the output link can correspond to one state of the drive links. Thus, to find the "correct" solution using the forward kinematics is even more difficult for such mechanisms.

The article proposes a numerical method for solving the forward kinematics of a parallel robot based on optimization methods. For the approximation of the system of nonlinear coupling equations describing the geometry of the robot, an optimization algorithm based on the concept of non-uniform covering, described in [4,5], was applied. This method has been successfully applied to the approximation of the working area of some types of parallel robots [6]. In the article, this method is used to solve the forward kinematics of the 3-RPR robot i.e., finding the coordinates of its output link.

2. Formulation of the problem

Consider a 3-RPR robot (figure 1), in which three rods of variable length are pivotally attached to a fixed base at the vertices of an equilateral triangle A_1,A_2 and A_3 [7]. The other ends of the rods are hinged at the vertices of the equilateral triangle B_1,B_2 and B_3 on the movable platform, which allows it to move in the plane of the base. The input coordinates of the robot are the rod lengths l_1,l_2 andl_3, the output coordinates are the coordinates of the geometric center of the moving platform x, y and the angle of rotation of the platform φ relative to an axis perpendicular to the plane of the base.

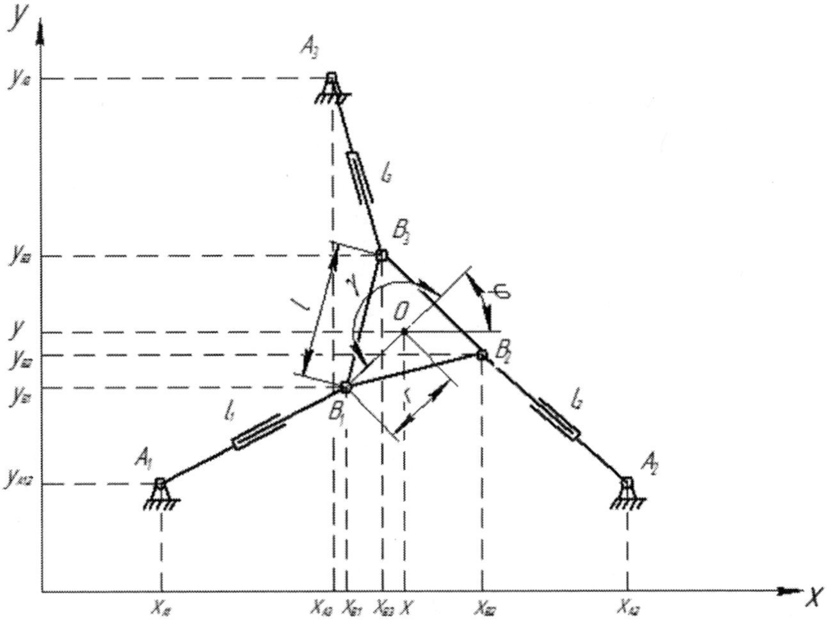

Figure 1. Flat 3-RPR robot

Changing rod lengths $l_i$has the form

$$l_i^2 = (x + r\cos(\gamma_i + \varphi) - x_{Ai})^2 + (y + r\sin(\gamma_i + \varphi) - y_{Ai})^2,$$

where γ_i is the angle determining the geometry of the output link, x_{Ai}, y_{Ai} are the coordinates of the point A_i and A_i – the hinge attaching the rod to the fixed base, r is the distance from B_i – the hinge attaching the rod to the moving platform to O – the geometric center of the platform, l_i - the length of the ith rod. If the points A_i and B_i are located at the vertices of equilateral triangles then the change in rod lengths is determined by

$$l_1^2 = \left(x + 0{,}5r\left(\sin\varphi - \sqrt{3}\cos\varphi\right) + 0{,}5\sqrt{3}R\right)^2 + \left(y - 0{,}5r\left(\sqrt{3}\sin\varphi + \cos\varphi\right) + 0{,}5R\right)^2, \quad (1)$$

$$l_2^2 = \left(x + 0{,}5r\left(\sin\varphi + \sqrt{3}\cos\varphi\right) - 0{,}5\sqrt{3}R\right)^2 + \left(y + 0{,}5r\left(\sqrt{3}\sin\varphi - \cos\varphi\right) + 0{,}5R\right)^2, \quad (2)$$

$$l_3^2 = (x - r\sin\varphi)^2 + (y + r\cos\varphi - R)^2, \quad (3)$$

where R is where the radii of the circumcircle of triangle $A_1A_2A_3$.

It should be noted that the solution of the forward kinematics of the robot is not unique, i.e. different positions of the movable platform correspond to the same rod lengths. In this case, an unambiguous determination of the position of the platform only by changing the lengths of the rods becomes impossible and there is an additional problem of identifying the "correct" solution of the forward kinematics to determine the current state of the control object. To solve the forward kinematics and determine all the possible positions of the platform for the given rod lengths, we write the equations of communication

$$\begin{cases}(x_{B1}-x_{A1})^2+(y_{B1}-y_{A1})^2-l_1^2=0\\(x_{B2}-x_{A2})^2+(y_{B2}-y_{A2})^2-l_2^2=0\\(x_{B3}-x_{A3})^2+(y_{B3}-y_{A3})^2-l_3^2=0\\(x_{B1}-x_{B2})^2+(y_{B1}-y_{B2})^2-l^2=0\\(x_{B3}-x_{B2})^2+(y_{B3}-y_{B2})^2-l^2=0\\(x_{B3}-x_{B1})^2+(y_{B3}-y_{B1})^2-l^2=0\end{cases} \quad (4)$$

Denote in the system of equations (4) the functions of changing the coordinates g_i,$i\epsilon$ 1,…,6 in the form: $g_1=(x_{B1}-x_{A1})^2+(y_{B1}-y_{A1})^2-l_1^2$, …, $g_6=(x_{B3}-x_{B1})^2+(y_{B3}-y_{B1})^2-l^2$. When these coordinates change, singularity zones appear that correspond to the special positions of the robot in which the loads on the links increase and uncontrollable mobilities appear. For a 3-RPR robot, singularities of only the second type are characteristic for which the Jacobian J_A is zero [8]

$$\det(J_A)=\begin{vmatrix}\frac{\partial F_1}{\partial x} & \frac{\partial F_1}{\partial y} & \frac{\partial F_1}{\partial \varphi}\\ \frac{\partial F_2}{\partial x} & \frac{\partial F_2}{\partial y} & \frac{\partial F_2}{\partial \varphi}\\ \frac{\partial F_3}{\partial x} & \frac{\partial F_3}{\partial y} & \frac{\partial F_3}{\partial \varphi}\end{vmatrix}, \quad (5)$$

where $\frac{\partial F_i}{\partial x}=2x-2x_{Ai}+2r\cos(\varphi+\gamma_i)$, $\frac{\partial F_i}{\partial y}=2y-2y_{Ai}+2r\sin(\varphi+\gamma_i)$,

$$\frac{\partial F_i}{\partial \varphi}=2r\left(y-y_{Ai}+r\sin(\varphi+\gamma_i)\right)\cos(\varphi+\gamma_i)-2r\left(x-x_{Ai}+r\cos(\varphi+\gamma_i)\right)\sin(\varphi+\gamma_i),$$

$F_i$is a function that characterizes the changes in rod lengths.

Then equation (5) takes the form:

$$\det(J_A)=12\sqrt{3}\,R\,r\,sin\,\varphi\,(R^2-2\,R\,r\,cos\,\varphi+r^2-x^2-y^2). \quad (6)$$

A singularity zones in the event that the Jacobian (6) is zero, that is, $\det(J_A)=0$, which is possible if the following conditions are fulfilled:

$$\sin\varphi=0 \quad (7)$$

$$x^2+y^2=R^2-2\,R\,r\cos\varphi+r^2 \quad (8)$$

We study the behavior of the robot at different positions of the platform taking into account the singularity zones described by conditions (7) and (8). Let us set the theoretically possible maximum boundaries for changing the coordinates of points of the platform B_1, B_2 and B_3

$$x_{A1}-l_{imax}-l\le x_{Bi}\le x_{A1}+l_{imax}+l \quad (9)$$

$$y_{A1}-l_{imax}-l\le y_{Bi}\le y_{A1}+l_{imax}+l \quad (10)$$

where l_{imax} is the maximum length of the ith rod.

We can estimate the lengths of the rods l_1,l_2 and l_3 from the equations (1) - (3) for some center of the platform *O*, given by the coordinates φ, x, y. We take the value of the angle φ in the range from – 119 ° to +119 °. Let us define the coordinates of the *O* - the center of the platform x and y, located both inside and outside the circle described by equation (8), ranging from 0 mm to 140 mm. Next, we

IOP Conf. Series: Materials Science and Engineering **489** (2019) 012060 doi:10.1088/1757-899X/489/1/012060

consider the forward kinematics i.e. determine the coordinates of the point B_i: x_{Bi}, y_{Bi} for the obtained values of l_1,l_2,l_3.

3. Algorithm for solving the forward kinematics

To solve this problem in connection with its ambiguity, we will use the numerical method of approximation of the system of nonlinear equations (4), based on the method of non-uniform covering [4, 5], with some given approximation accuracy δ. The detailed analysis of this method is as follows. If for some set of variable values of each functions g_i the following conditions are fulfilled:

$$\min g_i \leq 0 \quad (11)$$

$$\max g_i \geq 0, \quad (12)$$

then, therefore, each function takes the value 0 and there are solutions of nonlinear algebraic equations (4) within a given set of variables. It is necessary to obtain sets that have a diameter less than a given approximation accuracy δ, for which conditions (11), (12) are satisfied. The sets of values of six variables x_{Bi}, y_{Bi} are presented in the form of six-dimensional parallelepipeds. To calculate the minima and maxima of the functions, interval estimates are determined, which coincide with the extrema of the function. For this, the developed Snowgoose library in C++ is used.

An algorithm for determining the coordinates x_{Bi}, y_{Bi} of points of the moving platform B_1,B_2 and B_3 has been synthesized (figure 2). Let us introduce three lists of six-dimensional parallelepipeds $\mathbb{P}$,$\mathbb{P}_\mathrm{A}$ and $\mathbb{P}_\mathrm{E}$, where the list $\mathbb{P}$ includes a set of parallelepipeds containing intervals of variable values, the list $\mathbb{P}_\mathrm{A}$ includes a set of parallelepipeds containing intervals of values of variables less than δ for which the conditions (11), (12) are satisfied , $\mathbb{P}_\mathrm{E}$ includes a set of parallelepipeds containing intervals of values of variables for which at least one of the conditions (11), (12) is not satisfied. Each of the axes of the six-dimensional space corresponds to the changing parameters in the system of equations - the coordinates x_{Bi}, y_{Bi}. At the first step of the algorithm, the lists $\mathbb{P}_\mathrm{A}$ and $\mathbb{P}_\mathrm{E}$ are empty, and the list $\mathbb{P}$ consists of only one parallelepiped Q, which includes the entire range of boundaries of the platform points B_1,B_2 and B_3 (9) and (10).

The algorithm works as follows:

1. Input of geometrical parameters of mechanism and approximation accuracy δ.
2. The lists $\mathbb{P}_\mathrm{A}$ and $\mathbb{P}_\mathrm{E}$ are empty, and the list $\mathbb{P}$ consists of only one parallelepiped Q, which includes the entire range of theoretically maximum boundaries for finding points of the platform B_1, B_2 and B_3 (9) and (10).
3. Extract from the list $\mathbb{P}$ the parallelepiped $Q_j, j \in 1, n$.
4. We define for it a minimum of $\min g_i$ and a maximum of $\max g_i < 0$ of functions $g_i. i \in 1,..6$.
5. If at least one of the functions g_i satisfies at least one of the conditions $\min g_i > 0$ or $\max g_i < 0$, then the box is excluded from further consideration, falling into the list $\mathbb{P}_\mathrm{E}$, that is, $\mathbb{P}_E := \mathbb{P}_E \cup Q_j$
6. If the parallelepiped has a diameter less than or equal to the given parameter δ, that is, $d(Q_j) \leq \delta$ characterizing the accuracy of approximation, then it is added to the $\mathbb{P}_\mathrm{A}$ list, that is, $\mathbb{P}_\mathrm{A} := \mathbb{P}_\mathrm{A} \cup Q_j$
7. In other cases, the parallelepiped is divided into two equal parallelepipeds Q_{m+1} and Q_{m+2}along the edge with the greatest length. These parallelepipeds are entered at the end of the list $\mathbb{P}$, that is, $\mathbb{P} := \mathbb{P} \cup Q_j$.
8. If the list P becomes empty, that is $Q_j = \emptyset$, then the algorithm terminates its work.
9. In other cases, steps 3-9 are repeated.

The finiteness of the number of steps of the algorithm follows from the restriction on the minimum diameter of the parallelepiped.

IOP Conf. Series: Materials Science and Engineering **489** (2019) 012060 doi:10.1088/1757-899X/489/1/012060

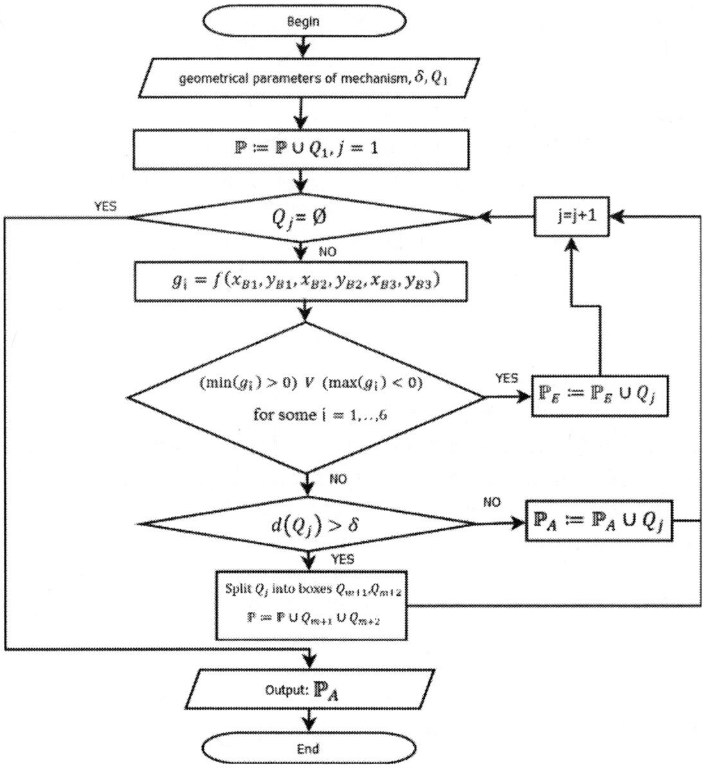

Figure 2. Algorithm for determining the coordinates x_{Bi}, y_{Bi} of the mobile platform points B_1,B_2 and B_3.

4. Simulation results

The following geometric parameters were used for modeling: R = 100 mm, r = 50 mm. Figure 3 presents the simulation results for condition (7). A singularity zone arises at the angle of rotation of the platform $\varphi = 0°$, at which the forward kinematics has one solution (figure 3, a). When changing rod lengths, two solutions arise, thus with one control signal besides the "correct" platform can move to another position (figure 3, b).

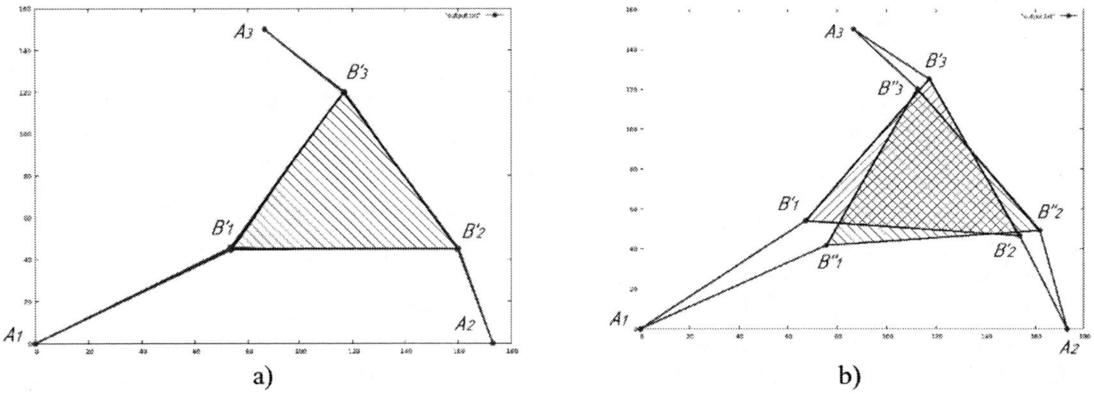

Figure 3. Simulation results a) subject to (7) the occurrence of special provisions *($\varphi = 0°$)*, b) without being in a singularity zone *($\varphi = 5°$)*.

Figure 4 presents the simulation results for the singularity zone that arises when the center of the platform is located on the circle described by condition (8). If the center of the platform is inside a circle, then the forward kinematics has two solutions (figure 4, a). When changing the lengths of the rods, providing for the exit of the center of the platform from the circle, the forward kinematics has four solutions. Thus, one control signal can move the platform to its "correct" or "incorrect" position (figure 4, b).

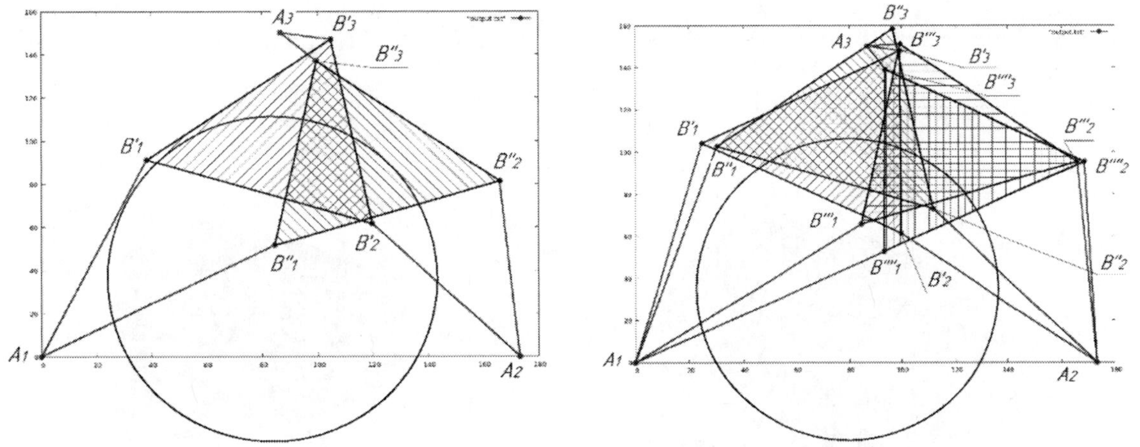

a) b)

Figure 4. Simulation results for finding the center of the platform a) inside the circle described by condition (7), b) outside the circle described by condition (7).

Figure 5 presents the simulation results for condition (7) with the position of the center of the platform outside the circle described by condition (8). A singularity zone occurs when the angle of rotation of the platform $\varphi = 0°$, at which the forward kinematics has three solutions (figure 5, a). When changing rod lengths, four solutions appear (figure 5, b).

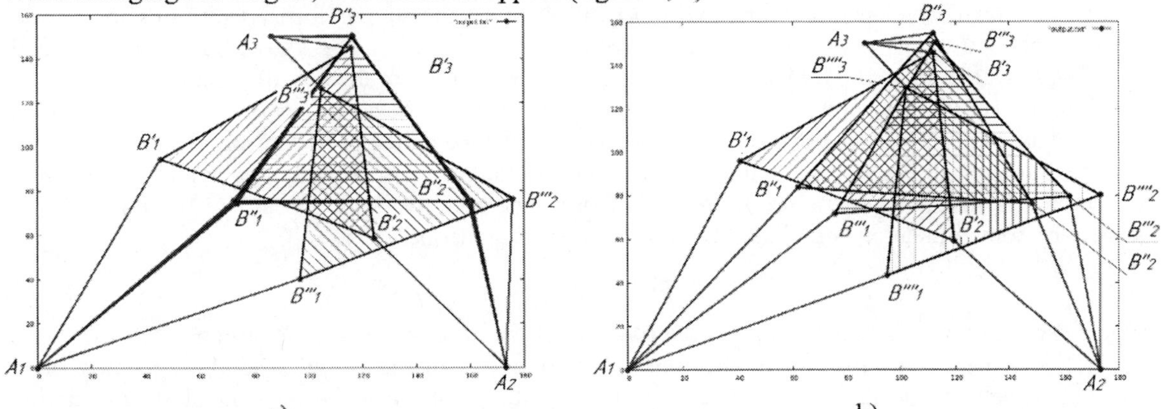

a) b)

Figure 5. Simulation results at the location of the center of the mobile platform outside the circle (8): a) if there are singularity zones *($\varphi = 0°$)*, b) without singularity zones *($\varphi = 5°$)*.

Thus, for the performance of technological operations, the center of the mobile platform must be within the circumference described by condition (8) and the angle φ must vary either within $-120° < \varphi < 0°$ or within $0° < \varphi < 120°$. When passing through an angle of 0°, duality of solutions arises in the problem of positions. The exit of the center of the platform beyond the circumference determined by the condition (8) also leads to ambiguity of decisions and the impossibility of controlling such a robot.

IOP Conf. Series: Materials Science and Engineering **489** (2019) 012060 doi:10.1088/1757-899X/489/1/012060

Let us determine the position of the singularity zones corresponding to the specific positions within the working area of the robot when performing a specific trajectory of the working platform. To determine the working area of the robot, we will also use the method of irregular covering, discussed earlier for a certain range of variation of rod lengths $l_{imax} = 130$ mm, $l_{imin} = 10$mm. The simulation results are presented in figure 6. It is clear from them that at the angle of rotation of the platform $\varphi = 20°$ (figure 6, a) the singularity zones described by condition (8) and representing a circle are inside the working area, for the angle of rotation $\varphi = 40°$ (figure 6, b) a significant part of the circle is located outside the working area and is unattainable.

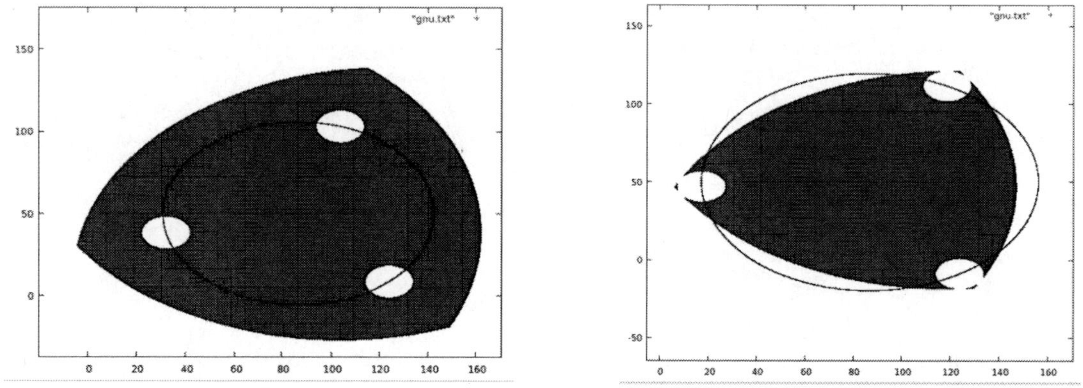

a) b)

Figure 6. Results of modeling the working area: a) for the angle of rotation of the platform $\varphi = 20°$, b) for the angle of rotation of the platform $\varphi = 40°$.

It should be noted that when solving management tasks, the position of the mobile platform should be determined regularly, and therefore the "correct" state of the output link always lies in a certain neighborhood from its previous state. The size of the neighborhood is known and depends on the speed of movement and the frequency of determining the position of the moving platform.

If, moreover, the trajectory of its movement passes far from the so-called special provisions of the robot, characterized by loss of degrees of freedom or the appearance of uncontrolled mobility, in such a neighborhood there will be only one solution to the forward kinematics, which is "correct", reflecting current position of the mobile platform. Accordingly, it is necessary to supplement the system of communication equations (4) with the following inequalities:

$$\Delta x_{Bi}^2 + \Delta y_{Bi}^2 \leq \delta_1, i \in 1,..,3, \quad (13)$$

where Δx_{Bi} and Δy_{Bi} are the change in the position of the coordinate point B_i along the *X* and *Y* axes, respectively, between the states of the robot, δ_1 is the size of the neighborhood, depending on the movement speeds and the frequency of determining the position of the moving platform. In fig. 6 shows the simulation results for: $l_1 = 70$ mm, $l_2 = 60$mm, $l_3 = 80$mm. As a result of additional testing for finding a platform in the vicinity of some given previous (13) position, the "correct" solution is selected from two solutions of the position problem (figure 7, a) (figure 7, b).

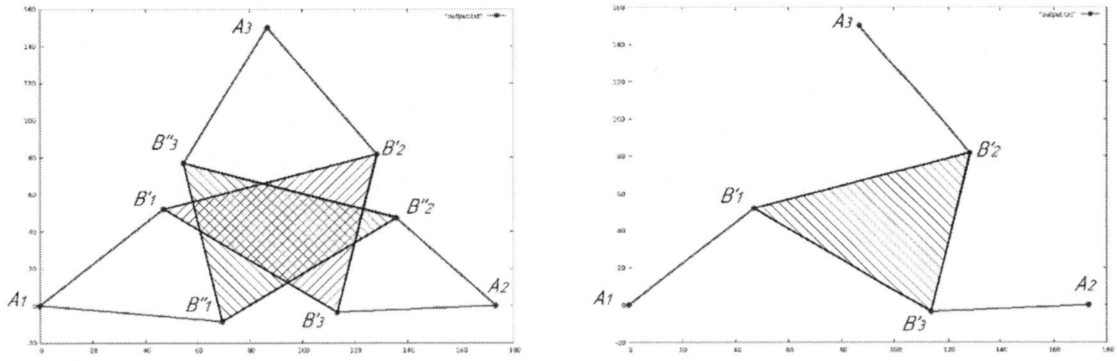

Figure 7. Simulation results a) without taking into account the condition (8), b) taking into account the condition (8).

5. Conclusion

In conclusion, we note that the method used to approximate nonlinear equation systems based on the concept of non-uniform covering can be successfully applied to achieve various tasks of the mechanics of parallel robots, including the problem of the positions of the robot and the definition of the working area. When such robots work, it is important to take into account the zones of singularities corresponding to singularity zones, when passing through which the ambiguity of solving the position problem takes place, which must be taken into account when constructing control algorithms for such robots. The method of non-uniform covering is applied to solve this problem. The effectiveness is characterized by the following indicators: the computation time for the accuracy of approximation $\delta = 0.001$, depending on the proximity to the singularity zones, was from 3 to 30 seconds. This is explained by the fact that as one approaches singularity zones, the number of parallelepipeds Q_j, describing the coordinate of the point B_i, increases, for example, 74 times for the singularity zone described by condition (7). This may be one of the indicators for assessing the closeness to singularities. This question requires further more detailed research. The obtained results can also be used in the selection of constructive and geometrical parameters of 3-RPR, as well as in solving planning problems and optimizing its trajectory taking into account the zones of singularity.

Acknowledgments

This work was supported by the Russian Science Foundation, the agreement number 16-19-00148.

References

[1] Laryushkin P A, Heilo S V, Chan K N, Glazunov V A 2011*MIKMUS - 2011 XXIII Int. Innovation-Oriented conf. of young scientists and students* (Mocsow) p 114

[2] Geng Z, Haynes L S 1992 *Robot Comput Integr Manuf.***9** 6 pp 485–95

[3] Yee C S, Lim K B 1997*Neurocomputing***16** 4 pp 333–49

[4] Evtushenko Yu G 1971*USSR Computational Mathematics and Mathematical Physics* **11** 6 pp 38–54

[5] Evtushenko Yu. G., Posypkin M. A 2013*Optimization Letters* **7** 4 pp 819–29

[6] Malyshev D I, Posypkin M A, Rybak L A, Usov A L 2018 *Int. J. of Open Information Technologies* **6** 7 pp 15–20

[7] Virabyan L G, Khalapyan S Y, Kuzmina V S 2018 *Bulletin of BSTU named after V.G. Shukhov* **9** pp 106–1

[8] Erastova K G, Laryushkin P A 2016 *The Engineering Bulletin. Electronic Scientific and Technical J. of Bauman MSTU.***7** pp 1–7

Synthesis planetary-layshaft transmission with single transition shifts

Konstantin B Salamandra

Mechanical Engineering Research Institute of the Russian Academy of Sciences (IMASH RAN), Maliy Kharitonievsky per., 4, Moscow 101990, Russia

ksalamandra@yandex.ru

Abstract. The more speeds in the automatic transmission, the better the fuel economy and the vehicle drivability. More speeds in conventional transmissions consisting only of planetary gear sets or only of gears with fixed axles lead to increase gears and shifting elements, which complicating the transmission design and growing the overall dimensions and weight. One of the ways to obtain simple transmission designs providing a great number of speeds with a relatively small number of gears and shifting elements is to use combinations of planetary gear sets and gears with fixed axles (planetary-layshaft transmissions). The main disadvantage that limits the use of planetary-layshaft transmissions is the presence of double and triple transition shifts in which 2 or 3 pairs of shifting elements are switched on and off at the same time. The problem to find out a set of speed ratios realized by the transmission with single transition shift sequences is solved in the paper on example of a planetary-layshaft transmissions consisting of two differentials and three sets of gears with fixed axles. The gear ratios magnitudes which satisfying obtained set of transmission speed ratios were found.

1. Introduction

One of the main ways to improve the fuel economy of vehicles is to ensure the engine operation in the best efficiency narrow range. The number of speeds in modern automatic transmissions has increased over the past 20 years from 4 to 8 - 10 for this reason.

Obviously, more speeds lead to a transmission design complication, an increase in overall dimensions and weight. The possible planetary-layshaft transmissions structures consisting of gears with fixed axles of gear wheels and two planetary differentials connecting in parallel were synthesized in paper [1]. It was also shown there that such transmissions make it possible to obtain a significantly more speeds at smaller number of gears and shifting elements used. In addition, a reduction in the loads on the internal parts of the transmission by more than 65% is achieved due to the division of power flow by means of differentials into three parallel branches [2]. The main disadvantage of the proposed planetary-layshaft transmissions with three power-flows is the presence of double and triple transition shifts, when not one pair of shifting elements is involved in the speed change, but two or three. Double and triple transition shifts lead to an increase in shifting time and, as a consequence, a decrease in efficiency, comfort and drivability of the vehicle [3].

The purpose of this work is to find a single transition shifts sequence of control elements for a planetary-layshaft transmission with three power-flows and to calculate gear ratios, providing set of transmission speed ratios close to a given.

IOP Conf. Series: Materials Science and Engineering **489** (2019) 012061 doi:10.1088/1757-899X/489/1/012061

2. Planetary-layshaft transmission IDDO

We use the planetary-layshaft structure IDDO shown in figure 1 as an example [1], in which I, O are the input and output links; D1, D2 - differentials; GB1, GB2, GB3 - sets of gears with fixed axles of gear wheels and shifting elements switching on these gears; m and n, p and q are the parts of the differentials D1 and D2, respectively; c1 – c6 – brakes and clutches of the differentials D1 and D2.

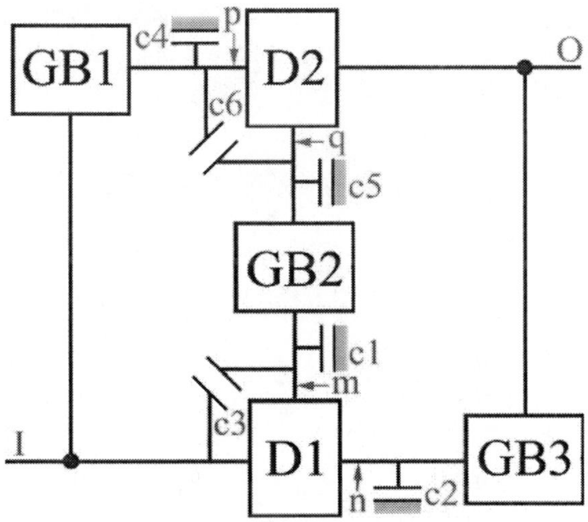

Figure 1. IDDO structure of planetary-layshaft transmission with three power-flows.

If in the IDDO structure (figure 1) the brakes c1, c2, c4 and c6, connecting the differentials parts m, n, p and q with a fixed body, are removed, then the structure will correspond, for example, to the transmission kinematic diagram presented in figure 2. Each GB contains 2 gears and a synchronizer, which switching on these gears. The parasitic gear wheels are additionally installed in GB3 to coordinate the rotation direction of the transmission inner links, since the sun gear n rotates in the opposite direction relative to the input shaft I. The shafts I and O rotate in different directions.

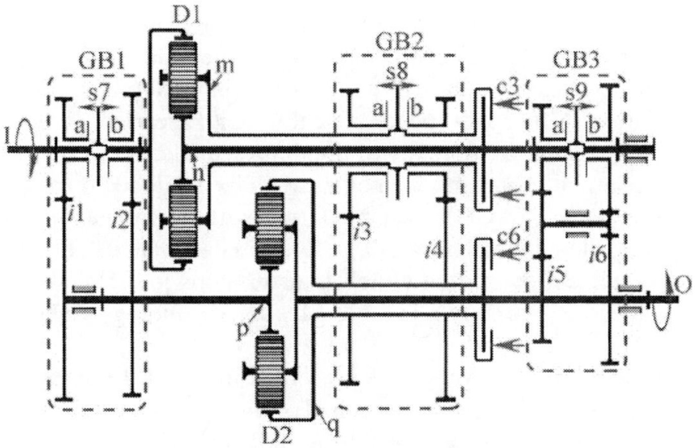

Figure 2. Kinematic diagram of planetary-layshaft transmission according to the IDDO structure.

In figure 2 $i1$, $i2$, $i3$, $i4$ are negative gear ratios of the respective gear pairs; i5, i6 – are positive gear ratios of gear pairs including the parasitic gears.

3. Transmission speed ratios

IOP Conf. Series: Materials Science and Engineering **489** (2019) 012061 doi:10.1088/1757-899X/489/1/012061

The transmission shown in figure 2 allows up to 20 forward and 2 reverse different speed ratios in accordance with the formula obtained in [1]. We distributed the twenty i_{IO} speed ratios implemented by the transmission by using the methodology from [4] from larger to smaller depending on the shifting elements switched-on. The results were entered into the table 1.

Table 1. Shifting elements sequence and speed ratios of the transmission shown in figure 2.

Speed	s7		s8		s9		c3	c6	i_{IO}	Sequences with single transition shifts		
	a	b	a	b	a	b				1 (x)	2 (Δ)	3 (o)
1			x		x			x	-8.345	1	1	1
2	x		x		x				-6.259	2		
3				o	o			o	-5.480			2
4	x			x	x				-5.129	3		3
5			Δ			Δ		Δ	-4.218		2	
6	Δ		Δ			Δ			-3.665		3	
7		x		x	x				-2.941	4		
8	Δ		(Δ)	(o)				Δ	-2.731		4	4
9	Δ		Δ				Δ		-2.401		5	
10			Δ				Δ	Δ	-2.285		6	
11		x	x		x				-1.977	5		
12	o			o		o			-1.529			5
13				–		–		–	-1.353			
14		x	x			x			-1.158	6		
15		Δ	Δ				Δ		-0.951		7	
16		x		x		x			-0.877	7		
17	o			o			o		-0.740			6
18				o			o	o	-0.565			7
19		x		x			x		-0.503	8	8	8
20		x		(x)				x	-0.400	9	9	9

Double and triple transition shifts occur as can be seen from table 1 between the following steps: 2–3, 4–5, 6–7, 7–8, 10–11, 11–12, 13–14, 15–16, 16–17. An analysis of table 1 shows that there are three possible single transition shift sequences in the resulting 20 speeds in which only one pair of shifting elements is switched on. The shifting elements for each of the three sequences are pointed with the corresponding symbol: "x" for the first sequence; "Δ" for the second; and "o" for the third. The numbers of speeds with single transition shifts for each sequence are shown in the three right columns of table 1. Each of the three sequences with single transition shifts has 9 speeds. The thirteenth speed in which the shifting elements are shown with the symbol "-", is not used in any of the three sequences.

The IDDO structure has 4 degrees of freedom, i.e. 3 shifting elements should be switched on at each speed. While in one-stream modes (in which only GB1 or GB3 is enabled), it is sufficient to enable only two shifting elements. Therefore, a third shifting element can be additionally switched on in these speeds and will not affect to the speed ratio being realized, as in dual-clutch transmissions [5]. The shifting elements additionally switched on in speeds 8 and 20 are shown in table 1 by corresponding symbols in brackets.

Thus, the planetary-layshaft transmission with three power-flows presented in figure 2 implements three dependent single transition shifts sequences with 9 speeds in each. To find out the best sequence from three obtained, requires the development of a criterion and optimization this criterion.

4. Optimization criterion

Let some speed ratios sequence $a_k = a_1, a_2, \ldots, a_k$ be preset, where a is the ratio i_{IO} of the corresponding speed; k is the speed number. The a_k sequence is ordered, the magnitudes of its members decrease in absolute value with increasing k, which is due to a decrease the speed ratios of the transmission when shifting from the lowest to the highest speed. It is unlikely that the members of the

a_k sequence will satisfy the speed ratios implemented by the transmission. Therefore, we will look for another sequence b_k, satisfying the speed ratio of the transmission closest to the given sequence a_k. The proximity of the sequences a_k and b_k will be evaluated by the criterion of the minimum of the function (1).

$$F = \sum_{k=1}^{9} \left(\frac{a_k - b_k}{a_k} \right)^2 \tag{1}$$

The best sequence will be evaluated by the smallest value of the function F after its minimization for each of the three sequences with single transition shifts.

5. Calculation results

Let the next sequence a_k [6] be preset, see table 2.

Table 2. Preset speed ratio sequence a_k.

Speed	Ratio	Step
1	-4.713	
2	-2.842	1.658
3	-1.909	1.488
4	-1.382	1.382
5	-1.000	1.382
6	-0.808	1.237
7	-0.699	1.156
8	-0.580	1.205
9	-0.480	1.209

5.1. Sequence "x" optimization

As a result of the criterion (1) minimization for the sequence "x" the next sequence b_k was obtained, see table 3.

Table 3. Optimization results of the sequence "x".

Speed	s7		s8		s9		c3	c6	i_{10}	Step
	a	b	a	b	a	b				
1			x		x			x	-4.796	
2	x		x		x				-2.629	1.825
3	x			x	x				-2.090	1.258
4		x		x	x				-1.182	1.768
5		x	x		x				-1.108	1.067
6		x	x			x			-0.904	1.226
7		x		x		x			-0.706	1.282
8		x		x			x		-0.535	1.318
9		x		(x)				x	-0.421	1.271
R1					x		x		2.109	
R2						x	x		0.160	

R1 and R2 are the ratios for the reverse speeds. The resulting sequence is implemented with the following magnitudes of gear ratios: $i1 = -1.342$, $i2 = -0.421$, $i3 = -2.034$, $i4 = -0.654$, $i5 = 2.109$, $i6 = 0.16$, $i_{D1} = -1.5$, $i_{D2} = -1.5$. Here and below, the gear ratios of the differentials from the sun gear to the ring gear under the assumption that the carrier has stopped.

The value of the function (1) of the sequence "x" is $F = 0.083$.

5.2. *Sequence "Δ" optimization*

As a result of the criterion (1) minimization for the sequence "Δ" the next sequence b_k was obtained, see table 4.

Table 4. Optimization results of the sequence "Δ".

Speed	s7		s8		s9		c3	c6	i_{IO}	Step
	a	b	a	b	a	b				
1			Δ		Δ			Δ	-4.415	
2			Δ			Δ		Δ	-2.646	1.668
3	Δ		Δ			Δ			-2.098	1.261
4	Δ		(Δ)					Δ	-1.326	1.582
5	Δ		Δ				Δ		-0.991	1.338
6			Δ				Δ	Δ	-0.848	1.168
7		Δ	Δ				Δ		-0.659	1.287
8		Δ		Δ			Δ		-0.574	1.148
9		Δ		(Δ)				Δ	-0.494	1.162
R1					Δ		Δ		4.501	
R2						Δ	Δ		1.848	

The resulting sequence is implemented with the following magnitudes of gear ratios: $i1 = -1.326$, $i2 = -0.494$, $i3 = -0.848$, $i4 = -0.644$, $i5 = 4.501$, $i6 = 1.848$, $i_{D1} = -1.5$, $i_{D2} = -1.5$.

The value of the function (1) of the sequence "Δ" is $F = 0.027$.

5.3. *Sequence "o" optimization*

As a result of the criterion (1) minimization for the sequence "o" the next sequence b_k was obtained, see table 5.

Table 5. Optimization results of the sequence "o".

Speed	s7		s8		s9		c3	c6	i_{IO}	Step
	a	b	a	b	a	b				
1			o		o			o	-4.732	
2				o	o			o	-2.310	2.048
3	o			o	o				-2.164	1.068
4	o			(o)				o	-1.342	1.612
5	o			o		o			-1.074	1.249
6	o			o			o		-0.743	1.446
7				o			o	o	-0.669	1.112
8		o		o			o		-0.617	1.083
9		o		(o)				o	-0.473	1.306
R1					o		o		3.131	
R2						o	o		0.16	

The resulting sequence is implemented with the following magnitudes of gear ratios: $i1 = -1.342$, $i2 = -0.473$, $i3 = -2.36$, $i4 = -0.669$, $i5 = 3.131$, $i6 = 0.16$, $i_{D1} = -2.314$, $i_{D2} = -4.0$.

The value of the function (1) of the sequence "o" is $F = 0.027$.

Thus, the best sequence with single transition shifts for the transmission presented in figure 2 in accordance with the criterion optimized, is the sequence "Δ".

6. Summary and results

IOP Conf. Series: Materials Science and Engineering **489** (2019) 012061 doi:10.1088/1757-899X/489/1/012061

One of the main problems of planetary-layshaft transmissions is the presence of double and triple transition shifts in which 2 or 3 pairs of shifting elements are involved. It leads to an increase in shifting time, reduce efficiency, worsen driving comfort and drivability. The problem of finding a sequence with single transition shifts and calculating the gear ratios that provide a range of speed ratios close to the given was solved using the example of a planetary-layshaft transmission consisting of two differentials and three sets of gears with fixed axles. The presented transmission has 5 shifting elements, provides 9 forward speeds with single transition shifts and 2 reverse speeds. The proposed planetary-layshaft transmission if compared with the 9-speed planetary transmission [6] has smaller number of gears and shifting elements and implements a close sequence of speed ratios.

Acknowledgments
The research was supported by Russian Science Foundation (project No. 17-79-10493).

References

[1] Salamandra K 2018 Transmissions with three power-flows for trucks and tractors *Engineering for Rural Development* 1955–61

[2] Antonov A S 1981 *Complex power transmissions: Theory of power flow and the calculation of transmitted systems* (Leningrad: Mashinostroenie)

[3] Bai Sh, Maguire J and Peng H 2013 *Dynamic analysis and control system design of automatic transmission* (Warrendale: SAE International)

[4] Salamandra K 2018 Static analysis and parameters synthesis of planetary-layshaft transmissions with three power flows *MATEC Web Conf.* **224** 02040

[5] Matthes B 2005 Dual Clutch Transmissions - Lessons Learned and Future Potential *SAE Tech. P.* 2005-01-1021

[6] Ebenhoch M, Greiner J, Gaertner L and Girres G 2013 9-speed transmission for front wheel drive applications *Proc. Int. VDI Cong.-Drivetrain for Vehicles* (Friedrichshafen) pp 57–80.

Manipulator-tripod on mounted parallelogram mechanism

E N Zakharov, I A Nesmiyanov and M E Nikolaev

Mechanics Department, Volgograd State Agrarian University, Volgograd, University avenue, 26, 400002, Russia

evgeny_zakharov@mail.ru

Abstract. The objects of development were a universal loader mounted on the base of the manipulator-tripod, the algorithm and method of the "copy" control, which allows to ensure the movement of cargoes along a predetermined path.

1. Introduction

In the modern world mobile robots have many various applications and their main working part is a manipulator. Robotic handling systems are used for loading and unloading, transport and storage operations in various technological processes [1, 2].

Manipulators equipped with actuators of positional tracking links and controlled by the human operator, allow you to mechanize a variety of labour-intensive loading and unloading operations. But even when working with the same type of cargoes, the operations vary from cycle to cycle, which is caused by a change in the coordinates of the initial and final points of the trajectory, i.e. the working operation is not completely clear in advance but varies widely [3, 4].

Such problems can be solved by improving and developing manipulators, which have ample opportunities for mechanization of various technological processes [5, 6, 7]. In such requirements it is possible to use the loading manipulators, in the kinematic chains of which there are spatial executive mechanisms that allow the movement of the cargo gripping arm along complex trajectories, and with a relatively simple design [8].

2. The design of the loader

To achieve the above goals a universal parallelogram platform with the attached manipulator-tripod was developed, as well as a control system for this mechanism.

The design of the loader is hinged and it is mounted on the spars of the self-propelled chassis (Fig.1).

The role of the loader boom is performed by the spatial parallel-program mechanism 1, which ensures the horizontal position of the platform 2 at different angles of rotation of the boom. On the platform 2 is fixed manipulator-tripod 3, equipped with a universal joint 4, with which the cargo's gripper 5 is attached. The manipulator-tripod provides movement of the cargo gripping device on width of a body of the self-propelled chassis.

IOP Conf. Series: Materials Science and Engineering **489** (2019) 012062 doi:10.1088/1757-899X/489/1/012062

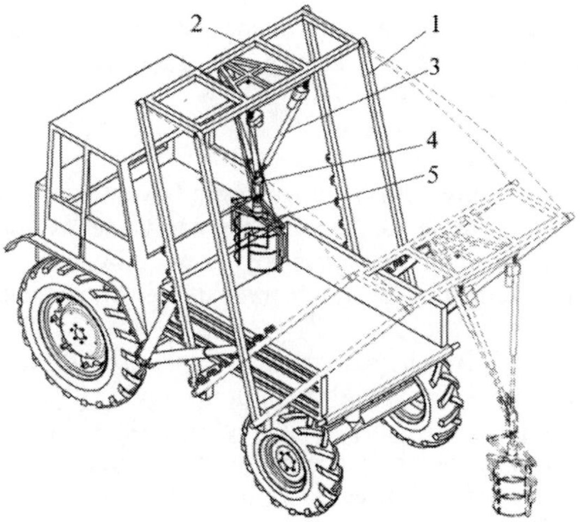

Figure 1. Mobile robot loader with cargo's gripper device based on T-16M self-propelled chassis.

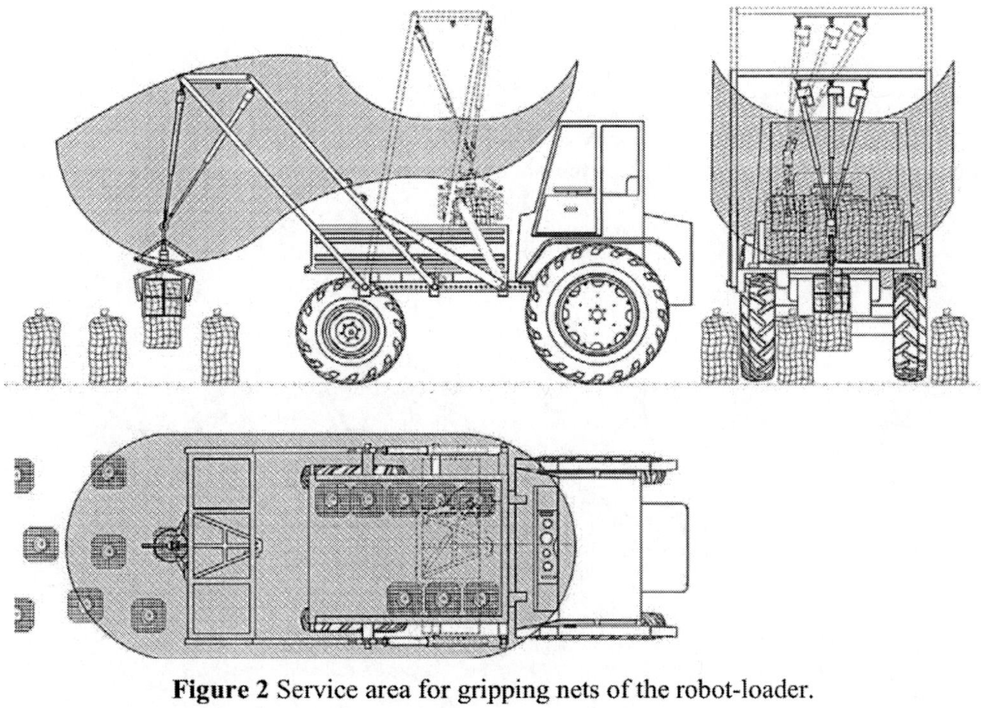

Figure 2 Service area for gripping nets of the robot-loader.

The geometric parameters of the mechanism are justified in such a way as to ensure the width of the load more than the width of the body of the self-propelled chassis (figure 2).

IOP Conf. Series: Materials Science and Engineering **489** (2019) 012062 doi:10.1088/1757-899X/489/1/012062

The calculations showed that each drive electric cylinder of the parallelogram mechanism should provide a force of P_{count}=3.28 kN. Thus, at operating speeds of electric cylinder rods up to V_{stock}=0.25 m·s^{-1} the required power of the actuator motor is $N_{needful}$=0.82 kW.

3. Control system

Manipulators of parallel structure along with the advantages have a significant drawback – the difficulty of manual control. This drawback is due to the need for simultaneous control of multiple drives. It is impossible to set the movement on a given trajectory, by partial movements of each degree of mobility of such manipulators, also, manual control of several degrees of mobility of the loading manipulator leads to rapid fatigue of the operator.

From the analysis of control systems it was revealed that the well-known and intuitive principle of manipulator's control with the participation of the operator is copying [9].

Control of the loading manipulator-tripod is a complex task, including planning and synthesis of trajectories, solving the problem of positioning the cargo at the initial and final trajectory points, setting and implementing the law of movement of the cargo along the trajectory [10, 11].

The algorithm of control of the gripping arm (p. M) of a manipulator-tripod is built basing on the conditions for the optimal trajectory tracking $M_0M_к$ (figure 3). The condition of optimal tracking of the trajectory of the point M assumes that the movement of the drive mechanisms of the control system and the rods of the working cylinders are connected by relations:

$$S_{M\text{м}} = S_{Mp} \cdot \lambda^{-1}; \qquad (1)$$
$$V_{M\text{м}} = V_{Mp} \cdot \lambda^{-1}.$$

where $S_{M\text{м}}$, $V_{M\text{м}}$ – displacement and velocity of point M of the manipulator; S_{Mp}, V_{Mp} – the corresponding displacement and velocity of point M of the handle of the control system, λ – the scale coefficient.

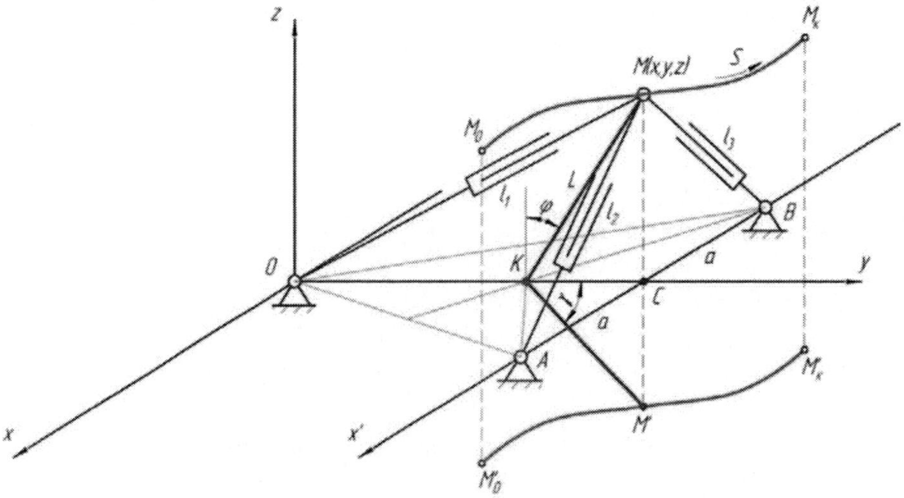

Figure 3. Calculation scheme for the formation of the control algorithm.

Let us suppose, the manipulator is characterized by the following geometric dimensions: $OM=\lambda l_1$; $AM=\lambda l_2$; $BM=\lambda l_3$; $OC=c$; $OK=d$; OA, AB, OB; $OA=OB$; $AC=CB=a$; $MK=L$.

The position of the point M can be described as equations of relations:

$$\begin{aligned} l_1^2 &= x_M^2 + y_M^2 + z_M^2, \\ l_2^2 &= (x_M + a)^2 + (y_M + c)^2 + z_M, \\ l_3^2 &= (x_M - a)^2 + (y_M + c)^2 + z_M, \\ L^2 &= x_M^2 + (y_M + d)^2 + z_M. \end{aligned} \tag{2}$$

The reproduced curve is represented as the function of $f_M = f(x_M, y_M)$, the initial position of the point M_0 of the driving mechanism is λx_0, λy_0, λz_0, (i.e. $M_0(x_0, y_0, z_0)$), besides the point M_0 coincides with one of the points of the trajectory of the reproduced curve and corresponds to the angles of the link $L - \varphi_0$ and γ_0.

Let us mark the starting lengths of the links of the master as l_{10}, l_{20}, l_{30}, and the courses of the rods, respectively, S_1, S_2 and S_3.

The algorithm of formation of the control program for electric cylinders of the manipulator is presented in figure 4.

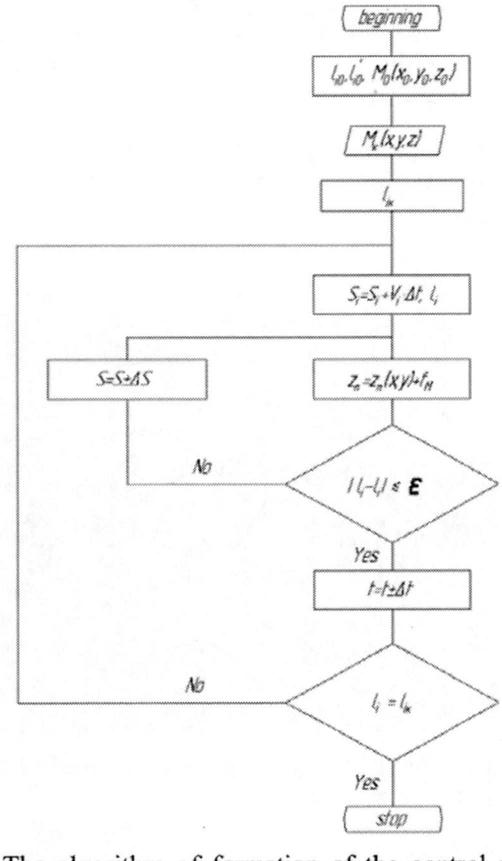

Figure 4. The algorithm of formation of the control program for electric cylinders of the manipulator.

IOP Conf. Series: Materials Science and Engineering **489** (2019) 012062 doi:10.1088/1757-899X/489/1/012062

The current readings of the position of sensors of the actuator rods l_{i0} and the positions of the rods of the control mechanism l'_{i0} are read in the body of the algorithm. According to the readings from equations (2) are Cartesian coordinates of the starting point of the tong position $M_0(x_0, y_0, z_0)$. Further, set the coordinates of the end point of positioning of the load $M_k(x, y, z)$ and the final values of the lengths of the executive links l_{ik} are considered. The lengths of actuators are related to the lengths of the links of the master by the scale factor λ, i.e.$l_i = \lambda \cdot l'_i$.

Further, we set the increment of the coordinates along the arc of the trajectory $S_i = S_i + V_i\Delta t$ and find the current values of l_i, after that the motion along the trajectory is effected. If the divergence between the actual and the specified trajectories is within the permissible error ε, then there is a further increment in time of movement t and a further increment of the arc coordinate S, if $|l_i - l^*_i| > \varepsilon$, then the arc coordinate is corrected. The movement along the trajectory is carried out until the current values of the lengths of the l_i actuators reach the specified final position of the l_{ik}.

As a result of the implementation of the proposed algorithm the trajectory of the load $\lambda z = f(\lambda_x, \lambda_y)$ is defined, as well as the sequence of switching on the working cylinder's engines. Depending on the specified error of the deviation ε from the reproducible trajectory, the values of the steps Δs and Δt are selected.

The presented above control algorithm was implemented on the developed device for manual control of the loading manipulator-tripod (Fig.5), which allows to control not each degree of mobility separately, but directly several generalized coordinates, setting the trajectory of the output link. Management of actuators is effected through manipulation of the handle, through which the operator acts on the positioning sensors [12].

In experimental studies of this control system control actions were set by the operator and the ideal trajectory tracking is possible only if the speed of the control action is less than the maximum possible speed of the actuator rod movement.

Figure 5. The model of the ergomonic copying control system of the manipulator.

4. Conclusion

The proposed method of controlling the electric cylinders of the loading manipulator is practically implemented and used in the manual control system of manipulators-tripods.

Experimental studies have shown that the deviation of the actual trajectory of the cargo's movement from the trajectory set by the operator does not exceed 3.4%, and the positioning error of the load is not more than 1.8%.

The developed control system allows to set the movement of the cargo directly, what could increase the productivity of loading and unloading operations with piece cargoes up to 18% when moving along straight trajectories.

When moving the cargo in a straight line with the simultaneous operation of the executive cylinders, the total work for the operation cycle is lower by 10-12% than when moving it from the same starting to the same final position with the work of each electric cylinder one-by-one.

Acknowledgments
This work was supported by the Russian Foundation for Basic Research, project No. 18-41-342002 p_mk.

References

[1] Ananiev A N 2015 Problems of application of robots in agricultural production. [Electronic resource.] Access mode: http://www.cckrb.ru/news/Innovatsii-v-APK/Problemy-primeneniya-robotov-v-selskokhozyaystvennom-proizvodstve-/?sphrase_id=2302

[2] Zavrazhnov A I, Zavrazhnov A A, Lantsev V Y, Zemlyanoy A A 2014 Development of robotic manipulator for 3D contour pruning of fruit trees *Robotics in agricultural technology: mater. International.scientific. prakt. Conf.* (Michurinsk) pp 304-12

[3] Afonin V L 2005 Mechanisms of parallel structure *Kinematics and dynamics. M.: IC MSTU "Stankin"* p 180

[4] Gerasun V M, Pindak V I, Nesmiyanov I A, Dyashkin-Titov V V, Pavlovsky V E 2012 Manipulators for mobile robots. Concepts and principles of design *Preprint IPM im. M. V. Keldysh* No. **44** p 24

[5] Gerasun V M, Zhoga V V, Nesmiyanov I A, Vorob'ova N S, Dyashkin-Titov V V 2013 To definition of a zone of service of the mobile manipulator-tripod M*echanical engineering and engineering education* **3** pp 2-8

[6] Belikov V A, V V Zhoga, V N Skakunov, S E Terekhov, I V Shamanov 2014 Computer vision Systems of mobile robots *Izvestia Volggtu. Ser. Actual problems of control, computer engineering and Informatics in technical systems* **22** pp 184-90

[7] Gerasun V M, Nesmiyanov I A 2010 Control systems of manipulators on the basis of space Executive mechanisms *Mechatronics, automation, control* **2** pp 24-8

[8] Nesmiyanov I A, Zhoga V V, Pavlovsky V E, Vorob'ova N S 2014 Agricultural robot manipulator control system *News of the lower Volga agro-University complex: Science and higher professional education* **3** pp 226-31

[9] Zhoga V V, Nesmiyanov I A, Dyashkin-Titov V V, Vorob'ova N S 2014 Control system of manipulator-tripod *Robotics and technical cybernetics* **4** pp 58-62

[10] Gerasun V M, Nesmiyanov I A 2010 Control systems of manipulators on the basis of space Executive mechanisms *Mechatronics, automation, control* **2** pp 24-8

[11] Zhoga V V, Dyaskin-Titov V V, Nesmiyanov I A, Vorob'ova N S 2016 The problem of positioning of the manipulator parallel-serial structure with a controlled gripper *Mechatronics, automation, control* **17** pp 525-30

[12] Gaivoronsky V A 1984 Control system of loading manipulators of special design *Sat. nauch. Tr. Volgograd agricultural Institute* **85** pp 72-9

Institute of Physics
Dirac House, Temple Back
Bristol BS1 6BE UK

ISSN: 1757-8981
ISBN 978-1-5108-8484-7

CURRAN ASSOCIATES INC.
proceedings.com